DIRECT CURRENT FUNDAMENTALS

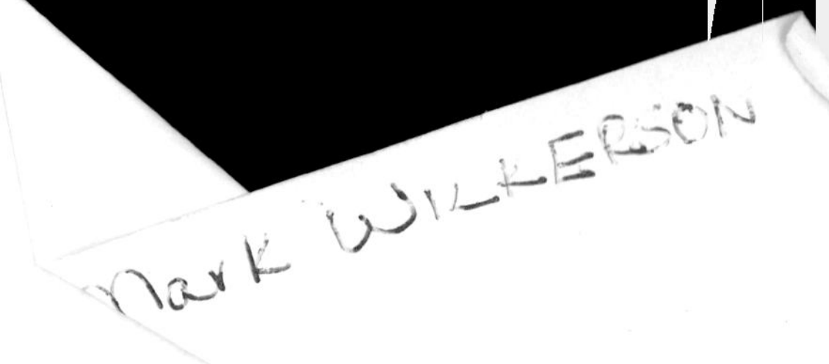

DIRECT CURRENT FUNDAMENTALS

THIRD EDITION

Orla E. Loper

Panama City, FL

Edgar Tedsen

Oakland Community College
Auburn Heights, MI

Delmar Publishers Inc.®

...e to dedicate this book to my son Mark.

Cover design by Patricia Pennington/Graphics

Delmar Staff
Administrative Editor: Mark Huth
Associate Editor: Jonathan Plant
Production Editors: Marilyn Hauptly, Cynthia Haller
Design Coordinator: John Orozco
Art Coordinator: Tony Canabush

For information, address Delmar Publishers Inc.
2 Computer Drive West, Box 15-015
Albany, New York 12212

Printed in the United States of America
Published simultaneously in Canada
by Nelson Canada
a division of International Thomson Limited

10 9 8 7 6 5 4 3

Library of Congress Cataloging in Publication Data

Loper, Orla E.
Direct current fundamentals.
Includes index.
1. Electric engineering. 2. Electric currents, Direct. I. Title.
TK1111.L66 1986 621.319'12 85-16234
ISBN 0-8273-2235-6
ISBN 0-8273-2234-8 (soft)
ISBN 0-8273-2236-4 (instructor's guide)

Contents

Preface

This book is a totally revised edition of a widely acclaimed textbook that, for nearly three decades, has helped thousands of students to launch their careers in electricity and electronics.

Like its predecessor, the revised edition is designed to help the beginning student at community colleges, vocational schools, technical institutes, and high schools in learning the fundamental concepts of DC electricity and magnetism. No prior knowledge of electricity is assumed, but the basic skills of algebra are required.

This revised edition has been totally reorganized to reflect current trends in presenting the traditional material. The content has been streamlined by shifting emphasis, consolidating related topics, and by eliminating the obsolete. Here are some of the specific new features.

- Every chapter is introduced with a set of learning objectives.
- For every chapter, the newly encountered vocabulary and technical terms are identified.
- Text material is organized in short sections that are limited to one specific topic and identified by number and subtitle.
- Wherever possible, illustrations have been placed in close proximity with the appropriate explanations.
- Every chapter is summarized by a series of brief, concise statements.
- The use of sample problems has been greatly improved. The step-by-step solution process has been standardized, and many new sample problems have been added.
- The Achievement Reviews at the end of each chapter have been expanded to offer a wide variety of practice problems, all of which have been field tested.
- The Instructor's Guide, available from the publisher, has been revised to contain solutions to all the new questions and problems.
- A list of basic electrical symbols has been added to the Appendix.
- A glossary has been added.
- Certain topics have been greatly expanded, including units related to numerical concepts, electrical measurements, and basic circuits.
- Different types of circuits are now grouped under their respective individual chapters.

- The concept of voltage dividers, both unloaded and loaded, has been added.
- Chapter 24, "Solving DC Networks," has been drastically changed to include well-explained solutions of the loop current method, superposition technique, and Thevenin's theorem.

Many well-proven features of previous editions have been retained. This includes the units related to DC machines, electrical heating, and lighting, which for some students represents a source of necessary and indispensable information. These chapters may be omitted or used at the discretion of the instructor.

The chapter dealing with motor starters and controllers does not ignore the modern approach to electronic controllers but purposefully retains some of the materials related to manual and magnetic control. Much of this type of equipment is still encountered throughout the smaller job-shops of industry, and the material also aids in studying the elementary concepts that are the necessary prerequisite for a future study of electronic controllers.

The presentation of the text material closely follows the author's lecture notes, which have been successfully used and refined over a period of nearly 20 years. It is hoped that this reflected experience will benefit other instructors and their students alike.

The companion volume to this book, Alternating Current Fundamentals, third edition, has also been newly revised. Together the two books offer a comprehensive introduction to electrical concepts, basic circuits, and electrical machinery.

ACKNOWLEDGMENTS

Special thanks go out to the following people, who reviewed the manuscript at various stages in its development:

Clifford Dickinson, Hudson Valley Community College
Roy Du Bose, New York Technical College
Ron Fusco, Mohawk Valley Community College
Larry Gazaway, Spokane Community College
Al Genest, New England Institute of Technology
Frank Griffin, Catawba Valley Technical College
Stephen Herman, Lee College
David Komola, Middlesex Vocational Technical High School
C. J. Lemmon, Renton Vocational Technical Institute
Thomas Roma, Jefferson State Vocational Technical School
Charles Thompson, Vermont Department of Labor and Industry

I'd like to thank my dear wife for her encouragement, patience, and support throughout this project.

My appreciation is extended to my friend and colleague David Braum of Oakland Community College for his valued suggestions and assistance in the completion of the manuscript.

Finally, I'd like to acknowledge the help and advice of the Delmar staff, especially Mr. Jonathan Plant, who has guided me in the task of making the revisions.

1

An Introduction to Electricity and Electronics

Objectives

After studying this chapter, the student should be able to

- Define and explain the new technical terms introduced in this chapter

alternating current (AC)	atom
direct current (DC)	electricity
electron	electronics
element	compound
molecule	proton
neutron	ion
positive ion	negative ion
single-phase AC	polyphase AC
rectifier	

- Give examples of elements and compounds
- Sketch a simple atom and label its parts
- State the law of attraction and repulsion related to electrical charges
- Name the three basic parts of an atom
- Explain the difference between atoms and molecules
- Distinguish between AC and DC

1–1 WHAT ARE ELECTRICITY AND ELECTRONICS?

Almost everyone is aware of the phenomenal developments in the field of electronics technology during the recent decades. The very term *electronics* evokes visions of exotic and complex devices that are quickly altering our individual and collective lifestyles.

The study of electricity and electronics has opened the door to rewarding careers for multitudes of people. The words electricity and electronics are part of everyone's vocabulary; yet a surprising number of people fail to make the proper distinction between these two words. As you begin your studies of these subjects, you should know how to differentiate between the two.

To begin, this is a book about electricity, not electronics. The study of electricity precedes the study of electronics. No one can hope to learn the concepts of electronics without having first mastered the principles of electricity. Then how do these two terms differ from one another?

Electricity is best thought of as a form of energy. Natural energy, of course, manifests itself in many different forms of which electricity is but one example.

You may recall one of the cardinal rules of science, which states that energy can neither be created nor destroyed; thus, mankind cannot create electricity. All we can do is produce and utilize electricity by converting various forms of energy.

Let us consider, by contrast, the word *electronics*. Electronics deals with specific applications of electrical principles that are earmarked by the following characteristics:

1. Electronics refers to the processing of informational signals. In other words, an electronic device is designed to convey, collect, or transmit informational data in the form of small variations in electrical voltages or currents.
2. Electronic equipment utilizes components such as electronic tubes or semiconductor devices.
3. The electronic signal does not necessarily require the use of metal conductors. The electrical energy may be wireless, or transmitted through space.

1–2 WHY THIS BOOK IS CALLED DIRECT CURRENT FUNDAMENTALS

Electricity is available to the consumer in two different forms: direct current (DC) and alternating current (AC). Of the two, alternating current is the more prevalent form. This kind of electricity is commercially generated and distributed by public utilities.

AC is available in two different versions: (1) polyphase AC, which is mainly used for industrial and commercial applications, and (2) single-phase AC, which is used in the home as well as in commerce.

Direct current, by contrast, is not commercially available to the average consumer. It is used in batteries, such as in mobile equipment; in all electronic devices; and for special industrial applications, such as adjustable speed drives and electroplating.

You may be curious about the difference between DC and AC. DC sources are distinguished by a fixed polarity, such as in a car battery, which has two terminals marked positive and negative. Current from DC sources flows steadily in one direction only.

AC sources do not have such polarity markings. Just think of an electrical wall outlet in your home. Current from such sources changes its direction continually, flowing back and forth in a conductor.

As stated before, direct current generally is not commercially available. If needed, it may be locally provided by use of

- DC generators
- Batteries
- Rectifiers (devices for the conversion of AC into DC)

This book, then, is concerned with the study and application of DC principles. One might ask: ''Why begin our studies with DC instead of the more common AC?'' The reason is that DC fundamentals are easier for the beginning student and, once learned, will afford an easy transition to AC fundamentals.

1–3 EARLY HISTORY OF ELECTRICITY

Our knowledge of electricity has been gathered over the years by experimenters in many areas: magnetism; batteries; current, through gases and through vacuum; studies of metals and of heat and light. Some of the simplest and most important ideas were discovered fairly recently. These recently discovered facts will be used in this discussion because they will be helpful in gaining an easier understanding of electricity.

The first written records describing electrical behavior were made 2,500 years ago. These records show that the Greeks knew that amber rubbed on cloth attracted feathers, cloth fibers, and other lightweight objects. The Greek name for amber was *elektron*. From elektron came our word *electric*, which at first meant ''being like amber,'' or, in other words, having the property of attraction.

A hard rubber comb and the plastic case of a pen both acquire a strange ability after being rubbed on a coat sleeve—the ability to attract other objects. Long ago, the name *charging* was given to the rubbing process that gives the plastic or hard rubber its ability to attract. After rubbing, the object was said to be charged. The *charge* given to the object was thought to be an invisible load of electricity.

About 300 years ago, a few men began a systematic study of the behavior of various charged objects. They soon found that *repulsion* was just as important as *attraction*. Their experiments showed that charged materials could be divided into the two groups shown in Figure 1–1.

- Any item from List A attracts any item from List B and vice versa. (Charged glass attracts charged rubber and vice versa.)
- Any item in List A repels any other item in List A. (Charged glass repels charged mica.)
- Any item in List B repels any other item in List B. (Charged rubber repels charged rubber.)

These results illustrate the *law of attraction and repulsion:*

> Unlike charges attract, like charges repel.

Various names were suggested to describe List A and List B. They could have been called by any pair of opposite-sounding names: Up and Down, or Black and

CHARGED MATERIALS	
List A	**List B**
Glass (rubbed on silk) Glass (rubbed on wool or cotton) Mica (rubbed on cloth) Asbestos (rubbed on cloth or paper) Stick of sealing wax (rubbed on wool)	Hard rubber (rubbed on wool) Block of sulfur (rubbed on wool or fur) Most kinds of rubber (rubbed on cloth) Sealing wax (rubbed on silk, wool, or fur) Glass or mica (rubbed on dry wood) Amber (rubbed on cloth)

FIGURE 1–1

White. The pair of names finally accepted by scientists was suggested by Benjamin Franklin: Positive for List A, and Negative for List B. The first item in each list was used as a standard and led to the original definition of the terms positive and negative: Anything that repels charged glass is like charged glass and has a *positive* charge; anything that repels charged rubber is like charged rubber and has a *negative* charge, Figure 1–2.

The frictional movement involved in rubbing the objects together is not of vital importance. Hard rubber simply pressed against wool (no rubbing) and then removed will get its negative charge although not as strongly as if it were rubbed. The only value of the rubbing is to bring the rubber into contact with more of the surface area of the wool fibers.

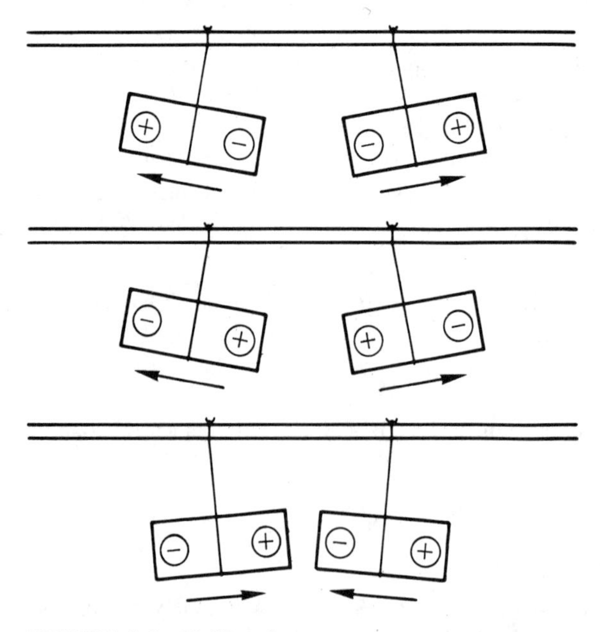

FIGURE 1–2 Unlike charges attract each other and like charges repel each other.

For a further understanding of what is occurring in materials when they are electrically charged, we need to review some facts about the internal structure and composition of all materials.

1–4 ONE HUNDRED ELEMENTS—BUILDING BLOCKS OF NATURE

All of the thousands of kinds of materials on the earth consist of various combinations of simple materials called *elements*. Carbon, oxygen, copper, iron, zinc, tin, chlorine, aluminum, gold, uranium, neon, lead, silver, nitrogen, and hydrogen are elements that most of us have heard of or have used. We do not often use the elements silicon, calcium, and sodium in the pure form, so their names may be less familiar. However, these three elements in combination with oxygen and other elements make up the largest part of the soil and rocks of our earth and help form many manufactured products of everyday use.

There are more than 100 elements. Some of them we never hear of, either because they are very scarce or because people have not yet developed industrial uses for them. Because germanium, beryllium, and titanium are now being used in the electronics and aircraft manufacturing industries, their names are more familiar than they were a few years ago, whereas in 1890 few people had heard of aluminum because it was then a rare and precious metal.

Since there are over 100 different elements, there are over 100 different kinds of atoms. The word *atom* is the name for the smallest particle of an element. We can talk about atoms of carbon, oxygen, and copper because these materials are elements. Single atoms are so small that there is no use wondering what one atom looks like. For example, it is estimated that there are about 30,000,000,000,000,000,000,000 atoms of copper in a penny and that the penny is about six million atoms thick. If an imaginary slicing machine sliced a penny into six million slices of copper, each slice one atom thick, then each slice would contain five million billion atoms.

We do not talk about an atom of water, because water is not an element; it is a compound. The smallest possible speck of a compound is properly called a *molecule,* Figure 1–3. Each molecule of water is made of two atoms of hydrogen and one atom of oxygen. The word *compound* is the name for a material composed of two or more different elements combined. Water is a compound, and the smallest particle of water is a molecule.

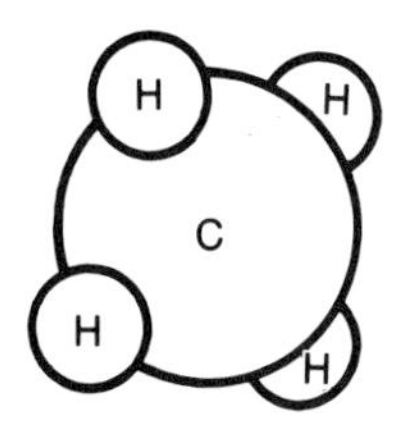

NATURAL GAS MOLECULE (METHANE)

4 HYDROGEN ATOMS COMBINED WITH 1 CARBON ATOM

WATER MOLECULE

2 HYDROGEN ATOMS COMBINED WITH 1 OXYGEN ATOM

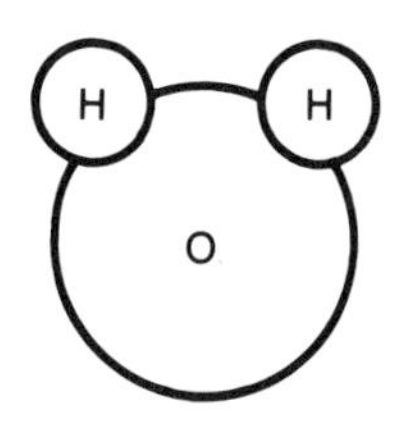

FIGURE 1–3 Molecules of compounds

1–5 THE ATOM ANALYZED—ELECTRONS, PROTONS, AND NEUTRONS

All of the more than 100 kinds of atoms are found to consist of still smaller particles. These particles are so completely different from any known material that any imaginative picture of them is sure to be inaccurate.

Atoms of hydrogen gas are the simplest in structure of all atoms. Hydrogen atoms consist of a single positively charged particle in the center, with one negatively charged particle whizzing around it at high speed. The positively charged particle has been given the name *proton;* the negatively charged particle is called an *electron*.

Figure 1–4 is not drawn to scale because the diameter of the atom is several thousand times greater than the diameters of the particles in it. To show relative dimensions, a more exact representation would have a pinhead-sized electron revolving in an orbit 150 feet across. However, the pinhead is not an exact representation either, for the electron is highly indefinite in shape. It is more like a fuzzy wisp that ripples, spins, and pulses as it rotates around the proton in the center. The mathematical equation that describes it best is the equation that describes a wave. An atom has no outer skin other than the surface formed by its whirling electrons. This is a repelling surface, comparable to the whirling ''surface'' that surrounds a child skipping a rope. There is as much relative open space within the atom as there is in our solar system.

The proton that forms the center of the hydrogen atoms is smaller than the electron but 1,840 times as heavy. The most important properties of the proton are its positive charge and its weight. The number of protons determines the identity of the element. For example, an atom containing 29 protons must be an atom of copper.

As we look at diagrams of other atoms, we need two new words to describe them. The *nucleus* of the atom is the name given to the tightly packed, heavy central core where the protons of the atom are assembled. Along with the protons are other particles called neutrons, Figure 1–5.

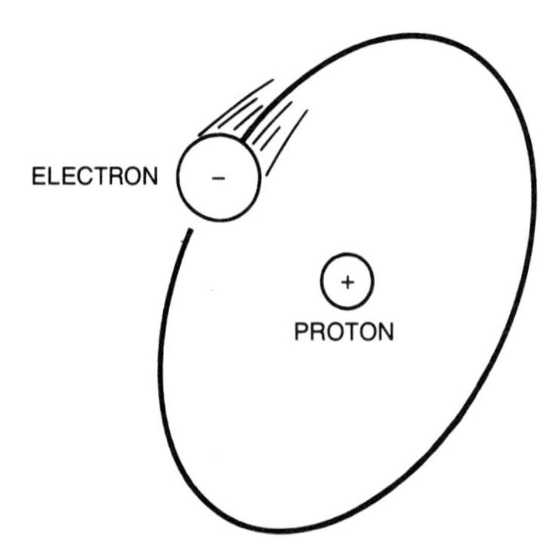

FIGURE 1–4 The hydrogen atom

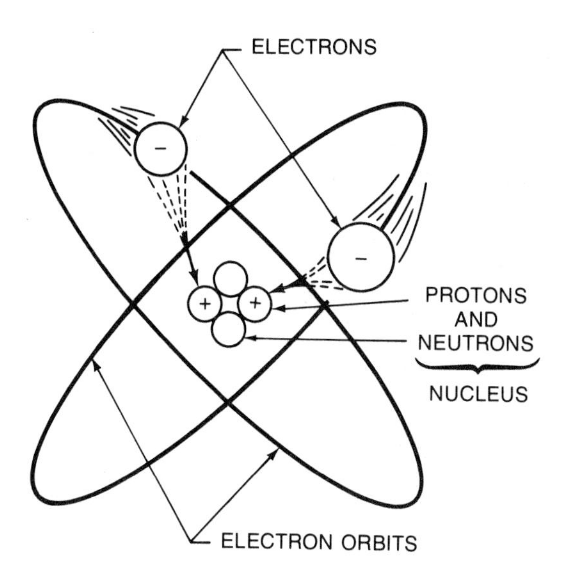

FIGURE 1–5 Atomic structure

The name *neutron* indicates that this heavy particle is electrically neutral; neutrality and weight are its most important properties. A neutron is probably a tightly collapsed combination of an electron and a proton.

At first, it may be hard to realize that these three particles—electrons, protons, and neutrons—make up all materials. All electrons are alike, regardless of the material from which they come or in which they exist, Figure 1–6. All protons are alike, regardless of the material in which they exist. Neutrons, too, are all alike.

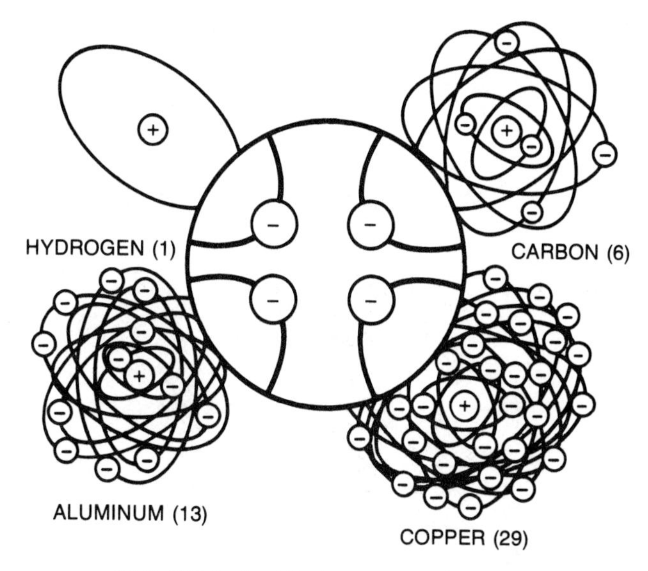

FIGURE 1–6 All electrons are identical.

1–6 THE ATOMIC THEORY—CORNERSTONE OF ELECTRICAL THEORY

The arrangement of electrons around the nucleus determines most of the physical and chemical properties and the behavior of the element. The electrons of the atom are often pictured in distinct layers, or *shells,* around the nucleus. The innermost shell of electrons contains no more than two electrons. The next shell contains no more than eight electrons; the third, no more than 18; and the fourth, 32. Let us consider the model of a copper atom, Figure 1–7.

The 29 electrons of the copper atom are arranged in four layers, or shells: two in the shell nearest the nucleus, eight in the next, and 18 in the third, for a total of 28 electrons. The single twenty-ninth electron circulates all alone in the fourth shell.

This outermost shell is known as the *valence shell,* and electrons occupying this orbit are known as *valence electrons*. When energy is applied to a valence electron, it may dislodge itself from its parent atom and is then known as a *free electron*. In this position (where it is relatively far from the positive nucleus and is screened from its attracting positive charge by the other electrons), this single electron is not tightly held to the atom and is fairly free to travel.

If we examine the electron arrangement in all kinds of atoms, we find that most of them have one, two, or three electrons in an outer shell, shielded from the positive nucleus by one or more inner shells of electrons. These elements are all called *metals*. Metals are fairly good conductors of electricity because they have many free electrons that can move from atom to atom.

Elements with five, six, or seven electrons in their outermost ring are classified as *nonmetals*. Diagrams of two such nonmetallic elements, sulfur and iodine, are shown in Figure 1–8. They are *not* good conductors for the following reasons:

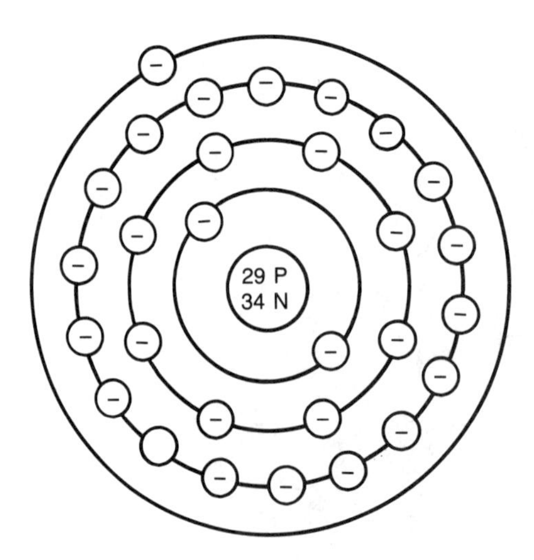

FIGURE 1–7 Copper atom

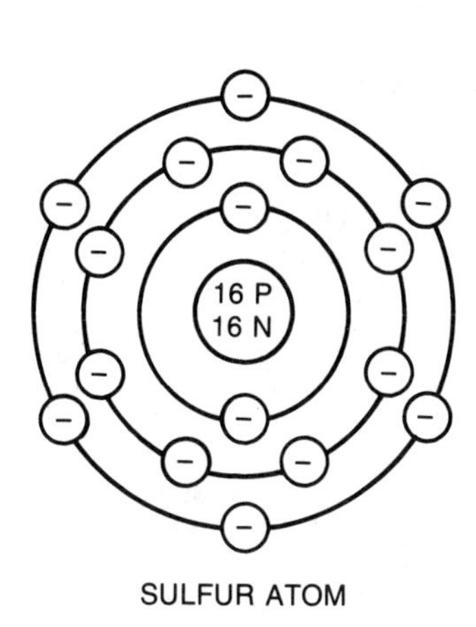

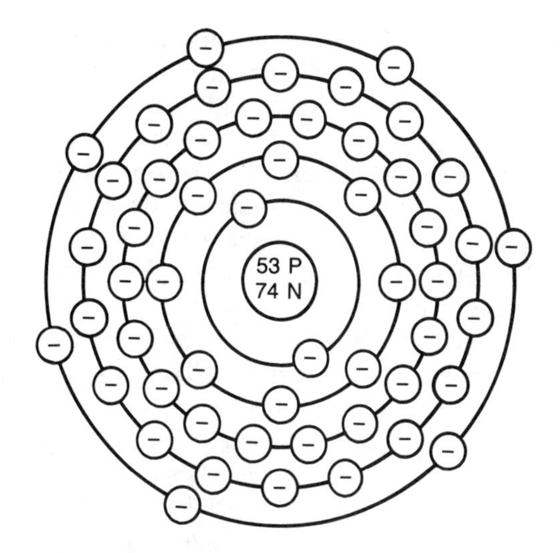

FIGURE 1–8 Nonmetallic elements

1. Their outside electrons are not as well shielded from the attracting force of the nucleus, because the atom has relatively fewer electrons in the inside shells helping to screen any individual outer electron from the attracting force of the nucleus.
2. A shell of eight electrons has a degree of energy stability. Atoms with seven, six, or five electrons in the outer shell will readily pick up and hold the one, two, or three electrons that will build the shell up to eight.

For example, if we try to push some electrons through a block of sulfur, we find that our electrons drop into the empty spaces in the outer shells of the sulfur atoms and are stuck there. This stable shell of eight electrons leaves sulfur with no free electrons ready to slide over to the next atom and with no room for a newcomer.

The word *ion* refers to an electrically unbalanced atom. Considering this statement, it may be concluded that a *positive ion* is an unbalanced atom that has lost some of its electrons, and conversely, a *negative ion* is an unbalanced atom that has gained some electrons.

SUMMARY

- Electricity is a form of energy.
- Electronics deals with specific applications of electrical principles.
- Electrical systems may be classified as being either direct current (DC) or alternating current (AC).
- AC can be converted into DC by the use of rectifiers.
- Unlike charges attract; like charges repel.

- An element is a single uncombined substance consisting of only one kind of atom. An atom is the smallest portion of an element.
- A compound is a substance that can be chemically separated into two or more elements. A molecule is the smallest portion of a compound.
- Atoms consist of various numbers of electrons, protons, and neutrons.
- Electrons are negatively charged and lightweight and move outside the nucleus.
- Electrons are arranged in layers, or shells, around the nucleus of the atom.
- The number of electrons in the outer shell of the atom determines most of the electrical properties of the element.
- Protons are positively charged and heavy and are contained within the nucleus.
- Neutrons are not charged, are heavy, and are contained within the nucleus.
- The number of protons determines the kind of element.
- A negatively charged object is one that has gained extra electrons.
- A positively charged object is one that has lost some of its electrons.
- Electricity is explained by the behavior of electrons.
- All materials can become electrically charged.
- The motion of electrons through a material is called the electric current.

Achievement Review

1. Using our knowledge of electrons, how do we now define the terms *positive charge* and *negative charge?*
2. Using what you know of electron theory, explain what must happen to give an object a positive charge. What happens to give an object a negative charge?
3. State the law of attraction and repulsion.
4. What kind of charge does an electron have?
5. Would two electrons repel or attract each other? Explain.
6. What do each of these words mean: atom, element, molecule, compound, proton, electron, neutron? (There is no point in memorizing definitions of such terms; you should try to understand their meaning so that you can use them correctly.)
7. Tell how atoms of metals differ from atoms of nonmetals in their electron arrangement. Why are metals good conductors?
8. There is an element called gallium. Its atoms have 31 electrons. Referring to the picture of a copper atom, Figure 1–7, how would you expect the electrons of an atom of gallium to be arranged? Is gallium a metal?
9. Explain the terms *AC* and *DC*. Tell how they differ from each other.
10. Complete the following sentences.
    a. All materials consist of over 100 simple substances called ________.
    b. The smallest particles of these simple substances are called ________.
    c. Atoms consist of three still smaller particles called ________, ________, and ________. Of these three, the one with least weight is

the __________; the one most readily movable is the __________; the positively charged particle is the __________; the negatively charged particle is the __________; the particle most responsible for the electrical behavior of materials is the __________. An atom with unbalanced electrical charges is known as a(n) __________. An atom with a surplus of electrons is said to be a __________ ion, and if it has a deficiency of electrons, it is called a __________ ion.

2
Electricity Production and Use

Objectives

After studying this chapter, the student should be able to

- Define and explain the new technical terms introduced in this chapter.

  photon
  photoconductive
  photovoltaic
  semiconductor
  solar cell
  P-type silicon
  N-type silicon
  thermoelectric effect
  thermocouple
  thermopile
  electrolyte
  electrode
  dry cell
  wet cell
  primary cell
  secondary cell
  ultrasonic
  piezoelectric effect
  electromagnetic induction

- Describe six different principles of energy conversions employed in the generation of electricity. An example of each conversion process should be given.
- Discuss and explain six different principles of energy conversion in the practical utilization of electrical energy.

2–1 ELECTRICITY PRODUCTION BY ENERGY CONVERSION

As we have seen in Chapter 1, electricity is a form of natural energy and, thus, can neither be created nor destroyed. What mankind has learned, however, is to derive the desired electrical effect from the conversion of any of the existing types of energy, such as heat, light, magnetism, chemical, and mechanical energy.

Sometimes the conversion process is a simple one, producing the desired electricity in a one-step operation. A photocell, for instance, delivers an electrical voltage as soon as light energy falls onto it.

More often than not, the conversion process is a complex one, requiring many intermediate steps. Consider, for example, how large power plants generate electricity for their customers.

The process begins with burning fossil fuels, either coal or gas, to create *heat*. The heat energy is used to produce steam for a turbine that, in turn, delivers *mechanical energy* to a generator. The generator then employs *magnetism* to produce the desired *electrical energy*. The flow chart that follows depicts this multiple conversion process.

Heat→Mechanical Energy→Magnetism→Electricity

Such multiple conversions are rather inefficient if one considers the losses involved in each conversion process. Remember, every machine (converter of energy) requires more input than it produces output.

Let us now have a closer look at some of the energy sources that are suitable to dislodge the valence electrons from their orbits, thereby creating the desired electrical effect.

2–2 ELECTRICITY FROM FRICTION

In Section 1–3 we discussed the well-known phenomenon of accumulating electrical charges on insulators, such as glass or rubber, by rubbing these substances intensely. The frictional heat causes the surface atoms to give up their valence electrons, giving rise to accumulated, nonmoving charges known as *static electricity*. Our next chapter, entitled "Electrostatics," is devoted to a more detailed description of this phenomenon.

2–3 ELECTRICITY FROM MAGNETISM

We have already mentioned that electrical generators operate on the principle of magnetism. Magnetism plays such an important part in the future study of electricity and electronics that we will devote a couple of chapters to this subject later on in this book.

For our present discussion it will suffice to know of the invisible force field that exists between the north pole and the south pole of a magnet.

Figure 2–1 illustrates how the force of this magnetic field can be used to push the free electrons of a conductor as it is being moved within a magnetic field. This principle of inducing a current flow in a wire when it is moved within a magnetic field is known as *electromagnetic induction*. It is the basis of every generator, however small or large.

2–4 ELECTRICITY FROM CHEMICAL ENERGY

When school children experiment with basic electrical principles, they sometimes make a battery from a grapefruit or lemon. They insert two dissimilar metals into the fruit and find a small, but demonstrable voltage between the two metal strips known as *electrodes*. The acetic acid of the fruit juice is the *electrolyte*, which interacts with the metals, causing a transfer of electrons. Thus, one electrode will accumulate a great

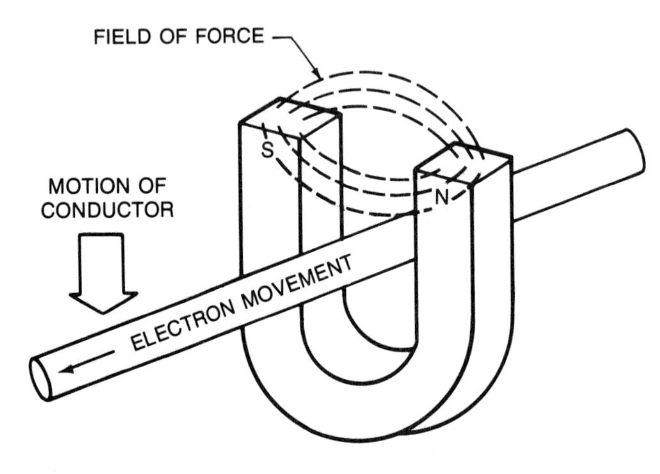

FIGURE 2–1

surplus of electrons, making it negative with respect to the other electrode, which is considered to be positive because it suffers a shortage of electrons.

The word *cell* refers to such a basic arrangement of a chemical substance, the electrolyte, interacting with two dissimilar electrodes. A *battery* is an arrangement of multiple cells.

By this definition, the ordinary flashlight battery should really be called a cell, or, more commonly, a dry cell. When new, such common dry cells yield 1.5 volts. By contrast a 12-volt car battery is a true battery, comprised of six wet cells, each providing 2 volts. Understand that all cells are basically wet. Dry cells, however, contain electrolyte as a moist paste. With a paste electrolyte and sealed construction, a dry cell provides the advantage of being functional in any position.

Some cells are classified as *primary cells,* because their chemical materials are used up as electrical energy is being produced. Primary cells are discarded after they are run down.

By contrast, *secondary cells* can be recharged, because the chemical reaction within the cell is reversible. Car batteries, for example, are comprised of secondary cells.

You should remember that cells and batteries have a fixed polarity, which gives rise to a unidirectional current flow; in other words, cells and batteries always deliver direct current. There is no such thing as an AC battery.

Cells and batteries are explained in greater detail in Chapter 14 of this book.

2–5 ELECTRICITY FROM LIGHT

Most of the life-sustaining energy encountered on our planet is derived from the sun in the form of heat and light energy.

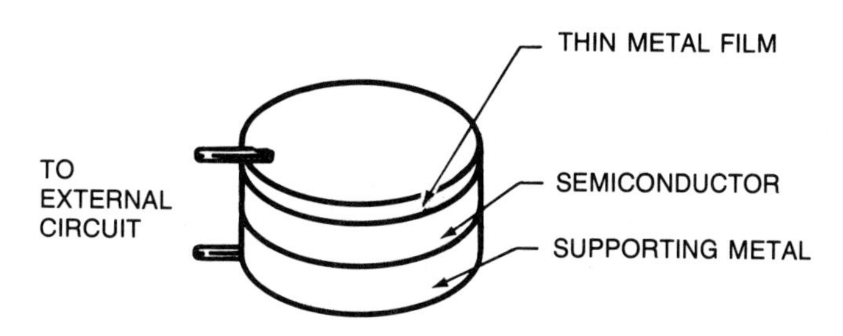

FIGURE 2–2 Photovoltaic cell

Light energy, according to one theory of physics, is transmitted by small particles known as *photons*. When light strikes the surface of certain materials, called *semiconductors,* the photons jar electrons loose from their low-energy state, and the semiconductor accumulates opposite charges similar to that of a battery. Such a device is known as a *photovoltaic cell,* Figure 2–2.

The *solar cell* is an improved type of photovoltaic cell. Specially treated silicon wafers, known as P-type silicon and N-type silicon, are fused together to form a junction. When photons strike such a device, opposite charges accumulate along the junction, producing electrical energy, Figure 2–3.

The photovoltaic devices described in this section must not be confused with *photoconductive* devices. Photoconductive devices do not produce electricity but change their internal resistance when struck by light. Such devices will be explained in a later section of this book.

The open-circuit voltage of a solar cell is about 0.5 volt. Voltage and efficiency (up to 25%) are fairly independent of the amount of illumination, but more light increases the current.

Solar batteries are used to energize electronic equipment in artificial satellites. The communication satellite Telstar is powered by 3,600 solar cells. Solar batteries are also used for maintaining a charge on storage batteries used for rural telephone systems.

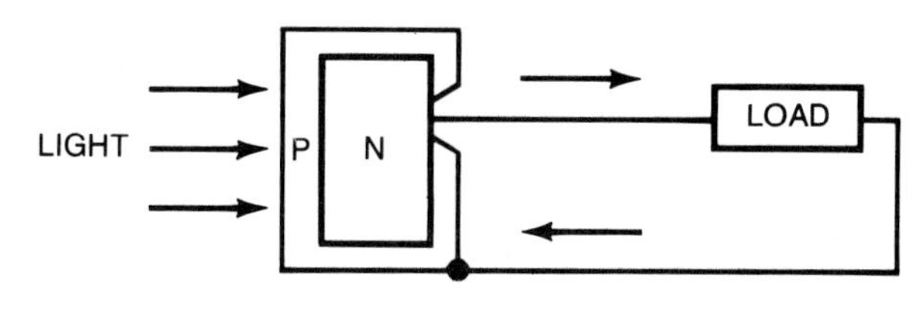

FIGURE 2–3 Solar cell

2–6 ELECTRICITY FROM HEAT

Thermoelectric effects utilizing temperature differentials to generate an electrical voltage have been known for a long time. As early as 1822, a German scientist named Seebeck showed that a circuit, such as in Figure 2–4, produces a steady current as long as the two junctions are at different temperatures.

The letters A and B in the drawing represent two different metals or, possibly, different semiconductors. This direct production of an electromotive force (emf) from heat is sometimes referred to as the *Seebeck effect*.

The explanation for the production of thermal voltage is found by a study of electron energies in conductors. When any two dissimilar metals, such as copper and iron, are in contact with each other, there is a tendency for a few electrons to drift out of one material and into the other. This slight accumulation of electrons causes a so-called *contact potential difference* between the materials.

It is a small voltage, difficult to measure and usually noticed only as a nuisance in delicate measurements. As shown in Figure 2–5, application of heat changes the contact voltage at the heated junction, and the difference in the contact voltage at the two junctions is the useful thermal emf.

Thermocouple is the name given to devices that produce a small emf when the junction of two dissimilar metals is being heated. The voltage output is a direct function of the amount of heat applied and can be used to measure temperature, especially temperatures beyond the range of liquid-containing glass thermometers. Thermocouples produce very small voltages rated in millivolts only. (Remember, 1 millivolt = 0.001 volt.)

Increased voltage output can be achieved when several thermocouples are placed in series, Figure 2–6. Such an arrangement is called a *thermopile*.

Gas-fired heating equipment is commonly controlled by an arrangement such as shown in Figure 2–7. The thermostat is a switch that closes when the room cools and turns on the gas supply to the furnace. The incoming gas is ignited by a pilot light. The pilot also heats a thermocouple, producing current so that the relay coil can hold switch

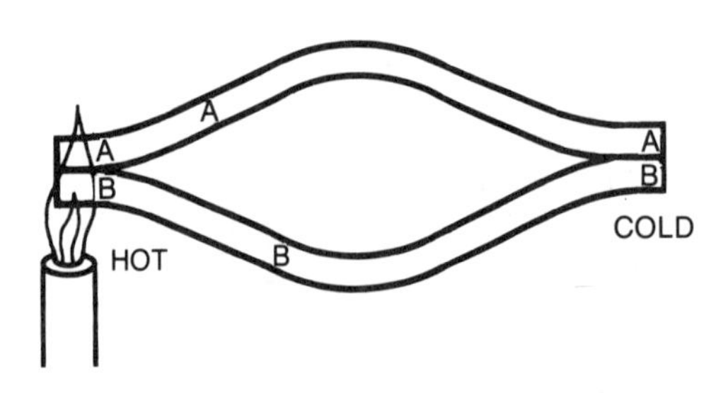

FIGURE 2–4 The Seebeck effect

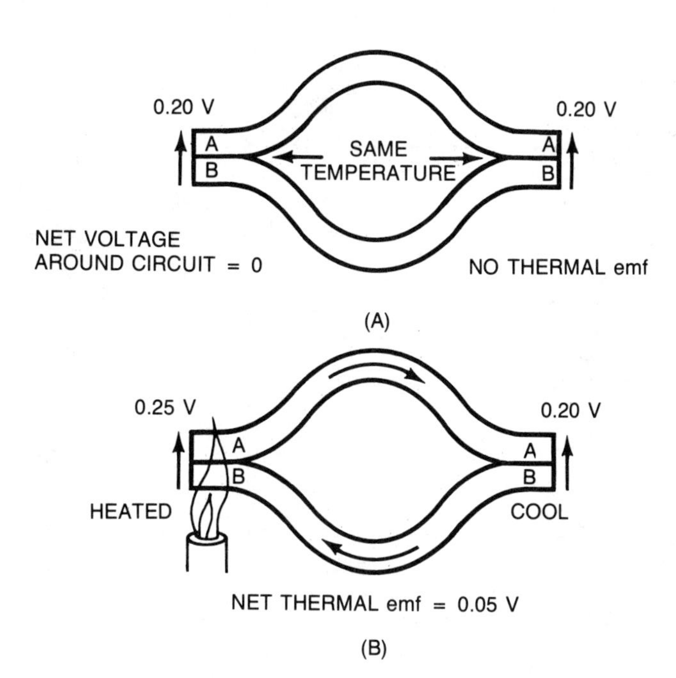

FIGURE 2–5 Thermocouple unit

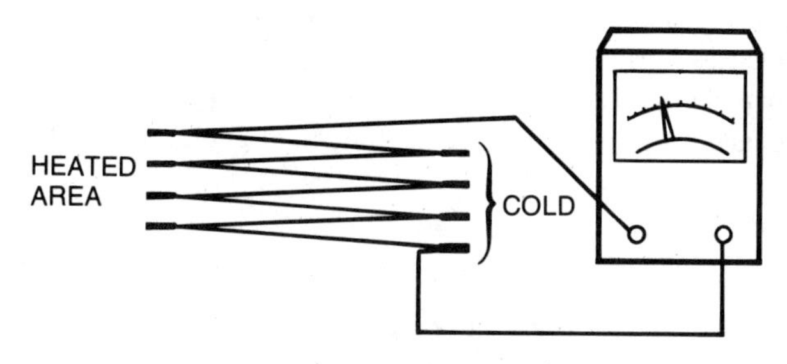

FIGURE 2–6 Thermopile

S closed. If the pilot flame fails, switch S opens so that the main gas valve cannot be opened by the thermostat. Thus, the thermocouple acts as a safety device, preventing an accumulation of unburned gas in the furnace area.

2–7 ELECTRICITY FROM MECHANICAL PRESSURE: PIEZOELECTRICITY

The term *piezo* (pronounced pee-ay′-zo) means pressure. Some materials develop opposite electric charges on opposite sides when they are compressed (or twisted, bent, or stretched). The most common application of this effect is in crystal microphones and crystal pickups for record players. Rochelle salt (sodium potassium tartrate) crystals are twisted back and forth by the sideways vibration of the phonograph stylus, producing

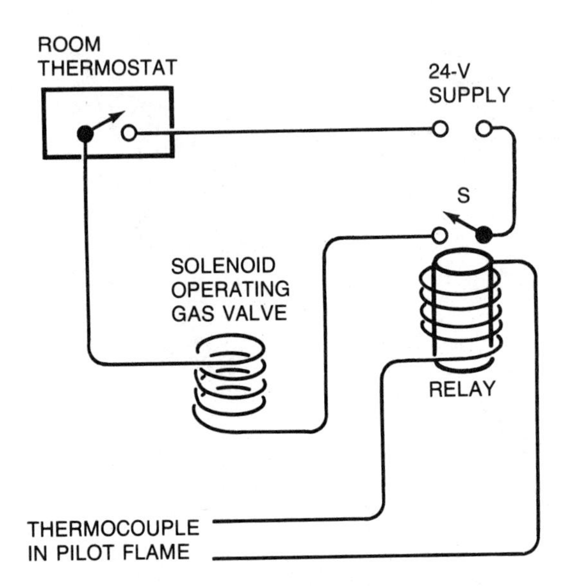

FIGURE 2–7 Gas-fired furnace control

alternating voltage on opposite sides of the crystals. That voltage, containing all the elements of speech or music, is used as the input signal voltage for an amplifier.

Some ceramic materials, such as barium titanate, show this piezoelectric property and can be used in record player pickups. A piece of barium titanate, when tapped with a heavy object, Figure 2–8, produces enough voltage to flash a small neon lamp of the

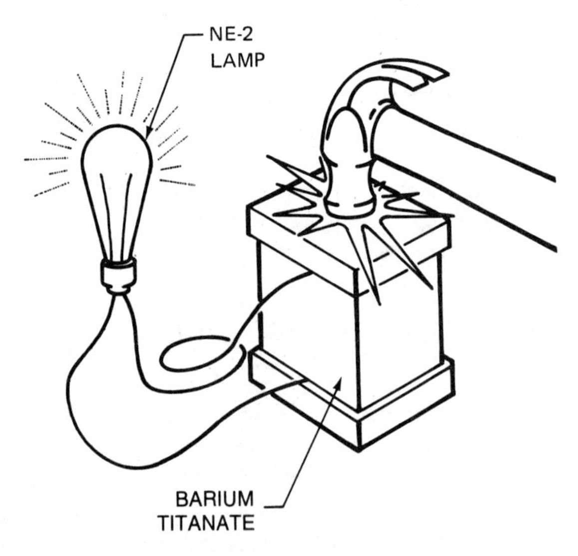

FIGURE 2–8 Piezoelectrical material

type NE-2 (0.04 W). Wires from the lamp contact each end of the piezoelectric material. As shown in the picture, a scrap of hard plastic is placed as a cushion between the hammer and the brittle barium titanate.

2–8 THE EFFECTS OF ELECTRICITY

Up to this point we have been concerned with the conversion of various energy sources into electrical energy. It should come as no surprise that this process is reversible at the consumer's end. After all, the production of light and heat from electrical power is so common in our daily lives that we need hardly mention it. But what of the other forms of energy that we discussed earlier? Some of their uses may not be obvious, so let us explain.

Chemical Reaction From Electricity

Certain industrial processes call for chemical reactions that can be initiated by an electrical current. This process, known as *electrolysis,* uses the electrical current to produce desirable chemical changes or variations in the properties of certain substances. An obvious application is found in the generation of voltages from chemical batteries, as explained in Section 2–4 and in Chapter 14. Electrolysis also finds application in the electroplating of metals, the production of chemicals, the refining of copper, and the extraction of aluminum and magnesium from their ores.

Mechanical Pressure From Electricity: The Piezoelectric Effect

We have seen, in Section 2–7, that the slight bending of a quartz crystal causes an alternating voltage to appear on its faces. This effect can also be reversed by applying opposite charges (voltage) to the faces of a slice of quartz so that the quartz crystals bend slightly. If the charges are reversed repeatedly (by an application of AC) at a frequency that is close to the natural mechanical vibration frequency of the crystal, the crystal begins to bend and vibrate rapidly back and forth. This mechanical oscillation, in turn, sustains the continued production of alternating charges on the faces of the crystal. Since a given crystal will oscillate at only one frequency, the quartz crystal is used to control the frequency of rapidly alternating voltages in radio transmitters. In other words, vibrating crystals are used in electronic circuits where high-frequency AC is being generated and stabilized. Crystals have found wide application in two-way radio communication sets.

Vibrating crystals are also used in small earphones, where the crystal vibrates a diaphragm, causing audible sound waves. Similarly, vibrating crystals have been used successfully by the recording industry in cutting phonograph records.

Vibrating crystals can also be used in the production of ultrasonic vibrations (approximately 50,000 cycles per second), which are used for medical purposes or in ultrasonic cleaners.

Industry uses piezo materials as transducers for pressure indicators or indicators of mechanical vibration of machine parts.

Magnetism From Electricity

Electricity and magnetism are two closely related phenomena. Whenever an electric current flows, magnetic forces are being created. This effect is known as *electromagnetism*.

Electromagnetism finds countless applications in a multitude of electrical and electronic devices and appliances. For example, every electrical machine or appliance producing motion is bound to have electromagnetic forces at work. Just think of all the household appliances driven by an electric motor. All electric motors operate on electromagnetism.

Without electromagnetism there would be no radio and television as we know it, because the sound from the loudspeaker and the picture on your TV screen are produced electromagnetically.

These are just a few examples highlighting the importance of magnetism produced by electricity. The subject of electromagnetism is covered more extensively in a later chapter of this book.

SUMMARY

- Electricity is produced by conversion from other energy sources, for example, friction, heat, light, magnetism, pressure, and chemical activity.
- Friction between certain insulating substances gives rise to static electricity.
- Electrical generators operate on electromagnetic principles.
- Photovoltaic devices convert light energy directly into electrical energy.
- Piezoelectricity (pressure emf) is a voltage developed by mechanically distorting certain crystals and ceramics. This emf is used in pickups and control elements, not as a power producer.
- Heat applied to the junction of two metals will displace electrons from one metal to the other. This emf is used in thermocouples for high temperature measurements.
- Batteries produce DC electricity from chemical energy.
- Cells and batteries are classified as being either primary or secondary.
- When electricity is used, its energy reverts back to any of the original source energies. In other words, the effects of electricity can be either heat, light, magnetism, pressure, or chemical activity.

Achievement Review

1. Name five methods (energy conversions) to generate electricity. Give an example of each.
2. What is meant by the term *electromagnetic induction?*
3. Explain the difference between the words *photovoltaic* and *photoconductive*.

4. Name at least two applications for
   a. Thermocouples
   b. Piezoelectricity
   c. Electrolysis
   d. Electromagnetism

3 Electrostatics

Objectives

After studying this chapter, the student should be able to

- Define and explain the new technical terms introduced in this chapter

  electrostatics
  electrostatic induction
  smoke precipitators
  potential energy
  selenium
  electroscope
  lightning arrester
  Van de Graaff generator
  joule
  foot-pound

- Recognize potentially dangerous situations involving electrostatic induction
- Give examples of nuisance aspects of static electricity
- Discuss some useful applications of static electricity
- Draw, from memory, the pattern of an electrostatic field between like and unlike charges

3–1 STATIONARY ELECTRONS

The word *static* means that something is at rest; it is not going anywhere. Therefore, static electricity is a charge that is stationary; it is not a current.

A good deal can be learned about electrons when they are stationary. Even though they are not traveling, electrons do not sit still but constantly jump from one place to another. The sparks and arcs they produce are often very important.

The word static in a discussion of electricity may first remind one of noise in a radio. When lightning flashes or an electric spark occurs, vibrating electrons broadcast some energy that is received as crackling in a radio receiver. The term static is now applied to any unexplained radio interference, whether the interference is of true electrostatic origin or not.

The material covered in this chapter is an extension of ideas introduced in Chapter 1. The Summary of Chapter 1 should be reviewed at this point, particularly these useful facts:

- Any object that has a negative charge has received some electrons that normally do not belong to it, and they will try to escape.
- Any object that has a positive charge has lost some of its electrons. When a positively charged object has electrons returned to it, it goes back to its original neutral state. The returned electrons do not have to be the same ones that were lost. (Remember that electrons are all alike, regardless of where they come from.) If a million electrons are lost and a different million electrons return, everything is still back to neutral.

Lists A and B in Figure 1–1 introduced insulators (nonconductors). Insulators easily acquire a static electric charge, because they are materials in which electrons cannot move readily. If a spot on the insulating material has a surplus of electrons, it keeps the electrons. If at another location on the material there is a lack of electrons, however, that surplus finds it difficult to slide over and fill the lack.

Materials in which electrons move freely are called *conductors*. Metals, with their loose electrons in the outer rings of the atoms, are good conductors. Conductors can be charged only if they are insulated from their surroundings by some nonconducting material.

3–2 ELECTROSCOPES

The *electroscope* was once widely used to compare electric charges and served as a crude voltmeter. It was also used in the detection of radioactive minerals. The device was an important aid in experiments that led to the discovery of electrons. For present-day use, it has been replaced by devices that are more accurate and sensitive but that are also more complex.

The electroscope, Figure 3–1, must first be charged using a known charge. The charging procedure is simple: (1) A hard rubber rod is charged by rubbing it on wool; (2) then the hard rubber rod is slid along the knob of the electroscope. Electrons from the negative hard rubber are transferred from the rod to the electroscope, where they repel each other and scatter all over the metal knob, metal rod, and metal leaves. The leaves repel each other because both are charged alike, in this case, negatively, Figure 3–1A.

The electroscope can also be charged positively, using a charged glass rod instead of the hard rubber rod. As the glass rod is wiped along the knob of a neutral electroscope, the rod attracts electrons from the electroscope. The electroscope now has a lack of electrons (since they have been transferred to the glass rod), and the electroscope is positively charged. The leaves again spread apart, Figure 3–1B, because the like positive charges repel.

Unknown Charges

A charged electroscope will indicate whether an unknown charge is positive or negative. For example, a pen is charged by rubbing it on a sleeve. The question now is whether the pen is negatively or positively charged. With a hard rubber rod, the electroscope is charged negatively, as described before. The pen is then brought slowly toward

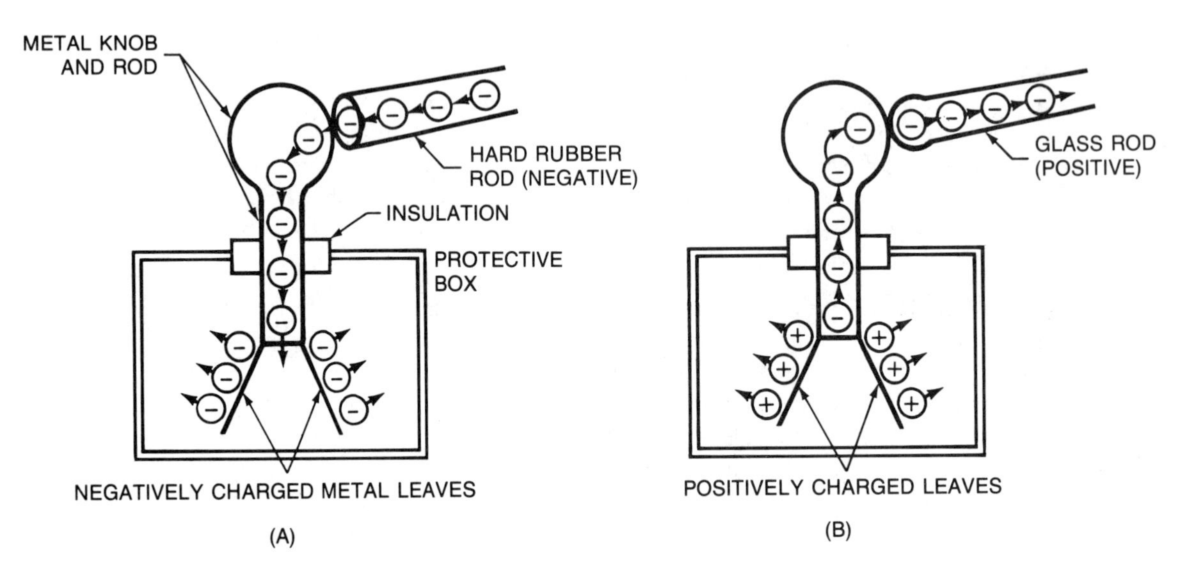

FIGURE 3–1 Charging the electroscope

the top of the electroscope, and the leaves are observed closely for their motion. The leaves may repel even more, Figure 3–2A. Why?

The leaves were made negative with the rubber rod. If they repel more, they must now be *more* negative. The leaves can become more negative only by having more electrons forced onto them. The unknown charge on the pen, therefore, must be negative, because more of the electroscope's electrons were repelled to the leaves.

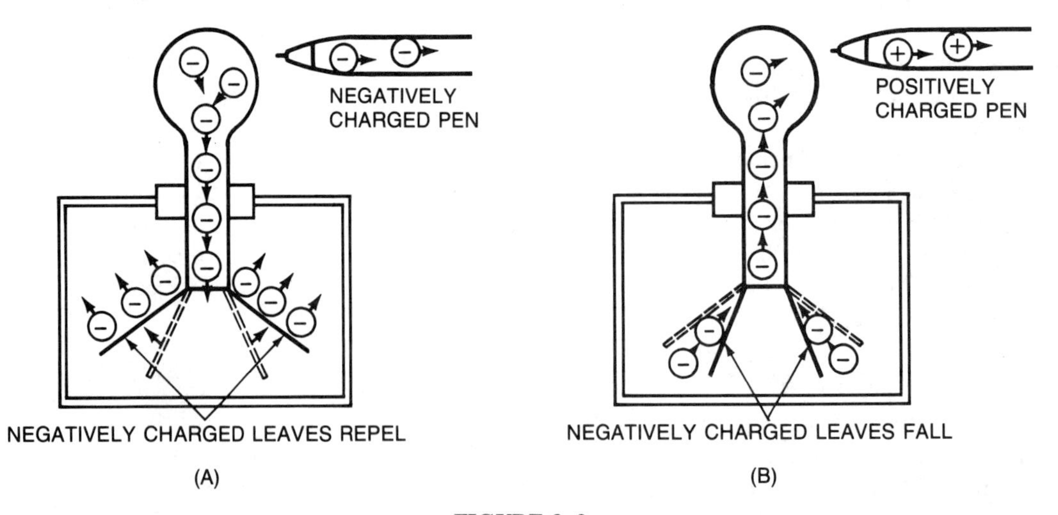

FIGURE 3–2

If the pen is charged positively, Figure 3–2B, what will happen? The positively charged pen will attract some of the excess electrons from the leaves to the knob. As these repelling electrons are removed, the leaves will fall together.

A positively charged electroscope can also be used to identify unknown charges. Keep in mind that the positively charged electroscope has relatively few of its electrons removed; there are still many electrons there. When a negative object comes near, it will repel the electrons in the electroscope toward the leaves. These electrons will neutralize the repelling positive charges on the leaves, causing the leaves to fall together. An approaching positive object will pull even more electrons from the positively charged leaves, causing them to repel each other more strongly.

In these instances, all of the electron motion is within the electroscope itself. The charged pen is not brought close enough to the knob to permit electrons to jump the gap between the two objects.

3–3 ELECTROSTATIC INDUCTION

A charged object always affects other objects in its neighborhood either by attracting or repelling electrons within the nearby object. *Electrostatic induction* is the process by which the neighboring object acquires a charge. In unusual circumstances, electrostatic induction can cause accidents. For example, on a warm, dry day, a truck driver speaks to the new gas station attendant, Joe, Figure 3–3. Joe lets go of the iron post he was sitting on near the gas pump. As the truck drives away, he steps over to remove the hose from another customer's gas tank. At this point, the gasoline catches fire. What causes this to happen?

A strongly charged truck, Joe's new rubber-soled shoes, the removal of his hand from the grounded post, electrostatic induction, and a spark in the presence of gasoline

FIGURE 3–3

fumes all contributed to the fire. If the truck was negatively charged, it repelled electrons from Joe through the post to ground (earth). Having lost some electrons, Joe was then positively charged. When he removed his hand from the post, Joe did not immediately regain the lost electrons since he was insulated by his rubber-soled shoes and the dry air. Electrons, attracted to his positive hand from the gasoline tank, jumped the gap, causing a spark. The spark ignited the gasoline vapor.

Let us illustrate this electrostatic induction process using the electroscope, Figure 3–4.

1. Bring a negatively charged rod (truck) near the electroscope (Joe). The experimenter's hand touching the electroscope corresponds to Joe's connection to the ground through the post. The rod will repel electrons from the electroscope knob. These electrons will move from the electroscope to the experimenter's hand touching the knob.
2. Remove the hand from the knob. This leaves the electroscope with a positive charge (lack of electrons).
3. Remove the negative rod.

If these steps are performed in the correct order, the electroscope will be left in a charged condition, shown by its repelling leaves. The fact that the charge is positive can be checked by its behavior when the negative rod is slowly brought near it again. The charge on the electroscope and the charge on Joe are examples of induced charges.

> An induced charge on one object is caused by the approach of another charged object, without contact between the two objects.

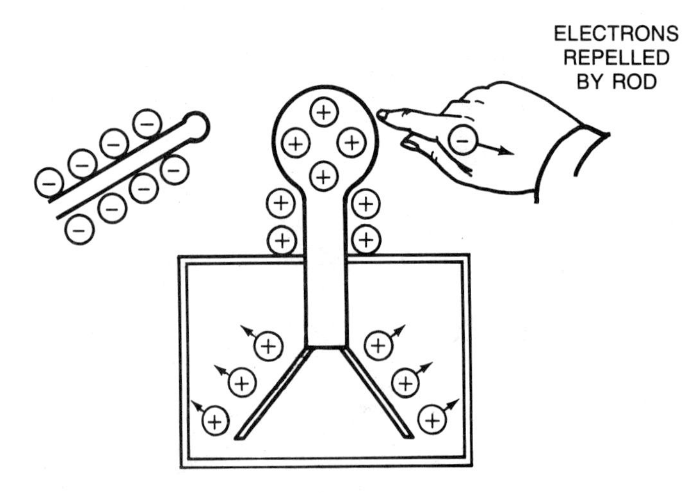

FIGURE 3–4

3–4 LIGHTNING

Benjamin Franklin showed that lightning is a large-scale performance of ordinary electrostatic behavior. (A few people who tried his kite experiment were killed by it.)

The distribution of electric charges in thunderclouds has been mapped, but as yet no one knows just how the charges are formed. Most of the lightning flashes occur within the cloud itself. Whether electrons go from cloud to ground or from ground to cloud depends on the part of the storm area that is over a particular location, Figure 3–5.

Objects, such as a house, that are under the small, positively charged area at the center of the storm are hit by electrons moving from the earth to the cloud. Objects, such as trees, under the large, negatively charged part of the cloud are hit by electrons driving from the cloud to the earth.

It is not possible to insulate against lightning. Protection can be provided only by giving it an easy path to ground. Well-grounded lightning rods are very effective. Ammunition sheds have been protected simply by a steel cable supported along the ridge of the shed and grounded at each end. Steel frames of large buildings, grounded during construction, serve as lightning protection for the building.

An automobile may be struck by lightning, but its occupants are not harmed because the steel body conducts the high-voltage electrons away from them. (Crawling under the car is definitely unsafe!) A *lightning arrester* for TV and radio antennas is simply a grounded wire brought close to, but not touching the antenna. Lightning arresters for power lines are made of materials with specialized resistance properties. The materials permit the high-voltage lightning discharge to pass to ground but stop the lower voltage energy of the power line from being grounded.

3–5 NUISANCE STATIC CHARGES

Besides lightning and the radio interference it causes, static charges are responsible for a variety of other nuisances and hazards.

- Power belting readily becomes charged. Grounded pulleys, combs or tinsel bars close to the belt, conductive belt dressing, and the use of conductive belt materials help prevent the buildup of static electricity.
- Charges that have accumulated on trucks and cars are usually grounded by a wire on the approachway to a tollgate to prevent electrical discharges (shocks) when coins change hands.
- Anesthetic gases are combustible. Precautions are taken to avoid static sparks, which may cause explosions during surgical operations. Grounded equipment, moist air, and conductive rubber help prevent the accumulation of charges.
- Grain dust, flour, wood dust, and cotton lint have produced disastrous explosions. Prevention of static charges helps avoid such dust explosions.
- In printing, sheets of paper may fail to feed into the press or to stack properly if

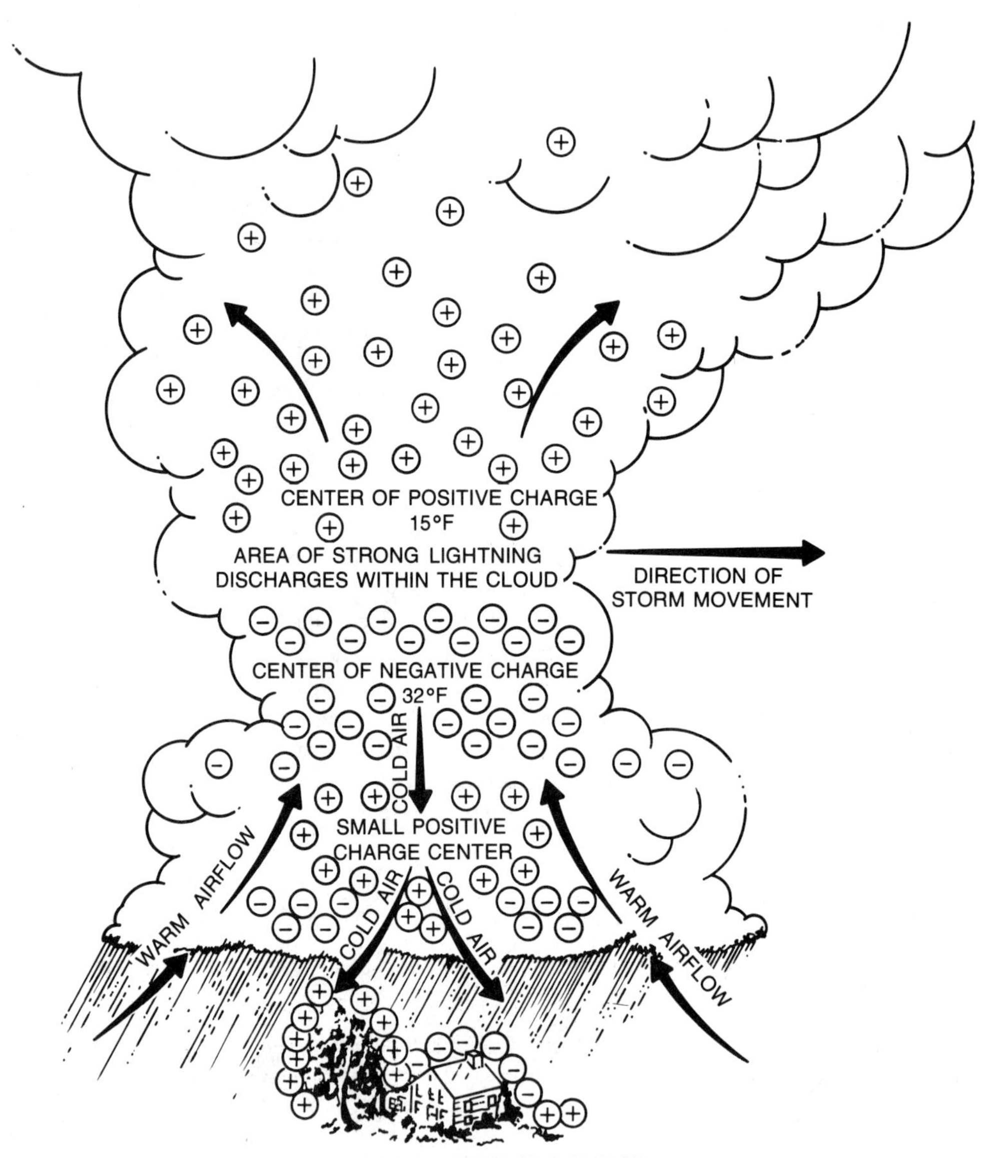

FIGURE 3–5

they repel each other or are attracted to nearby objects. When printing is done from a continuous sheet, the sheet may become charged and ignite if combustible solvent vapors are present. Similar problems occur in the cloth and plastic industries. Some devices that ground or neutralize these charges are the tinsel bar, flame, or a long metal comb connected to an alternating current power supply.

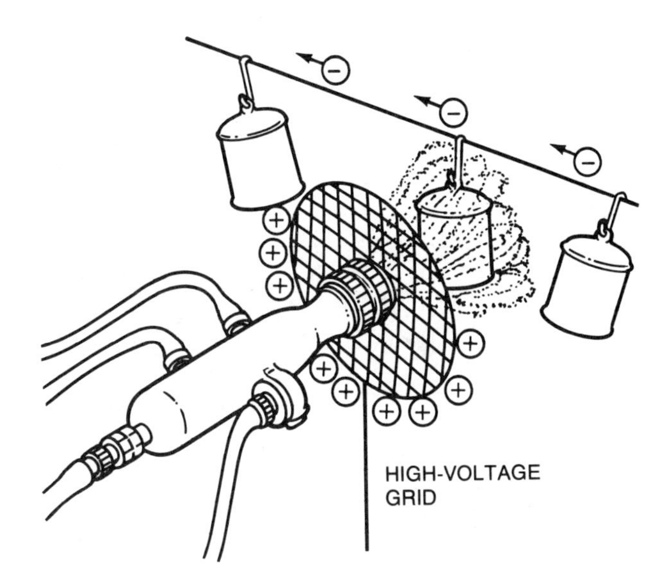

FIGURE 3–6 Electrostatic painting

3–6 USEFUL STATIC CHARGES

Painting

Although some industries must prevent it, other industries use the static charge as a tool. In paint spraying applications, paint particles, given a charge after they leave the spray gun, are attracted to the oppositely charged object receiving the paint, Figure 3–6. This method produces an even coat without wasting paint, and paint particles do not accumulate in holes or openings since the charge there is no greater than that at the outer surface.

Sandpaper Manufacture

Sandpaper grit can be made to stand up as it is being applied to the paper backing by giving the sand particles a positive charge and the glue-covered paper backing a negative charge, Figure 3–7. Since the individual sand grains have like charges, they repel each other and stand apart. This method produces a sharper sandpaper.

Smoke Precipitators; Rug Manufacture

Smoke precipitators charge the smoke particles, which are then collected on oppositely charged screens, so that the particles are not released to pollute the atmosphere. Similarly, charged rug fibers are attracted to oppositely charged, glue-covered backings to form new types of rugs and fabrics.

Xerography (Dry Copying)

After 15 years of developmental work, Carlson's application of electrostatic charging in an office copying machine appeared on the market in 1950. The core of the

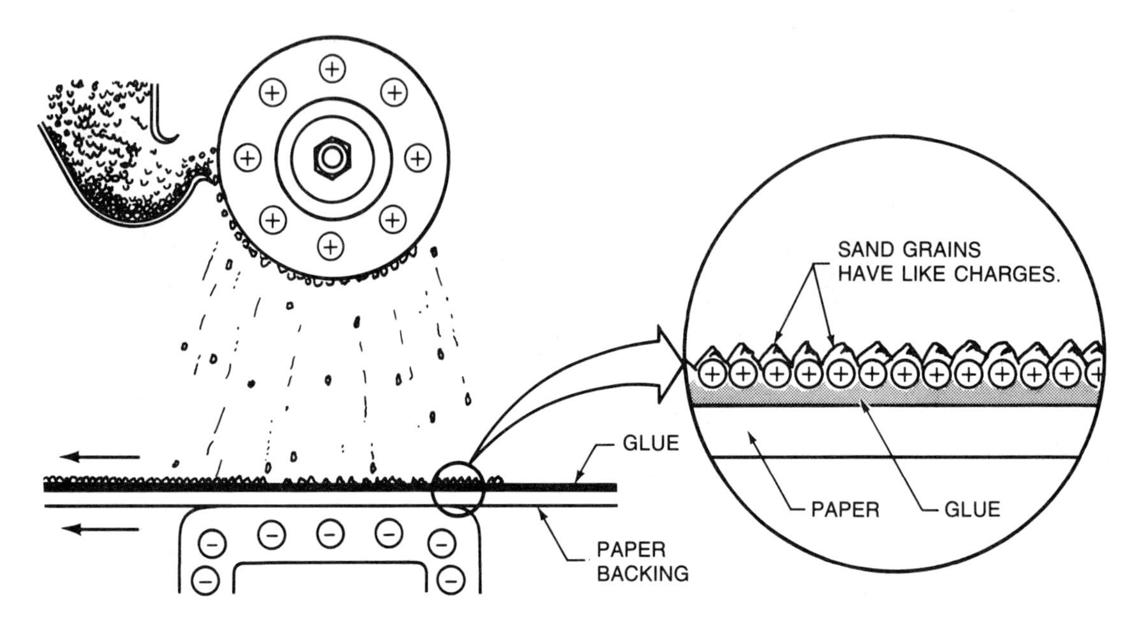

FIGURE 3–7 Sandpaper manufacture

machine is a rotating aluminum drum coated with a thin film of the element selenium, a semiconducting material, Figure 3–8. The conductivity of selenium is increased by light. The machine operates in the following manner:

1. As the selenium surface rotates past a positively charged wire, it loses electrons to the wire. This charging of the selenium is done in darkness.
2. Light, reflected from the sheet of material to be copied, is projected by lenses and mirrors onto the positively charged selenium on the drum. The black form of the letter R, for example, reflects no light to the drum. However, the white paper background does reflect. This reflection forms an image there, just as a camera or projector forms an image. Where the light surrounding the letter strikes the selenium, the selenium becomes a better conductor. Electrons flow from the aluminum drum to this lighted area and neutralize its positive charge. The dark area of the drum, where the letter R was, remains positively charged.
3. A small conveyor belt, not shown in Figure 3–8, pours a mixture of positively charged glass beads and negatively charged black powder over the previously illuminated surface of the drum. The black particles are attracted to the positively charged areas left on the drum in Step 2. The black particles are not attracted to the neutralized areas of the polished selenium surface. Any excess black powder toner and glass beads fall away.
4. The piece of paper that is to receive the finished print now passes under the drum. The paper has been positively charged by a wire so that it attracts the negatively charged black powder from the surface of the drum.
5. Finally, the paper carrying the black powder image passes either under a radiant

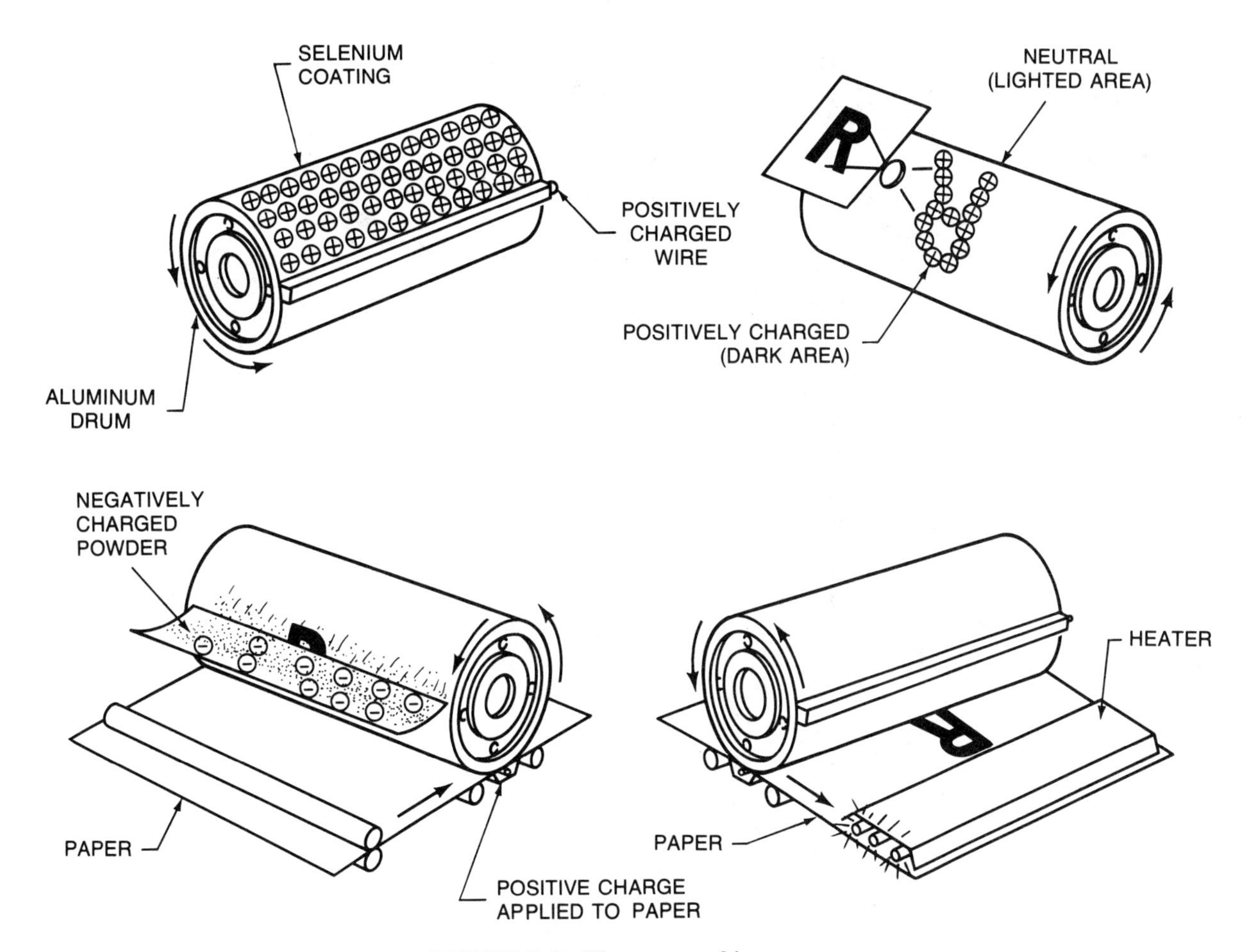

FIGURE 3–8 The xerographic process

heater or over a heated roller, where the black particles are melted into the paper to form a permanent copy. A brush (not shown) removes any particles still on the drum, and the selenium surface is ready for recharging.

Electrostatic Generators

Many types of static generators constructed in the past produced tiny currents at high voltage. A much more efficient device is the Van de Graaff generator, Figure 3–9. This device is used to create high voltages for speeding up charged particles in atom smashing experiments. This generator can also be used to test lightning protection equipment. Small models of the Van de Graaff generator are available for small-scale laboratory work.

In Figure 3–9, the sphere at the top of the device charges to a few hundred thousand volts, but the number of electrons accumulated there is small enough so that the spark is harmless.

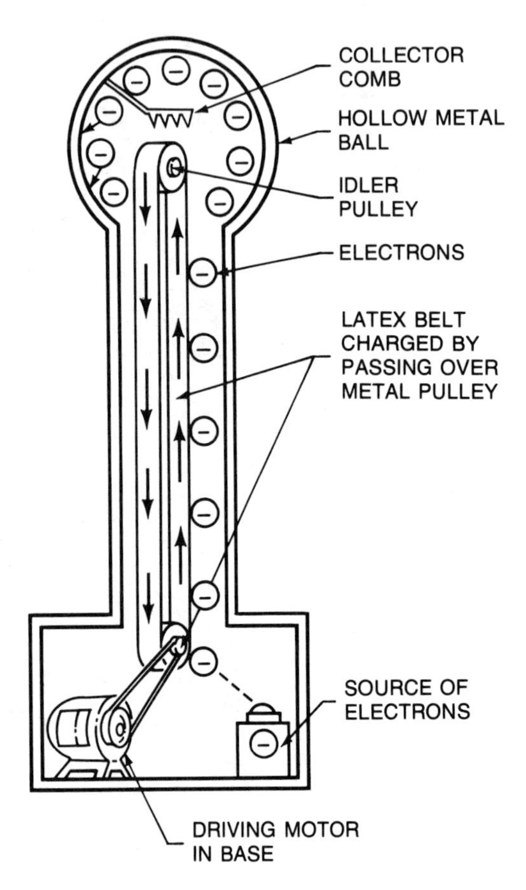

FIGURE 3–9 Van de Graaf generator

3–7 POTENTIAL ENERGY OF ELECTRONS

An electron forcibly taken away from one neutral object and put on another neutral object has gained *potential energy,* Figure 3–10. Force is needed to pull the electron away from Object A because the electron is attracted back to A by the positive charge that remains on A. If several electrons are transferred to Object B, they repel additional electrons from A. These electrons possess potential energy. Given the opportunity, the electrons will return to A, producing heat as they return. If these electrons are permitted to return to A through an electric motor, their potential energy is converted to mechanical energy.

Figure 3–11 illustrates another example of potential energy. Water can be taken from the pond and put into the tank, but it cannot get there by itself. In other words, energy must be used to carry the water up to the tank. The water, however, then has *potential energy*. Given the opportunity, the water in the tank will run back down to the ground, releasing energy on whatever it hits; that is, it will do useful work if it passes over a water wheel, or it will produce heat by friction.

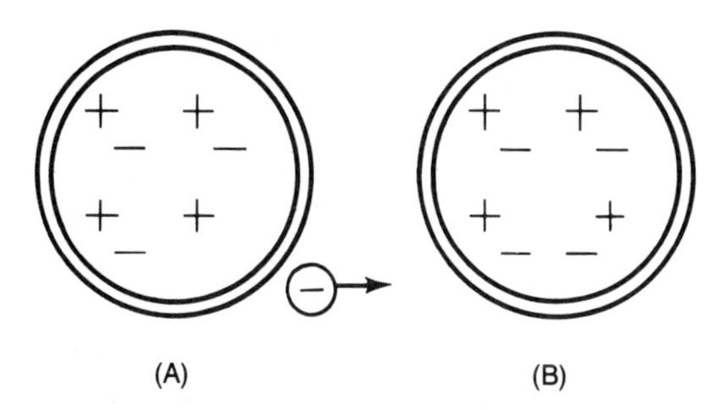

FIGURE 3–10 Electron with potential energy

The energy used to carry the pond water up to the tank is not wasted. Each gallon of water in the tank has more potential energy than a gallon of water in the pond. There is a *potential energy difference* between the water in the tank and the water in the pond.

In a similar manner, there is a potential energy difference between the electrons in Object B, Figure 3–10, and the electrons in Object A. In other words, there is a *potential difference* between B and A.

The potential energy of water is measured in a unit called a *foot-pound*. (See Chapter 9 for a more detailed discussion of energy and its measurement.) The potential energy difference between the water in the tank and the water in the pond can be expressed as a number of foot-pounds of energy per gallon of water.

The potential energy of electrons is often measured in a metric system unit called the *joule*. The transfer of each coulomb of electrons from A to B requires the use of a

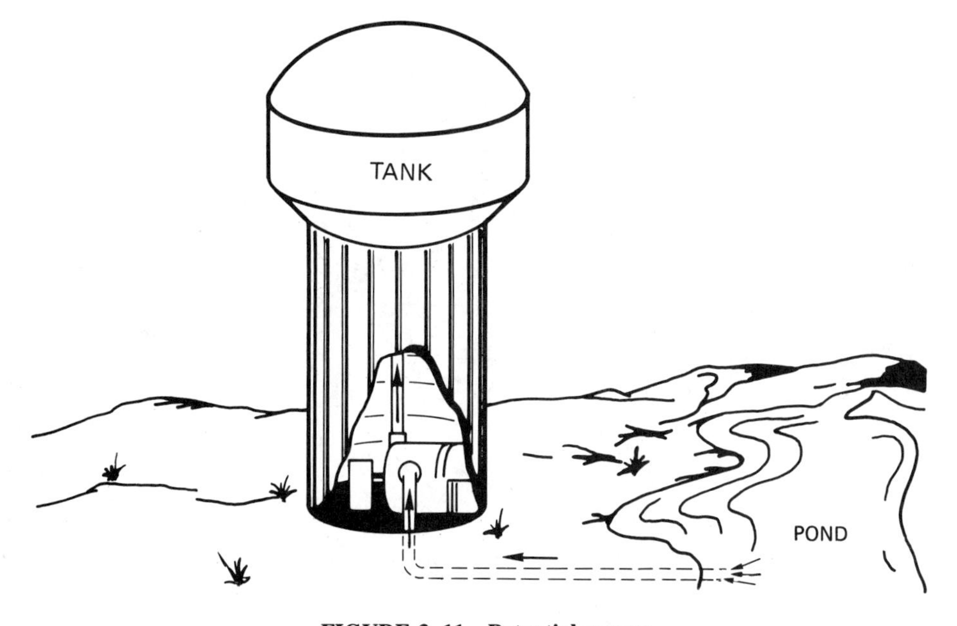

FIGURE 3–11 Potential energy

certain number of joules of energy. The potential energy difference between A and B can be expressed as a number of *joules* of energy required *per coulomb* of electrons. (This is comparable to the energy per gallon of water.) This expression, *joules of energy per coulomb,* is brought into this discussion because it is the accurate definition of a volt.

We may continue to think of a volt as a measure of electrical pressure or electromotive force because these terms may be a little easier to understand. However, the exact meaning of the *volt* is a measure of the *potential energy difference* between two points.

Volts = Joules/Coulomb

3–8 ELECTRIC LINES OF FORCE

Many years ago, people were puzzled as to just how two objects could exert force on each other when they were some distance apart with no material connection between them. How do two oppositely charged objects attract each other in a vacuum in the absence of light and heat? Even today, this question is still puzzling. We have no evidence that any small particles pass back and forth between them to pull them together or push them apart. There is no evidence that any sort of wave passes from one object to another.

Years ago, a solution to this puzzle was proposed by picturing invisible lines of force, like ropes or rubber bands, pulling two opposite charges together, Figure 3–12. The pattern of these lines may actually be seen by scattering splinters or dust of some nonconducting material between two strongly charged objects, Figure 3–13. Shredded

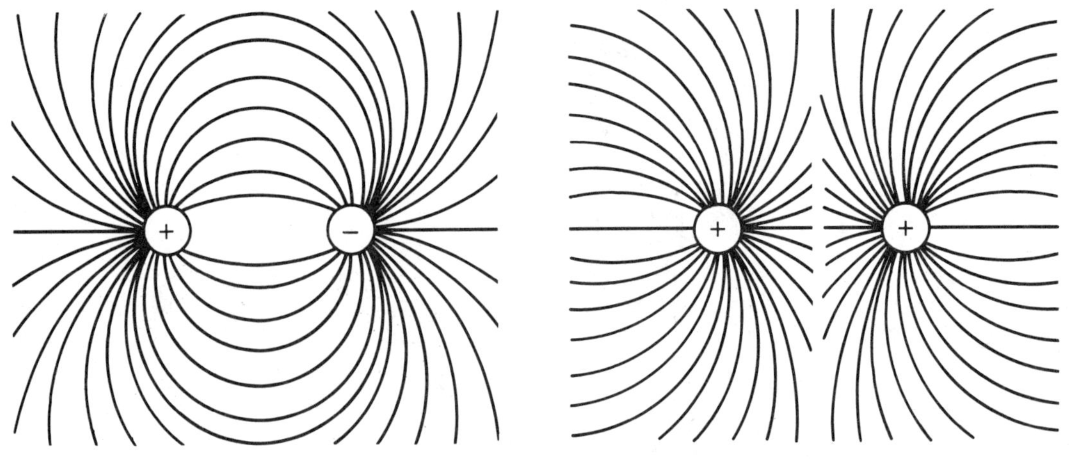

LINES OF FORCE

FIGURE 3–12

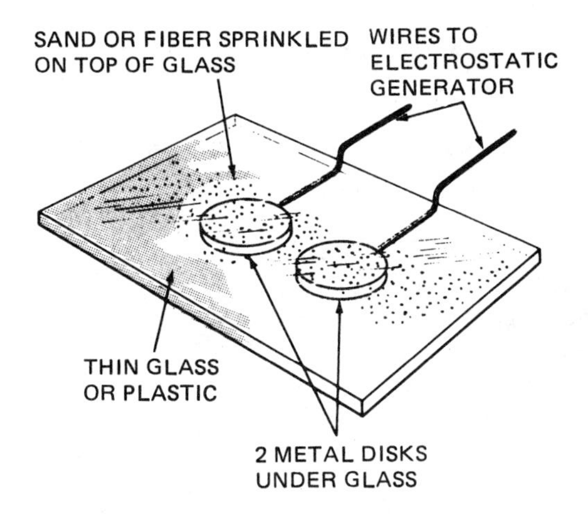

FIGURE 3–13

wheat, short fibers, grass seed, or wood splinters can be used. The short fibers become charged by induction and tend to become aligned in patterns like those shown in Figure 3–12. These patterns graphically show the existence of a very real force between the objects.

SUMMARY

- A negatively charged object has a surplus of electrons; a positively charged object lacks some electrons.
- A charged electroscope identifies unknown charges.
- Objects may become charged by being near a charged object, without contact. This is known as *electrostatic induction*.
- Potential difference is measured in volts. It is the energy difference between electrons in different locations.
- Charged objects are surrounded by an electric field that consists of lines of electric force extending from the charged object.

Achievement Review

1. What is static electricity?
2. Is static electricity more often noticed on conductors or insulators? Why?
3. Can conductors be charged?
4. Describe what happens when a positively charged object comes near but does not touch a positively charged electroscope.

5. What happens when a negatively charged object comes near a positively charged electroscope?
6. What is electrostatic induction?
7. When a negatively charged object approaches a neutral object, what sort of charge is induced on the neutral object?
8. Charged objects are said to possess energy. Explain.
9. A volt is a joule per coulomb. Explain.
10. What are electric lines of force?
11. How does a lightning arrester stop lightning?
12. State a few industrial uses of static charge.
13. What industries try to avoid static charges?

4
Basic Circuit Concepts

Objectives

After studying this chapter, the student should be able to

- Define and explain the new technical terms introduced in this chapter

voltage *(E* or *V)*	current *(I)*
resistance *(R)*	electromotive force
emf	volt
electron current	conventional current
ampere (A)	ohm (Ω)
source	load
control	schematic
closed circuit	open circuit
short circuit	fuse
circuit breaker	coulomb

- Draw a schematic diagram of a simple circuit
- Use correct circuit notation to describe the values of voltage, current, and resistance
- Differentiate between conventional current flow and electron current flow
- Explain the concept of a short circuit
- Observe proper safety precautions in working with electrical circuits

4–1 THREE MEASURABLE CIRCUIT QUANTITIES

In dealing with any useful quantity, whether vegetables, steel bars, or electrons, a system of measurement must be used to keep track of the production, transfer, and use of the commodity. So it is with electrical circuits that are arranged to obtain practical use of electrical energy. There are three fundamental concepts that constitute the elements of an electrical circuit, namely

Voltage, Current, and Resistance

Since these words represent abstractions, quantities that cannot be directly perceived by one's senses, it is very important that you develop a correct mental image of the key concepts. The following sections are designed to help you gain a clear understanding of these terms.

4–2 VOLTAGE

Let us consider a simple source of DC, a common flashlight battery, which is technically known as a dry cell, Figure 4–1. The little minus signs in the drawing represent a huge quantity of electrons. This should suggest that the shell, accessible at the bottom plate, has a vast surplus of electrons as compared with the number of electrons at the top cap of the cell.

Remember that the electrons have a negative charge, making the bottom plate with its surplus of electrons more negative than the top cap with its electron deficiency. It is this difference in potential (difference in the number of electrical charges) that is known as *voltage*. Frequently, you will find the term electromotive force (emf) used to describe the same condition.

It is important to note that protons do not enter into our discussion when we define the positive pole of the dry cell in Figure 4–1. (Remember, protons are the carriers of positive charges.) Both the positive and the negative poles can be defined in terms of their relative number of electrons. In this context the word *positive* simply means *less negative,* as compared to some other reference point.

Voltage is always measured between two points of different potential. The unit of measurement used for this is called the *volt*. Thus we can say that our dry cell has an emf of 1.5 volts. Compare this with the two terminals of a car battery. Between these two points we would expect to measure 12 volts. To measure such a voltage, we use an instrument known as a *voltmeter*.

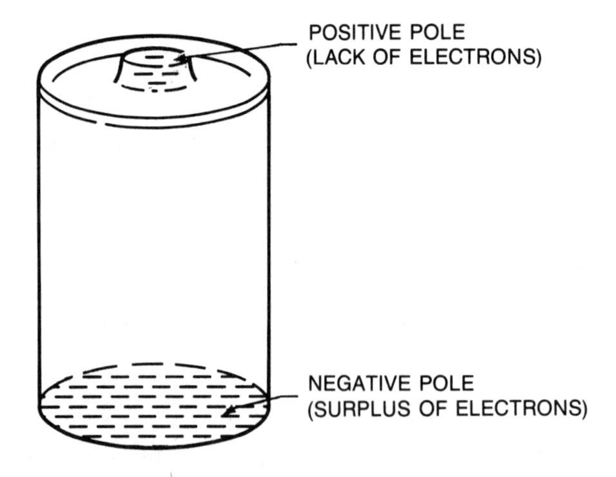

FIGURE 4–1

4–3 CURRENT

For discussion purposes, let us pretend that we connect a heavy wire between the poles of our battery. In reality this must never be done because it constitutes an undesirable condition known as a *short circuit*. Figure 4–2 shows how such a wire provides a path by which the electron surplus can drain off toward the point of electron deficiency. It is this motion of electrons that we refer to as *current*. Notice that this definition of current considers the motion of free electrons from a point of electron surplus to a point of electron shortage. In other words, the electron theory defines current flow as a motion of free electrons from the negative pole of a source, through the outer circuit path, and back to the positive pole. This is the definition of electron current that we will use throughout this book.

You should know, however, that there is yet another theory that leads to the conclusion that current flows from positive to negative. This is known as *conventional current* flow. The concept of conventional current can be very useful and will be reintroduced to you at a later date when you begin your study of electronics and semiconductors.

Current as a Rate of Flow

In measuring the rate of electron flow, we are concerned with a quantity rate rather than simple speed. In ordinary electrical devices, the number of electrons passing through the device each second is the important consideration, not their speed in miles per hour. Water pumps are rated in gallons per minute, ventilation fans in cubic feet per minute, and grain-handling equipment in bushels per hour. All of these are quantity rates.

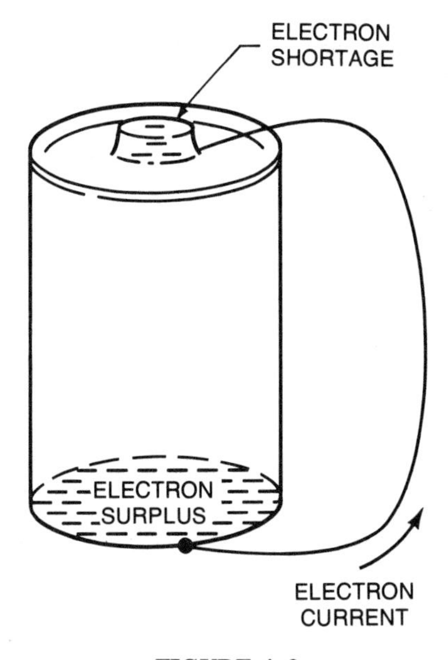

FIGURE 4–2

To establish such a rate for electrons, we must first decide on a measure of quantity. We could use the number of electrons passing a point per second, but so many pass by a point in one second that the number is too great to use. Instead, we lump together 6,250,000,000,000,000,000 electrons and call this quantity a *coulomb* of electrons (in honor of Charles Coulomb, a French scientist). To show just how large a number 6,250,000,000,000,000,000 is, assume that this many flies are in New York State in the summer. If all of the flies were killed, they would cover the state to a depth of 6 feet, packed so that each cubic inch contained 500 flies.

We can measure the rate of electron flow in coulombs per second. This is comparable to measuring the flow of traffic in cars per hour or measuring air current in cubic feet per second. It can also be compared to measuring water flow in gallons per second, Figure 4–3.

To express the rate of electron flow, the phrase *coulombs per second* is seldom used. Rather, we use one word that means coulombs per second. This word is *ampere* (A), again named in honor of a French scientist.

One ampere is a flow rate of 1 coulomb/second.

To measure the current flow in an electrical circuit, we use an *amperemeter,* or *ammeter*.

Current Speed

People sometimes get into discussions of what is meant by the speed of electricity. A more exact term than speed of electricity is needed to distinguish between (1) the average speed of individual electrons as they drift through the wire and (2) the speed of the impulse. We realize that when we turn on a light, the light is *on* immediately. In a house wired with #12-gauge wire, calculations show that there are so many electrons in the wire that the average speed of individual electrons is only about 3 inches per hour when the current is 1 ampere. Three inches per hour is the speed of the electron drift through the copper wire. However, keep in mind that since the wire is full of electrons

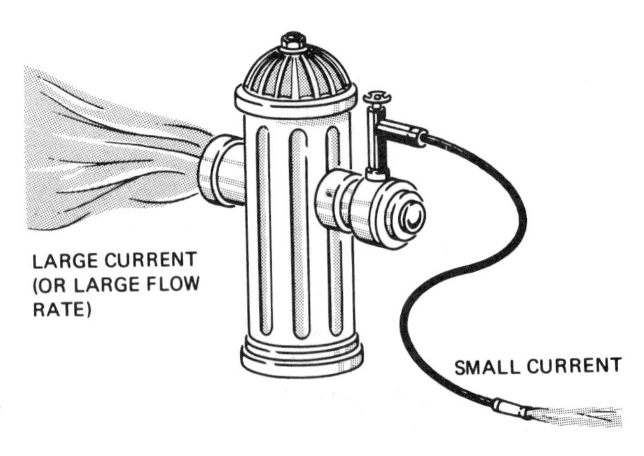

FIGURE 4–3 Flow rates

to begin with, they start moving everywhere at once when the switch is turned on. The actual speed of this impulse depends on the arrangement of the wires and may be anything from a few thousand miles per second to the speed of light, 186,000 miles per second, as a theoretical top limit.

4–4 RESISTANCE

A single stroke of the oars will not keep a rowboat moving indefinitely at the same rate of speed. Neither will voltage keep electrons moving indefinitely at the same rate. Friction slows the movement. This internal friction, which retards the flow of current (electrons) through a material, is called *electrical resistance*.

Electrons slide through a copper wire easily, like a boat through water. Electrons also move through iron and some metal alloys fairly easily, although not as easily as they do through copper. But there are many materials through which electrons can move hardly at all, even if a lot of pressure (high voltage) is applied. Trying to move electrons through sulfur, glass, plastic, or porcelain, for example, is about as effective as trying to row a boat on a concrete road or on plowed ground.

The list in Figure 4–4 compares the resistance of common materials. Those of highest resistance (so high it is difficult even to measure) are the best insulators. Those of lowest resistance are the best conductors. In the range between the two extremes are the materials that are poor conductors yet do not have quite enough resistance to be called insulators.

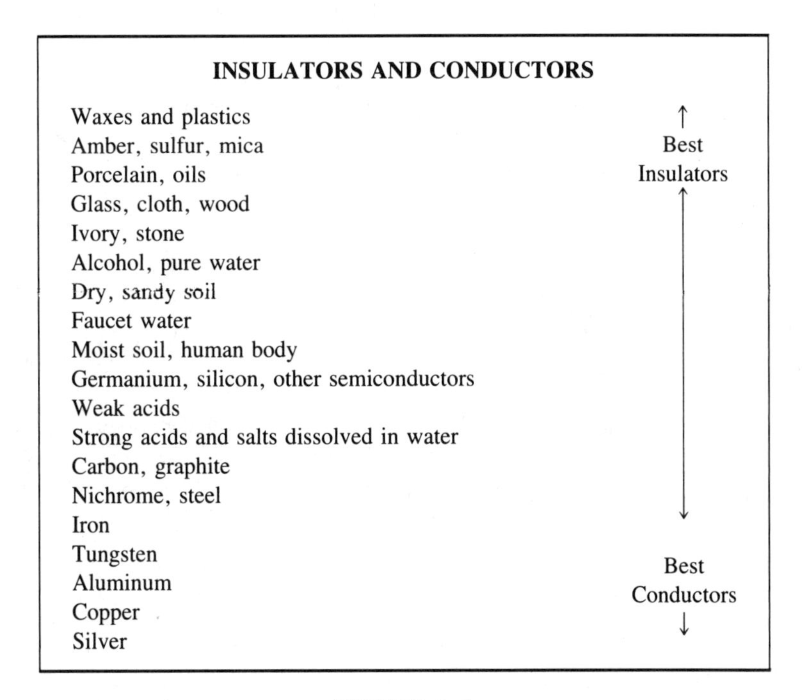

INSULATORS AND CONDUCTORS

Waxes and plastics
Amber, sulfur, mica
Porcelain, oils
Glass, cloth, wood
Ivory, stone
Alcohol, pure water
Dry, sandy soil
Faucet water
Moist soil, human body
Germanium, silicon, other semiconductors
Weak acids
Strong acids and salts dissolved in water
Carbon, graphite
Nichrome, steel
Iron
Tungsten
Aluminum
Copper
Silver

FIGURE 4–4

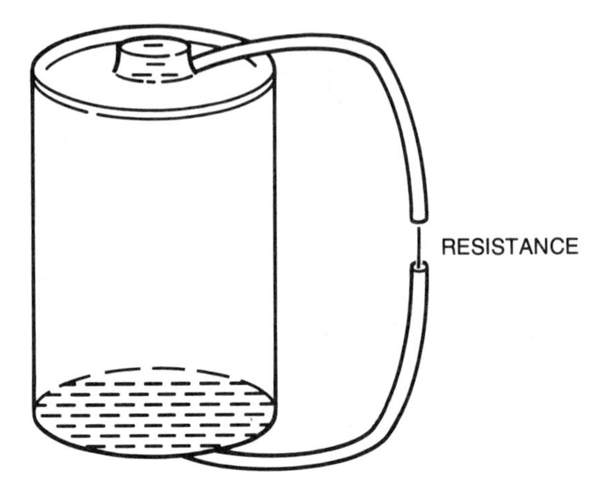

FIGURE 4–5

Recall from Chapter 1 that elements whose atoms have only one, two, or three electrons in the outer shell (or orbit) are conductors, because these electrons are free to move. Elements whose atoms have only five, six, or seven electrons in the outer shell are insulators, because there are no free electrons.

This unit measure of resistance is the *ohm,* which is denoted by the Greek letter omega (Ω). This unit is named after a German scientist, G. S. Ohm. Electrical resistance can be measured with an *ohmmeter*. Many multimeters (multipurpose instruments), such as a VOM, have an ohmmeter function built in.

The thickness of a conductor also affects the resistance of a circuit. Assume that the heavy wire we placed across the dry cell in Figure 4–2 has a short section of very thin wire inserted, as shown in Figure 4–5. This thin section represents a resistance to the flow of current in the same manner that the bottleneck of a one-lane detour represents a reduction in the flow of traffic on a broad freeway.

It is important to note that heat is being developed whenever an electrical current is forced through a resistance. It is conceivable that a resistance wire, like the one shown in Figure 4–5, becomes so hot that it begins to glow intensely and gives off light. This is called *incandescence*. Ordinary light bulbs, called *incandescent lamps,* operate on this principle.

Most of the heat-generating appliances that you know contain an electrical resistance wire. Think, for instance, of the cigarette lighter in the dashboard of your car. It, too, has a built-in resistance wire that glows from heat whenever a voltage forces current through the resistance.

4–5 A SIMPLE ELECTRIC CIRCUIT

Figure 4–6 represents a simple, functional circuit. The necessary elements of such a circuit are described as source, load, conductors, and control. The first three of these terms are closely linked with the three important concepts discussed in the foregoing section. The *source* represents the voltage that forces the electrons through the circuit.

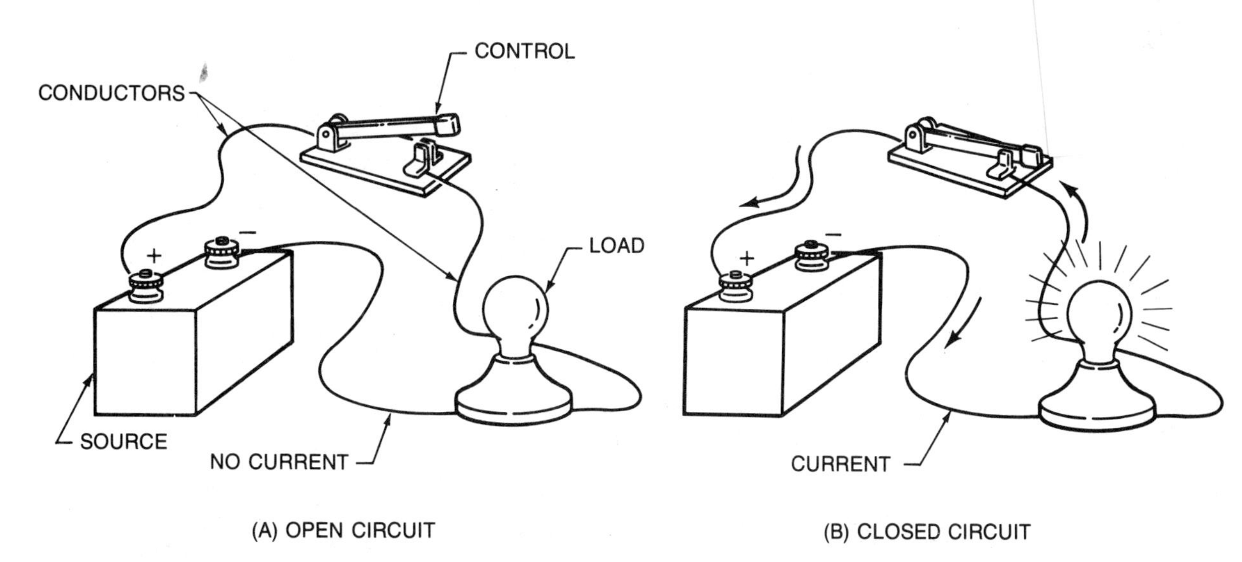

FIGURE 4–6 Current requires a complete circuit path.

The battery, in this example, represents the source. DC electrical sources are very often color coded to designate polarity of the terminals, positive or negative. By convention (widespread agreement and usage) the negative terminal of a source will be identified by the color black and the positive terminal will be marked by red. The *load* (or load resistance) utilizes the electrical energy. The lamp represents the load. The *conductors* provide the path for the electron current; and the *control* is provided by the switch, which can be operated to turn the lamp on or off.

4–6 OPEN CIRCUITS AND CLOSED CIRCUITS

Careful comparison of the two circuits shown in Figure 4–6 will reveal the difference between open and closed circuits. You see, if electrons are to start moving and keep moving, a complete path for them to follow must exist. In other words, a closed circuit is required to allow the current to flow whenever electrical energy is needed by the load. Figure 4–6B represents this condition.

Now, let us look at Figure 4–6A. The switch has been opened. This simple switch is known as a *knife switch*. Its operation resembles that of a drawbridge. When the drawbridge is opened, the infinite resistance of the air between the switch contacts stops all current motion. This condition is known as an *open circuit*. This term applies to a circuit containing an open switch, a burned-out fuse, or any separation of wires that prevents current from flowing. If an appliance fails to operate when connected but does not blow a fuse, an open circuit in the device may be the problem.

4–7 THE SCHEMATIC DIAGRAM

Pictures of electrical circuits and their components, such as shown in Figure 4–6, are seldom used in the practice of electrical trades. It is more practical to convey ideas by means of graphic symbols known as *schematic diagrams,* or *schematics*. Figure 4–7

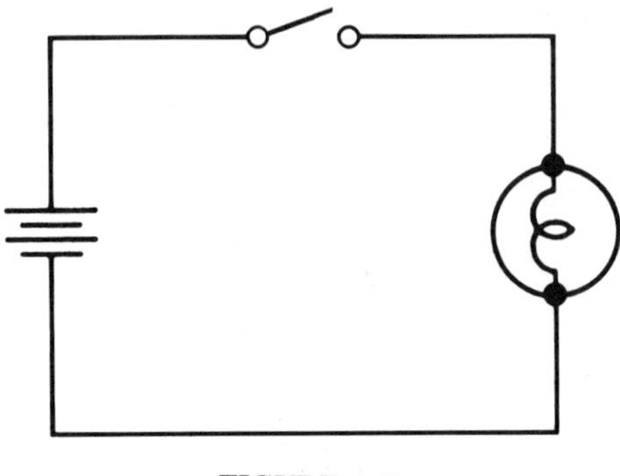

FIGURE 4–7

is a schematic representation of the circuit shown in Figure 4–6A. Compare the two drawings carefully. It is important that you learn how to draw and interpret such schematic diagrams. This type of skill requires some time, of course, and necessitates that you learn many graphic symbols. A chart of such symbols is shown in the Appendix, Figures A–12 and A–13. It is suggested that you frequently refer to this list as you progress in your studies.

4–8 BASIC CIRCUIT NOTATION

In the preceding sections you became acquainted with some of the basic units of electrical measurement. At this time you should recall that the voltage (also called emf) is measured in units called volts.

Let us assume that the battery symbol in Figure 4–7 represents a 12-volt car battery. We might then use a circuit notation to show this specific value, as illustrated in Figure 4–8, where the notation reads $E = 12$ V. The letter E, in this case, stands for electromotive force (emf).

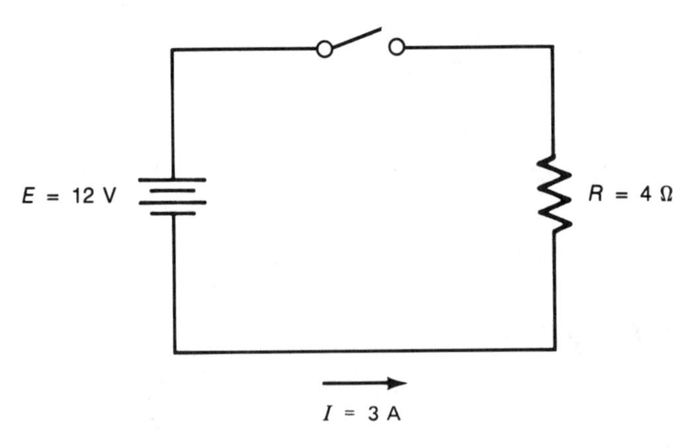

FIGURE 4–8

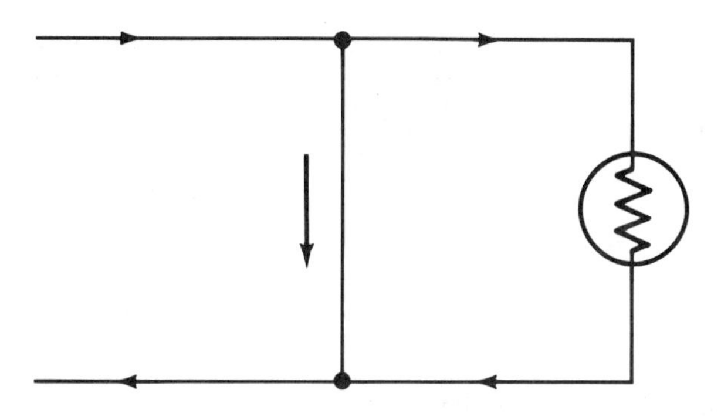

FIGURE 4–9 Circuit of low resistance

Alongside the wire in Figure 4–8, you see the notation $I = 3$ A. This means the current in this circuit is equal to 3 amperes. Do you remember that current is measured in amperes? The letter I stands for current.

In the same circuit diagram, the symbol of a resistor has been used to replace the symbol of a lamp. This symbol is often used to indicate the general resistance of any load, without specifying its exact nature. The adjoining legend, $R = 4\ \Omega$, means this load has a 4-ohm resistance. The word *ohm* is denoted by the Greek letter omega (Ω).

4–9 THE SHORT CIRCUIT

A *short circuit* is a parallel path of very low resistance, often caused accidentally, Figures 4–9 and 4–10. In Figure 4–10 the frayed insulation on the lamp cord may

FIGURE 4–10 Short circuit

permit the two wires to touch each other. If the wires do touch, they form a path of nearly 0 ohms resistance. As a result, a large amount of current appears in the wires leading to the place of contact. The wires can overheat and start a fire. To prevent such an outcome, fuses or circuit breakers are used in series with each house circuit. If a short circuit occurs, the excessive current melts the fuse wire or trips the circuit breaker, opening the circuit.

4–10 ELECTRICAL SAFETY

All personnel working with electrical circuits must be made aware of the special hazards they may encounter when they come in contact with live circuit components. A firm knowledge of electrical principles and safety practices will aid in combating irrational fear and help to develop good safety attitudes.

Consider the following safety precautions:

1. Electrical shock can be harmful, if not fatal. Voltage levels are not the only determining factor. One hundred twenty volts is sufficient to kill. On the other hand, one can experience high voltages, say 25,000 volts, without lasting damage. Instead, it is the effect of current through the body that is harmful. Generally, currents as low as 0.15 ampere are fatal.
2. Therefore, to avoid or minimize electrical shock, one should
   a. Never touch two wires at the same time.
   b. Never touch one wire and ground (earth) at the same time.
3. Instead, be sure to
   a. Shut power off and, if possible, lock it out before working on a circuit.
   b. Insulate the body from the ground.
   c. Properly ground all metal enclosures of electrical equipment. A green-colored wire usually serves this purpose. This green wire is the *equipment ground* and attaches to the third prong of a male attachment cap (plug).
   d. Discharge capacitors before touching their terminals.
   e. Avoid use of metal measuring tapes in the vicinity of live conductors.
   f. Use a type C extinguisher to combat electrical fires. Any other type of fire extinguisher may cause fatal shock.
4. At times it may be necessary tc test live circuits in operation. Under such conditions it is advisable to put one hand in your pocket or behind your back. This prevents contact between two hot wires or between one hot wire and ground. Remove metal jewelry, such as rings and watches.
5. Hot soldering irons can cause severe burns. Be sure hot soldering irons have sufficient time to cool before you store them.
6. Hot solder splatters easily. Keep this in mind before you shake your soldering tool. Molten solder on someone's body may cause serious injury.
7. Wear safety glasses for the following activities:
   a. Using rotating power tools, such as grinders and drills
   b. Chiselling or chipping
   c. Handling chemical electrolytes

8. When charging batteries, remember that hydrogen gas is highly volatile. A small electrical spark from a wire is sufficient to start an explosion.
9. Ventilate batteries when you are charging them. Be sure to loosen the cell caps to allow gases to escape.
10. To prevent oxidation, make electrical connections and splices well and be sure they are tight. Poor connections introduce resistance at the junction, which inevitably causes heat and fire hazards.
11. Do not walk away from a machine that is in motion. When shutting down a machine, stay with it until it has completely stopped.
12. Report all injuries to your instructor, regardless of how slight or unimportant you think they might be.
13. Some hand tools, such as pliers or screwdrivers, have insulated handles. Such insulation may not always be sufficient to protect you from accidental shock. You may have to take additional safety precautions.
14. If contact is made with metal parts, causing a short circuit, live wires may cause serious burns and blinding flashes.
15. Some electrical components, especially resistors, develop considerable heat. Allow them to cool before touching.

There are a few more specialized safety precautions that will be pointed out to you as you progress in your studies.

SUMMARY

- The three key concepts of an electrical circuit are voltage, current, and resistance.
- Voltage may be thought of as a pressure (or force) that exists between two points of different potential.
- Voltage is also known as electromotive force (emf).
- Voltage is measured in units called volts.
- Electron current is a motion of free electrons from negative to positive. This is the standard adopted for this book.
- Conventional current is said to flow from positive to negative.
- Current is measured in units called amperes.
- When measuring amperes, we measure the rate of flow of electrons (1 ampere = 1 coulomb per second).
- The effect of the electrical impulse travels with the speed of light, while individual electrons move relatively slowly.
- The opposition to current flow is called resistance.
- Resistance is measured in units called ohms.
- Resistance always generates heat when a current flows through it.
- An electrical circuit consists of a load connected to a source by means of conductors. An optional control is often added to such a circuit.
- Schematic diagrams are the blueprints of the electrical trades.
- Closed circuits provide a complete path for the electrons.

- There can be no current flow in an open circuit.
- Short circuits are caused when a low-resistance path bypasses the load.
- Short circuits cause excessive current flow and create potential fire hazards.
- Fuses and circuit breakers are used to protect against excessive currents.
- Electrical safety rules must be followed to insure the safety of personnel and equipment.

Achievement Review

1. What is the difference between the negative pole and the positive pole of a battery or power source?
2. What colors are used to identify (a) the negative pole and (b) the positive pole?
3. What is voltage? Define the word.
4. What are some other names for voltage?
5. What letter symbol is used for voltage?
6. Define the word *current*.
7. What letter symbol is used for current?
8. Explain the difference between conventional current flow and electron current flow.
9. What is an ohm, and what is its letter symbol?
10. What is the unit of measurement for current?
11. What kind of instrument is used for measuring resistance?
12. What are amperemeters used for?
13. What is an emf? Explain.
14. How fast does the electron current travel?
15. What is meant by potential or potential difference?
16. Draw a simple schematic with a DC source of 6 volts connected to a 1.5-ohm load. Indicate a current of 4 A flowing, as well as the voltage and resistance values.
17. Explain the purpose of fuses and circuit breakers.
18. Explain how fuses differ from circuit breakers.

THREE FUNDAMENTAL CONCEPTS

Voltage, current, and *resistance* are the basic concepts underlying all electrical and electronic principles.

1. Write brief but concise definitions for the following terms: voltage, current, and resistance.

2. Be sure to complete the chart below as directed during lecture.

Name of the Electrical Concept	Letter Symbol for This Concept	Unit of Measurement	Letter Symbol for This Unit of Measurement	Name of Instrument Used for Measurement
Voltage				
Current				
Resistance				

5

Scientific Notation and Metric Prefixes

Objectives

After studying this chapter, the student should be able to

- Define and explain the new technical terms introduced in this chapter

milli-	kilo-
micro-	mega-
nano-	giga-
pico-	

- Express decimal numbers in powers of 10 and vice versa
- Convert any number into scientific notation
- Round off numbers to three-digit accuracy
- Express any number with an appropriate metric prefix
- Convert numbers with metric prefixes into decimal numbers or scientific notation
- Perform the basic arithmetic operations (add, subtract, multiply, and divide) with numbers expressed in powers of 10

5–1 RATIONALE FOR STUDYING THIS UNIT

Personnel working in the field of electricity and electronics are frequently called upon to perform mathematical calculations in order to predict or confirm proper circuit operation. The numbers involved in such calculations are often extremely large or extremely small. For example, you may remember the electrical unit *coulomb* (defined in Section 4–3), consisting of 6,250,000,000,000,000,000 electrons. Many people feel ill at ease about pronouncing such numbers, let alone manipulating the numbers by multiplication or division.

For more than a generation, electrical technicians have made use of a tool known as the slide rule, which requires knowledge and application of the *powers of 10* principle. The invention and widespread use of modern electronic calculators has not relieved us of this need to manipulate numbers in scientific notation. In fact, the ability of a hand-held calculator to process and display very large (or very small) numbers depends upon the use of scientific notation.

It is assumed that all students of this chapter will use an electronic calculator in working out the various exercises. If you own a calculator but do not fully understand all of its operations, you should consult your calculator's instruction manual or seek help from someone within your study group.

A word of encouragement: If the topics of this chapter are new to you and you suffer frustrations in your attempt to solve the assigned problems, do not despair! You are not expected to become an expert at this merely by finishing just one more chapter. The proficiency will come by repeated exposure to the ideas and principles presented here. You may have to refer back to this chapter many times as you progress in your studies.

5–2 POWERS OF 10

Surely, you know that 10^2 means 10×10, or 100. Likewise, 10^3 means $10 \times 10 \times 10 = 1{,}000$, and so forth. Armed with this knowledge, you should easily construct a chart like the one shown in Figure 5–1.

Let us work a few examples to show how this knowledge can be used to expresss any number of powers of 10.

EXAMPLE 5–1

Given:

$$25{,}000 = 25 \times 1{,}000$$
$$10^3 = 1{,}000$$

Find: 25,000 in terms of powers of 10.

Power of 10	Decimal Equivalent	Name
10^6	1,000,000	one million
10^5	100,000	one hundred-thousand
10^4	10,000	ten thousand
10^3	1000	one thousand
10^2	100	one hundred
10^1	10	ten
10^0	1	one unit
10^{-1}	0.1	one tenth
10^{-2}	0.01	one hundredth
10^{-3}	0.001	one thousandth
10^{-4}	0.0001	one ten-thousandth
10^{-5}	0.00001	one hundred-thousandth
10^{-6}	0.000001	one millionth

FIGURE 5–1

SOLUTION

$$25{,}000 = 25 \times 10^3$$

EXAMPLE 5–2

Given:

$$5{,}000{,}000 = 5 \times 1{,}000{,}000$$
$$10^6 = 1{,}000{,}000$$

Find: 5,000,000 in terms of powers of 10.

SOLUTION

$$5{,}000{,}000 = 5 \times 10^6$$

EXAMPLE 5–3

Given:

$$0.007 = 7 \times 0.001$$
$$10^{-3} = 0.001$$

Find: 0.007 in terms of powers of 10.

SOLUTION

$$0.007 = 7 \times 10^{-3}$$

EXAMPLE 5–4

Given:

$$470{,}000 = 47 \times 10{,}000$$
$$10^4 = 10{,}000$$

Find: 470,000 in terms of powers of 10.

SOLUTION

$$470{,}000 = 47 \times 10^4$$

5–3 SCIENTIFIC NOTATION

The term *scientific notation* refers to a special way of using powers of 10. The method demands that a decimal point be placed behind the first significant digit when the number is changed into powers of 10. Let us use Example 5–4 to demonstrate this principle.

In Example 5–4 we concluded that

$$470{,}000 = 47 \times 10^4$$

The thoughtful student will recognize that we could have equally well stated the results as

$$470 \times 10^3$$

In fact, we have many more ways to write our answer. Here is a list of alternatives.

$4{,}700 \times 10^2$
$47{,}000 \times 10^1$
4.7×10^5

Of these choices, only the last number qualifies as scientific notation, because the decimal point appears behind the first significant digit.

Suggested Procedure

Count the number of spaces that the decimal point has to be moved in order to appear behind the first digit. This count, the number of spaces moved, becomes your exponent in the power of 10.

Let us express the number 6,750.0 as scientific notation. Note that the decimal point must be moved three places to the left in order to appear behind the numeral 6. The three places moved determines the exponent. Thus, we write our answer

6.75×10^3

5–4 SIGNIFICANT DIGITS

Most practical calculations are generally limited to only three significant digits. What does this statement mean? An example will make this clear.

Assume that some friends you know want to brag about the performance of their new automobile. They tell you that they clocked their speed at 104.6784935 miles per hour. (That is downhill with a tailwind blowing.) How would you receive such information short of calling them a liar? Obviously, speedometers do not give 10-digit readouts. Neither do electrical measuring instruments. Most such instruments (called *meters*) are limited to three-digit accuracy.

Does this mean that electricity is a sloppy science? Quite to the contrary! Electronic measurements can be extremely precise; however, practical considerations would recommend that all numbers be limited to three-digit accuracy. Thus, your friends should have reported their speed as 105 miles per hour.

5–5 ROUNDING OFF TO THREE SIGNIFICANT DIGITS

The last statement may have puzzled you. ''Why 105 miles per hour?'' you might ask. ''Why not 104 miles per hour?'' The answer to this is determined by the *fourth* digit. If the fourth digit is 5 or greater, the third digit should be rounded off to the next higher number. On the other hand, if the fourth digit is smaller than 5, it would be dropped along with all successive digits.

It must be pointed out that 0's are not considered to be significant unless they are preceded by a number other than 0. Let us consider, for instance, the number 0.00020543. In rounding off such a number, it is important to recognize that the numeral 2 is the first significant digit to be retained. Thus, the answer would be reported as 0.000205, or in scientific notation as

$$2.05 \times 10^{-4}$$

5–6 METRIC PREFIXES

As stated earlier, in electrical theory we are often confronted with very large or very small numbers. To overcome the inherent inconvenience of dealing with such awkward numbers, it is customary to modify the basic measuring units by attaching a prefix to their multiples or submultiples.

For instance, 1,000 volts can be called 1 kilovolt. This example shows that the word *kilo* stands for a multiple of 1,000, or 10^3. In other words, *1 kilo means 1,000 times a unit.*

Can you see that with this convention we can express a number like 27,000 ohms as $27 \times 10^3 = 27$ kilohms, or just 27 kΩ?

Now that we understand the word kilo, let us have a look at Figure 5–2 to become acquainted with some other prefixes.

Note that both uppercase letters and lowercase letters are used for letter symbols. This is especially important to distinguish between *M* for mega and *m* for milli.

The letter symbol for micro is the Greek letter μ (pronounced: myoo).

EXAMPLE 5–5

Given:

milli (m) = 0.001
micro (μ) = 0.000001

Find:

a. 0.0035 amperes in milliamperes
b. 0.0035 amperes in microamperes

Prefix	Numerical Equivalent
giga (G)	1,000,000,000 $= 10^9$
mega (M)	1,000,000 $= 10^6$
kilo (k)	1,000 $= 10^3$
milli (m)	0.001 $= 10^{-3}$
micro (μ)	0.000001 $= 10^{-6}$
nano (n)	0.000000001 $= 10^{-9}$
pico (p)	0.000000000001 $= 10^{-12}$

FIGURE 5–2

SOLUTION

a. $0.0035\ \text{A} = 3.5 \times 10^{-3}\ \text{A} = 3.5\ \text{mA}$
b. $0.0035\ \text{A} = 3{,}500 \times 10^{-6}\ \text{A} = 3{,}500\ \mu\text{A}$

EXAMPLE 5–6

Given:

kilo (k) = 1,000
mega (M) = 1,000,000

Find:

a. 470,000 in kilohms
b. 470,000 in megohms

SOLUTION

a. $470{,}000\ \Omega = 470 \times 10^{3}\ \Omega = 470\ \text{k}\Omega$
b. $470{,}000\ \Omega = 0.47 \times 10^{6}\ \Omega = 0.47\ \text{M}\Omega$

More examples are shown in conjunction with the practice problems at the end of the chapter.

5–7 MULTIPLICATION AND DIVISION WITH POWERS OF 10

Even though it is assumed that all students of electricity have a basic knowledge of algebra, it may be helpful to briefly refresh our memory.

Rule # 1 (Multiplication): When powers of 10 are to be multiplied, add their exponents. For instance

$$10^{5} \times 10^{3} = 10^{5+3} = 10^{8}$$

Rule # 2 (Division): When powers of 10 are to be divided, subtract their exponents. For instance

$$10^{5}/10^{3} = 10^{5-3} = 10^{2}$$

EXAMPLE 5–7

Given: Rule # 1

Find: 3×10^{6} times 2.5×10^{3}

SOLUTION

$$3 \times 2.5 \times 10^{6} \times 10^{3} = 7.5 \times 10^{6+3} = 7.5 \times 10^{9}$$

EXAMPLE 5–8

Given: Rule # 2

Find: 24×10^{5} divided by 8×10^{2}

SOLUTION

$24/8 \times 10^{5-2} = 3 \times 10^3$

EXAMPLE 5–9

Given: Rule # 2

Find: 6×10^3 divided by 2×10^{-2}

SOLUTION

$6/2 \times 10^{3-(-2)} = 3 \times 10^{3+2} = 3 \times 10^5$

SUMMARY

- Powers of 10 are useful in making computations involving very large or very small numbers.
- Scientific notation mandates that a decimal point be placed behind the first significant digit of a number expressed in powers of 10.
- Three-digit accuracy is generally sufficient for quantities related to practical applications. Numbers should be rounded off accordingly.
- Rounding off to three digits demands a look at the fourth digit to determine whether to round up or down.
- Metric prefixes are used in the electrical trades to describe multiples or submultiples of basic units. Examples are kilovolt, milliampere, and megohm.
- In multiplying powers of 10, the exponents are added.
- In dividing powers of 10, the exponent of the divisor must be subtracted.

Achievement Review

ROUNDING OFF MATHEMATICAL ANSWERS

Key Ideas

1. It will be assumed that electric circuit quantities are sufficiently accurate to justify solutions containing three significant digits.
2. A sequence of significant figures never begins with 0.
3. The fourth significant digit determines whether the third figure should be rounded off.
4. If a number has less than three digits, fill in the rest with 0's. Thus: 75 = 75.0.

Examples

28.4623 = 28.5
799.5238 = 800
1,947,268 = 1,950,000

0.000203728 = 0.000204
4.6924 = 4.69
0.05 = 0.0500

Exercises

Round off the following numbers:

1. 746.23
2. 0.0050629
3. 84,896
4. 27.452
5. 0.023465
6. 2.012345
7. 8
8. 17
9. 468,642
10. 4,672
11. 89.072
12. 73,141
13. 11,569
14. 37,732
15. 4,711
16. 6,789.0
17. 0.42632
18. 31
19. 29.862
20. 29,862
21. 0.29862
22. 852,100
23. 5,280,362
24. 373,806
25. 7,380,650
26. 0.002468
27. 146.43
28. 7,896,245
29. 0.0026793
30. 39.05

POWERS OF 10

Exercises

1. Write the following numbers as powers of 10:
   a. 100
   b. 1
   c. 100,000
   d. 10,000,000
   e. 1,000,000,000
   f. 1,000,000
   g. 1,000
   h. 0.0001
   i. 0.1
   j. 0.001
   k. 0.000 0001
   l. 0.000 000 001
   m. 0.000 000 000 001
   n. 0.000 001
   o. 10,000
   p. 0.01
2. Write the following numbers in decimal form:
   a. 10^0
   b. 10^1
   c. 10^3
   d. 10^5
   e. 10^8
   f. 7.31×10^5
   g. 9.60×10^{-2}
   h. 17.4×10^{-3}
   i. 10^{-1}
   j. 10^{-3}
   k. 10^{-5}
   l. 10^{-8}
   m. 10^{-12}
   n. 6.183×10^2
   o. 3.101×10^6
   p. 5.68×10^{-4}
3. Write the following numbers in scientific notation:
   a. 438
   b. 9,800
   c. 0.1006
   d. 27.10
   e. 77×10^4
   f. 38.6×10^{-5}

g. 0.000 082
h. 7,610,000
i. 0.0002
j. 6.72
k. 381×10^2
l. 11.7×10^{-5}

METRIC CONVERSIONS

Prefix Kilo

The prefix kilo means 1,000 times a unit. For instance, 220 kV means 220 kilovolts, or 220,000 volts.

Examples

270 kV = 270,000 V
15 kV = 15,000 V
3 kΩ = 3,000 Ω
0.8 kV = 800 V
39,000 Ω = 39 kΩ
4,700 Ω = 4.7 kΩ
500 Ω = 0.5 kΩ

Exercises

Perform the indicated changes below.

1. 1,200 Ω = __________ kΩ
2. 3.3 kΩ = __________ Ω
3. 68,000 Ω = __________ kΩ
4. 47 kΩ = __________ Ω
5. 220,000 Ω = __________ kΩ
6. 150 kΩ = __________ Ω
7. 18,000 V = __________ kV
8. 24 kV = __________ V
9. 150 Ω = __________ kΩ
10. 0.27 kΩ = __________ Ω
11. 2.7 kV = __________ V
12. 330 Ω = __________ kΩ
13. 3,200 Ω = __________ kΩ
14. 2.56 kΩ = __________ Ω
15. 0.39 kΩ = __________ Ω
16. 24.6 kV = __________ V
17. 500,000 Ω = __________ kΩ

Prefix Milli

The prefix milli means 1/1,000 of a unit. For instance, 5 milliamperes means 5/1,000 ampere, or 0.005 ampere.

Examples

620 mA = 0.62 A
37 mA = 0.037 A
3.2 mA = 0.0032 A
1.2 A = 1,200 mA
0.65 A = 650 mA
0.034 A = 34 mA
0.0063 A = 6.3 mA

Exercises

Perform the indicated changes below.

1. 0.002 A = ________ mA
2. 6 mA = ________ A
3. 0.37 A = ________ mA
4. 16 mA = ________ A
5. 0.052 A = ________ mA
6. 64 mA = ________ A
7. 1.7 A = ________ mA
8. 3,400 mA = ________ A
9. 0.0029 A = ________ mA
10. 8.6 mA = ________ A
11. 25 mA = ________ A
12. 3.2 mA = ________ A
13. 0.45 A = ________ mA
14. 0.027 A = ________ mA
15. 0.64 mA = ________ A
16. 3.8 A = ________ mA
17. 3.8 mA = ________ A

Prefix Mega

The prefix mega means 1,000,000 times a unit. For instance, 5 MΩ means 5 megohms, or 5,000,000 Ω.

Examples

450 MΩ = 450,000,000 Ω
27 MΩ = 27,000,000 Ω
3.2 MΩ = 3,200,000 Ω
0.5 MΩ = 500,000 Ω
470,000 Ω = 0.47 MΩ
2,500,000 Ω = 2.5 MΩ
68,000,000 Ω = 68 MΩ

Exercises

Perform the indicated changes below.

1. 3,200,000 Ω = ________ MΩ
2. 60 MΩ = ________ Ω
3. 2,500 V = ________ MV
4. 98,000 V = ________ MV
5. 0.21 MV = ________ V
6. 7 MΩ = ________ Ω
7. 18 Ω = ________ MΩ
8. 200 MV = ________ V
9. 4,800 V = ________ MV
10. 0.73 MV = ________ V
11. 700 kΩ = ________ MΩ
12. 0.47 MΩ = ________ kΩ
13. 2.5 MΩ = ________ Ω
14. 17,500,000 Ω = ________ MΩ
15. 10 MΩ = ________ Ω
16. 2.7 MΩ = ________ Ω
17. 0.033 MΩ = ________ kΩ
18. 39 kΩ = ________ MΩ

Prefix Micro

The prefix micro means 1/1,000,000 of a unit. For instance, 5 μA = 0.000,005 A.

Examples

27 μA = 0.000,027 A
490 μA = 0.00049 A
6,800 μA = 0.0068 A

72,000 μA = 0.072 A
260,000 μA = 0.26 A
5.22 A = 5,220,000 μA
0.16 A = 160,000 μA
0.034 A = 34,000 μA
0.0002 A = 200 μA
0.000,008 A = 8 μA

Exercises

Perform the indicated changes below.

1. 0.000001 V = ________ μV
2. 50 μV = ________ V
3. 0.00069 V = ________ μV
4. 7,200 μA = ________ A
5. 250 μA = ________ A
6. 0.1753 V = ________ μV
7. 0.0083 A = ________ μA
8. 72,695 μV = ________ V
9. 62 V = ________ μV
10. 95 μV = ________ V
11. 58 μA = ________ A
12. 32 μV = ________ V
13. 100 A = ________ μA
14. 381,500 μV = ________ V
15. 0.0095 V = ________ μV
16. 0.0635 A = ________ μA
17. 83 μA = ________ A
18. 0.123 V = ________ μV
19. 83,560 μV = ________ V
20. 592,600 μA = ________ A

ELECTRICAL UNITS, SCIENTIFIC NOTATION, CONVERSION, AND ABBREVIATIONS

Express each of the following electrical terms, units, or numbers as integers or decimal numbers, as well as powers of 10 (scientific notation).

Note: Some answers have been provided throughout this assignment, so that you, the student, may have guidelines for this task. Be sure to study *all* these examples before you begin.

Electrical Quantity	Integer or Decimal Number in Basic Units	Scientific Notation in Powers of 10
1. kilo means		
2. mega means		
3. milli means		
4. micro means		
5. micro-micro means	0.000,000,000,001	10^{-12}

Electrical Quantity	Integer or Decimal Number in Basic Units	Scientific Notation in Powers of 10
6. pico means		
7. nano means		
8. 10 microamperes	0.000,01	1×10^{-5}
9. 1 megohm		
10. 10 kilohms		
11. 25 milliamperes	0.025 A	2.5×10^{-2}
12. 1,000 kilohertz		
13. 500 milliwatts		
14. 10 milliamperes		
15. 0.01 microfarad	0.00000001	10^{-8}
16. 2.5 millihenries		
17. 900 milliamperes		
18. 35 milliamperes		
19. 75 microvolts		
20. 2,000 millivolts		
21. 35 picofarads	0.000,000,000,035	3.5×10^{-11}
22. 1,000 micro-microfarads		
23. 47 kilohms		
24. 5 megohms		
25. 0.05 millivolt		

continued

Electrical Quantity	Integer or Decimal Number in Basic Units	Scientific Notation in Powers of 10
26. 1 milliwatt		
27. 0.01 kilovolt		
28. 62 nanoseconds	0.000,000,062	6.2×10^{-8}
29. 400 microseconds		
30. 150 volts	150 V	1.5×10^{2}
31. 3,000,000 volts		
32. 275 milliamperes		
33. 300 microamperes		
34. 0.05 millivolt		
35. 100 kilohms		
36. 2,000 microseconds	0.002	2×10^{-3}
37. 11 nanoseconds		
38. 0.001 kilovolts		
39. 0.05 megohm		
40. 200,000 kilohms		
41. 0.1 ohm		
42. 0.05 volt		
43. 0.001 second		
44. 0.000 001 ampere	0.000,001 A	1×10^{-6} A
45. 50 millihenries		
46. 500 megahertz		

Conversions

47. 35 milliamperes is the same as ________ microamperes.
48. 75 microamperes is the same as ________ amperes.
49. 0.001 milliamperes is the same as ________ microamperes.
50. 5,000,000 ohms is the same as ________ kilohms.
51. 1,000 micro-microfarads is the same as ________ microfarads.
52. 2,000 seconds is the same as ________ milliseconds.
53. 200 millivolts is the same as ________ microvolts.

Abbreviations

54. μa means
55. mA means
56. μH means
57. mV means
58. mW means
59. μV means
60. mH means
61. kV means
62. kWh means
63. meg means
64. kHz means

6
Electrical Quantity Measurement

Objectives

After studying this chapter, the student should be able to

- Define and explain the new technical terms introduced in this chapter

ammeter	analog meter
digital meter	DVM
range switch	VOM
voltmeter	VTVM
ohmmeter	linear
scale	range
multiplier	nonlinear

- Choose the proper instrument for the measuring task
- Connect a meter correctly to the circuit
- Read and interpret meter scales accurately

In Chapter 4 you were introduced to the three basic circuit concepts: voltage, current, and resistance. The electrical instruments used to measure these quantities are called: voltmeters, amperemeters (ammeters), and ohmmeters, respectively.

The purpose of this chapter is to explain the proper use of such meters, the proper connection of the meters, and the interpretation of the meter scales.

6–1 TYPES OF METERS

The traditional meters have a pointer moving across the face of a scale (similar to the speedometer in your car). Such instruments are classified as *analog* meters. Readings from analog meters are generally limited to three-digit accuracy. The third digit must often be estimated and, therefore, is uncertain.

Digital meters, by contrast, directly display the numerical results of the measurement. In spite of their apparent superiority over analog meters, they will never displace analog meters completely. Therefore, students are urged to practice the reading of meter scales.

What sometimes complicates the interpretation of meter readings is the multiple use of one scale for different ranges and applications. Figure 6–1, for instance, portrays the face of a *multimeter,* commonly known as a *VOM* (volt-ohm-milliammeter). Similar multimeters have been known as *VTVM,* or vacuum tube voltmeters (now nearly obsolete), and *FET meters* (field-effect transistor meters).

6–2 AMMETER AND VOLTMETER SCALE INTERPRETATION

A portion of an easy-to-read meter scale is shown in Figure 6–2. Note that the space between whole number values is divided into 10 parts so that each small division is equal to 0.1 (one tenth). If the pointer is at position 1, the indicated amount is 0.3. At position 2, the pointer is halfway between 1.70 and 1.80, so the indicated reading is 1.75. At position 3, the pointer is at the 2.6 mark.

The student should always be careful to note the value by the small divisions on the scale. On the meter in Figure 6–3, each small division is equal to 0.2. At position 1, the pointer is at 0.6; at position 2, the pointer is at 3.5; and at position 3, the pointer indicates 5.2.

It is customary to use decimals rather than common fractions to record the values of meter readings. For example, a scale reading is given as 8.75 rather than 8 3/4, or as 7.4 rather than 7 2/5.

The scale of Figure 6–4 represents a dual-range ammeter. Such an instrument would generally have three terminals for making the connections labelled COM, 3 A, and 30 A.

For the 30-ampere range (top scale), the test leads are connected to the COM and 30 A terminals. For the 3-ampere range (bottom scale), connections are made to the COM and the 3 A terminals.

The *range* of a meter is the maximum amount of the instrument at full-scale deflection. For the protection of the meter, the range should never be exceeded. Therefore, it is wise to measure unknown quantities on the higher range first before switching to a lower range for the sake of greater accuracy.

Notice that in our example, Figure 6–4, each pointer position has two different interpretations.

Position 1
- on the 3-ampere scale ----- 0.65 A
- on the 30-ampere scale ----- 6.5 A

Position 2
- on the 3-ampere scale ----- 2.05 A
- on the 30-ampere scale ----- 20.5A

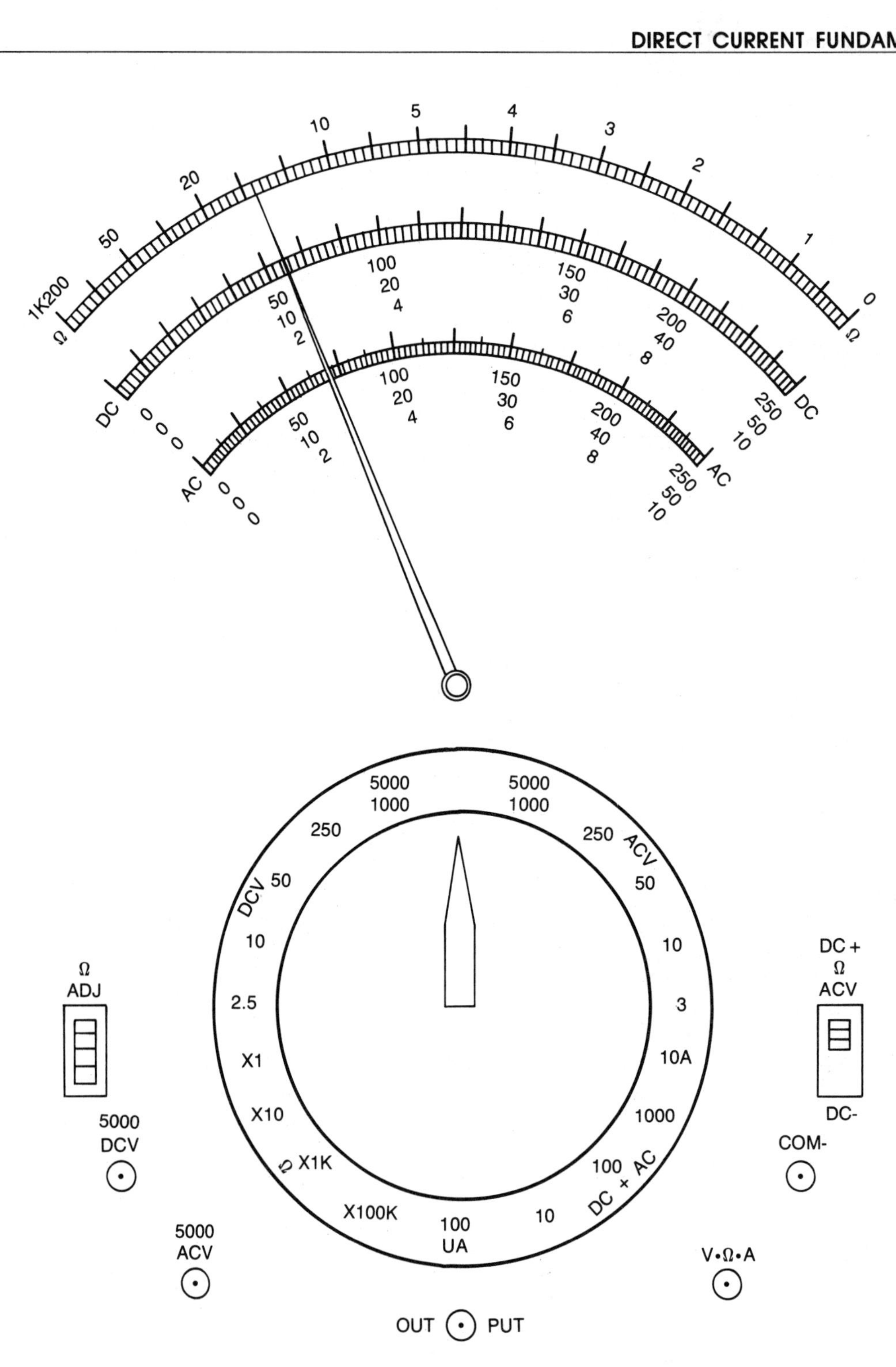
Ω ADJ
5000 DCV
5000 ACV
DCV
ACV
DC + AC
UA
X1K
X100K
DC +
Ω
ACV
DC-
COM-
V•Ω•A
OUT
PUT

FIGURE 6–1

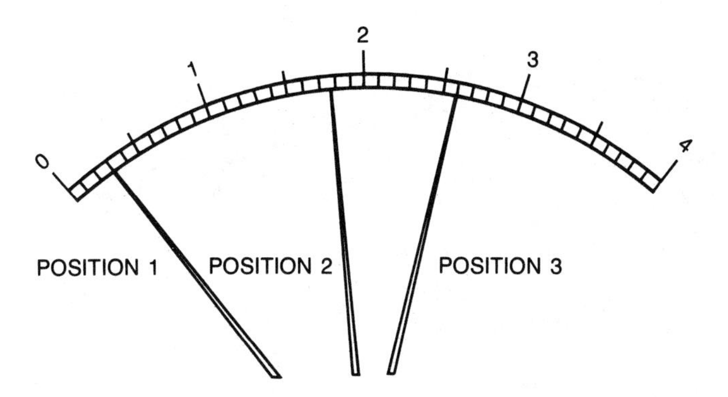

FIGURE 6–2 Meter scale, 0.1 intervals

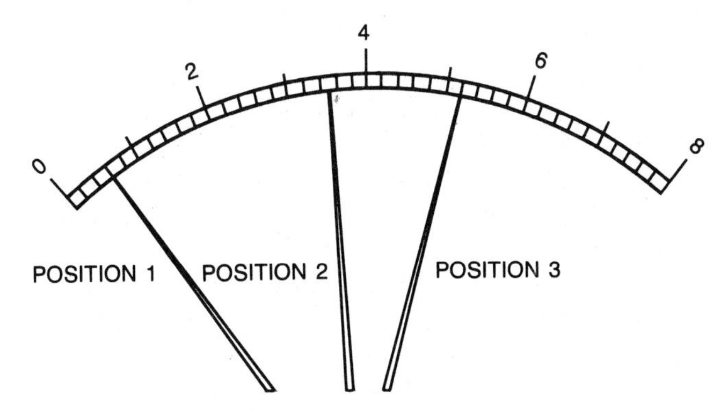

FIGURE 6–3 Meter scale, 0.2 intervals

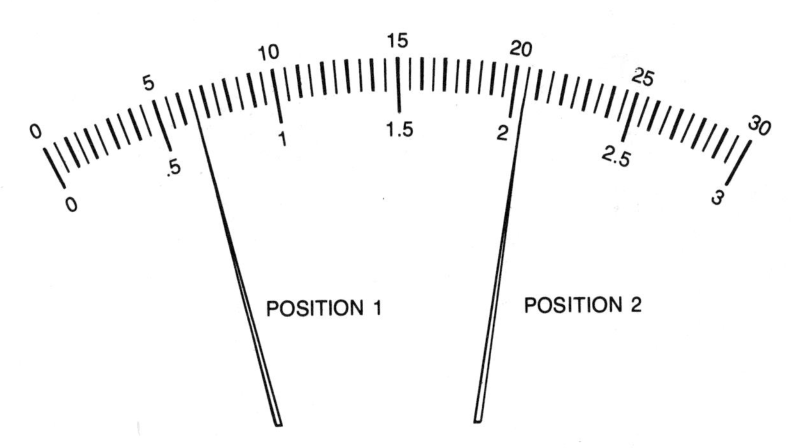

FIGURE 6–4 Two-range ammeter scale

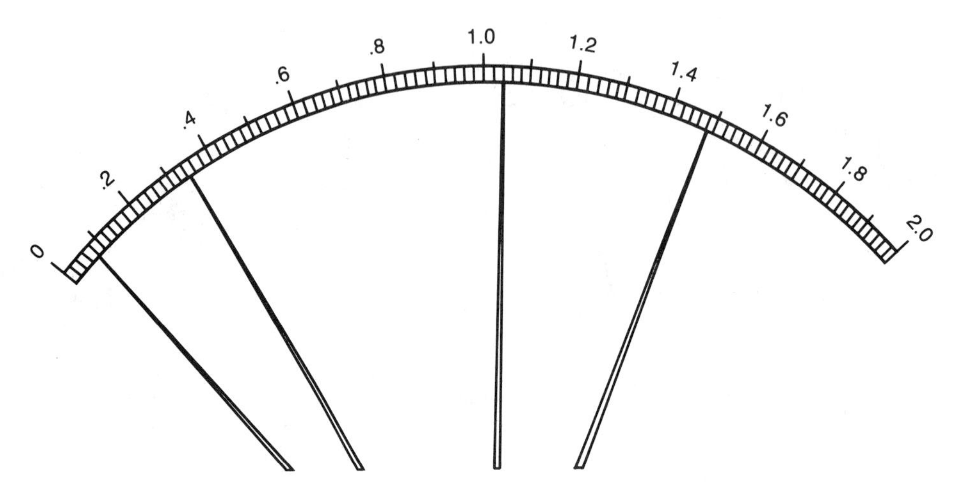

FIGURE 6–5 Two-ampere ammeter

The scale in Figure 6–5 shows pointers indicating values at 0.08, 0.35, 1.04, and 1.49, the smallest divisions representing 0.02 ampere.

Figure 6–6 represents the scale of a 50-ampere ammeter. The smallest divisions here represent 0.5 ampere. The first position of the pointer indicates 22 amperes. Depending on the accuracy of the meter mechanism, the next indicated position may be called 38.25 because the pointer is halfway between 38.0 and 38.5. However, if the accuracy of the meter is to within 2% of the full-scale reading, then the meter is accurate to within only 1 ampere. Therefore, there is no point in worrying about determining the meter reading to an accuracy of 0.5 ampere. To justify a meter reading of 38.5, the meter must have an accuracy of 1% or better, and to justify the 38.25, an accuracy of 0.5%.

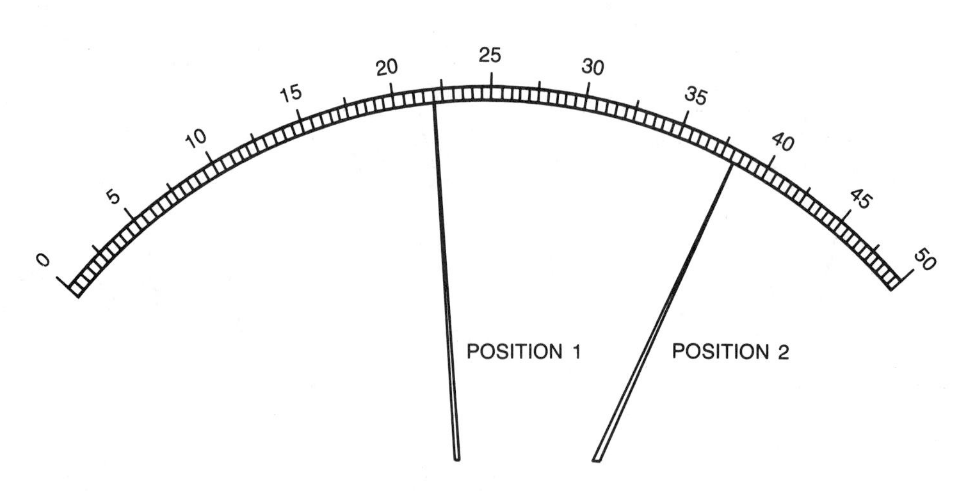

FIGURE 6–6 Fifty-ampere ammeter

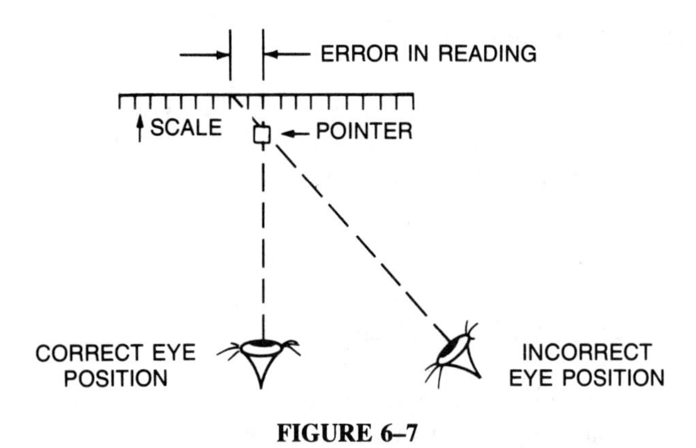

FIGURE 6–7

Highly accurate meters have a narrow mirror strip along the scale to help prevent reading errors. On these scales, the image of the pointer disappears behind the pointer itself. With the ordinary nonmirror scale, the line of sight should be perpendicular to the surface of the scale, Figure 6–7.

6–3 OHMMETER SCALE INTERPRETATION

We are accustomed to read analog meters with their numbers arranged from left to right and with the 0 digit on the left-hand side of the scale.

Most ohmmeters have the numbers of their scales arranged from right to left, with the 0 on the right-hand side of the scale, Figure 6–8.

Note that this is a nonlinear scale where equal spaces do *not* represent equal values. Note, for instance, that the space from 0 to 1 is greater than the space between 10 and 20. This makes the meter readings much more accurate toward the lower end of the scale (the right-hand side).

Ohmmeters generally have a range switch indicating *multiplier* values, such as × 1, × 100, × 1K, × 100K. This means that each reading taken on the scale must be multiplied by the value indicated on the range switch.

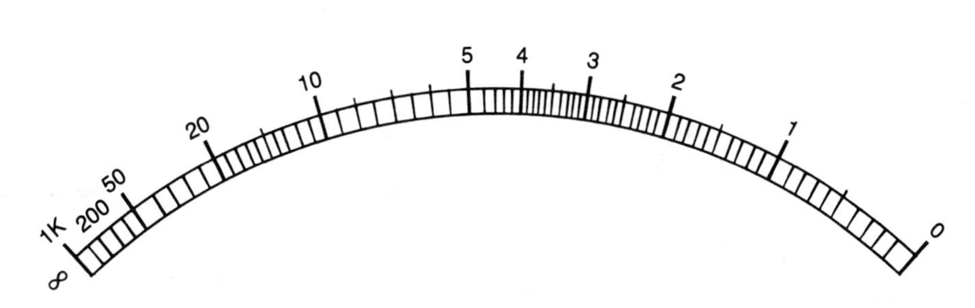

FIGURE 6–8

6–4 ELECTRICAL METER CONNECTION

The proper techniques of connecting electrical meters should be demonstrated and practiced in a laboratory setting. It is assumed that the reader has access to such facilities. However, the following guidelines are offered for your review and consideration:

1. Always connect *voltmeters parallel* to a load source, Figure 6–9.
2. Always connect *amperemeters* in *series*. To establish such a connection, break the circuit and then insert the meter into the break, Figure 6–10.
3. DC meters are polarized: positive (+) and negative (−). Take care to connect the positive (red-colored) test lead toward the positive end of the source. If mistakes are made in the polarization, the pointer will deflect downhill, that is, to the left of 0.
4. When *ohmmeters* are applied to a circuit, be sure that the *power is turned off*. Better yet, disconnect one side of the resistance to be measured.
5. When meters with *multiple ranges* are employed, protect the meter by switching to the *highest range first* to prevent overloading.

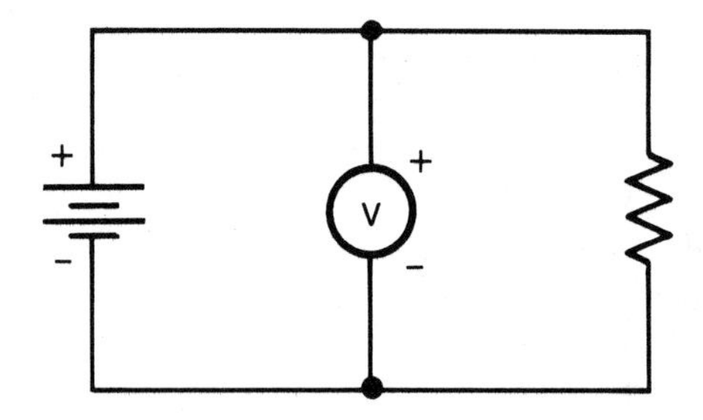

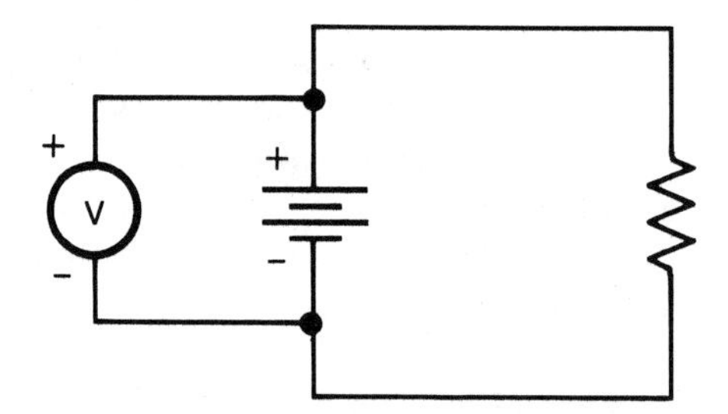

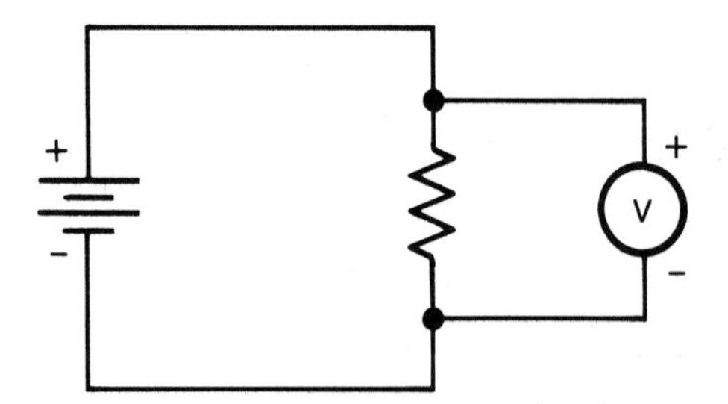

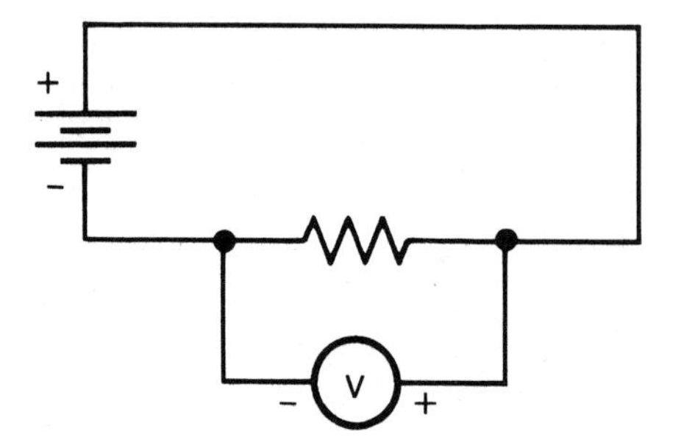

FIGURE 6–9 Examples of voltmeter connections

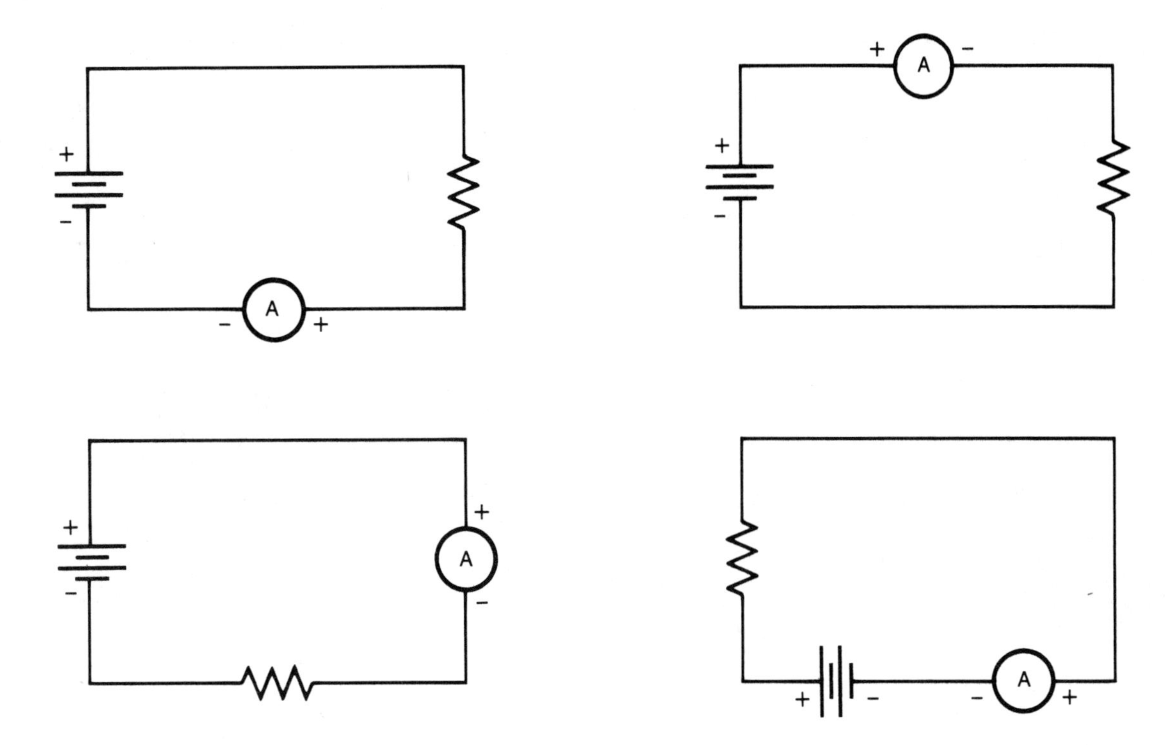

FIGURE 6–10 Examples of ammeter connections

SUMMARY

- Analog meters are instruments in which numerical data are represented by the position of a moving pointer over a scale.
- Multimeters, such as VOMs, serve different functions with multiple ranges.
- The range of a meter at maximum deflection must not be exceeded.
- DC meters must be correctly polarized.
- Voltmeters are connected in parallel.
- Ammeters are connected in series.
- Resistors must be disconnected from the circuit before an ohmmeter is attached.

Achievement Review

DC VOLTMETER READING

Look at the scale of a DC voltmeter as shown on page 73. Then assume that the pointer deflects to the individual settings as shown in the chart on page 72 and that the range selector is in the position indicated. The instructor will give further instruction as necessary.

DC VOLTMETER READING					
	Reading When the Range Selector Switch is at				
The Needle Points to	5 V	25 V	50 V	250 V	500 V
1. 2 divisions past the number 3					
2. 4 divisions past the number 10					
3. 3 divisions past the number 2					
4. 5 divisions past the number 5					
5. 7 divisions past the number 10					
6. 6 divisions past the number 15					
7. 2 divisions past the number 10					
8. 4 divisions past the number 1					
9. 5 divisions past the number 10					
10. 1 division past the number 2					
11. 8 divisions past the number 3					
12. 6 divisions past the number 10					
13. 4 divisions past the number 5					
14. 5 divisions past the number 2					
15. 2 divisions past the number 15					
16. 6 divisions past the number 5					
17. 4 divisions past the number 20					
18. 2 divisions past the number 5					
19. 6 divisions past the number 4					
20. 4 divisions past the number 2					

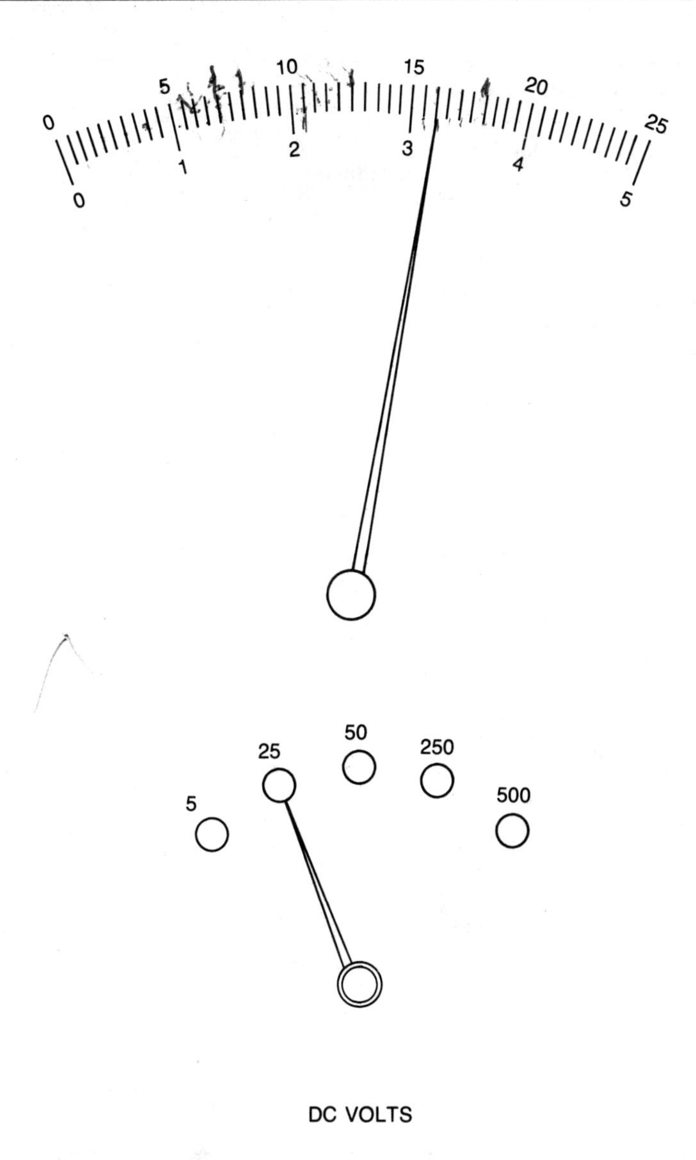

DC AMMETER READING

Look at the scale of a DC ammeter as shown on page 75. Next, assume that the pointer deflects to the individual settings shown in the chart on page 74 and that the range selector is in the position indicated. The instructor will give further explanation if necessary.

DC AMMETER READING				
Range Selector Switch at		Pointer in the Following Position	Answer	
			In Milliamperes	In Amperes
1.	.01	.2		
2.	.10	.2		
3.	1 A	.2		
4.	.01	5 divisions past .4		
5.	.001	3 divisions past .6		
6.	.0001	2 divisions left of .8		
7.	.1	3 divisions past .2		
8.	.01	4 divisions past .6		
9.	1 A	5 divisions past .4		
10.	.01	5 divisions past 0		
11.	.0001	3 divisions left of .4		
12.	.001	2 divisions past .6		
13.	.1	7 divisions past .2		
14.	.01	2 divisions left of .8		
15.	.001	3 divisions past .2		
16.	1 A	exactly between .2 and .4		
17.	.001	exactly between .6 and .8		
18.	.1	exactly between .8 and 1		
19.	.01	4 divisions past .4		
20.	.10	4 divisions left of .4		

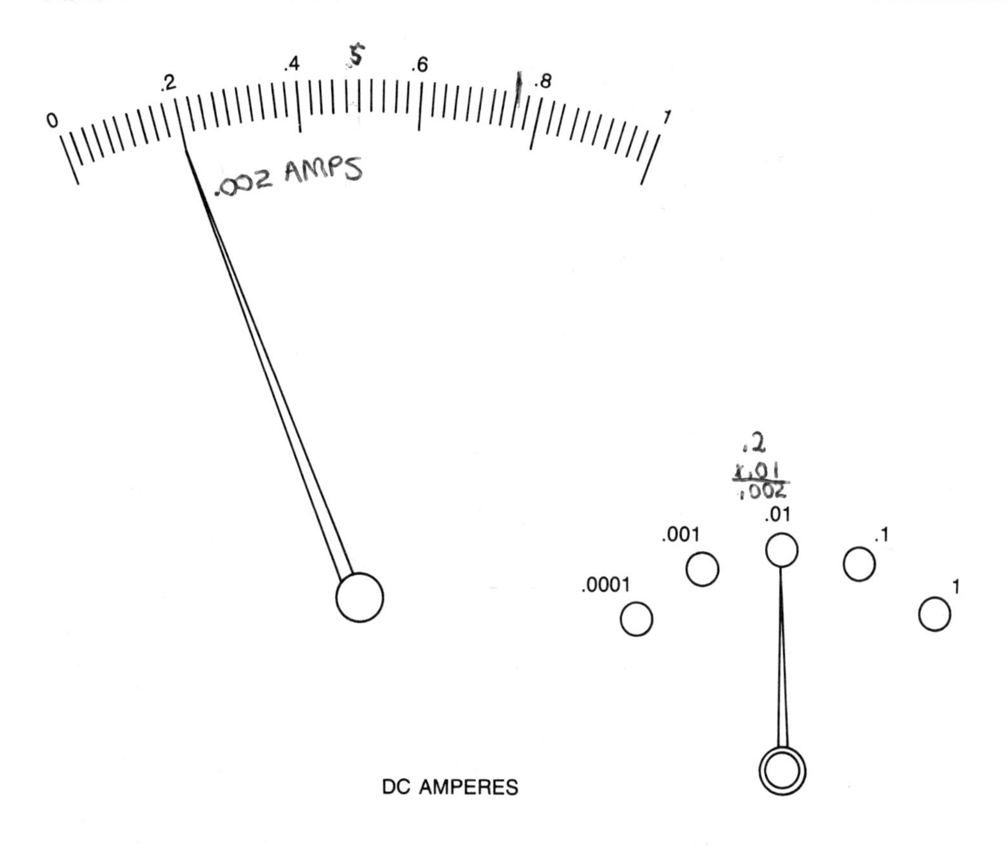

CIRCUIT CHECKS

On the next pages you will see a number of circuits with instruments attached. Some of these circuits have mistakes incorporated.

Your task is to check each circuit for mistakes. Circle the mistake(s) as you see them, and briefly explain what is wrong. If no mistakes are found, label the drawing OK and state what is being measured, for example R_T, I_2, E_1.

1. **2.**

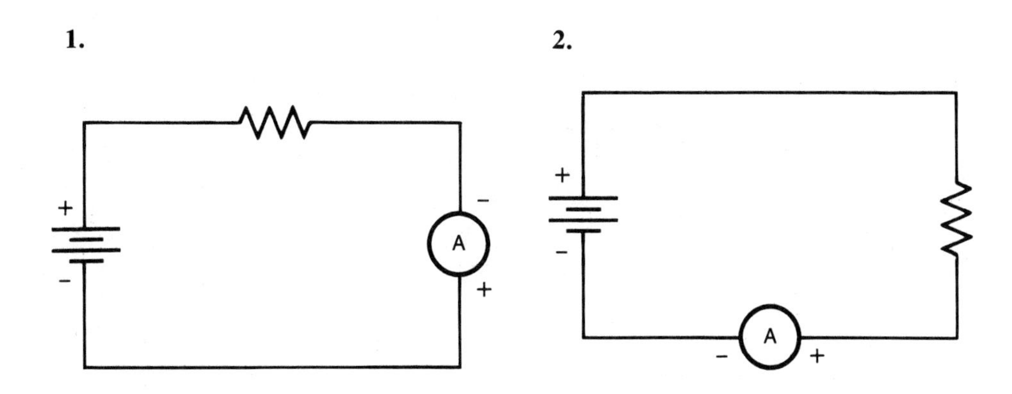

3.

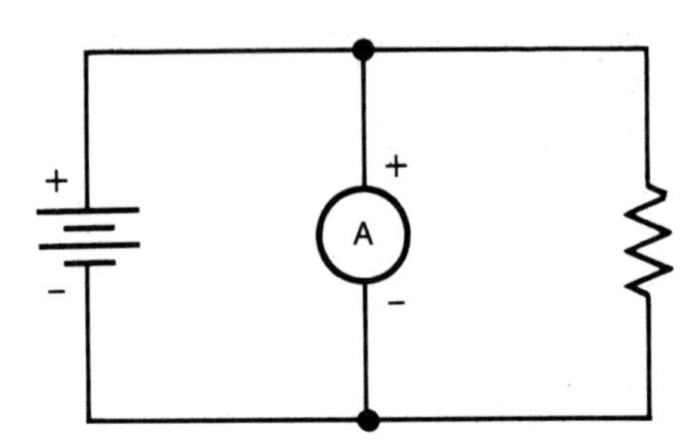

4.

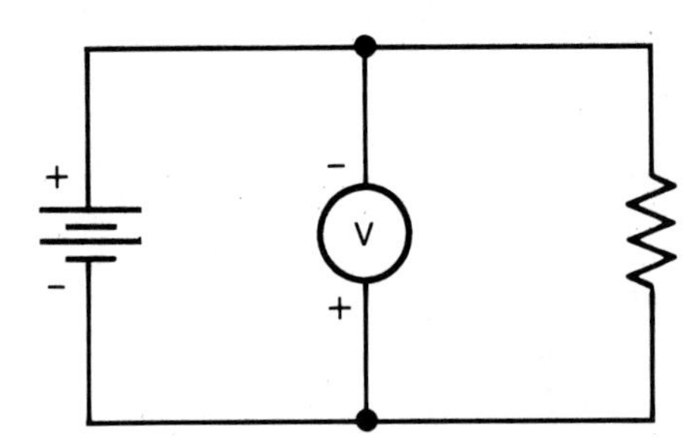

5.

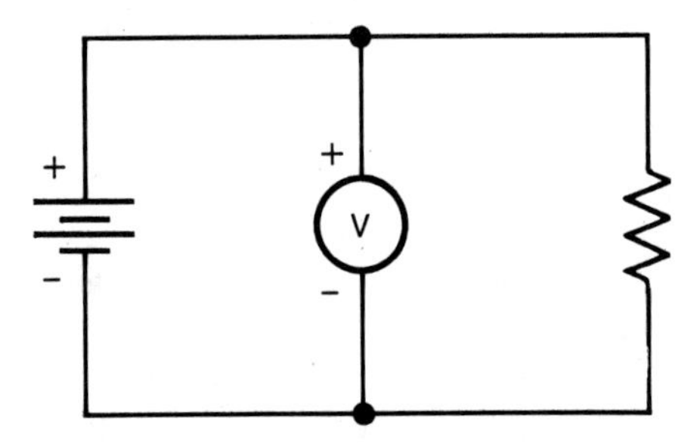

6.

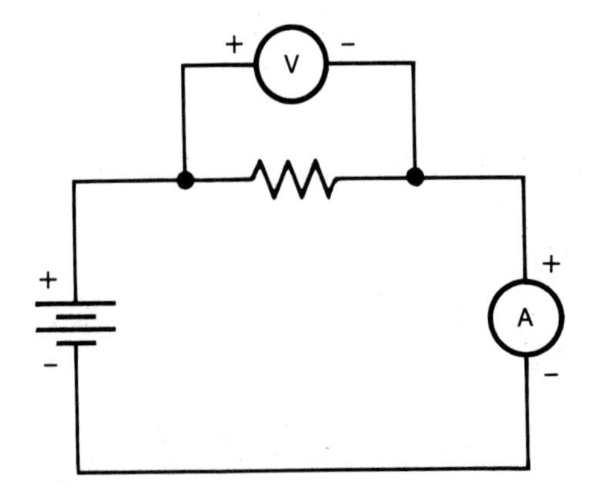

7.

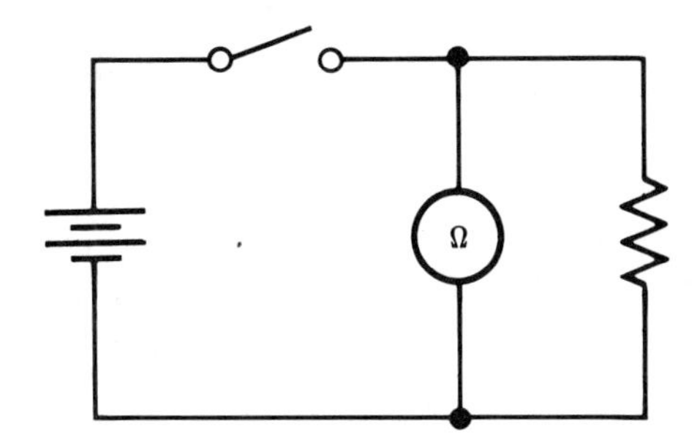

8.

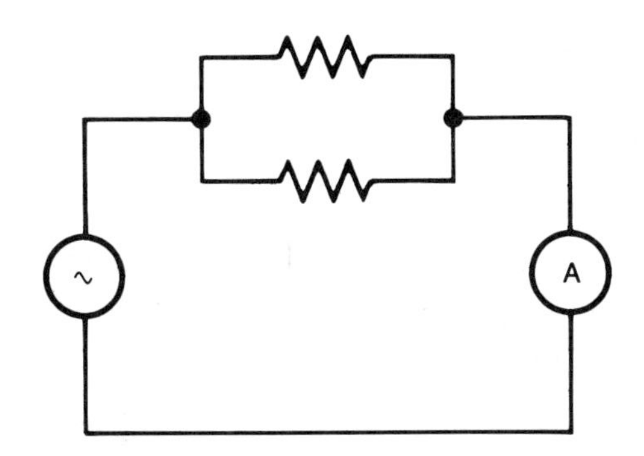

9.

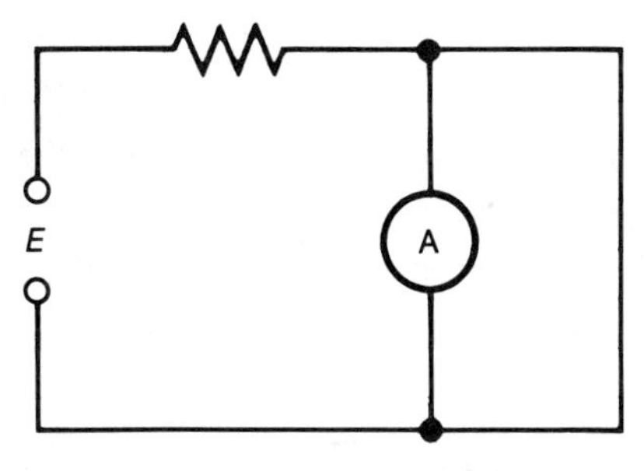

IN PARALELL WRONG.
SHORT CIRCUIT

10.

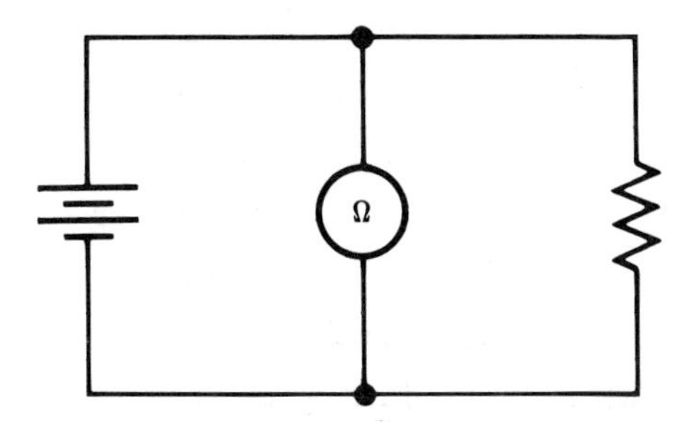

11.

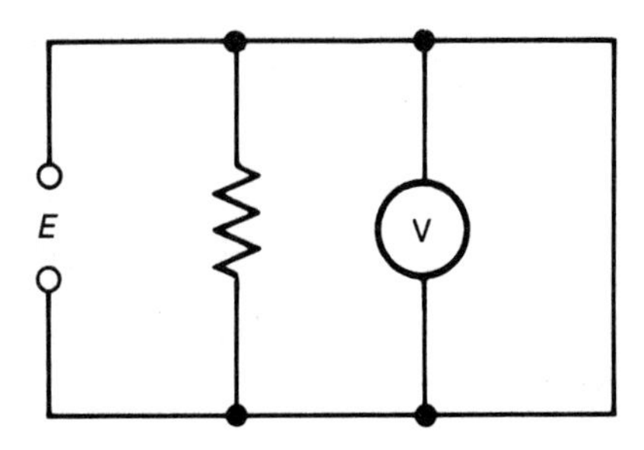

12.

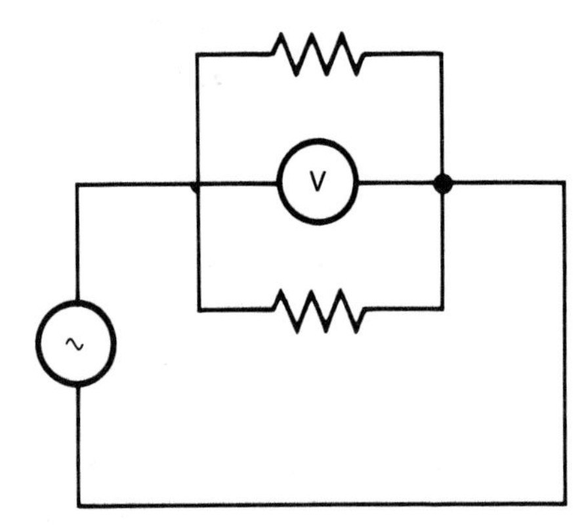

METER SCALE READING

Four actual-size meter scales are shown on the next page. Record the indicated reading for each of the three positions of the pointer on each scale. Note that two sets of readings are required for meter scale C and that three sets are required for meter scale D.

Scale		First Position	Second Position	Third Position
A				
B				
C	0–300			
	0–150			
D	250–0–250			
	50–0–50			
	10–0–10			

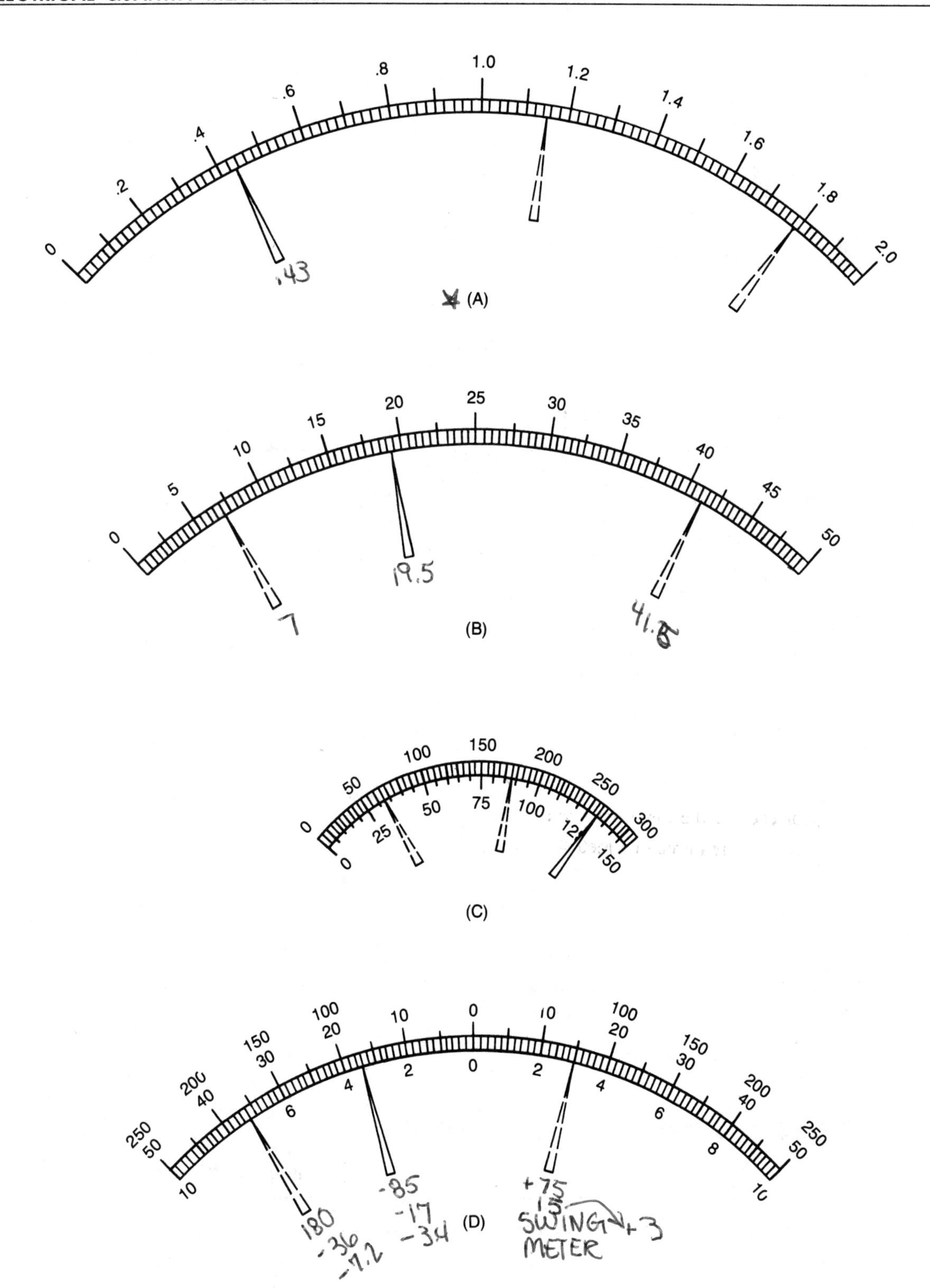
0
.2
.4
.6
.8
1.0
1.2
1.4
1.6
1.8
2.0
(A)
0
5
10
15
20
25
30
35
40
45
50
(B)
0
50
100
150
200
250
300
0
25
50
75
100
125
150
(C)
250
50
10
200
40
150
30
100
20
10
0
10
100
20
150
30
200
40
250
50
10
4
2
0
2
4
6
6
8
(D)

7

Resistance

Objectives

After studying this chapter, the student should be able to

- Define and explain the new technical terms introduced in this chapter

conductivity
conductance
insulators
specific resistance
circular mil
American Wire Gauge (AWG)
temperature coefficient
potentiometer
color code
resistivity
conductor
semiconductors
mil
mil-foot
nichrome
wattage rating
rheostat

- Name various materials classified as either good conductors, poor conductors, or insulators
- Explain the four factors that determine the amount of resistance in a wire
- Discuss the aspects of conductivity and resistivity
- Perform computations involving the equation

$$R = \frac{K \times l}{A}$$

- Use the charts and tables in this chapter
- Compare wire sizes using the American Wire Gauge
- Determine the correct values of resistors by using the color code

7–1 CONDUCTANCE VS. RESISTANCE

It is well known that some materials, especially metals, permit electrical currents to easily flow through them. These materials, whose atomic structures readily provide free electrons, are referred to as good *conductors*.

Silver is the best of such conductors, but it is seldom used because of its high cost. Copper is almost as good a conductor as silver, is relatively inexpensive, and serves for most types of wiring. Aluminum, a fairly good conductor, is used where weight reduction is an important factor.

By contrast, other materials tend to prevent current from flowing through them. Such substances are known as *nonconductors* or *insulators*. Examples of insulators are glass, rubber, and nylon.

It is important to realize that not all materials can be so easily classified. The degree of conductivity varies over an extremely wide range. Conductors and insulators merely represent two extremes between which a great variety of materials are classified as poor conductors or semiconductors.

Among these poor conductors are metals and alloys that find use as resistance wire in heat-producing appliances. Nickel-chromium, often called *nichrome,* is an example of such wire.

But even the extremes do not represent absolute perfection. Every conductor, no matter how good, still offers some resistance. Just how much resistance a conductor has depends on four factors, namely

1. The type of material from which it is made
2. The conductor's length
3. Its cross-sectional area (thickness)
4. Its temperature

These factors will be explored in the following sections.

7–2 RESISTIVITY OF MATERIALS

The resistance rating of different materials is based on a comparison of the number of ohms measured in a standard sized sample of the material. In the metric system, for instance, the standard is a 1-centimeter cube whose resistance is measured in millionths of an ohm (microhms). Look at Figure A–1 in the Appendix, and compare the resistivity in microhm-centimeters. This resistivity is also known as *specific resistance*.

In the English system of measurement, the standard of resistivity is called the mil-foot. This standard is based on the resistance of a piece of wire 1 foot long with a diameter of 0.001 inch (1 mil = 1/1,000). Figure 7–1 illustrates the concept of a mil-foot, and the chart in Figure 7–2 shows some characteristic values of resistivity.

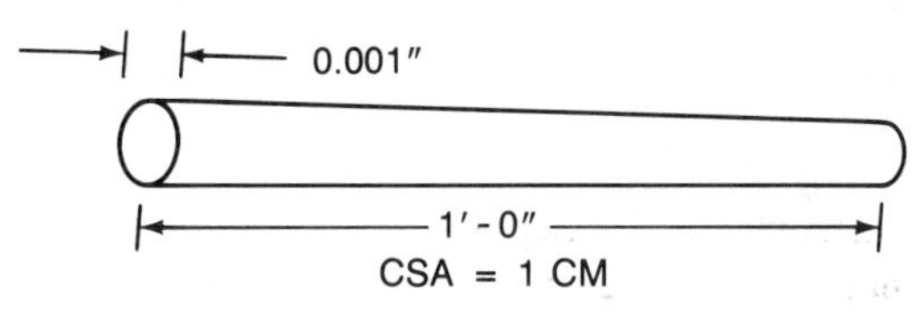

FIGURE 7–1

K = Ohms Resistance per Mil-foot (at 70°F)	
Aluminum	17
Brass	42
Cadmium Bronze	12
Copper	10.4
Copperclad Aluminum (20% Cu)	15.2
Copperweld	26–34
Iron	60
Nichrome	600
Silver	9.6
Steel	75
Tungsten	33
Ohms/mil-ft resistance for other metals can be calculated from Figure A–1 in the Appendix.	

FIGURE 7–2

7–3 LENGTH OF A CONDUCTOR

The resistance of any conductor is directly proportional to its length. For example, if 50 feet of wire has a resistance of 1 ohm, then 100 feet of the same wire will have a resistance of 2 ohms. In other words, the longer the wire, the more resistance it has. This simple fact must be taken into account whenever electricians plan the installation of a long supply line.

7–4 CROSS-SECTIONAL AREA (CSA) OF A CONDUCTOR IN CIRCULAR MILS

Resistance is inversely proportional to the cross-sectional area of the wire. This is a way of saying that the thicker the wire, the lower its resistance. A thick wire, Figure 7–3, allows many electrons to move through it easily, just as a wide road can carry many cars per hour or a large pipe can allow a large volume of water to flow through it.

Because the use of square inches or square feet results in complicated calculations and inconveniently small numbers for ordinary sizes of wire, a more convenient unit of area (thickness) is used. The circle in Figure 7–4 represents the end of a wire that is

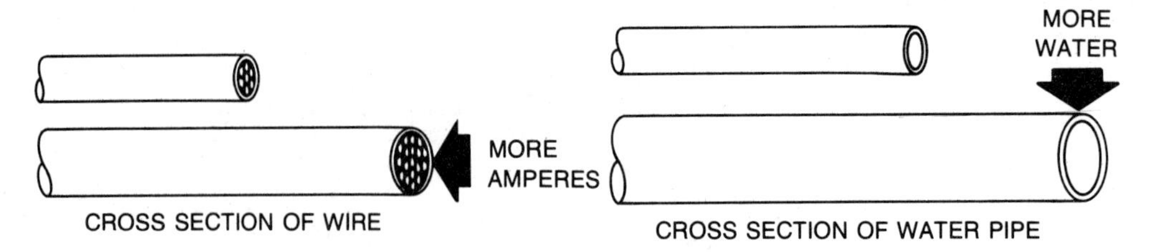

FIGURE 7–3

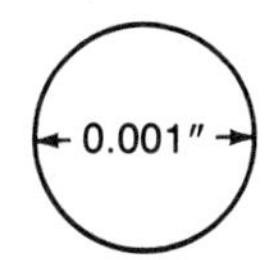

FIGURE 7–4 One circular mil cross section

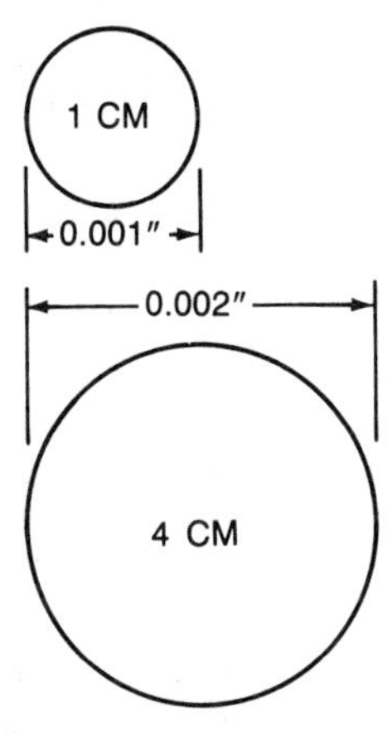

FIGURE 7–5 Comparative areas

1/1,000 of an inch thick. This distance, 0.001 inch, is 1 *mil*. The area of this circle, which is 1 mil in diameter, is 1 *circular mil* (CM). A circular mil is a unit of area (A) measurement. It is the same type of measurement as a square foot or an acre but more convenient to use.

For example, what is the area of a circle with a diameter of 0.002 inch? Such a circle has an area exactly four times larger than a circle with a diameter of 0.001 inch, Figure 7–5. If either the radius or the diameter of any flat surface is doubled, then the area will increase four times. Let us look at an example using the formula for finding the area of a circle, $A = \pi r^2$. If the radius r of circle X is 5, then the area r^2 is 25π. If the radius r of circle Y is 10 (*double* that of circle X), then the area of circle Y is 100π (*four times* that of circle X). Therefore, to answer the question about the area of a circle that has a diameter of 0.002 inch, we can say that the area is four times greater than that of a 1-mil circle, or 4 circular mils.

What is the area of a circle that has a diameter of 0.003 inch? Three-thousandths of an inch is 3 mils: 3^2 is 9. This circle is nine times as large in area as a 1-mil circle. The area of the circle, therefore, is 9 circular mils. The area of a circle, using circular mils, can be found by the following steps:

1. Write the diameter of the circle in mils.
2. Square this number (multiply it by itself); the result is the area of the circle in circular mils.

EXAMPLE 7–1

Given: A conductor of 1/2 inch diameter.

Find: Its cross-sectional area in circular mils.

SOLUTION

1/2 in. = 0.5 in. = 500 mils
$(500)^2$ = 250,000 CM

Note: 250,000 CM is also known as 250 MCM.

If the circular mils of a wire are known, it is a simple matter to reverse the mathematical procedure and find the diameter.

EXAMPLE 7–2

Given: A conductor with cross-sectional area of 100 circular mils.

Find: Its diameter in inches.

SOLUTION

$d_{mil} = \sqrt{100} = 10$ mils
10 mils = 0.01 in. = 1/100 in.

Not all conductors are circular in shape. Solid rectangular conductors *(bus bars)* are used to carry large currents. Bus bars are often easier to assemble and take up less space than the large sized round conductors required to carry the same current. In addition, the flat shape of the bus bar provides more surface area from which heat can radiate.

The cross-sectional area of such a bus bar is figured in *square mils* rather than in circular mils. A square mil is defined as the area of a square with sides of 0.001 inch, Figure 7–6.

EXAMPLE 7–3

Given: A bus bar 4 inches wide and 0.25 inch thick, Figure 7–7.

Find: Its cross-sectional area in square mils.

SOLUTION

0.25 in. = 250 mils
4 in. = 4,000 mils
250 × 4,000 = 1,000,000 sq mil

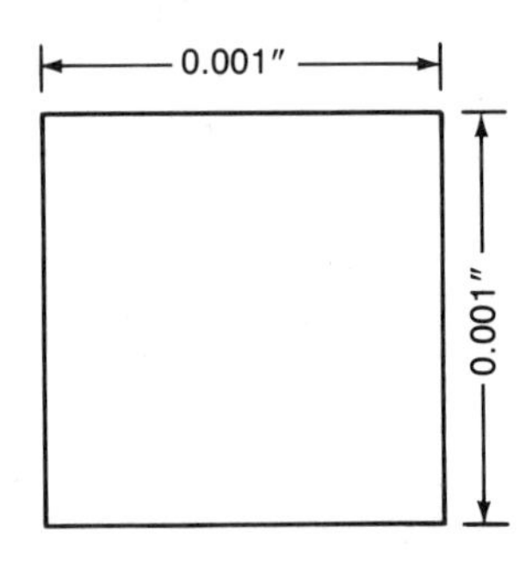

FIGURE 7–6

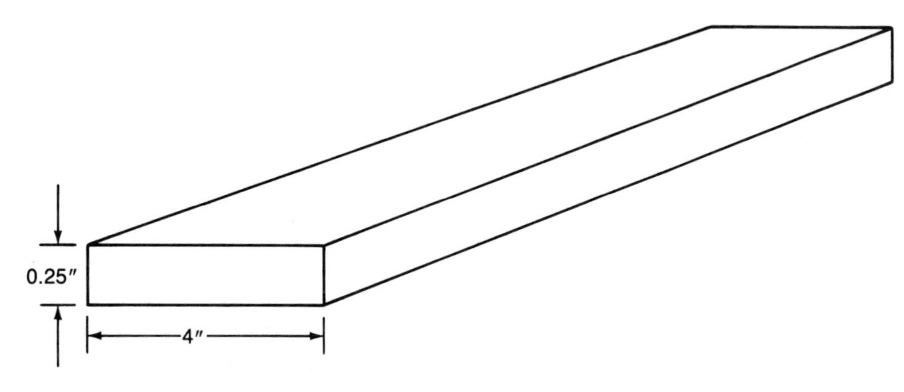

FIGURE 7–7

For purposes of comparison, square mils may be converted into circular mils by a multiplying factor of 0.7854.

1 CM = 0.7854 sq mil

For example, using the answer found in the problem above,

1,000,000 sq mil = 785,400 CM

Calculating the Resistance of a Wire

As previously stated, the resistance of a wire is a function of its resistivity *(K)*, Figure 7–2. Furthermore, it is directly proportional to its length *(l)* and inversely proportional to its cross-sectional area *(A)*. This statement can be written mathematically as

$$R = \frac{K \times l}{A}$$

where

R = resistance of the wire
K = resistivity of the material (ohms per mil-ft)
l = length in feet
A = cross-sectional area in CM

This formula enables us to calculate the resistance of any given length of wire.

EXAMPLE 7–4

Given: A copper wire with a cross-sectional area of 4,107 circular mils.

Find: The resistance of 175 feet cut from the same wire.

SOLUTION

$$R = \frac{K \times l}{A}$$

$$R = \frac{10.4\ (175)}{4{,}107}$$

$$R = 0.443\ \Omega$$

Note: K is taken from Figure 7–2.

This given equation can also be transposed to yield an expression for unknown variable l and unknown variable A.

$$l = \frac{R \times A}{K}$$

and

$$A = \frac{K \times l}{R}$$

Note: Remember to use Figure 7–2 for K values and Figure A–2 for cross-sectional area values.

EXAMPLE 7–5

Given: A spool of nichrome wire of gauge #20.

Find: The length of a 30-ohm wire to be cut from this stock.

SOLUTION

$$l = \frac{30 \times 1{,}020}{600}$$

$$l = 51 \text{ ft.}$$

EXAMPLE 7–6

Given: A two-wire power line is to be erected between poles 1,200 feet apart (the length of the wire is 2,400 feet). Resistance of the line should not be more than 1.7 ohms.

Find: The size of the aluminum wire to be used.

SOLUTION

$$A = \frac{17 \times 2{,}400}{1.7}$$

$$A = 24{,}000 \text{ CM}$$

This value is close to #6 wire (from Figure A–2).

7–5 EFFECT OF TEMPERATURE ON RESISTANCE

Resistance depends not only on length, area, and kind of material, but also on the temperature of the material. Figure A–2 of the Appendix gives two sets of values for the resistance of copper at two different temperatures. (The resistance values given for the higher temperature are used when calculating motor and transformer windings since

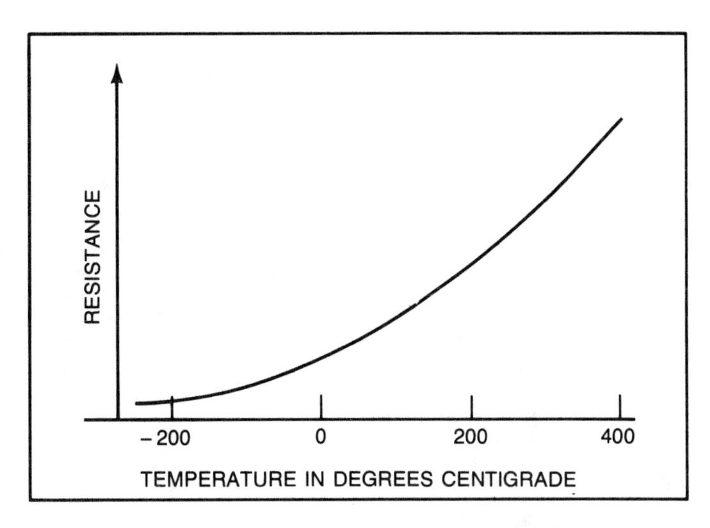

FIGURE 7–8 Resistance of most metals

these windings are intended to operate warm.) These values in Figure A–2 show that *as the temperature rises, the resistance of metals increases.*

The graph in Figure 7–8 shows, in a general way, how the resistance of most metals increases as the temperature increases. Not only is the resistance greater, but it is also increasing at a faster rate as the temperature rises.

The resistance of carbon, however, decreases slightly with a rise in temperature. The resistance of conducting liquid solutions decreases rapidly with a temperature rise. Finally, the resistance of semiconductors, such as germanium and metal oxides, decreases very rapidly as the temperature goes up.

Not all metals increase in resistance at the same rate. The rate at normal temperatures is given as the *temperature coefficient of resistance* in Figure A–1 of the Appendix.

As the temperature approaches absolute 0 (−273°C), many metals and alloys lose all of their electrical resistance. This behavior is called *superconductivity*. When the niobium-alloy wire coils of extremely powerful electromagnets are cooled (by liquid helium) to temperatures about 4°C above absolute 0, the coils have a very low resistance. Thus, they can carry huge currents without an appreciable energy loss. Research scientists are developing super-cooled power lines that can be buried underground.

Using Temperature Coefficient to Find Resistance at Higher Temperature

The resistance increase due to a higher temperature is equal to the original resistance multiplied by the temperature coefficient and by the rise in temperature in degrees. Expressed in equation form, this statement becomes

$$\text{Resistance increase} = R \times \text{coefficient} \times \text{degree rise}$$

This resistance increase is added to the original resistance to give the resistance at the higher temperature.

Let us find the resistance of 10 feet of #30 platinum wire to be used in an oven at 500°F.

1. The resistance of 10 feet of #30 platinum wire at room temperature is found by comparison with copper: Figure A–2 (Appendix) gives the resistance of 1,000 feet of #30 copper wire as 103 ohms. Therefore, 10 feet of #30 copper wire has a resistance of 1.03 ohms. Since platinum has 5.80 times as much resistance as copper, 10 feet of platinum has a resistance of

   $1.03 \times 5.80 = 5.97 \text{ or } 6\ \Omega$

   Thus, the original resistance in the formula for finding the amount of resistance increase is the value of 6 ohms at 68°F.
2. The temperature coefficient given in Figure A–1 (Appendix) is 0.003 per degree Celsius.
3. The rise of temperature in degrees must be found and expressed in Celsius degrees. From 68°F to 500°F, there is a difference of 432°F. Since 9°F is required to equal 5°C, 432 is divided by 9 and multiplied by 5. Therefore, the number of Celsius degrees equivalent to 432°F is 240°C.
4. The three quantities needed to calculate the amount of change in resistance are now known. Thus, resistance increase is

   $6\ \Omega \times 0.003 \times 240°\text{C} = 4.3\ \Omega$

   This value is added to the original resistance of 6 ohms. As a result, the resistance of 10 feet of #30 platinum wire at 500°F is 6 ohms + 4.3 ohms, or 10.3 ohms.

7–6 THE AMERICAN WIRE GAUGE (AWG)

It would be impractical having to calculate resistance for any given length and wire diameter. Such information is more readily available from the American Wire Gauge (AWG), also known as the Brown & Sharpe (B&S) Gauge.

The American Wire Gauge assigns numbers from 0 to 44, Figure A–2. The table is self-explanatory, but the following observations may be useful:

- The larger the number, the smaller the wire.
- A #2 wire is nearly the size of a standard wooden pencil.
- A #44 wire is about the thickness of a fine hair.
- Wires larger than 0-gauge are labelled 00 (2/0), 000 (3/0), and 0000 (4/0), respectively.
- Wires larger than 0000 (4/0) are classified by cross-sectional area in CM, the next larger size being 250,000 CM = 250 MCM (see example 7–1).
- Odd-numbered conductors are rarely used and seldom stocked by wholesalers.
- The most prevalent wire sizes used in house wiring are #14 and #12.
- A #14-gauge wire is the smallest wire permissible by the National Electric Code to be used for permanent installations.

- No. 8-gauge wire can be used for installing electric kitchen ranges and other heavy duty appliances.
- No. 22-gauge wire may be used for electronic circuits.
- Ordinary lamp cords are generally made of #16 or #18 wire.
- A stranded wire has the same amount of copper and the same current-carrying capacity as a solid wire of the same size.

7–7 STRANDED WIRE AND CABLE

Bunch Stranding

Bunch-stranded wire is a collection of wires twisted together. These wires are not placed in specific geometrical arrangement. For example, #18 lamp cord contains 16 #30 wires that are loosely twisted together in the same direction.

Concentric Stranding

For this type of wire stranding, a center wire is surrounded by one or more definite layers of wires. Each wire layer contains six wires more than the layer immediately beneath it, Figures 7–9 and 7–10. When several wire layers are used, each layer is twisted in a direction opposite to that of the layer under it.

The size of the wire strands used in a cable depends on the flexibility required. For example, #00 cable may consist of seven strands of #7 wire, or 19 strands of #12 wire, or 37 strands of #24 wire. If #24 wire is used, the resulting cable is rated as extra flexible.

However, since even fine wire will break if bent back and forth many times, stiffer wire may be used with the copper wire in a cable to increase the durability of the cable.

Number of Wire Layers over Center Wire	1	2	3	4	5	6	7
Total Number of Wires in Cable	7	19	37	61	91	127	169

FIGURE 7–9

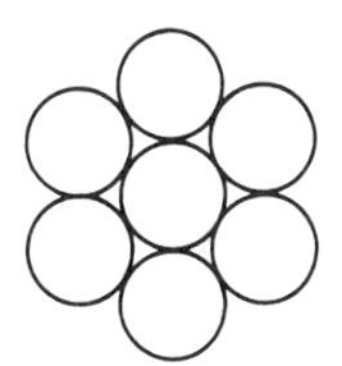

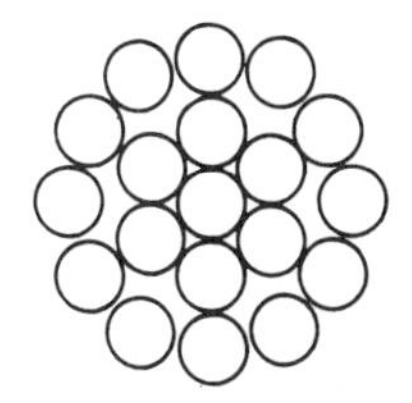

FIGURE 7–10 Concentric stranding

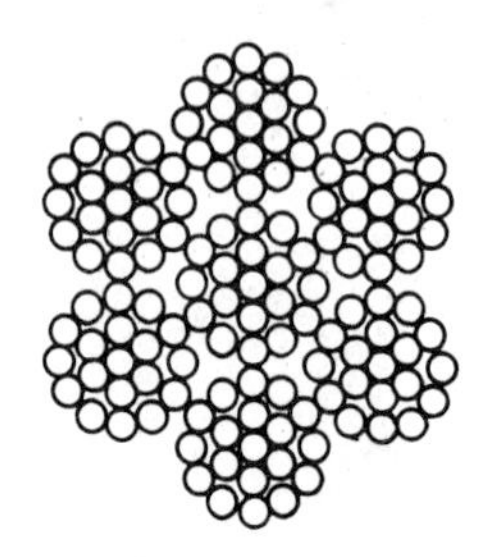

FIGURE 7–11 Rope stranding

One type of #25 microphone cable consists of four strands of #33 copperweld wire added to three strands of #33 copper wire. Steel as well as copperweld wire may be used to supply extra strength to large cables.

Rope Stranding

A rope-stranded cable consists of several concentric cables twisted together. The 7 × 19 rope-stranded cable shown in Figure 7–11 consists of seven 19-strand conductors twisted together.

Cables larger than #4/0 are rated in circular mils. Standard sizes range from 200,000 CM to 5 million CM. These cables may be covered with any of a great variety of insulation materials.

Cables are not always circular. Flat braid is available in many shapes and sizes. The battery grounding strap on an automobile is made from wire in the form of a flat braid.

7–8 COMMERCIAL RESISTORS

A *resistor* is a circuit component that has electrical resistance and is used for the control of current or the production of heat. Resistance elements, made of nickel-chromium alloys, have been mentioned before in this chapter. They are designed to develop heat in appliances such as toasters, irons, and hot-water tanks. Incandescent lamps (ordinary light bulbs), too, have a fine resistor element made of tungsten, which produces light when it glows from heat.

The next few sections will acquaint you with resistors designed for the control of current. Such resistors can be classified as being *fixed* or *variable,* Figures 7–12 and 7–13. The words fixed and variable refer, of course, to their electrical resistance in ohms.

Fixed Resistors

Electronic devices, such as radios and television sets, contain multitudes of fixed resistors. These small, color-banded resistors are cylindrical in shape with axial wire leads molded into them, Figure 7–12B. They are generally known as *carbon resistors,* because they are manufactured from a carbon-based resistance material. Figure A–1 reveals that carbon has a resistance of up to 2,000 times that of copper, so it takes only

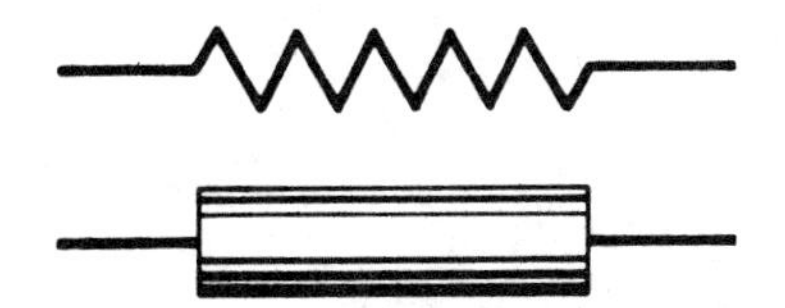

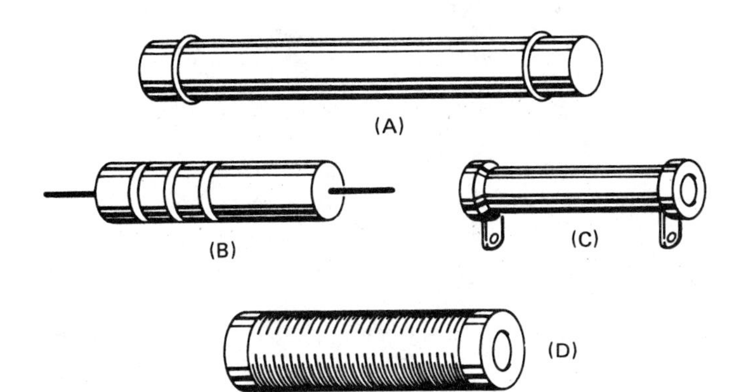

FIGURE 7–12 Types of fixed resistors

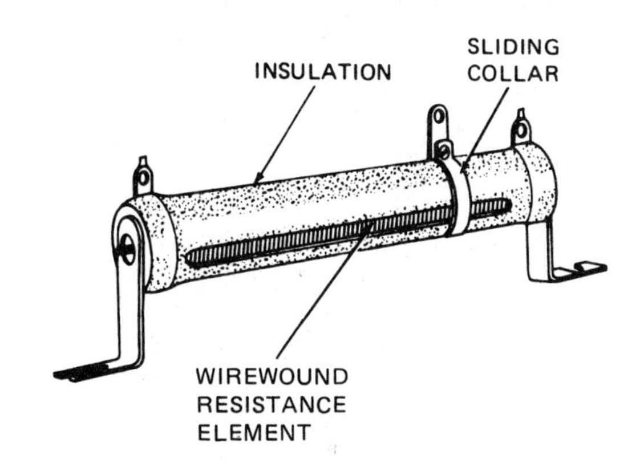

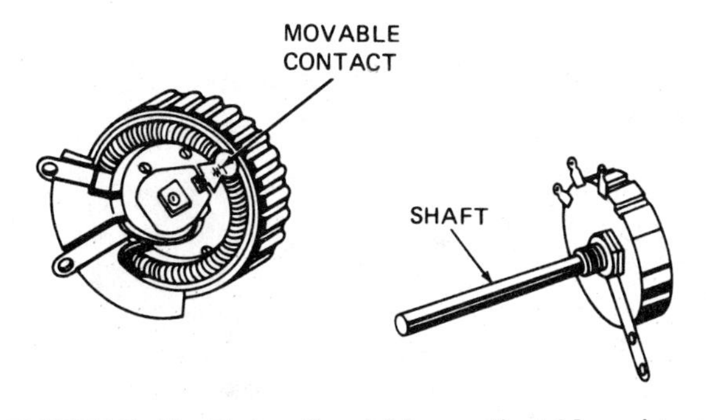

FIGURE 7–13 Types of variable or adjustable resistors

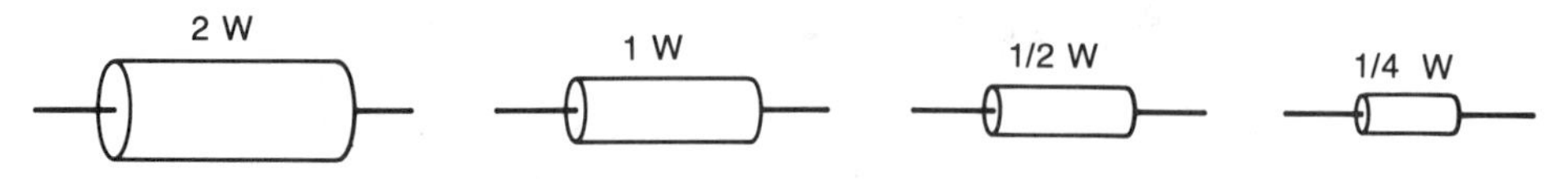

FIGURE 7–14 Typical resistor sizes

small amounts of carbon to obtain high resistance values. It is not uncommon to encounter resistors with values of many millions of ohms (megohms) measuring only fractions of an inch, Figure 7–14.

The size of a resistor, in fact, has nothing to do with its ohmic value. Instead, it is related to its power rating in *watts* (W). The wattage rating, as you will learn in Chapter 9, is an indication of the resistor's ability to dissipate heat without becoming damaged. Obviously, the larger the surface area of the resistor, the better it dissipates the heat.

Resistors with power ratings larger than 2 watts are generally wirewound on a ceramic base. In addition, high-wattage resistors often have a hollow core to provide greater surface area for heat dissipation, Figures 7–12C and D.

For low-wattage applications, carbon resistors have the advantage of being small, inexpensive, and rugged, while being reasonably accurate. However, carbon resistors are not without disadvantages. Besides their limited current-carrying capacity, they tend to change their value with age or overheating.

Color-coded bands reveal the ohmic value of most carbon resistors. This system of color coding is explained later in this chapter.

Variable Resistors

In many practical applications, it becomes desirable to vary the performance characteristics of an appliance. Examples of this include the volume control on your radio, the brightness control on a TV set, a light dimmer, or a speed control of an electric motor. They all have one common link: variable resistors.

Variable resistors generally have three terminals, one on each end and one movable contact in between. In many cases the resistance element is circular in shape and has a movable contact sliding across it. The movable contact is attached to a shaft that can be rotated to select the desired amount of resistance. (Remember the volume control on your radio.)

Variable resistors such as these are generally called *potentiometers* or *rheostats*, depending on their function in the circuit. In general, a potentiometer is used to vary a *voltage* by utilizing all three of its terminals. By contrast, a rheostat is used to vary a *current* by utilizing only two of the three terminals, Figure 7–15.

The potentiometer circuit shown in Figure 7–15A can be regulated to show any voltage, from zero to maximum, on the attached voltmeter. The theory of this operation will be covered in a later chapter.

Compare this with the rheostat circuit in Figure 7–15B in which the brightness of the lamp is controlled by varying the current through it. The schematic diagram in Figure 7–15B represents the same idea as the pictorial illustration in Figure 7–16.

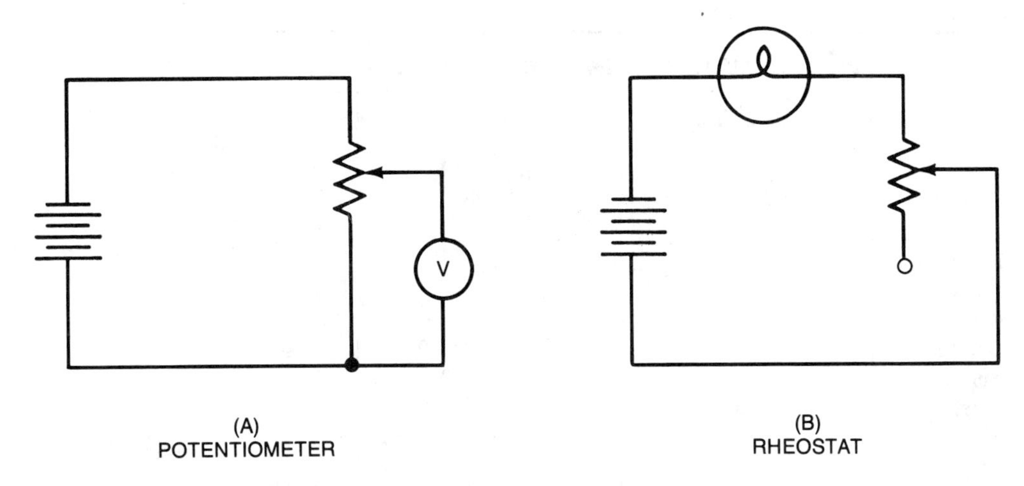

FIGURE 7–15 Potentiometer and rheostat hookups

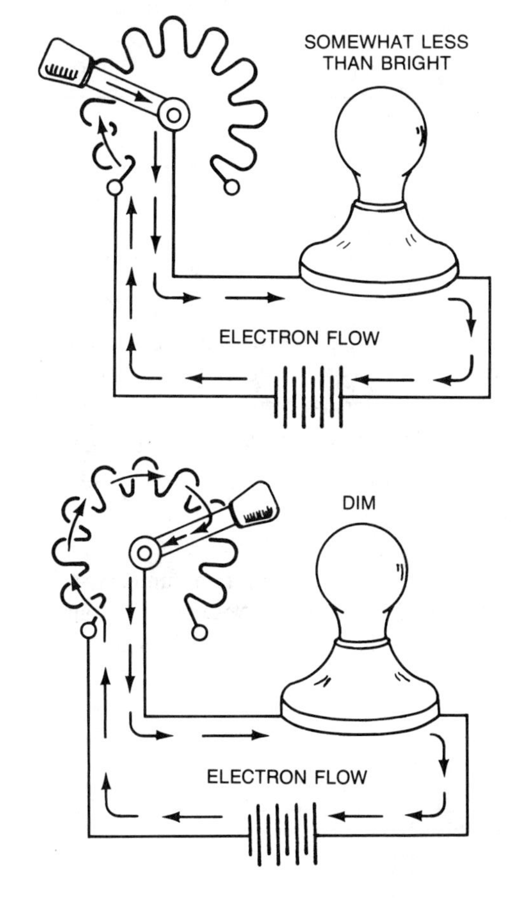

FIGURE 7–16 A rheostat

OHMS RESISTANCE COLOR SYSTEM			
Color	Numerical Value of 1st and 2nd Figures	Multiplying Value	Percentage Tolerance (Fourth Band)
Black	0	1	
Brown	1	10	
Red	2	100	
Orange	3	1,000	
Yellow	4	10,000	
Green	5	100,000	
Blue	6	1,000,000	
Violet	7	10,000,000	
Gray	8	100,000,000	
White	9	1,000,000,000	
Gold	→	0.1	5%
Silver	→	0.01	10%
No color			20%

FIGURE 7–17

The Resistor Color Code

A system of indicating the number of ohms resistance for small resistors is shown in Figure 7–17. Carbon resistors cannot be made as accurate as wirewound resistors; thus, the color also shows the percentage accuracy of the resistor.

The following examples show how to determine the resistance values of color-banded carbon resistors.

1. First band, red, first figure of number: 2.
   Second band, green, second figure of number: 5.
   Third band, yellow, how many 0's to place after the first two numbers: 4.
   In other words, 2 5 0000. Resistance is 250,000 ohms. Absence of a fourth color means that the actual ohms may be as much as 20% more or less than the rated value of 250,000.

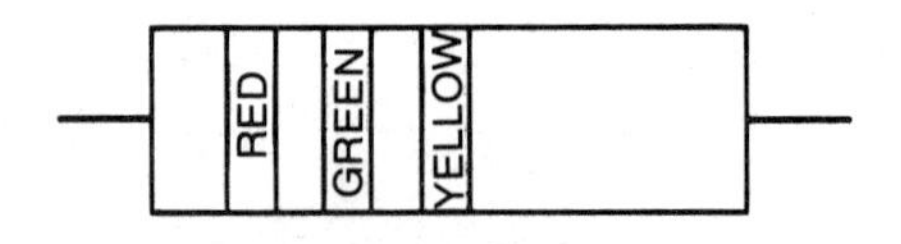

FIGURE 7–18

2. First band, yellow: 4.
Second band, violet: 7.
Third band, brown: 1. (Add one 0 to the 4 and 7.)
Resistance is 470 ohms.
Fourth band, gold: 5% tolerance. Actual resistance in ohms within 5% of the 470.

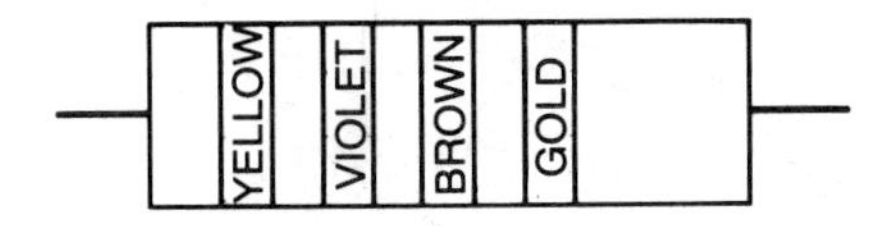

FIGURE 7–19

3. First band, green: 5.
Second band, black: 0.
Third band, green: 5 (zeros). 5 0 00000 = 5,000,000 ohms.
Fourth band, silver: 10% tolerance.
Resistance measures between 4,500,000 and 5,500,000 ohms.

FIGURE 7–20

4. First band, red: 2.
Second band, black: 0.
Third band, gold: multiplying value of 0.1.
$20 \times 0.1 = 2$ ohms
Resistance is within 5% of 2 ohms.

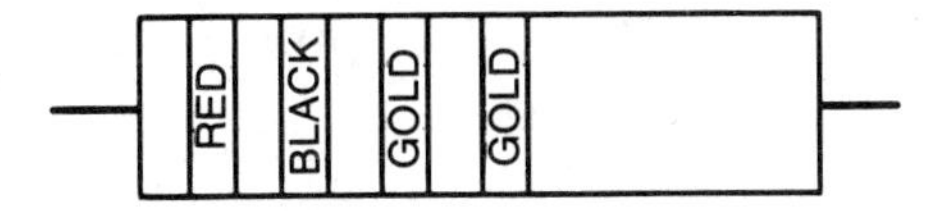

FIGURE 7–21

SUMMARY

- Every conductor, no matter how good, has some resistance.
- The amount of resistance a wire has depends on its material composition, length, cross-sectional area, and temperature.
- The statement above is mathematically expressed by the equation

$$R = \frac{K \times l}{A}$$

- Cross-sectional area (CSA) of a round conductor is figured in circular mils.
- Cross-sectional area of a rectangular bus bar is figured in square mils.

- 1 CM = 0.7854 sq mil.
- Resistance is more for a long wire than a short one, a thin wire than a thick one, a hot wire than for a cold one, and iron than for copper.
- A change in resistance due to temperature change is equal to the original resistance times temperature coefficient times degrees change.
- Commercial resistors must be specified by resistance value and wattage rating.
- Variable resistors are known as potentiometers or rheostats, depending on their circuit function.
- The wire table gives the resistance values in ohms per 1,000 feet for copper and aluminum.

Achievement Review

1. Write definitions for the words (a) resistance, (b) mil, (c) circular mil, (d) mil-foot.
2. There is no perfect conductor. Every wire has at least some resistance. Just how much resistance it has depends on four reasons. Explain.
3. What is the name of the special resistance wire used in heating appliances?
4. If you want to obtain a lot of heat from such heating wire, should it have a lot of resistance or just a little bit? Explain. (Hint: Heating power is equivalent to $P = E \times I$ or $P = I^2R$.)
5. Which has more resistance, a wire 20 feet long or a wire 20 inches long, if both are taken from the same stock? Explain.
6. Which has more resistance, a given length of #12 wire or #14 wire? Explain.
7. What happens to the resistance of a lamp as it heats up? Explain.
8. Does a hot lamp draw as much current as a cold lamp? Explain.
9. Two pieces of wire are cut off the same coil. One is 17 feet long, the other is 204 inches long. Which of the two has more resistance? Explain.
10. The diameter of #14 copper wire is 64 mils. Find the cross-sectional area of the wire in circular mils (CM).
11. What is the cross-sectional area of
    a. A wire 0.012 inch in diameter
    b. A wire 0.0155 inch in diameter
12. Find the diameter of a wire that has a cross-sectional area of 81 CM.
13. Find the resistance of the wires in parts a–d, using

    $$R = \frac{K \times l}{A}$$

    a. 100 feet of #14 aluminum
    b. 25 feet of #20 nichrome
    c. 1 mile of #8 iron
    d. 6 inches of #18 copper
14. A power line requires 12,000 feet of wire. The total resistance of this wire is not to exceed 5 ohms.

a. What size copper wire meets these requirements? What will 12,000 feet of the wire weigh?
b. What is the smallest size aluminum wire that meets the requirements? What will this wire weigh?

15. In question 14, the specific resistances for copper and aluminum are 10.4 and 17 at 68°F (20°C). Find the resistance of the wire at
    a. 0°F (−18°C)
    b. 104°F (40°C)
16. A resistance of 0.0005 ohm is required in the connecting wire of an ammeter. What length of #10 copper wire must be used?
17. The field magnet winding of a 230-volt DC generator has a resistance of 54.5 ohms at 20°C. Find the resistance of the copper winding when the temperature rises to 50°C.
18. A coil of copper wire has 150 ohms resistance at 20°C. After several hours of operation, the resistance of the coil is 172 ohms. Find the temperature of the coil. Of what use is a calculation of this type?
19. Find the diameter (in inches) of a wire whose cross-sectional area is 4,096 CM.
20. A flexible copper cable has 74 strands each 8.75 mils in diameter. Compute the cross-sectional area. (Hint: Compute first the CSA of one strand.)
21. A cable with a cross-sectional area of 1,200 MCM is made up of 19 strands. Find the diameter of each strand.
22. The cross-sectional area of a #10 wire is 10,382 CM. Find its diameter.
23. A 300 MCM cable is composed of 37 strands of copper wire of equal size. Find the diameter of each strand.
24. Find the length of a copper wire that has 4,000 CM and has 2.5 ohms.
25. Find the resistance of a 0.1-inch-diameter aluminum wire that is 100 feet long.
26. Assume that 25 feet of #24 manganin wire is used as a resistance element on a 120-volt circuit. Find the current flowing in this circuit. (*Note:* Manganin has a resistivity of 260.)
27. Use your wire table to find the resistance of 200 feet of #23 wire at 68°F.
28. How heavy (in pounds) is a #2 wire, 600 feet long?

COLOR CODE EXERCISE

Part A

For each of the following resistors, state the value and tolerance.

1. Yellow, Purple, Black, Silver
2. Brown, Green, Red, Gold
3. Red, Purple, Red, Silver
4. Orange, Purple, Orange, Silver
5. Brown, Black, Green, Silver
6. Blue, Gray, Orange, Gold
7. Brown, Green, Orange, Silver
8. Red, Purple, Green, Silver
9. Brown, Black, Black, Gold
10. Brown, Black, Blue, Silver
11. Red, Red, Orange, Silver
12. Orange, Orange, Black, Silver
13. Yellow, Purple, Yellow, Silver
14. Brown, Black, Yellow, Silver

15. Red, Red, Brown, Gold
16. Brown, Black, Red, Silver
17. Red, Red, Yellow, Silver
18. Blue, Gray, Silver, Gold
19. Green, Blue, Gold, Silver
20. Green, Blue, Brown, Silver

Part B

For each of the following resistors state the color code.

1. 270 kΩ ± 5%
2. 33 Ω ± 10%
3. 4.7 MΩ ± 20%
4. 68 kΩ ± 10%
5. 2,700 Ω ± 20%
6. 100 kΩ ± 10%
7. 22 kΩ ± 5%
8. 2.2 MΩ ± 10%
9. 1,500 Ω ± 5%
10. 10 Ω ± 10%
11. 4.7 Ω ± 5%
12. 39,000 Ω ± 20%
13. 0.15 MΩ ± 10%
14. 10 MΩ ± 10%
15. 1,000 Ω ± 20%
16. 0.1 MΩ ± 10%
17. 15 kΩ ± 10%
18. 10,000 Ω ± 20%
19. 3.3 Ω ± 5%
20. 0.68 Ω ± 5%

8
Ohm's Law

Objectives

After studying this chapter, the student should be able to

- Describe the interrelationship of voltage *(E)*, current *(I)*, and resistance *(R)*
- Apply Ohm's law in calculating an unknown circuit quantity
- Use Ohm's law in conjunction with metric prefixes

8–1 VOLTAGE, CURRENT, AND RESISTANCE

The three quantities E, I, and R relate to one another in a very significant way. This relationship, first formulated by the German scientist Georg S. Ohm, states that the current *(I)* is directly proportional to the voltage *(E)* but inversely proportional to the resistance *(R)*. In other words,

- Current goes up when voltage goes up.
- Current goes down when voltage goes down.

} (with constant R)

- Current goes up when resistance goes down.
- Current goes down when resistance goes up.

} (with constant E)

In the language of mathematics this is simply stated as

$$I = \frac{E}{R}$$

The expression above is known as *Ohm's law* and can be algebraically transposed and rewritten into other forms, namely

$$E = I \times R \text{ and } R = \frac{E}{I}$$

Ohm's law is one of the most important formulas in electrical theory. It enables us to compute unknown circuit quantities, thereby helping us to predict or confirm proper circuit operation. Memorize Ohm's law, because you will encounter it time and again in your future studies of electricity and electronics.

A popular memory device, for some students, has Ohm's law arranged in a circle as shown in Figure 8–1. To use this circle, place a finger on the letter representing the unknown quantity. The other two letters will show themselves in the proper arrangement to indicate either multiplication or division.

EXAMPLE 8–1

Given: A resistance *(R)* of 4 ohms connected to a 12-volt battery *(E)*.

Find: The current *(I)*.

SOLUTION

Covering the letter I in the memory circle suggests the correct formula, namely

$$I = \frac{E}{R} = \frac{12}{4} = 3\ A$$

EXAMPLE 8–2

Given: A current *(I)* of 6 amperes flowing through a 4-ohm resistance *(R)*.

Find: The voltage *(E)* needed to force the current through the resistance.

SOLUTION

Wanting to find the voltage *(E)*, cover the letter E within the memory circle. This would suggest that

$$E = I \times R$$
$$E = 6 \times 4$$
$$E = 24\ \text{V}$$

EXAMPLE 8–3

Given: A source of 6 volts *(E)* forcing a current of 1.5 amperes *(I)* through a resistor *(R)*.

Find: The value of the unknown resistor.

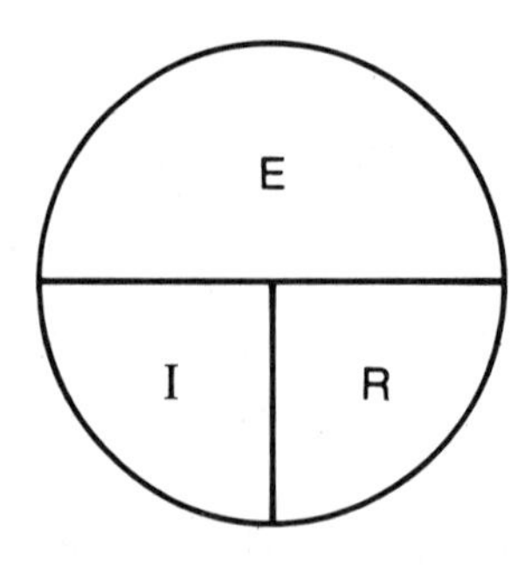

FIGURE 8–1

SOLUTION

Placing your finger on the unknown value *(R)* of your memory circle will suggest that

$$R = \frac{E}{I}$$

$$R = \frac{6}{1.5}$$

$$R = 4\ \Omega$$

8–2 OHM'S LAW WITH METRIC PREFIXES

Ohm's law is designed for use with basic units only. That means

- Current must be expressed in amperes.
- Voltage must be expressed in volts.
- Resistance must be expressed in ohms.

Special precaution must be observed when given values are stated with metric prefixes instead of basic units. In such cases it is recommended to convert the metric prefixes to powers of 10.

EXAMPLE 8–4

Given: An electromotive force of 100 volts and a current of 20 milliamperes.

Find: The resistance.

SOLUTION

$$R = \frac{E}{I}$$

$$R = \frac{100}{20 \times 10^{-3}}$$

$$R = \frac{100 \times 10^{3}}{20}$$

$$R = 5 \times 10^{3}$$

$$R = 5{,}000\ \Omega \text{ or } 5\ \text{k}\Omega$$

EXAMPLE 8–5

Given: A 3-kilohm resistor is connected to a 12-volt power source.

Find: The current.

SOLUTION

$$I = \frac{E}{R}$$

$$I = \frac{12}{3 \times 10^{3}}$$

$$I = \frac{12 \times 10^{-3}}{3}$$
$$I = 4 \times 10^{-3}\ \text{A}$$
$$I = 0.004\ \text{A or } 4\ \text{mA}$$

EXAMPLE 8–6

Given: A current of 2 milliamperes is flowing through a 5-kilohm resistance.

Find: The voltage.

SOLUTION #1

$$E = I \times R$$
$$E = 0.002\ \text{A} \times 5{,}000\ \Omega$$
$$E = 10\ \text{V}$$

SOLUTION #2

$$E = I \times R$$
$$E = 2 \times 10^{-3} \times 5 \times 10^{3}$$
$$E = 10\ \text{V}$$

Solution #2 of the preceding problem demonstrates that multiplication of milli × kilo = 1, and the powers of 10 drop out. Also, since

$$\frac{1}{10^{-3}} = 10^{3}$$

it follows that

$$\frac{1}{\text{milli}} = \text{kilo}$$

and vice versa

$$\frac{1}{10^{3}} = 10^{-3}$$

it follows that

$$\frac{1}{\text{kilo}} = \text{milli}$$

Look, once more, at Examples 8–4 and 8–5 to confirm this idea.

SUMMARY

- Ohm's law states that
  1. Current and voltage are directly proportional
  2. Current and resistance are inversely proportional

- Mathematically, Ohm's law may be expressed by the equations

$$E = I \times R \qquad I = \frac{E}{R} \qquad R = \frac{E}{I}$$

- When solving Ohm's law problems, numerical values must be stated in basic units (volts, amperes, and ohms).

Achievement Review

1. Given voltage = 120 volts and current = 4 amperes, find resistance.
2. Given resistance = 6 ohms and current = 3 amperes, find voltage.
3. Given voltage = 24 volts and resistance = 8 ohms, find current.
4. Given voltage = 150 volts and resistance = 450 ohms, find current.
5. Given current = 5.5 amperes and voltage = 440 volts, find resistance.
6. Given voltage = 6 volts and resistance = 0.05 ohms, find current.
7. Given current = 0.022 ampere and voltage = 660 volts, find resistance.
8. Given voltage = 60 volts and current = 0.01 ampere, find resistance.
9. Given resistance = 48 ohms and current = 2.5 amperes, find voltage.
10. Given voltage = 12 volts and current = 0.002 ampere, find resistance.
11. Given resistance = 50,000 ohms and voltage = 250 volts, find current.
12. Given current = 0.01 ampere and resistance = 3,000 ohms, find voltage.
13. Given resistance = 7.5 ohms and current = 4.5 amperes, find voltage.
14. Given resistance = 31.4 ohms and current = 3.99 amperes, find voltage.
15. Given resistance = 480 ohms and current = 220 milliamps, find voltage.
16. Given current = 50 milliamps and resistance = 2 kilohms, find voltage.
17. Given resistance = 30,000 ohms and current = 3 milliamps, find voltage.
18. Given voltage = 100 volts and resistance = 10 kilohms, find current.
19. Given voltage = 240 volts and current = 8 milliamps, find resistance.
20. Given current = 0.002 ampere and voltage = 6.62 volts, find resistance.

9

Electrical Power and Energy

Objectives

After studying this chapter, the student should be able to

- Define and explain the new technical terms introduced in this chapter

force	power (P)
work	horsepower (hp)
potential energy	efficiency
kinetic energy	watt (W)
foot-pound	I^2R losses
joule (j)	watt-hour (Wh)
Btu	kilowatt-hour (kWh)

- Calculate the power dissipation of a resistor
- Use the power equation in conjunction with Ohm's law to find an unknown circuit quantity
- Explain the derivation of the formulas embodied in the PIRE wheel
- Calculate the cost of electrical energy expended by a load

9–1 ENERGY

In Section 2–8 we discussed the conversion of electrical energy to perform work. The type of work or change accomplished involves force and motion. The following changes all require energy: mechanical movement, the production of heat or light, the production of sound, the conversion of one chemical compound into another, and the production of radio waves. The amount of energy required for these changes can be measured, although we cannot see it. How this is accomplished is the subject of this chapter.

In common usage the words *work* and *energy* have broad meanings. The physical work or energy with which we are concerned does not include such things as the work done by someone counting the cars that pass a corner, the work done in getting people to change their minds, or the energy with which one tackles an arithmetic problem. Instead, words like work and energy have precise, scientific meanings, which should be understood before we discuss the concept of *power*. The following sections will help to clarify these ideas.

Mechanical Energy

The lifting of a weight illustrates the meaning of a *unit of measurement of energy*. One foot-pound of energy, or 1 foot-pound of work, is required to lift a 1-pound weight a distance of 1 foot. (The words *work* and *energy* can be interchanged where measurement is concerned.)

A *foot-pound* is defined as the energy used when a 1-pound force moves an object a distance of 1 foot. The 1-foot movement is in the *same direction* as the applied force.

How much work is done in lifting a 20-pound weight 5 feet vertically, Figure 9–1?

Work = Force × Distance

A 20-pound force traveling 1 foot accomplishes 20 foot-pounds of work. If the force must travel 5 feet, then 100 foot-pounds of work is accomplished.

Foot-pounds = Feet × Pounds

Potential Energy

What happens to the 100 foot-pounds of energy? It is saved up or conserved. When the 20-pound weight is lifted 5 feet, it has 100 foot-pounds of energy that it did not have when it was on the ground. This energy is called *potential energy*. When the weight is permitted to fall back to earth, it delivers 100 foot-pounds of energy to the earth, Figure 9–2.

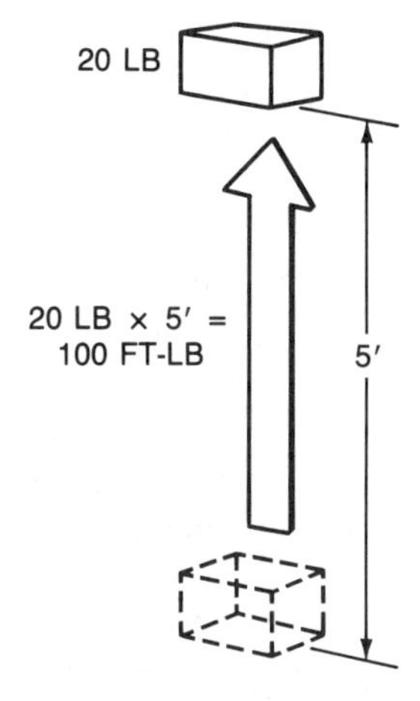

FIGURE 9–1 Energy required to lift weight

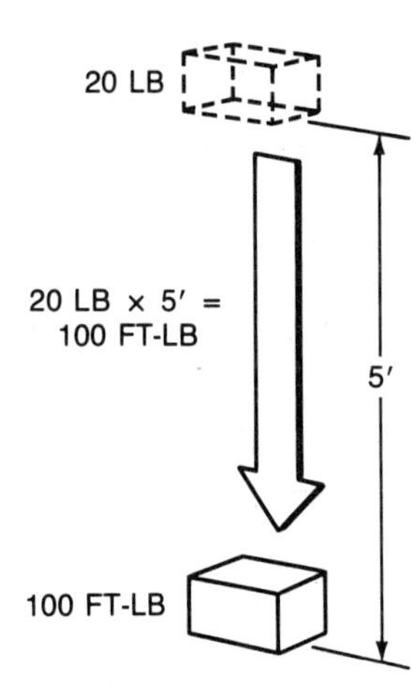

FIGURE 9–2 Energy delivered when weight falls

How much work is done when a 200-pound box is dragged horizontally along the floor a distance of 6 feet? The question cannot be answered when asked in this manner. Since 200 pounds is a vertical force, it is not in the same direction as the motion. Using a spring scale, however, we find that the horizontal force is 50 pounds. The amount of work done can now be computed.

$$50 \text{ lb} \times 6 \text{ ft} = 300 \text{ ft-lb}$$

What becomes of this 300 foot-pounds of energy? The energy is converted into heat by the process of friction against the floor. The box does not gain potential energy.

Kinetic Energy

How much work is done when a force of 8 pounds is applied to a 3/4-pound ball to throw it a distance of 6 feet?

$$\text{Work} = \text{Force} \times \text{Distance} = 8 \text{ lb} \times 6 \text{ ft} = 48 \text{ ft-lb}$$

What becomes of this energy? It exists as the energy of motion of the ball and is called *kinetic energy*. The ball in flight has 48 foot-pounds of energy it will deliver when it strikes its target.

Let us return to the example of potential energy, Figure 9–2. What happens when the 20-pound weight is allowed to fall the 5-foot distance to the floor? The potential energy of the weight is 100 foot-pounds. This energy becomes the kinetic energy of motion as the weight falls through 5 feet.

The purpose of these mechanical examples is to illustrate the meaning of such terms as work, energy, potential energy, and kinetic energy. These terms are often used in electrical energy discussions, but they are easier to visualize in mechanical energy examples.

9–2 UNITS OF ENERGY

The foot-pound is the energy unit commonly used in the British system of measurements. The metric unit of energy is called a joule. Most common electrical units are based on the joule as the unit of energy. The *joule* is the work done when a force of 1 newton is exerted through a distance of 1 meter. (A newton is 100,000 dynes, which is about 3 1/2 ounces; a meter is about 39 inches.)

1 j = 0.738 ft-lb
1 ft-lb = 1.35 j

There are two units of measurement that are used to express heat energy. The unit in the British system for heat is called Btu (British thermal unit). A *Btu* is the amount of heat needed to raise the temperature of a pound of water 1 Fahrenheit degree.

How much heat is needed to warm 10 pounds of water from 50°F to 65°F? The temperature increase is 65° − 50° = 15°F. Therefore, 10 lb × 15°F = 150 Btu.

The metric system unit of heat is called a calorie or a gram-calorie. The *calorie* is the energy needed to raise the temperature of 1 gram of water 1 Celsius degree. (The calorie used in food energy calculations is a kilogram-calorie; this amount of energy will raise 1,000 grams of water 1 degree.)

1 Btu = 778 ft-lb
1 Btu = 252 cal

9–3 POWER

The word power (P), as commonly used, means a variety of things. In technical language, *power* means how fast work is done or how fast energy is transferred. Two useful definitions of the term power are as follows:

Power is the rate of doing work.
Power is the rate of energy conversion.

The student should understand what is meant by the statement: *power is a rate.* We do not buy or sell electrical power. What we buy or sell is electrical energy. Power indicates how fast the energy is used or produced. A mechanical example will illustrate the meaning of power and energy.

EXAMPLE 9–1

Given: An elevator lifts 3,500 pounds a distance of 40 feet in 25 seconds.

Find:

a. The work done in 25 seconds (foot-pounds).
b. The rate of work (foot-pounds per second).

SOLUTION

a. $40 \times 3{,}500 = 140{,}000$ ft-lb

b. $\dfrac{140{,}000 \text{ ft-lb}}{25 \text{ sec}} = 5{,}600$ ft-lb/sec

Naturally, power can also be expressed in foot-pounds per minute. The next example will make this clear.

EXAMPLE 9–2

Given: A pump takes 20 minutes to lift 5,000 pounds of water 66 feet.

Find: The rate of doing work.

SOLUTION

$$\text{rate} = \frac{5{,}000 \text{ lb} \times 66 \text{ ft}}{20 \text{ min}}$$

$$\text{rate} = \frac{330{,}000 \text{ ft-lb}}{20 \text{ min}}$$

$$\text{rate} = 16{,}500 \text{ ft-lb/min}$$

The Horsepower

When James Watt started to sell steam engines, he needed to express the capacity of his engines in terms of the horses they were to replace. He found that an average horse, working at a steady rate, could do 550 foot-pounds of work per second. This rate is the definition of 1 horsepower (hp), Figure 9–3.

$$1 \text{ hp} = 550 \text{ ft-lb/sec}$$

Considering that 1 minute has 60 seconds, it follows that

$$1 \text{ hp} = 33{,}000 \text{ ft-lb/min}$$

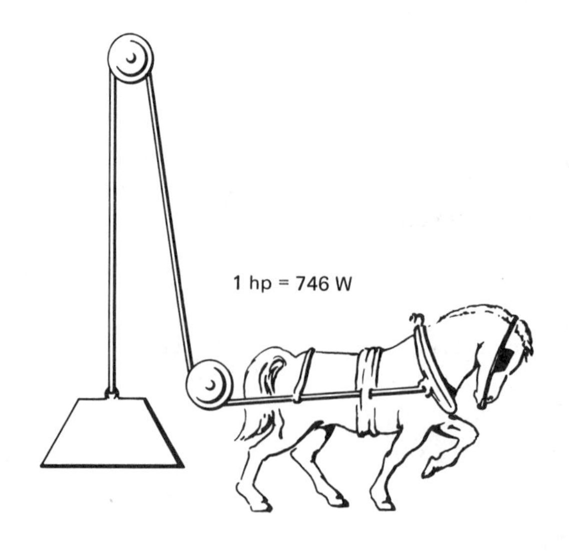

1 hp = 550 FT-LB PER SECOND

FIGURE 9–3

1 hp	= 746 W
1 W	= 3.42 Btu/hr
1 Btu/sec	= 1,055 W
1 cal/sec	= 4.19 W
1 ft-lb/sec	= 1.36 W

FIGURE 9–4 Comparison of power units

Can you see that the elevator in Example 9–1 is doing work at the rate of approximately 10 horsepower? Remember, its rate of work was found to be 5,600 foot pounds per second. To convert this to horsepower, we compute

$$5{,}600 \text{ ft-lb/sec} \times \frac{1 \text{ hp}}{550 \text{ ft-lb/sec}} = 10.18 \text{ hp}$$

Now try to compute the horsepower rating of the pump described in Example 9–2. Your answer should be approximately 0.5 horsepower. Do you agree?

Horsepower can also be expressed in units of electrical power called watts (W).

1 hp = 746 W

The Watt—Unit of Electrical Power

Recall that the metric unit of energy is called a joule. In the metric system, power is measured in *joules per second*. This unit of measurement corresponds to foot-pounds per second in the British system. Because the unit *joules per second* is used so often, it is replaced with the single term *watt*. (Figure 9–4 gives a comparison of various power units with the watt.)

One watt is a rate of 1 joule per second

Another way of explaining the term *watt* is to say that 1 watt of power is dissipated (in the form of heat) when 1 volt of electrical pressure forces 1 ampere of current through the resistance of 1 ohm, Figure 9–5. This relationship can be expressed by the equation

$$P = E \times I$$

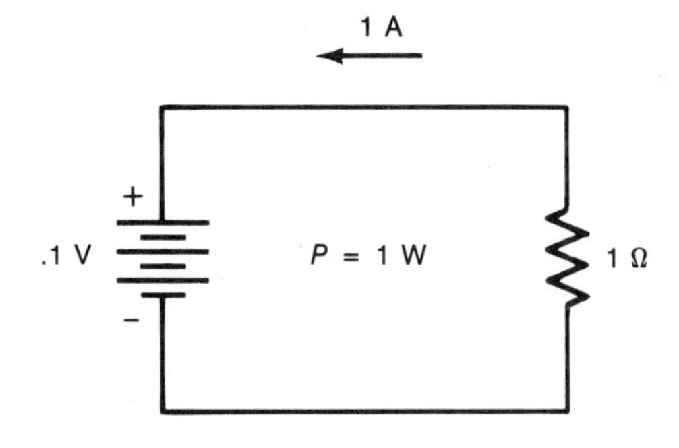

FIGURE 9–5

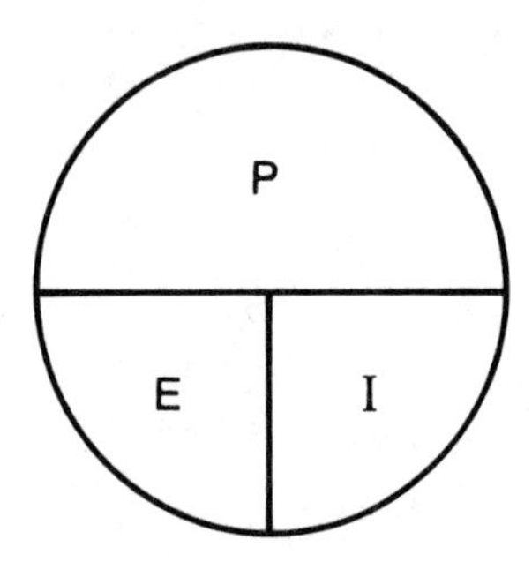

FIGURE 9–6

You should recall the Ohm's law circle used as a memory device. The power equation shown previously also lends itself to another such memory circle often referred to as a PIE diagram, Figure 9–6. Cover the unknown quantity to be found and read off the equation. Thus,

$$E = \frac{P}{I} \qquad P = EI \qquad I = \frac{P}{E}$$

EXAMPLE 9–3

Given: A 120-volt circuit protected by a 15-ampere fuse.

Find: Maximum power.

SOLUTION

$$P = E \times I$$
$$P = 120 \times 15$$
$$P = 1{,}800 \text{ W}$$

This value of 1,800 watts is the maximum amount of energy available to the appliances attached to the circuit.

EXAMPLE 9–4

Given: A radio rated at 36 watts.

Find: How many volts of electromotive force are needed to force 3 amperes through the radio.

SOLUTION

$$E = \frac{P}{I}$$
$$E = \frac{36}{3}$$
$$E = 12 \text{ V}$$

EXAMPLE 9–5

Given: The label of a television that states 720 W/120 V.

Find: The current flowing through the supply line to the TV.

SOLUTION

$$I = \frac{P}{E}$$

$$I = \frac{720\ \text{W}}{120\ \text{V}}$$

$$I = 6\ \text{A}$$

A Second Equation for Power

The power equation $P = E \times I$ may be combined with Ohm's law to yield the hybrid equation $P = I^2R$. Let us see how this is derived.

Assume that we need to calculate the electrical power of a circuit for which the voltage is unknown. From Ohm's law we know that

$$E = I \times R$$

Substituting $I \times R$ for E into the power equation

$$P = E \times I$$

we get

$$P = I \times R \times I$$

or simply

$$P = I^2R$$

$P = I^2R$ is an important formula to remember. The synonymous word *power* and the term I^2R have found their way into the vocabulary of people in the electrical trades. Thus, we speak of I^2R losses (power losses in the form of heat), I^2R heating, and I^2R ratings.

Mathematically, the quantities of this equation may be transposed to yield two other expressions, namely

$$R = \frac{P}{I^2} \quad \text{and} \quad I^2 = \frac{P}{R} \quad \text{or} \quad I = \sqrt{\frac{P}{R}}$$

The following examples will explain the application of these formulas in finding an unknown circuit quantity.

EXAMPLE 9–6

Given: A current of 2 amperes flowing through a 250-ohm resistance.

Find: The electrical power converted into heat by the resistor. (Remember, all resistors develop heat when a current flows through them.)

SOLUTION

$$P = I^2R$$
$$P = (2)^2(250)$$
$$P = 4(250)$$
$$P = 1{,}000 \text{ W} \quad (1{,}000 \text{ W can also be expressed as 1 kW.})$$

EXAMPLE 9–7

Given: A 60-watt light bulb with a 0.5-ampere current flowing through it.

Find: The resistance of the lamp.

SOLUTION

$$R = \frac{P}{I^2}$$
$$R = \frac{60}{(0.5)^2}$$
$$R = \frac{60}{0.25}$$
$$R = 240\ \Omega$$

EXAMPLE 9–8

Given: A 40-ohm resistor dissipating 360 watts of power.

Find: The current.

SOLUTION

$$I = \sqrt{\frac{P}{R}}$$
$$I = \sqrt{\frac{360}{40}}$$
$$I = \sqrt{9}$$
$$I = 3 \text{ A}$$

A Third Equation for Power

It is possible to develop yet a third formula by combining the power equation with Ohm's law. Let us see how this is accomplished.

Beginning with $P = E \times I$, let us assume that the quantity I is unknown. From Ohm's law, we substitute the equality

$$I = \frac{E}{R}$$

Instead of $P = E \times I$ we now have

$$P = E \times \frac{E}{R}$$

which is more commonly stated as

$$P = \frac{E^2}{R}$$

This, then, is our third equation for electrical power, which may be transposed to yield two more expressions.

$$R = \frac{E^2}{P} \qquad \text{and} \qquad E = \sqrt{PR}$$

The following three examples will demonstrate the use of these equations.

EXAMPLE 9–9

Given: A heating appliance with a 30-ohm resistance connected to a voltage source of 120 volts.

Find: The amount of power converted to heat.

SOLUTION

$$P = \frac{E^2}{R}$$

$$P = \frac{120 \times 120}{30}$$

$$P = 120 \times 4$$

$$P = 480 \text{ W}$$

EXAMPLE 9–10

Given: A resistor rated at 1,500 ohms, 2 watts.

Find: The greatest potential difference *(E)* that can be applied across the resistor.

SOLUTION

Since P and R are given,

$$E = \sqrt{PR}$$

$$E = \sqrt{2 \times 1{,}500}$$

$$E = \sqrt{3{,}000}$$

$$E = 54.8 \text{ V}$$

EXAMPLE 9–11

Given: A lamp rated at 300 watts, 120 volts.

Find: The resistance of the lamp while it is operating.

SOLUTION

$$R = \frac{E^2}{P}$$

$$R = \frac{(120)\ (120)}{300}$$

$$R = 48\ \Omega$$

An alternate method of finding R is to determine

$$I = \frac{P}{E}$$

and then substitute I in Ohm's law to find R.

The PIRE Wheel

Up to this point we have become acquainted with twelve mathematical expressions that relate to Ohm's law and electrical power. For your convenience, these twelve equations are summarized in the circular chart in Figure 9–7.

The four letters P, I, R, and E, shown in the inner circle, represent the unknown quantities that may need to be found. Radiating outward from each of these letters are three choices of equalities that can be used for calculating the unknown.

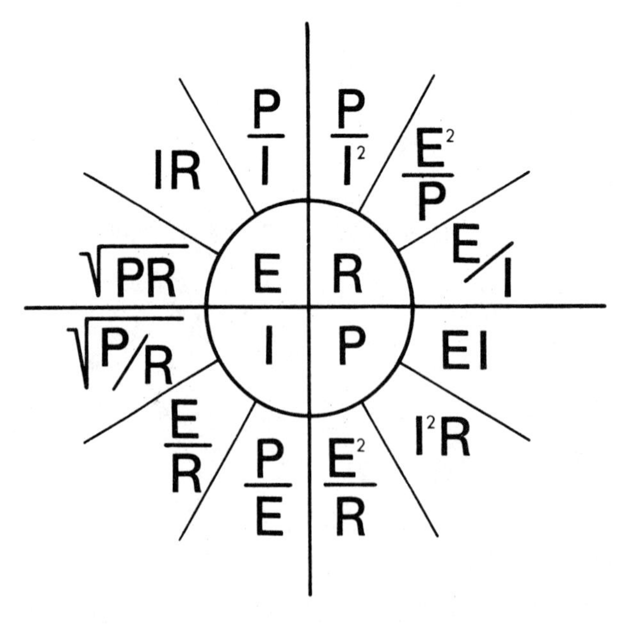

FIGURE 9–7

9–4 ENERGY AND COST CALCULATIONS

Electrical energy is commercially sold in units called *kilowatt hours*. The consumption of energy is not only a function of the power rating of the appliance, but also of the amount of time for which it is employed. Obviously, it makes a difference on your electrical bill whether you burn a lamp for 1 hour or for 10 hours. Take, for instance, a 100-watt light bulb. If you burn it for 1 hour, you have expended 100 watts × 1 hour = 100 watt-hours, which is equivalent to 0.1 kilowatt-hour (kWh). Burning it for 10 hours will require 10 times as much energy, namely

$$100 \text{ W} \times 10 \text{ h} = 1{,}000 \text{ Wh} = 1 \text{ kWh}$$

Note: Energy = Power × Time
(kWh) (kW) (h)

EXAMPLE 9–12

Given: A 500-watt heater operated for 10 hours.

Find: The cost of operating the heater at $0.06 per kilowatt-hour.

SOLUTION

$$500 \times 10 = 5{,}000 \text{ Wh} = 5 \text{ kWh}$$

$$5 \text{ kWh} \times \frac{\$0.06}{\text{kWh}} = \$0.30$$

EXAMPLE 9–13

Given: A 120-volt, 10-ampere iron operated for 3 hours.

Find: The cost of operation at $0.064 per kilowatt-hour.

SOLUTION

$$120 \text{ V} \times 10 \text{ A} = 1{,}200 \text{ W}$$
$$1{,}200 \text{ W} \times 3 \text{ h} = 3{,}600 \text{ Wh} = 3.6 \text{ kWh}$$
$$3.6 \text{ kWh} \times \$0.064 = \$0.23$$

EXAMPLE 9–14

Given: Five 60-watt lamps and four 100-watt lamps burning daily for 5 hours.

Find: The cost of operating the lamps for a billing period of 30 days, at the cost of $0.056 per kilowatt-hour.

SOLUTION

$$5 \times 60 \text{ W} = 300 \text{ W}$$
$$4 \times 100 \text{ W} = 400 \text{ W}$$
$$\text{total: } 700 \text{ W} = 0.7 \text{ kW}$$

$$0.7 \text{ kW} \times \frac{5 \text{ h}}{\text{day}} \times 30 \text{ days} = 105 \text{ kWh}$$

$$105 \text{ kWh} \times \frac{\$0.056}{\text{kWh}} = \$5.88$$

The Kilowatt-Hour Meter

The electrical energy delivered to consumers is measured by an instrument known as the *kilowatt-hour meter*. Such a meter generally has four dials geared to each other with a ratio of 10 to 1. Thus, when the *unit dial* makes ten revolutions, the *tens dial* makes one revolution. Similarly, the *tens dial* makes ten revolutions for every one revolution of the *hundreds scale*. The dials are driven by a small motor with a turning effect proportional to the product of voltage and current ($E \times I$). The reading of each dial represents a whole number, with any fractional part of that number being found on the next lower dial.

Figure 9–8 represents the four dials of such a kWh meter. In reading such a meter, some people start with the *unit dial* and then read from right to left; others read the dials from left to right, taking care to record the last number the pointer has passed. No matter how you do it, remember this

If the pointer is between two digits, select the lower of the two numbers.

Let us put this knowledge to use and read the meter in Figure 9–8. If you do it correctly, the answer should be: 4,294 kWh.

Now let us look at the dials in Figure 9–9. These dials represent the same meter read one month later. Do you agree that the new value is 4,579 kWh? By subtraction we find the difference between the two meter readings: 4,579 − 4,294 = 285 kWh. Thus, we have determined the amount of energy used during that month.

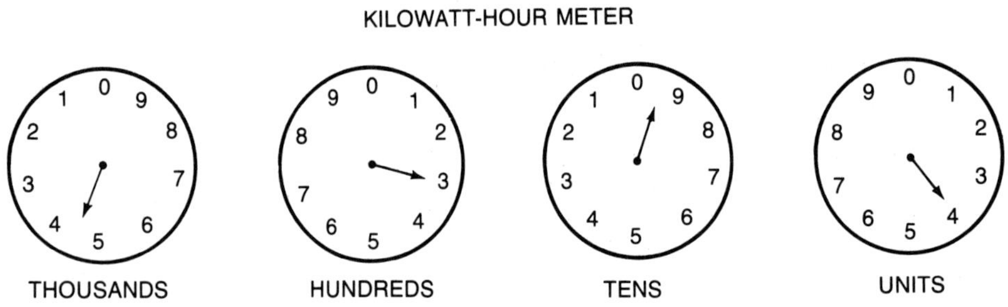

KILOWATT-HOUR METER

THOUSANDS HUNDREDS TENS UNITS

FIGURE 9–8

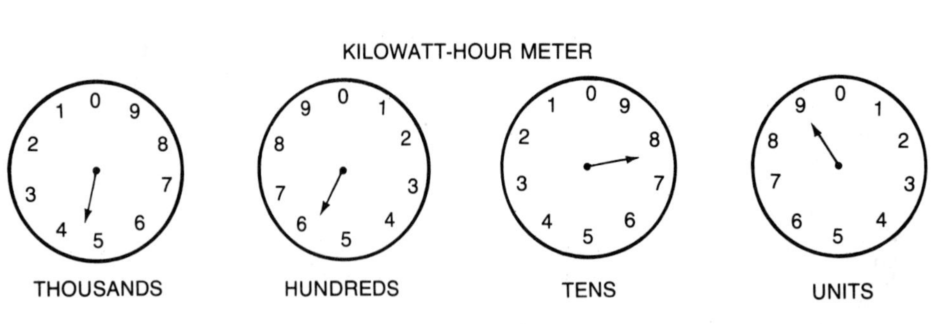

FIGURE 9–9

9–5 EFFICIENCY OF ENERGY CONVERSION

"Energy is never created (from nothing) and energy never vanishes" is one way of stating a principle long known as the *law of conservation of energy*. Each kilowatt-hour used comes from the burning of a fuel or from the release of some stored water. The energy of coal is still responsible for batteries used to light lamps; that is, coal is oxidized to release the zinc or lead used in the battery from other elements when the metal is refined. The electrical energy that we use daily is soon converted to heat by one process or another. Electrical energy can be stored briefly by charging batteries or by pumping water into a storage tank. However, examples of energy storage are few. In general, energy is converted from one form to another. The *efficiency* of this energy conversion is a way of measuring how well the energy-converting device accomplishes its task.

$$\text{Efficiency} = \frac{\text{Useful energy obtained}}{\text{Total energy used}}$$

or

$$\text{Efficiency} = \frac{\text{Power output}}{\text{Power input}}$$

EXAMPLE 9–15

Given: A DC motor taking 4.2 amperes on a 120-volt line and delivering 0.5 horsepower.

Find: The efficiency of the motor.

SOLUTION

The power output is 0.5 hp, or 373 W (1 hp = 746 W).
The power input is 120 V × 4.2 A = 504 W.
The efficiency of the motor is 373/504 = 0.74 or 74%.

The efficiency of any device can be no greater than 100%. In other words, the device cannot give out more energy than it takes in. The efficiency of all electrical heating devices is 100%, because the electrical production of heat is easy. Heating devices may vary, however, in how effectively they deliver the heat from the coils in which it is produced to the place where it is to be used. In Example 9–15, we found that the motor is 74% efficient. The other 26% of the energy used appears as heat. If the motor stalls and produces no mechanical power, then 100% of the energy is converted to heat.

EXAMPLE 9–16

Given: An electrical generator with a 10-horsepower input producing 50 amperes at 100 volts.

Find: The efficiency of the generator.

SOLUTION

Power output: 50 A × 100 V = 5,000 W
Power input: 10 hp = 7,460 W
Efficiency = 5,000/7,460 = 0.67 or 67%

You may ask yourself, "Can I drive an electrical generator with an electric motor and let the generated current run the motor?" The answer is, "This scheme will not work very well." The reason is apparent from the previous discussion and Example 9–16. Both the motor and the generator waste some of the energy applied with the result that one device is not going to produce enough energy to run the other device.

SUMMARY

- Work = Force × Distance.
- Energy is the ability to do work.
- Power is the rate of using energy.

$$\text{Power} = \frac{\text{Work}}{\text{Time}}$$

- Power × Time = Energy.
- Watts = Volts × Amperes.
- I^2R is synonymous with electrical power.
- Watts and kilowatts measure power, which is a rate.
- Watt-hours and kilowatt-hours measure energy.
- $\text{Efficiency} = \frac{\text{Power Output}}{\text{Power Input}}$

Achievement Review

1. The word *rate* implies division by a unit of time. For example, the rate of speed for a car that travels 200 miles in 4 hours is 200 ÷ 4 or 50 miles per hour. Electric power, too, is defined as a rate. What is this definition?
2. A certain motor consumes 1,492 watts. What is the power rating in
   a. Horsepower?
   b. Kilowatts?
3. If it was desired to obtain more light in a room, should a bulb with a smaller or greater wattage rating be used? Why?
4. What is meant by I^2R losses?
5. What is the equation for finding P, if E and I are given?
6. What is the equation for finding E, if P and R are given?
7. What is the equation for finding I, if R and P are given?
8. What is the equation for finding R, if P and E are given?
9. Calculate the wattage for a 60-volt, 10-ampere arc lamp.
10. Find the power used by a 12-ampere, 110-volt heater.

11. What is the wattage rating of a 100-ohm resistor carrying 0.5 amperes?
12. Calculate the watts for a 2-ohm resistor on a 6-volt line.
13. Calculate resistance of a 60-watt, 120-volt lamp when operating.
14. Find operating current for an 800-watt, 115-volt toaster.
15. Find the cost of operating the toaster of question 14 for 5 hours per month if energy costs $0.135 cents per kilowatt-hour.
16. A 4-ohm resistor is rated at 144 watts. What is the maximum current it can carry without burning up?
17. Another resistance is rated at 25 watts. This particular resistor has 400 ohms. How much current can be pushed through this resistor before it might burn up?
18. A 750-watt electric iron, when connected to 120 volts, will take how much current?
19. How much current will flow through a wire supplying a 1-horsepower motor at 120 volts?
20. What would be the power in watts consumed if a heater rated at 7.5 amperes were connected to a 220-volt circuit?
21. A small pilot light is rated at 5.4 watts. What is the voltage needed to produce a current of 0.3 amperes?
22. How much will it cost to operate a washing machine for 15 hours a month if its motor is rated at 1/3 horsepower and the price of electric energy is $0.12 cents per kilowatt-hour? Assume 60% efficiency.
23. What is the resistance of a toaster whose label reads: 1,200 W/120 V?
24. A long power transmission line suffers a power loss of 3.2 megawatts when 800 amperes are flowing through it. What is the resistance of the cable?
25. Two hundred forty milliamperes are flowing through a 4.7 kilohm resistance. What should be the minimum wattage rating of the resistor, including a 10% margin for safety?
26. A 6-volt battery is charged at the rate of 5 amperes for 24 hours. What is the amount of energy charge in kilowatt-hours?
27. One 1/4-horsepower DC motor is 70% efficient. What is the current through the motor, in amperes, when it is operating on a 120-volt line and delivering 1/4 horsepower?
28. A DC motor takes 5 amperes on a 110-volt line and is 60% efficient. Find the horsepower output. Find the heating rate in watts.
29. One pound of coal releases 12,000 Btu of energy when it is burned. What is this value expressed in foot-pounds?
30. If the energy of a pound of coal is converted into electrical energy (kilowatt-hours) by means of equipment that has an overall efficiency of 30%, what is the energy in kilowatt-hours obtained from a pound of coal?
31. Assuming that 50,000 Btu per hour are required to heat a house in cold weather, how many watts of electrical heat are needed to produce heat at this rate? (Electrical heat is 100% efficient.)
32. Calculate the power in watts that must be supplied to a 40-gallon electric water heater to raise the temperature of a tank of water from 50°F to 150°F in 2

hours. Assume that no heat is wasted. (1 gallon = 8 1/3 lb) Calculate the cost of heating the water at $0.04 cents per kilowatt-hour.

33. Find the amount of time required for a 1,600-watt heating element to warm 30 gallons of water from 50°F to 150°F.
34. Find the time required to heat 1,900 grams of water (about 2 quarts) from 10°C to boiling (100°C) on a 660-watt hotplate.
35. Find the time required for the heating operation in question 34 if the process is 90% efficient; that is, only 90% of the heat input is delivered to the water.

10

Series Circuits

Objectives

After studying this chapter, the student should be able to

- Define and explain the new technical terms introduced in this chapter

series circuit	voltage drop
voltage divider	series-aiding
series-opposing	ground

- Describe and explain the behavior of voltage, current, resistance, and power in a series circuit
- Write four mathematical statements describing the relationships of E, I, R, and P in a series circuit
- Calculate unknown components and/or circuit quantities in a given series circuit
- Identify correct polarity with respect to a common reference point (ground)
- Determine the net voltage from series-connected voltage sources

10–1 CHARACTERISTICS OF SERIES CIRCUITS

When any number of devices are connected so that there is only a single circuit path for electrons, the devices are in *series*. Each device has the same amount of current in it, Figure 10–1.

Most of us are familiar with a series string of Christmas tree lights in which all of the lamps fail to light when one lamp burns out. When any one lamp goes out, its filament is removed from the circuit. The effect is the same as opening a switch in the circuit. When electrons cannot flow through one lamp, they cannot flow through the rest of the lamps either. If the defective lamp is replaced, then electrons can flow through the entire circuit. Each device in a *series* circuit has the *same current* through it as each other device in the circuit, since there is only one path for the electrons to travel.

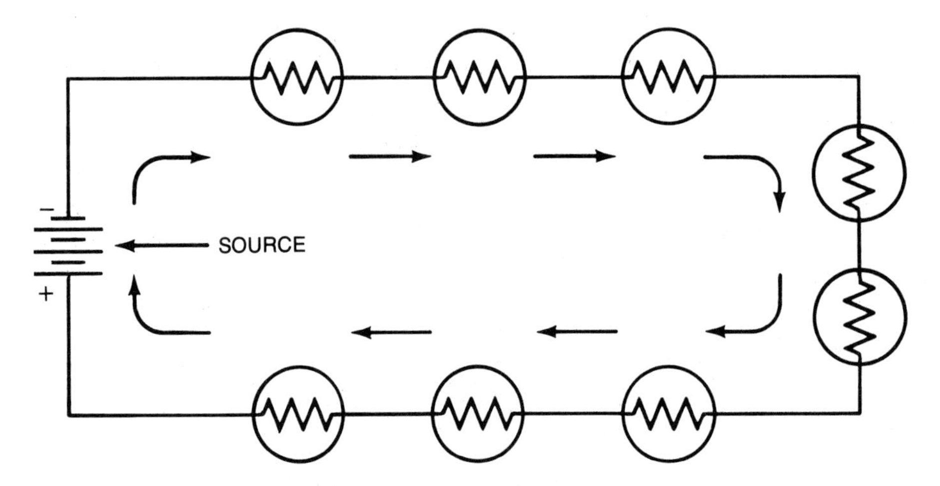

FIGURE 10–1 A series circuit

10–2 THE VOLTAGE DROP

The eight lamps shown in Figure 10–1 share the 120 volts supplied and, assuming all lamps in the circuit to be of equal size, they will share alike. In other words, each lamp has a potential difference of 15 volts (120/8) across its terminals. These individual voltages appearing across each resistance of a series circuit are known as *voltage drops*.

To reinforce this idea, let us consider the circuit of Figure 10–2. Let us first examine the nomenclature used in this schematic diagram. Note that all resistors have been labelled with subscripts: R_1, R_2, and R_3. Since all the resistors are equal in size, we may assume that they will share the supply voltage of 24 volts equally among themselves. These voltage drops of 8 volts each are labelled correspondingly: E_1, E_2, and E_3. After 8 volts are dropped at R_1, only 16 volts are left over for the remainder of the circuit.

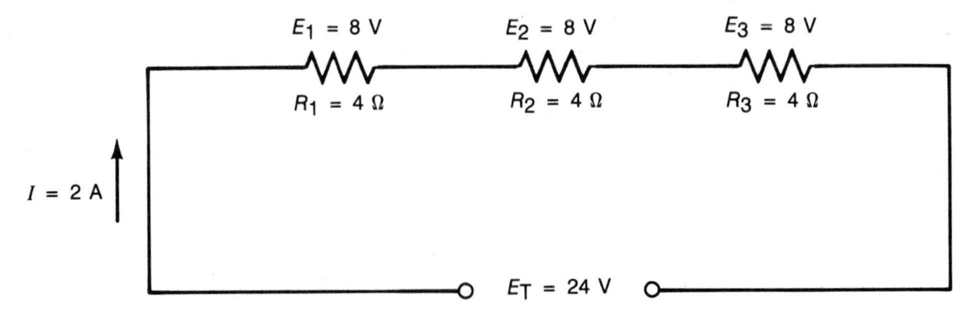

FIGURE 10–2

These examples should confirm that

The total voltage of a series circuit is equal to the sum of all the voltage drops.

Mathematically stated

$$E_T = E_1 + E_2 + E_3 + \cdots + E_n$$

10–3 RESISTANCE AND CURRENT IN SERIES CIRCUITS

Compare the circuit in Figure 10–2 with that of Figure 10–3. Can you detect the similarities and the differences? Both circuits consist of three resistors attached in series to a 24-volt supply. Both circuits have 2 amperes flowing through the resistors. By Ohm's law, we compute the *total resistance* (R_T) that is connected to the battery.

$$R_T = \frac{E_T}{I_T}$$

$$R_T = \frac{24}{2}$$

$$R_T = 12\ \Omega$$

This may have been obvious to you right along. After all, if you add all the resistances in each circuit, you obtain the same result for R_T, namely 12 ohms. In other words

The total resistance of a series circuit is equal to the sum of all individual resistors.

Mathematically stated

$$R_T = R_1 + R_2 + R_3 + \cdots + R_n$$

This idea of adding individual values *does not* apply to the current. You may remember our earlier statement, in Section 10–1, that the current in a series circuit has

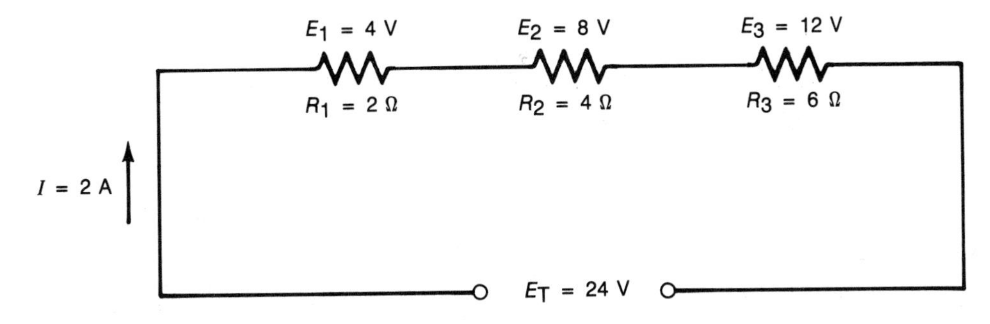

FIGURE 10–3

the same value everywhere. The current in these two circuits is 2 amperes at any given point. Yet it is customary to label the current with different subscripts as it flows through different components. We refer to it as I_1 when it flows through resistor R_1 and, correspondingly, I_2 when it flows through R_2. The current in the supply line is dubbed I_T to correspond to the supply voltage E_T. Some people prefer to use the subscript S for the word supply, and write I_S and E_S. Just remember, by whatever name you call it,

The current in a series circuit is the same everywhere.

Mathematically stated

$$I_T = I_1 = I_2 = I_3 = \cdots = I_n$$

So much for the similarities. Now let us have a quick look at the difference between the two circuits. Have you noticed? The voltage drops are different when the resistors are different. In fact, the voltage drops are proportional to the values of the resistors. For instance, if one resistor is twice as large as another, its voltage drop, too, will be twice as large. This, of course, can be confirmed by use of Ohm's law; so, we compute for the circuit in Figure 10–3.

$$E_1 = I_1 \times R_1 = 2 \times 2 = 4 \text{ V}$$
$$E_2 = I_2 \times R_2 = 2 \times 4 = 8 \text{ V}$$
$$E_3 = I_3 \times R_3 = 2 \times 6 = 12 \text{ V}$$
$$E_T = I_T \times R_T = 2 \times 12 = 24 \text{ V}$$

From now on, as you can see, we must use *matching* subscripts whenever we use Ohm's law or the power equations. This idea will be illustrated by the solved sample problems in the next two sections.

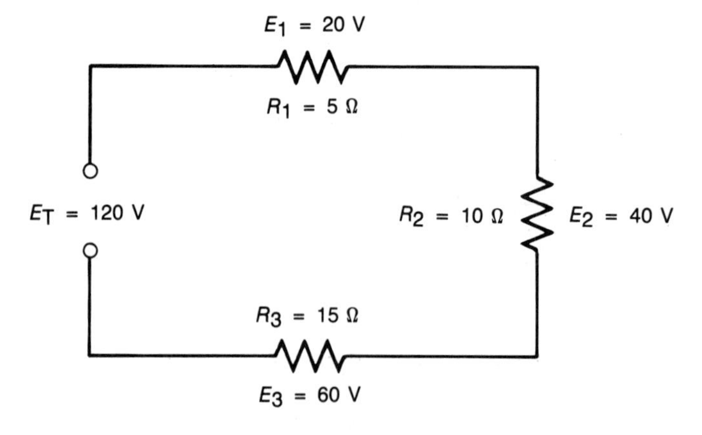

FIGURE 10–4

10–4 POWER CONSUMPTION IN SERIES CIRCUITS

You should recall, from our discussion in Chapter 7, that the wattage rating of resistors is an important specification. With our knowledge of the power equations, determining the power dissipation of an individual resistor, as well as an entire circuit, becomes an easy task. The following example will illustrate.

EXAMPLE 10–1

Given: A series circuit as shown in Figure 10–4.

Find:

a. The power dissipation of each individual resistor.
b. The total power used by the complete circuit.

SOLUTION

a. The power dissipated by R_1 is called P_1. We have two choices to solve for P_1. Either

$$P_1 = E_1 \times I_1$$
$$P_1 = 20 \times 4$$
$$P_1 = 80 \text{ W}$$

or

$$P_1 = I^2R_1$$
$$P_1 = 16(5)$$
$$P_1 = 80 \text{ W}$$

Likewise

$$P_2 = E_2 \times I_2$$
$$P_2 = 40 \times 4$$
$$P_2 = 160 \text{ W}$$

or

$$P_2 = I^2R_2$$
$$P_2 = 16(10)$$
$$P_2 = 160 \text{ W}$$

And finally

$$P_3 = E_3 \times I_3$$
$$P_3 = 60 \times 4$$
$$P_3 = 240 \text{ W}$$

or

$$P_3 = I^2R_3$$
$$P_3 = 16(15)$$
$$P_3 = 240 \text{ W}$$

b. To find the total circuit power, the same two methods can be used. Either we compute

$$P_T = E_T \times I_T$$
$$P_T = 120(4)$$
$$P_T = 480 \text{ W}$$

Or we use

$$P_T = I^2R_T$$
$$P_T = 16 \times 30$$
$$P_T = 480 \text{ W}$$

In addition to the two methods shown above, there is yet a third option. Consider the fact that

The total power dissipation is equal to the sum of all the individual power ratings.

Mathematically stated

$$P_T = P_1 + P_2 + P_3 + \cdots + P_n$$

For this particular problem then

$$P_T = 80 \text{ W} + 160 \text{ W} + 240 \text{ W}$$
$$P_T = 480 \text{ W}$$

10–5 CALCULATION OF SERIES CIRCUIT QUANTITIES

This section presents a number of solved problems to demonstrate the proper techniques of solving series circuit problems. Study them carefully.

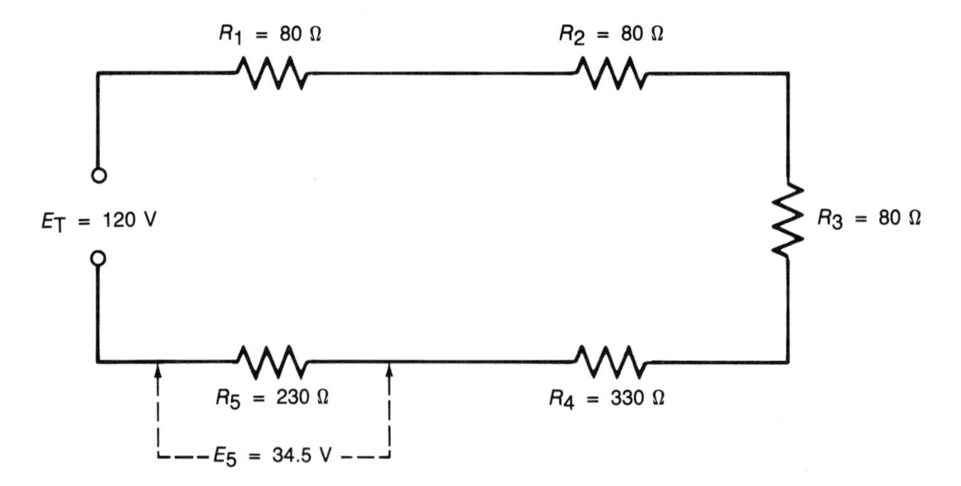

FIGURE 10–5

EXAMPLE 10–2

Given: An electronic circuit in which the heating elements of five radio tubes are connected as shown in Figure 10–5 and a voltmeter connected across R_5 that reads 34.5 V.

Find:

a. The current flowing through each resistor.
b. The voltage drop across each resistor.

SOLUTION

a. *Method #1*

$R_T = R_1 + R_2 + R_3 + R_4 + R_5$
$R_T = 80 + 80 + 80 + 330 + 230$
$R_T = 800\ \Omega$

$I_T = \frac{E_T}{R_T}$
$I_T = \frac{120}{800}$
$I_T = 0.15\ \text{A}$

Method #2

$I_5 = \frac{E_5}{R_5}$
$I_5 = \frac{34.5}{230}$
$I_5 = 0.15\ \text{A}$

$I_1 = I_2 = I_3 = I_4 = I_5 = 0.15\ \text{A}$

b. $E_1 = I_1R_1$
$E_1 = 0.15(80)$
$E_1 = 12\ \text{V}$

$E_2 = I_2R_2$
$E_2 = 0.15(80)$
$E_2 = 12\ \text{V}$

$E_3 = I_3R_3$
$E_3 = 0.15(80)$
$E_3 = 12\ \text{V}$

$E_4 = I_4R_4$
$E_4 = 0.15(330)$
$E_4 = 49.5\ \text{V}$

$E_5 = 34.5\ \text{V (given)}$

Although solid-state electronic devices have replaced vacuum tubes in many applications, vacuum tubes still are used for specific circuits. The model numbers of some commonly used tubes are 12BA6, 12BE6, 12AV6, 35W4, and 50C5. Does the above problem give an indication of the meaning of the numbers 12, 35, and 50 in the tube model numbers?

EXAMPLE 10–3

Given: The circuit shown in Figure 10–6.

Find:

a. The current flowing through R_2, (I_2).
b. The voltage across R_2, (E_2).
c. The resistance R_2.

SOLUTION

a. $I_2 = I_1$

$$I_2 = \frac{E_1}{R_1}$$

$$I_2 = \frac{20}{10}$$

$$I_2 = 2 \text{ A}$$

b. Knowing that

$$E_T = E_1 + E_2$$

it follows that

$$E_2 = E_T - E_1$$
$$E_2 = 120 - 20$$
$$E_2 = 100 \text{ V}$$

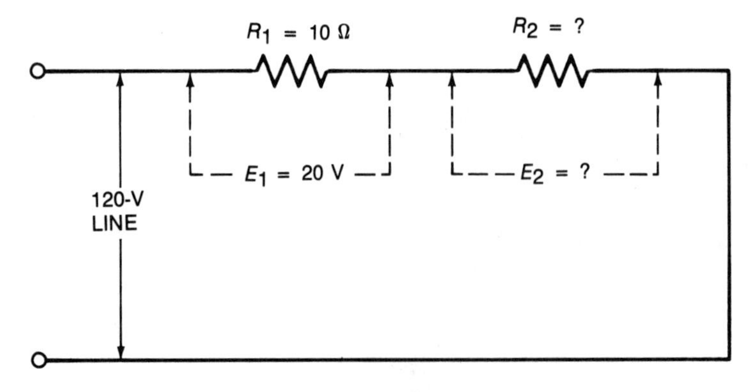

FIGURE 10–6

c. $R_2 = \frac{E_2}{I_2}$

$R_2 = \frac{100}{2}$

$R_2 = 50\ \Omega$

EXAMPLE 10–4

Given: The circuit shown in Figure 10–7.

Find:

a. E_T
b. P_1

SOLUTION

a. $I_2 = \frac{E_2}{R_2}$

$I_2 = \frac{12}{24}$

$I_2 = 0.5\ \text{A}$

$R_T = R_1 + R_2$
$R_T = 12 + 24$
$R_T = 36\ \Omega$

$E_T = I_T R_T$
$E_T = 0.5(36)$
$E_T = 18\ \text{V}$

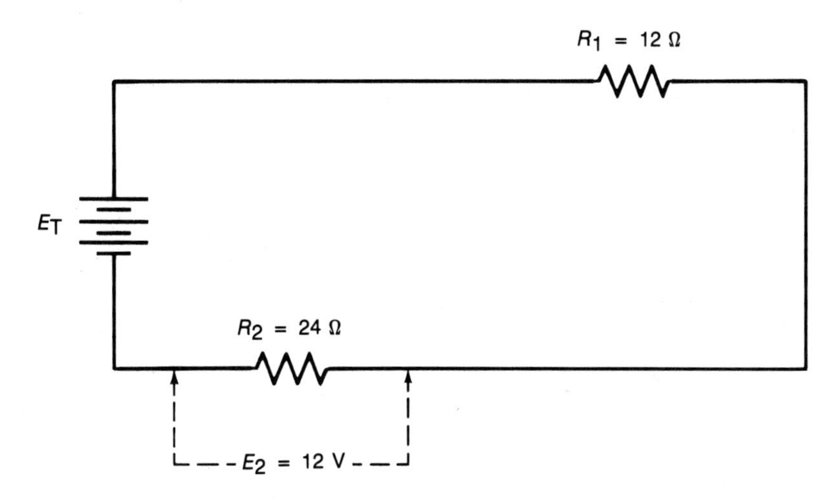

FIGURE 10–7

b. *Method #1*

$P_1 = I_2R_1$
$P_1 = (0.5)^2(12)$
$P_1 = 3\ \text{W}$

Method #2

$E_T = E_1 + E_2$

therefore

$E_1 = E_T - E_2$
$E_1 = 18 - 12$
$E_1 = 6\ \text{V}$

Substituting $E_1 = 6$ V into

$P_1 = E_1I_1$
$P_1 = 6(0.5)$
$P_1 = 3\ \text{W}$

EXAMPLE 10–5

Given: A circuit as shown in Figure 10–8.

Find:

a. E_T
b. P_2

SOLUTION

a. $R_T = R_1 + R_2$
$R_T = 3\ \text{k}\Omega + 6\ \text{k}\Omega$
$R_T = 9\ \text{k}\Omega$

$E_T = I_TR_T$
$E_T = 40 \times 10^{-3} \times 9 \times 10^3$
$E_T = 360\ \text{V}$

b. *Method #1*

$P_2 = I_2R_2$
$P_2 = (0.04 \times 0.04)(6{,}000)$
$P_2 = 0.0016(6{,}000)$
$P_2 = 9.6\ \text{W}$

Method #2

$E_2 = I_2R_2$
$E_2 = 40\ \text{mA} \times 6\ \text{k}\Omega$
$E_2 = 240\ \text{V}$

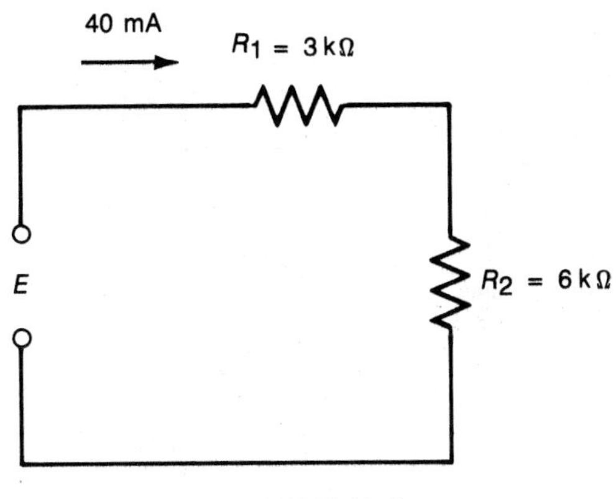

FIGURE 10–8

Substituting the above value into

$$P_2 = E_2 I_2$$
$$P_2 = 240 \text{ V} \times 40 \text{ mA}$$
$$P_2 = 240 \times 40 \times 10^{-3} \text{ A}$$
$$P_2 = 9.6 \text{ W}$$

EXAMPLE 10–6

Given: A circuit as shown in Figure 10–9.

Find:

a. E_2
b. R_1
c. R_2

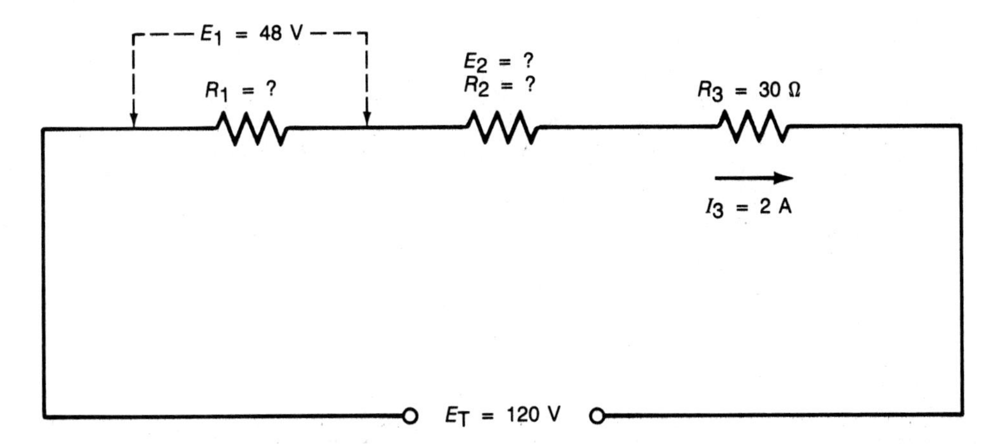

FIGURE 10–9

SOLUTION

a. $E_3 = I_3R_3$
$E_3 = 2(30)$
$E_3 = 60 \text{ V}$

Since

$E_T = E_1 + E_2 + E_3$

it follows

$E_2 = E_T - (E_1 + E_3)$
$E_2 = 120 - (48 + 60)$
$E_2 = 120 - 108$
$E_2 = 12 \text{ V}$

b. $R_1 = \dfrac{E_1}{I_1}$

$R_1 = \dfrac{48 \text{ V}}{2 \text{ A}}$

$R_1 = 24\ \Omega$

c. $R_2 = \dfrac{E_2}{I_2}$

$R_2 = \dfrac{12 \text{ V}}{2 \text{ A}}$

$R_2 = 6\ \Omega$

10–6 VOLTAGE DROP ON A LINE

Why do the lights in a house dim when a motor starts? The question is answered by Ohm's law and the use of the series circuit principle.

The sum of individual voltages equals the total applied voltage.

Assume that each wire leading to the house, Figure 10–10A, has 0.5-ohm resistance and that lamps in the house cause a 2-ampere current in the line. We then have a series circuit and can calculate the voltage at the house.

Each line wire is, in effect, a 0.5-ohm resistor with 2 amperes through it.

$E = IR$
$E = 2 \times 0.5$
$E = 1 \text{ V}$

which is the potential energy used to maintain the 2-ampere current in the 0.5-ohm resistance of the wire. One volt is used on each wire, since the potential difference between the wires at the house is equal to 120 − 2 or 118 volts, Figure 10–10B.

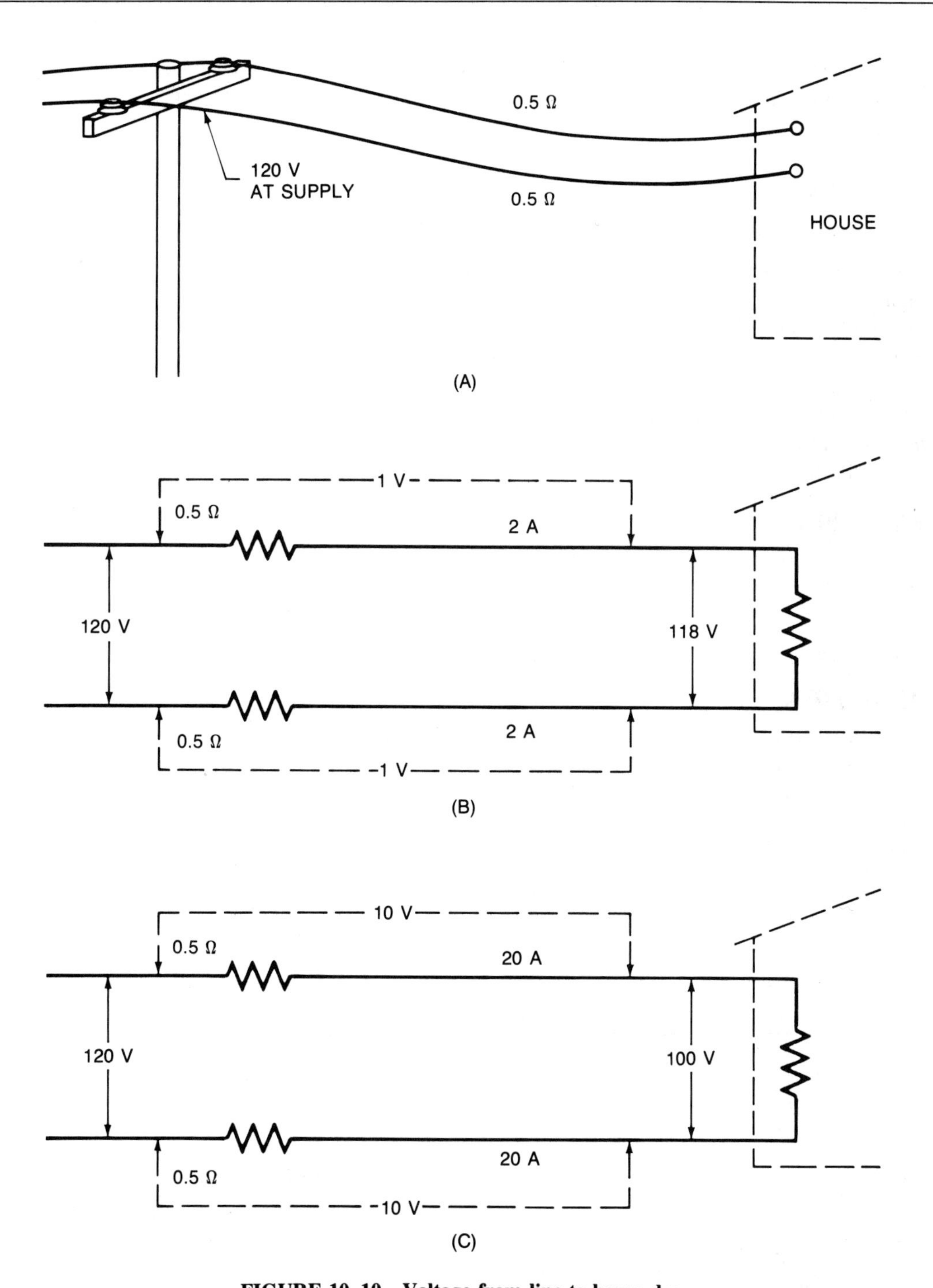

FIGURE 10–10 Voltage from line to house drop

If a motor is turned on so that the current in the line becomes 20 amperes instead of 2 amperes, more voltage will be required to maintain the current in the line leading to the house.

$$E = IR$$
$$E = 20 \times 1/2$$
$$E = 10 \text{ V for each wire}$$

Therefore, subtracting 20 volts from 120 volts gives 100 volts delivered at the house, Figure 10–10C.

With 2 amperes in the line, the voltage at the house is 118 volts; and with 20 amperes in the line, the voltage at the house is 100 volts. Lighting in the house is dimmer on 100 volts than on 118 volts, since the decreased voltage means that there is less current in the lamps. This 2-volt or 20-volt loss is called the *voltage drop on the line*. According to Ohm's law, this voltage depends on the resistance of the line and the current in the line.

EXAMPLE 10–7

Given: A certain electric motor requires at least 12 amperes at 110 volts to operate properly, Figure 10–11, and the motor is to be used 500 feet from a 120-volt power line.

Find: The size copper wire needed for the 500-foot extension.

SOLUTION

Since the source is 120 volts and 110 volts must be delivered to the motor, the difference, or 10 volts, can be used by the wire when the current is 12 amperes. Since E and I are known, we can find the resistance of the extension wires.

$$R = E/I$$
$$R = 0.83\ \Omega$$

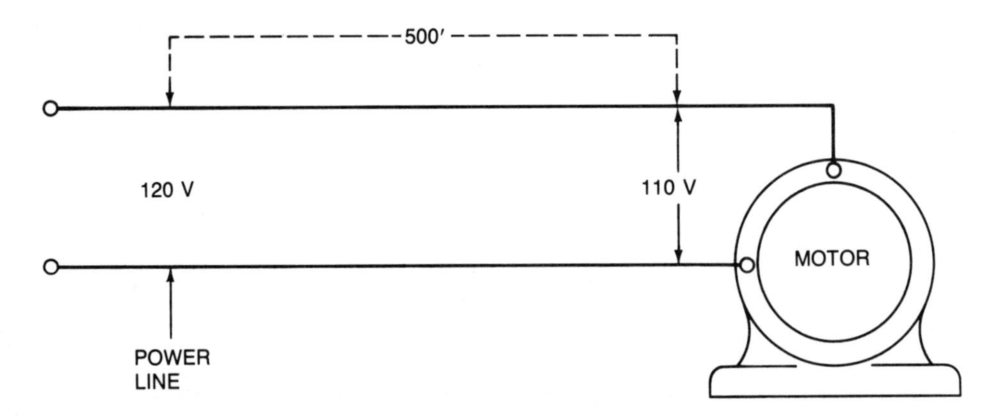

FIGURE 10–11 Electric motor

(These wires are a resistance in series with the motor. Using Ohm's law, we used the *10 volts* to find the resistance of the *wire* that is *responsible for the 10-volt* potential difference.)

The wire size can now be found from tables. (See Figure A–2 in the Appendix: American Wire Gauge.) The 500-foot extension requires 1,000 feet of wire with no more than 0.83 ohms of resistance. Using the "Ohms per 1,000 ft." column in Figure A–2, we find that the closest wire size is #9. (If #9 copper wire were not available, then the extension would be made of #8 copper wire.)

10–7 VOLTAGE AT AN OPEN ELEMENT

In Figure 10–12 there were seven lamps in the series circuit, but one lamp has been removed. How much voltage, if any, exists across the open socket?

One way of arriving at the answer is to consider how much of the 220-volt source is being used on each of the six remaining lamps, using $E = IR$. The current in each lamp is zero. As a result, each lamp has 0 volts across it; therefore, the entire 220-volt potential appears at the open socket.

The same reasoning applies to voltage at an open switch. The pressure is there, even though it may not be causing any current.

10–8 SERIES CIRCUITS AS VOLTAGE DIVIDERS

When series circuits are used for the purpose of obtaining different voltages from one voltage source, they are known as *voltage dividers*. The principle of voltage division was implied earlier, in Section 10–3, where it was stated that ". . . voltage drops are proportional to the values of the resistors. . . ." In other words, voltage dividers are series circuits in which the source voltage is divided among the resistors proportionally to their ohmic values.

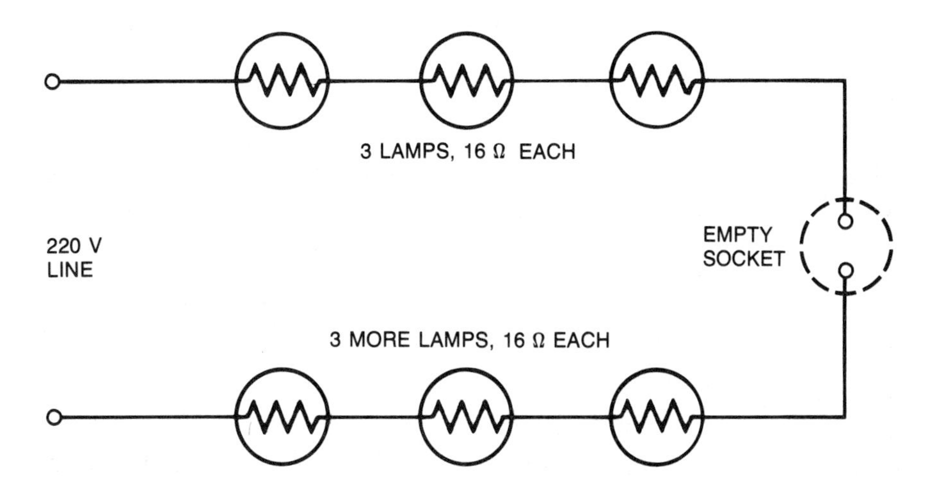

FIGURE 10–12 Series circuit

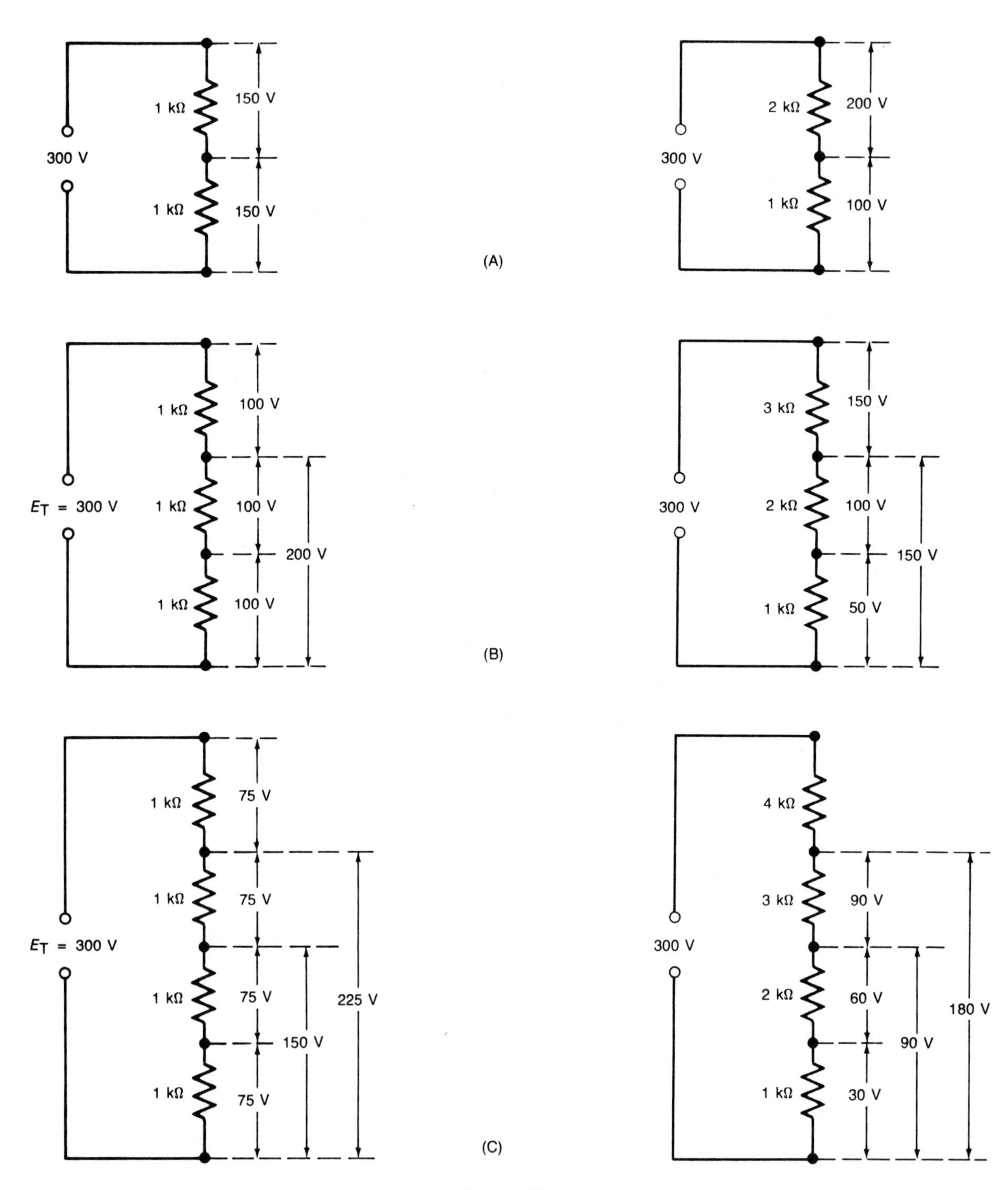
1 kΩ
150 V
300 V
1 kΩ
150 V
2 kΩ
200 V
300 V
1 kΩ
100 V
(A)
1 kΩ
100 V
E_T = 300 V
1 kΩ
100 V
200 V
1 kΩ
100 V
3 kΩ
150 V
300 V
2 kΩ
100 V
150 V
1 kΩ
50 V
(B)
1 kΩ
75 V
1 kΩ
75 V
E_T = 300 V
1 kΩ
75 V
225 V
150 V
1 kΩ
75 V
4 kΩ
3 kΩ
90 V
300 V
2 kΩ
60 V
180 V
90 V
1 kΩ
30 V
(C)

FIGURE 10–13

The six circuit diagrams shown in Figure 10–13 will help to clarify this point, especially if you make the effort to confirm the mathematical accuracy by use of Ohm's law.

In some applications, voltage division is achieved by means of *potentiometers*. You may wish to refresh your memory on that subject by rereading Section 7–8, dealing with variable resistors.

Remember, a potentiometer is a three-terminal device having the same effect as the two-resistor circuit in Figure 10–13A. In fact, the movable contact of the potentiometer divides the resistance element into two series-connected resistors. By varying the setting of the movable contact, any voltage drop from zero to maximum can be obtained, Figure 10–14.

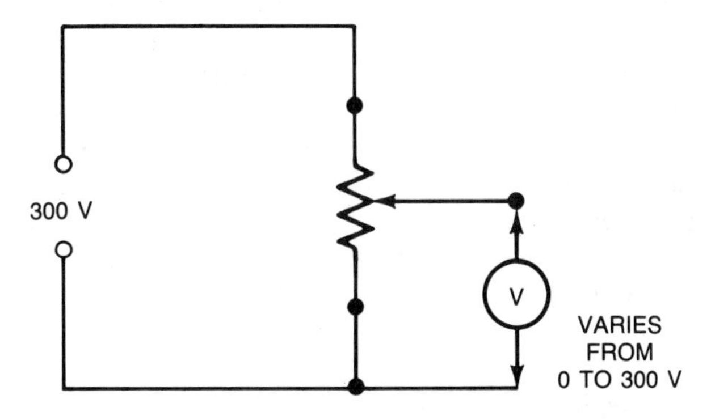

FIGURE 10–14 Potentiometer as voltage divider

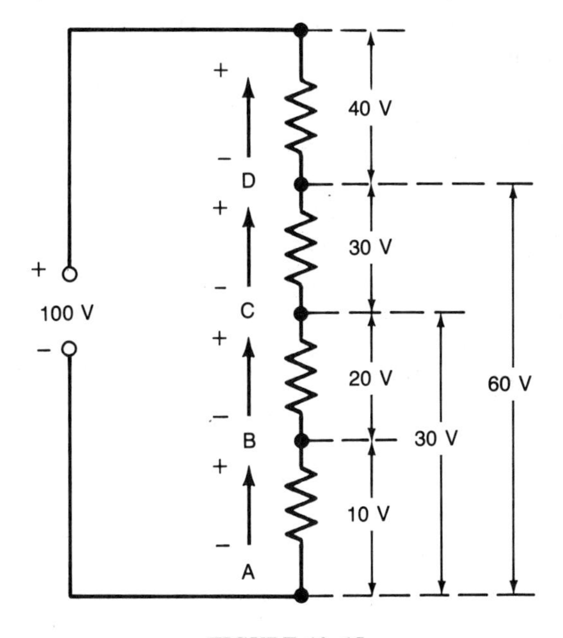

FIGURE 10–15

10–9 POLARITY CONSIDERATIONS

In discussing the voltage distribution of a circuit, it is not good enough to merely refer to the magnitude of the voltages. Equal consideration should be given to the polarity of the voltages.

To illustrate this point, let us have another look at a voltage divider circuit in which we have identified the terminal points of the resistors with letters, Figure 10–15.

To determine the polarity of a voltage drop, we simply consider the direction of the current. Remember, electron current moves from negative to positive, as indicated by the arrows in our drawing. Since the current flows from point A toward point B, it follows that point A is negative with respect to point B. (We could have said, instead, that point B is positive with respect to point A.) Similarly, point B is negative with respect to point C (or point C is positive with respect to point B).

Do not be confused by the two polarity markings (+ and −) at points B, C, and D. Some students may wonder how any given point can be assigned both positive *and* negative markings at the same time. This simply depends on your reference point. Point C is positive with respect to point B, but it is negative with respect to point D. This is somewhat analogous to describing someone's age. A person can be described as being young *and* old at the same time. Your age is relative to the point of reference. Compared to your grandfather, you are a young person; but compared to a baby brother, you are the old one.

10–10 GROUND AS A REFERENCE POINT

The concept of a reference point is important in describing not only polarity markings but voltage levels as well. In electrical theory such reference points are often called *ground,* regardless of whether or not they are actually connected to the earth. Automobiles, for instance, are said to have a negative ground. That simply means that the chassis, connected to the negative terminal of the battery, is the common reference point. But certainly the car is not grounded to the earth. If anything, it is well insulated from the earth by its rubber tires.

Two graphic symbols are used to denote a ground, Figure 10–16. The first of the two, Figure 10–16A, is used to describe a chassis ground. Let us employ that symbol to denote a reference point in the voltage divider circuit of Figure 10–15.

We are going to place this reference point first on point A, then on point B, then on to points C, D, and E. The schematics shown in Figure 10–17 will show how the voltage levels on each point change with every change of the reference ground.

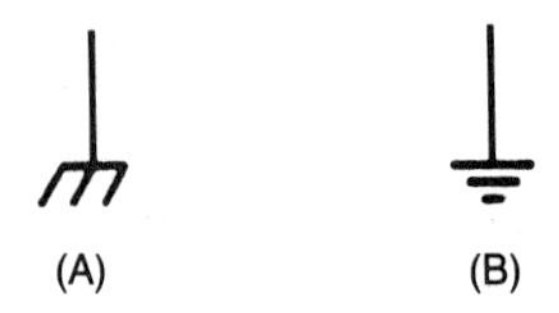

FIGURE 10–16

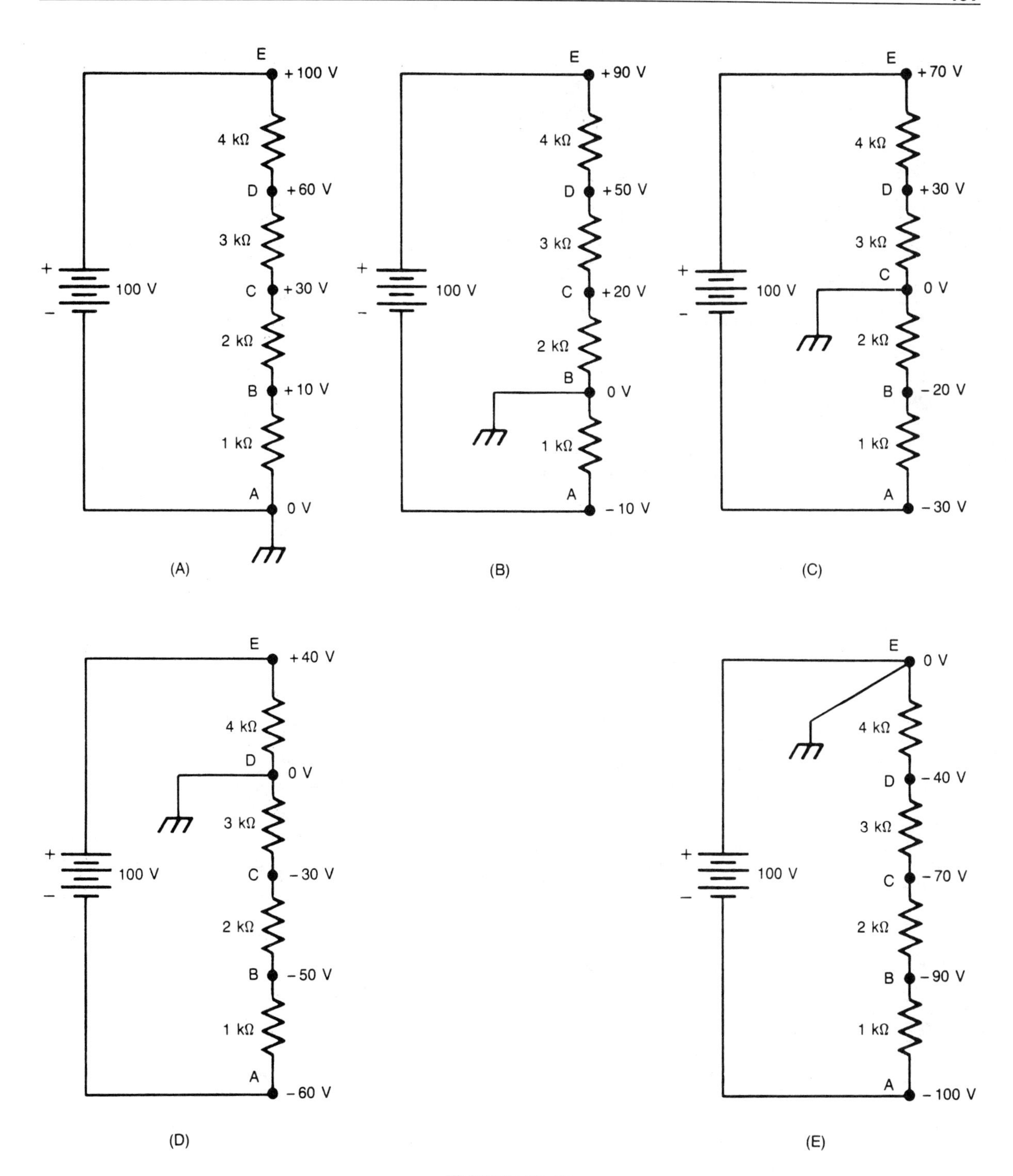
E
+100 V
4 kΩ
D
+60 V
3 kΩ
+
100 V
–
C
+30 V
2 kΩ
B
+10 V
1 kΩ
A
0 V
(A)
E
+90 V
4 kΩ
D
+50 V
3 kΩ
+
100 V
–
C
+20 V
2 kΩ
B
0 V
1 kΩ
A
–10 V
(B)
E
+70 V
4 kΩ
D
+30 V
3 kΩ
+
100 V
–
C
0 V
2 kΩ
B
–20 V
1 kΩ
A
–30 V
(C)
E
+40 V
4 kΩ
D
0 V
3 kΩ
+
100 V
–
C
–30 V
2 kΩ
B
–50 V
1 kΩ
A
–60 V
(D)
E
0 V
4 kΩ
D
–40 V
3 kΩ
+
100 V
–
C
–70 V
2 kΩ
B
–90 V
1 kΩ
A
–100 V
(E)

FIGURE 10–17

Be sure to study these drawings to gain understanding of this important principle. The same concept will be presented in the following section, as well as in the chapter entitled "Series-Parallel Circuits and Loaded Voltage Dividers."

10–11 VOLTAGE SOURCES IN SERIES

Cells and other voltage sources are often connected in series to obtain higher voltages. The voltage produced by several cells connected in series is the total of the individual cell voltages, Figure 10–18. This method of connecting voltage sources is known as *series-aiding*. Note that with this method, the positive pole of one source always connects with the negative pole of another source. A familiar example of this is the loading of a flashlight with multiple dry cells.

Consider, by contrast, two voltage sources interconnected with identical poles, Figure 10–19. Such connections are known as *series-opposing*. When voltage sources are connected like this, their voltages subtract from one another. If the individual source voltages are equal, their combined output will equal 0 volts, Figure 10–19A.

If the individual source voltages are of different magnitude, their net difference will represent the output voltage, with the larger of the two determining the polarity of the output. (Compare Figures 10–19B and C.)

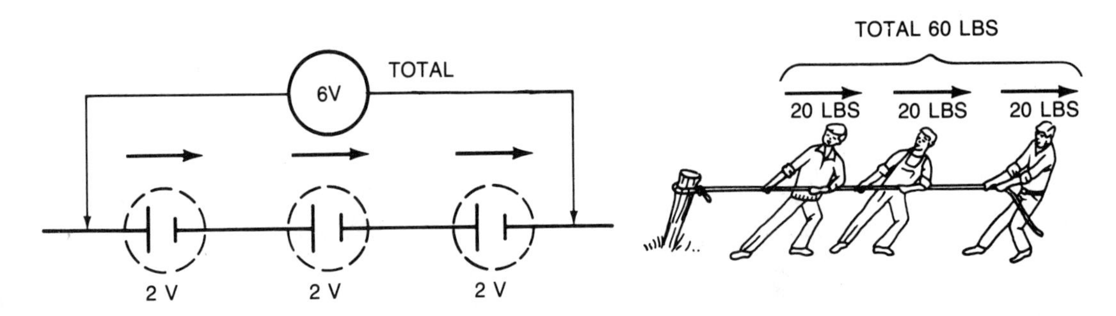

FIGURE 10–18 Three 2-volt lead storage cells in series

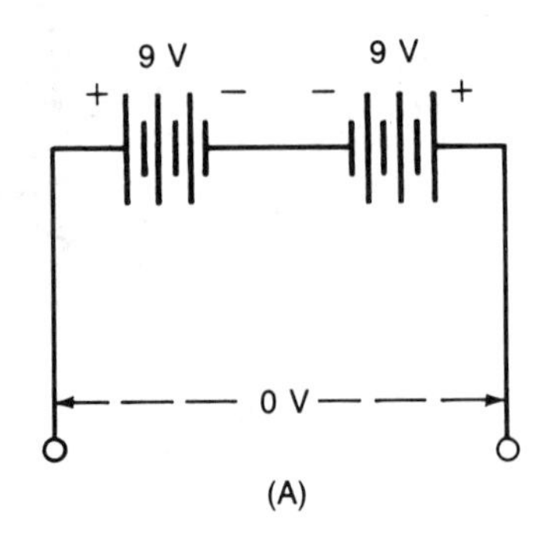

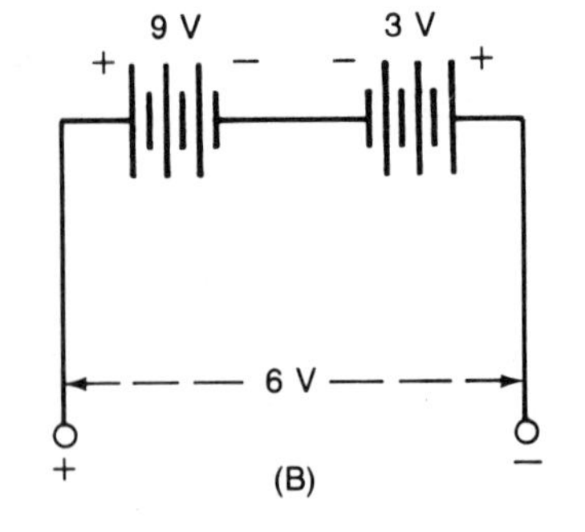

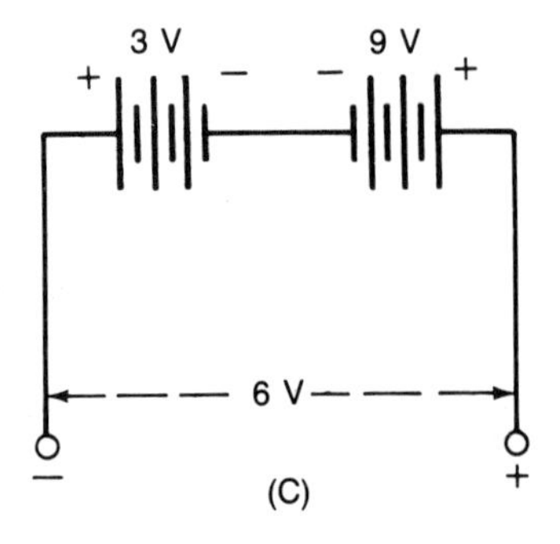

FIGURE 10–19

SUMMARY

- In series circuits there is only one path for the current.
- In series circuits the current is the same for all components.

$$I_T = I_1 = I_2 = I_3 = \cdots = I_n$$

- In series circuits the sum of all voltage drops is equal to the applied voltage.

$$E_T = E_1 + E_2 + E_3 + \cdots + E_n$$

- In series circuits the total circuit resistance is equal to the sum of all individual resistors.

$$R_T = R_1 + R_2 + R_3 + \cdots + R_n$$

- In series circuits the total circuit power is equal to the sum of all the individual resistors' power dissipation.

$$P_T = P_1 + P_2 + P_3 + \cdots + P_n$$

- Long power lines supplying a load represent resistors in series with the load causing a voltage drop in the line.
- When one resistor element in a series circuit opens up, the source voltage will appear across its terminals.
- Series circuits may be considered as voltage dividers.
- Voltage division is achieved proportionally to the values of the resistors.
- The polarity of voltage drops is negative at the point where the electron current enters the resistor and positive where the current exits.
- The word ground denotes a common reference point.
- Voltage sources may be connected series-aiding or series-opposing.

Achievement Review

Note: For each problem, draw a circuit diagram to gain practice in schematic work and to aid in visualizing each problem. All answers *must include units of measurement.*

1. In a television receiver, two resistors are in series. The first resistor (R_1) has a voltage drop of 200 volts across it. The second resistor (R_2) has a drop of 50 volts across it. Current in the circuit is 20 milliamperes (0.02 A). What is the value of each resistor? What is the total resistance?
2. In question 1, what wattage is dissipated by each resistor?
3. Across a 300-volt power supply in an electronic device there are four resistors in series. Resistor values are 20,000, 50,000, 500, and 4,500 ohms. What is the current, in milliamperes, that flows through this network?
4. In a radio there are three resistors in series: R_1 has a 1.5-volt drop across it; R_2 has a 3-volt drop across it; and R_3 has a 0.6-volt drop across it. R_3 is 20 ohms. What is the current in the circuit, and what are the resistance values of R_1 and R_2?

5. Consider a barn located a considerable distance from a farmhouse. The two wires supplying the power to the barn have a resistance of 0.3 ohm each. The resistance of these wires forms a series circuit with the load (machinery) in the barn. The voltage at the farmhouse is 120 volts. How many volts will be available at the barn when 20 amperes is supplied?
6. Explain how the answer would change in question 5 if the current were increased to 35 amperes. Specifically, what effect would this have on the machinery to be operated in the barn?
7. Two resistors are connected in series across a 24-volt power supply. A voltage drop of 9 volts can be measured across the first resistor. The current through the second resistor is 1.5 amperes. What are the values of the two resistors?
8. Vacuum tubes such as the ones used in older radios or TVs require different voltages for their heaters. An efficient method of connecting such heaters is in a *series* configuration, providing that all tubes require the same heater current. Consider the following four tubes connected in such a manner to a 120-volt supply:

   Tube #1 requires 25 V/0.3 A
   Tube #2 requires 12 V/0.3 A
   Tube #3 requires 6 V/0.3 A
   Tube #4 requires 12 V/0.3 A

   If the tubes were to be connected to a 120-volt circuit, an additional series dropping resistor would be needed to compensate for the difference in voltages. What must be the ohms and wattage rating of such a resistor?
9. A motor rated for use on a 230-volt line can generally tolerate 7% voltage fluctuation.
   a. What is the lowest voltage it will tolerate?
   b. Assuming that such a motor will draw 12 amperes when run at 214 volts, if the motor is located 800 feet away from the power source, what size wire must be installed? Let us find out! First, draw a simple schematic diagram like this.

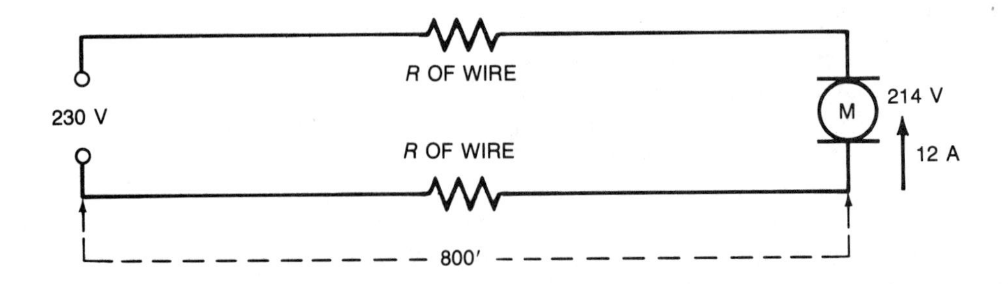

Note that the resistances of the wires are connected in series with the motor, dropping a voltage along the way that is lost to the motor.

c. How much is the voltage drop in the wire?
d. Using this answer (and Ohm's law) find the *maximum resistance* permissible *in the wire*.
e. How many feet of wire are needed for this installation?
f. Use the preceding two answers to compute the number of ohms per 1,000 feet.
g. Now consult your wire table to find the proper wire size for this installation.

10. An advertising sign is to be constructed with low-voltage light bulbs rated at 6 V/25 W. The bulbs will be connected to a source of 114 volts.
a. How many lamps should be used?
b. If only 16 bulbs were used, how much voltage would each lamp receive?
c. What effect will this have on the current?
d. How will this affect the lifespan of the bulb?

By contrast, *if twenty or more* lamps were to be used, what effect would this have on
e. The light intensity?
f. The power consumption?

11. Find E_T and P_2.

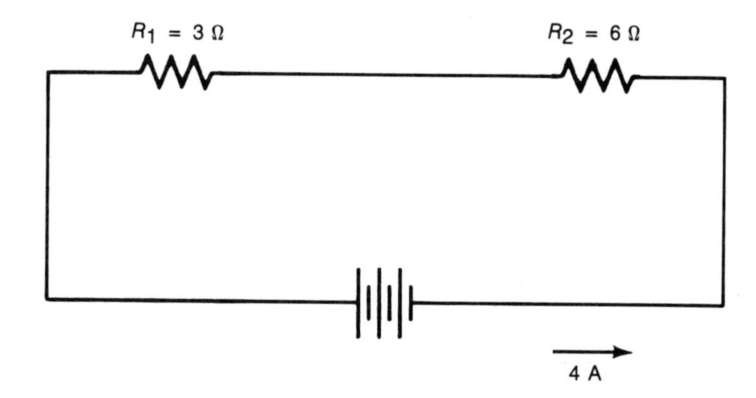

12. Find E_T and P_1.

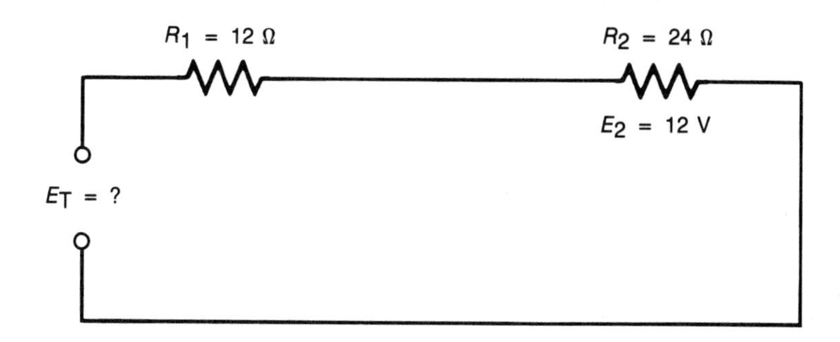

13. Find I and R_3.

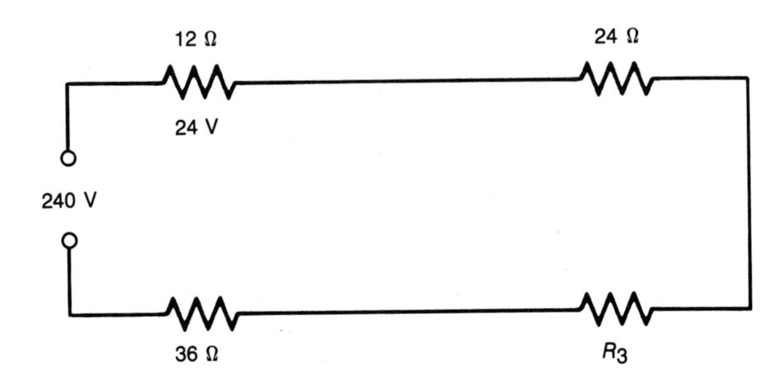

14. Find E_1 and R_2.

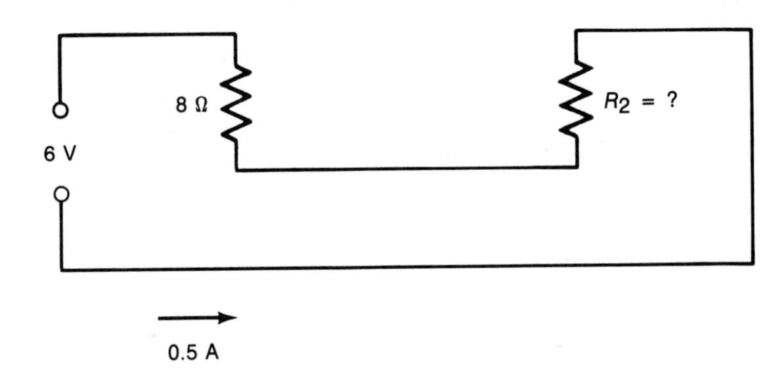

15. Find E_2 and R_1.

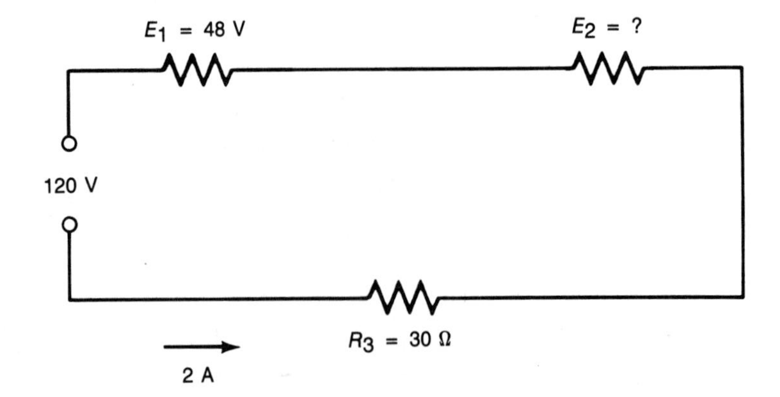

16. Find I and P_1.

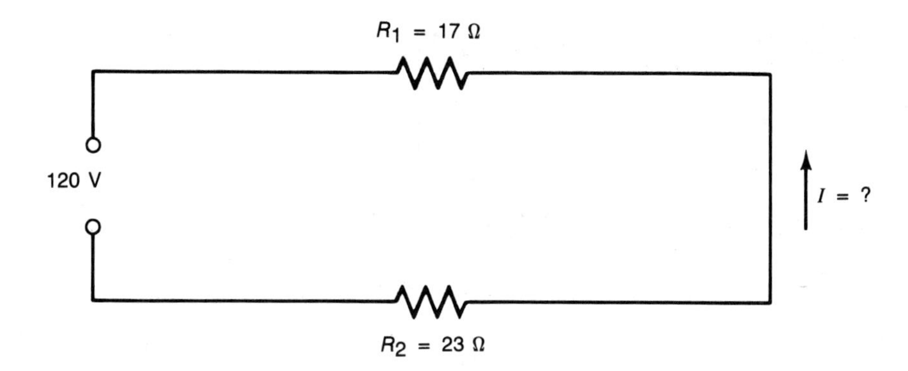

17. Find E_T and P_T.

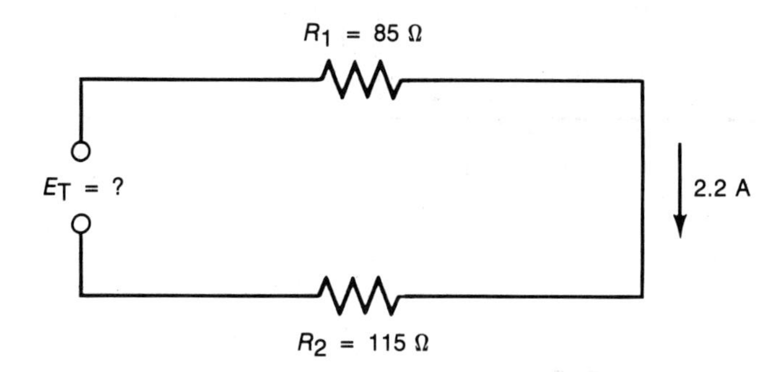

18. Find R_2 and P_1.

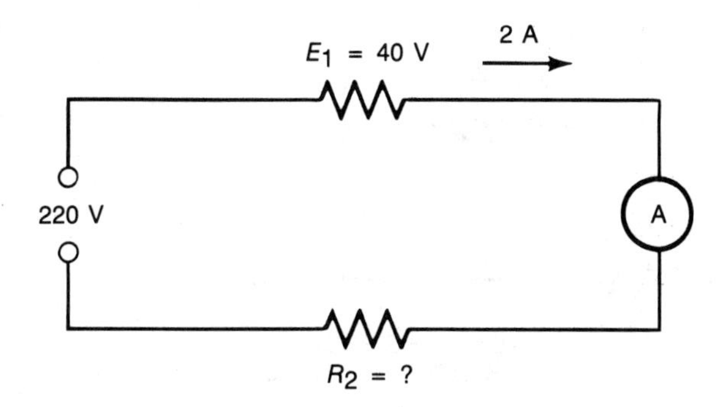

19. Find I_T and P_2.

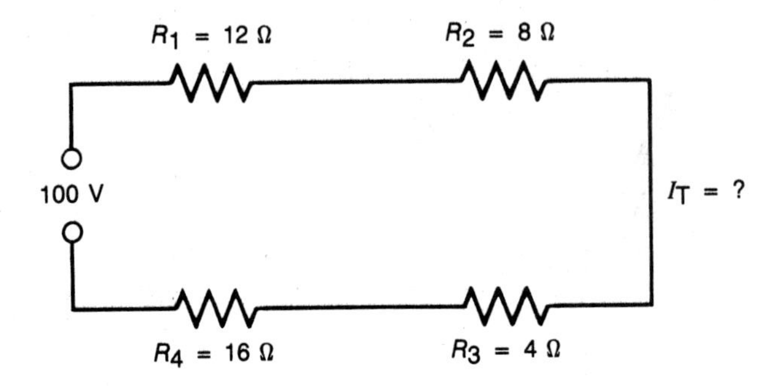

20. Find R_2 and P_3.

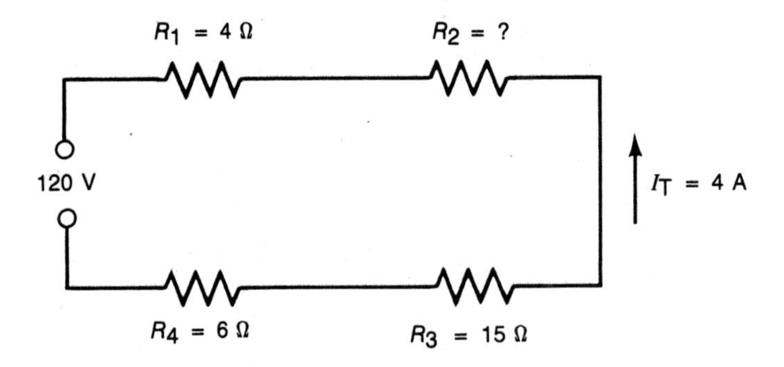

21. Find E_4, E_T, P_2, and P_T.

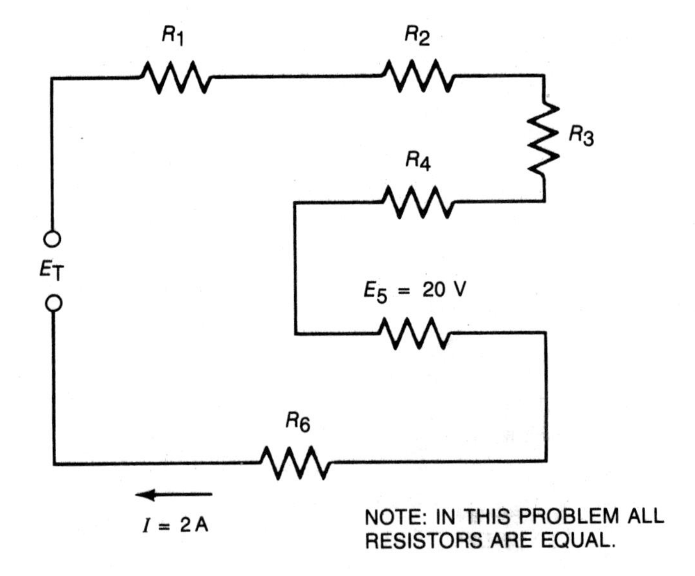

22. Solve the series circuit shown below for the following quantities:

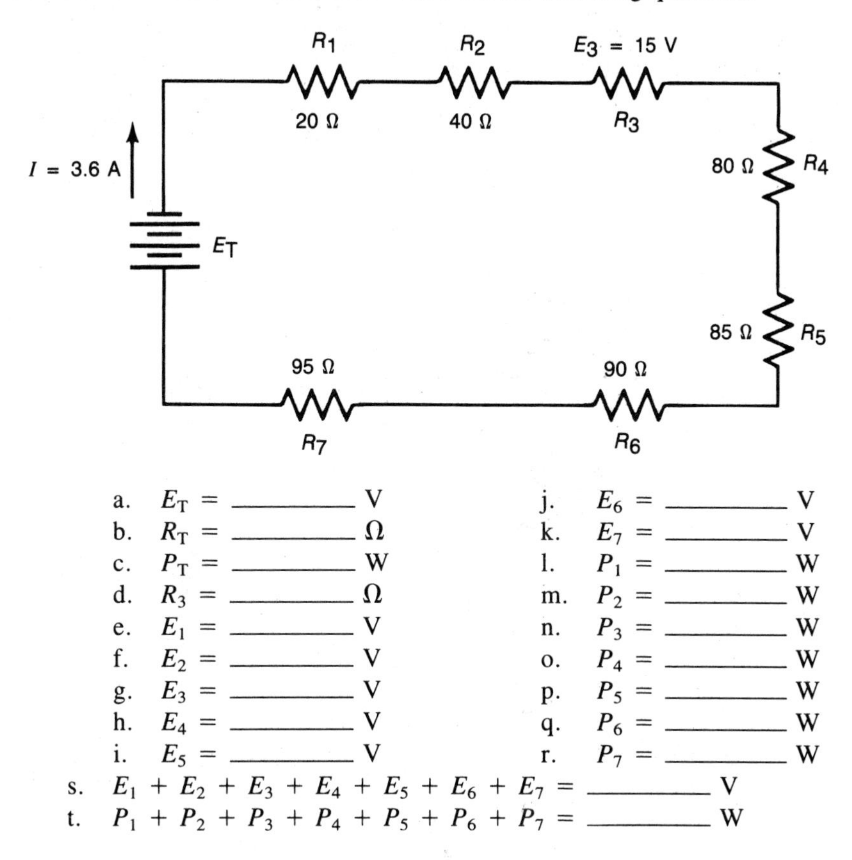

a. E_T = ________ V
b. R_T = ________ Ω
c. P_T = ________ W
d. R_3 = ________ Ω
e. E_1 = ________ V
f. E_2 = ________ V
g. E_3 = ________ V
h. E_4 = ________ V
i. E_5 = ________ V
j. E_6 = ________ V
k. E_7 = ________ V
l. P_1 = ________ W
m. P_2 = ________ W
n. P_3 = ________ W
o. P_4 = ________ W
p. P_5 = ________ W
q. P_6 = ________ W
r. P_7 = ________ W
s. $E_1 + E_2 + E_3 + E_4 + E_5 + E_6 + E_7$ = ________ V
t. $P_1 + P_2 + P_3 + P_4 + P_5 + P_6 + P_7$ = ________ W

23. Solve the series circuit shown below for the following quantities:

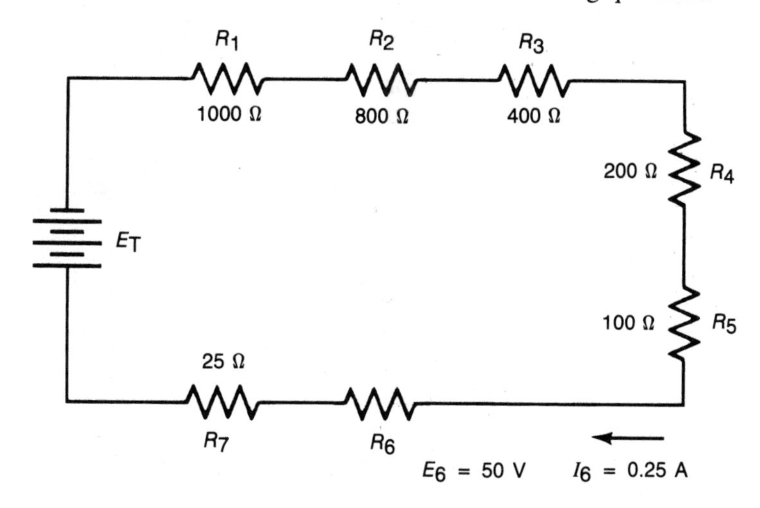

a. E_T = ________ V
b. I_1 = ________ A
c. I_7 = ________ A
d. E_5 = ________ V
e. P_4 = ________ W
f. R_T = ________ Ω
g. I_2 = ________ A
h. P_6 = ________ W
i. I_T = ________ A
j. I_5 = ________ A
k. E_4 = ________ V
l. P_3 = ________ W
m. E_1 = ________ V
n. E_7 = ________ V
o. P_5 = ________ W
p. P_T = ________ W
q. I_3 = ________ A
r. E_2 = ________ V
s. P_1 = ________ W
t. R_6 = ________ Ω
u. I_4 = ________ A
v. E_3 = ________ V
w. P_2 = ________ W
x. P_7 = ________ W
y. $E_1 + E_2 + E_3 + E_4 + E_5 + E_6 + E_7$ = ________ V
z. $P_1 + P_2 + P_3 + P_4 + P_5 + P_6 + P_7$ = ________ W

11
Parallel Circuits

Objectives

After studying this chapter, the student should be able to

- Explain five different methods of computing the total resistance of a parallel circuit
- Compute any unknown circuit quantity in a parallel circuit having sufficient data

11–1 THE NATURE OF PARALLEL CIRCUITS

You should recall that series circuits were defined as having one, and only one, path along which the current can flow. By contrast, a parallel circuit has more than one path through which the current can flow. Figure 11–1 is such a parallel circuit. This is the standard representation where all resistors are drawn parallel to one another, resembling the ties between railroad tracks.

Now look at the two schematics in Figure 11–2, which are representations of the *same* parallel circuit shown in Figure 11–1. The common feature of these parallel circuits can be detected by carefully tracing each possible current path. Then it will become clear that each individual resistor is connected to the negative and positive pole of the battery. If you have difficulty in perceiving this fact, it is recommended that you trace the negative and positive supply lines with two different colored pencils from the source to each resistor.

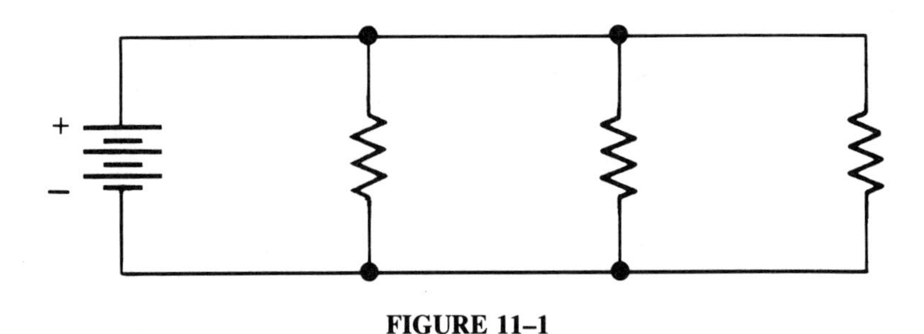

FIGURE 11–1

What all these drawings convey is the fact that *there is only one voltage* serving all load resistors. (Compare this with the characteristics of a series circuit, where every resistor may have a different voltage drop.) Mathematically stated

$$E_T = E_1 = E_2 = E_3 = \cdots = E_n$$

Also, as you trace the current path from its power source, you will notice that the current divides at some junction point, flows through the individual branch resistors, and then recombines at some other junction point before it returns to the power source.

Now, carefully check the three drawings just mentioned. They all show identical current flow to demonstrate that the *total line current* (supplied by the source) *is equal to the sum of all the individual branch currents*. (Compare this with the characteristics of a series circuit, where we dealt with one common current in each part of the circuit.) Mathematically stated

$$I_T = I_1 + I_2 + I_3 + \cdots + I_n$$

We have compared the voltage and current distribution of parallel circuits with that of series circuits. Now it is time to compare the total resistance of such circuits.

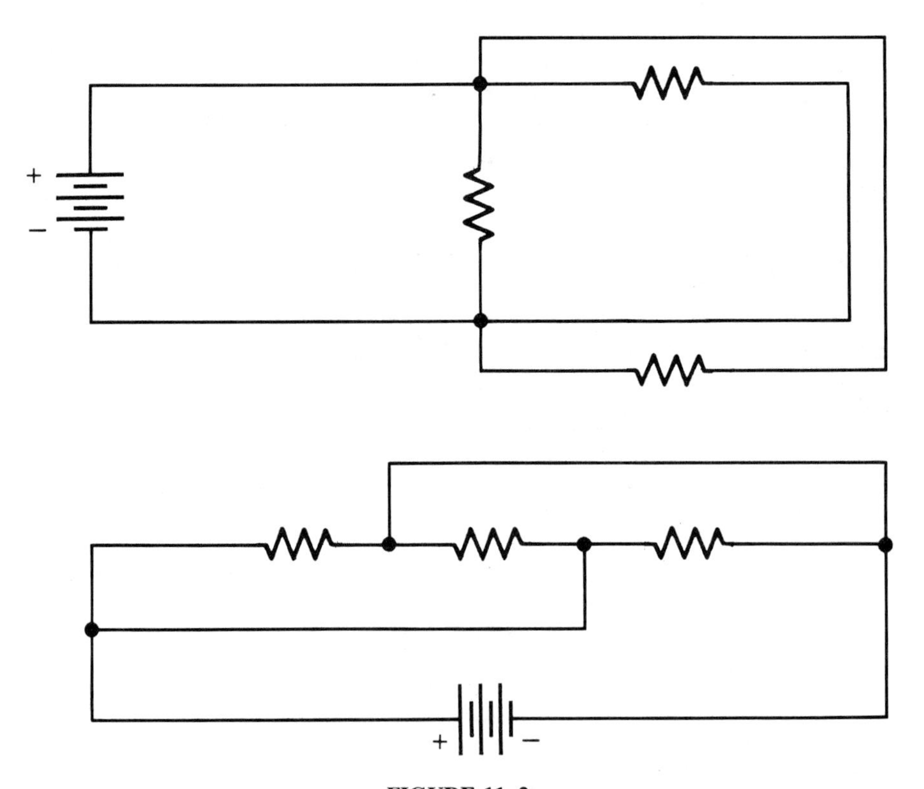

FIGURE 11–2

In a series circuit, as you should recall, the total resistance increases with every additional resistance. In parallel circuits, by contrast, we encounter exactly the opposite. In other words, the total resistance of a parallel circuit decreases whenever additional resistors are connected to the circuit. In fact, *the total resistance is always smaller than any of the individual branch resistors*.

There are five different methods by which the total resistance of a parallel circuit may be computed. How to employ these methods will be the subject of the next few sections.

11–2 USE OF OHM'S LAW FOR COMPUTING R_T

An easy way of determining the total resistance is by applying Ohm's law:

$$R_T = \frac{E_T}{I_T}$$

This necessitates, in some cases, finding the total current first. The following example should make this clear.

EXAMPLE 11–1

Given: The circuit shown in Figure 11–3.

Find:

a. The branch current I_1, I_2, and I_3.
b. The total current supplied by the power source (I_T).
c. The total circuit resistance.

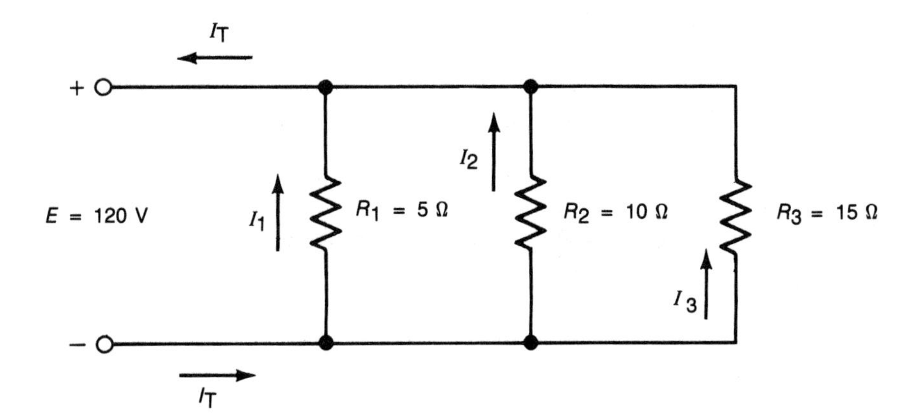

FIGURE 11–3

SOLUTION

a. $I_1 = \frac{E_1}{R_1} = \frac{120}{5} = 24\text{ A}$

$I_2 = \frac{E_2}{R_2} = \frac{120}{10} = 12\text{ A}$

$I_3 = \frac{E_3}{R_3} = \frac{120}{15} = 8\text{ A}$

b. $I_T = I_1 + I_2 + I_3$

$I_T = 24 + 12 + 8$

$I_T = 44\text{ A}$

c. $R_T = \frac{E_T}{I_T} = \frac{120}{40}$

$R_T = 2.73\ \Omega$

Note: In the event that a voltage has not been specified, you may assume any convenient number of your choice.

11–3 USE OF THE RECIPROCAL EQUATION

One of the most common methods of solving for the total resistance of a parallel circuit involves the use of the reciprocal formula shown here

$$\frac{1}{R_T} = \frac{1}{R_1} + \frac{1}{R_2} + \frac{1}{R_3} + \cdots + \frac{1}{R_n}$$

EXAMPLE 11–2

Given: The circuit shown in Figure 11–4.

Find: Total resistance using the reciprocal equation.

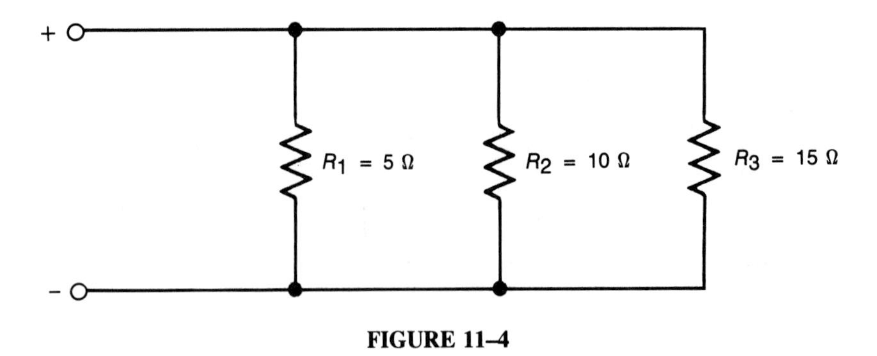

FIGURE 11–4

SOLUTION

$$\frac{1}{R_T} = \frac{1}{5} + \frac{1}{10} + \frac{1}{15}$$
$$\frac{1}{R_T} = \frac{6}{30} + \frac{3}{30} + \frac{2}{30}$$
$$\frac{1}{R_T} = \frac{11}{30}$$
$$R_T = \frac{30}{11}$$
$$R_T = 2.73\ \Omega$$

Compare our answer with that of the previous section. They are identical.

11–4 A PRACTICAL METHOD

The reciprocal equation inevitably requires the finding of the lowest common denominator. This procedure could, at times, be cumbersome. The following method is offered as an alternate practical solution. This method works in all cases.

EXAMPLE 11–3

Given: The circuit shown in Figure 11–5.

Find: The total resistance.

SOLUTION

Pick the largest resistance value in the circuit; then divide it by *all* the parallel resistors (including itself).

$12 \div 12 = 1$
$12 \div 4 = 3$
$12 \div 6 = 2$

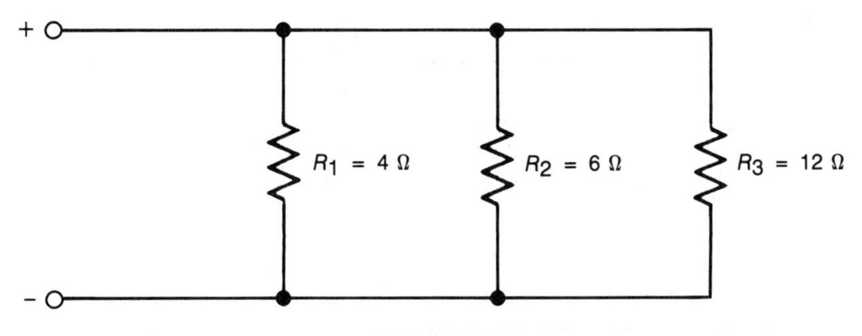

FIGURE 11–5

Now add the results.

$$1 + 3 + 2 = 6$$

Divide that result again into the largest resistor.

$$12 \div 6 = 2$$

2 Ω is the answer.

Now that you have been given an example, try to do it yourself. Substitute the numbers 20, 30, and 60 for the resistance values shown in Figure 11–5. When you are done, you should get 10 ohms for an answer. If not, better try again.

11–5 THE PRODUCT OVER THE SUM FORMULA

This is a quick way of getting your answer, provided that *only two* resistors are given. Study the example shown here. The equation to be used is

$$R_T = \frac{R_1 \times R_2}{R_1 + R_2}$$

EXAMPLE 11–4

Given: The circuit shown in Figure 11–6.

Find: The total circuit resistance.

SOLUTION

$$R_T = \frac{3 \times 6}{3 + 6}$$

$$R_T = \frac{18}{9}$$

$$R_T = 2\ \Omega$$

If you think you understand it, solve the above problem once again, substituting 15 ohms for R_1 and 30 ohms for R_2. Your answer should be 10 ohms.

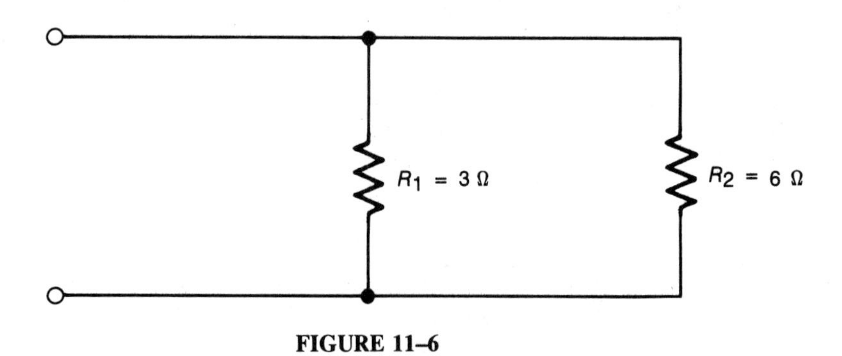

FIGURE 11–6

11–6 A SPECIAL CONDITION: ALL EQUAL RESISTORS

Occasionally, you may encounter circuits where all load resistors are of identical value. That makes our computation really easy. We merely take the resistance value of one device and divide it by the number of branch resistors. Mathematically, we might state it like this

$$R_T = \frac{R_1}{n}$$

where n represents the number of resistors shown.

EXAMPLE 11–5

Given: The circuits shown in Figure 11–7.

Find: Total resistance.

SOLUTION

a. $R_T = \frac{R_1}{n}$

$R_T = \frac{60}{2}$

$R_T = 30\ \Omega$

b. $R_T = \frac{R_1}{n}$

$R_T = \frac{60}{3}$

$R_T = 20\ \Omega$

c. $R_T = \frac{R_1}{n}$

$R_T = \frac{60}{4}$

$R_T = 15\ \Omega$

d. $R_T = ?$

e. $R_T = ?$

What answers did you get for the circuits shown in D and E? Do you agree with 12 and 10 ohms, respectively?

11–7 POWER DISSIPATION IN PARALLEL CIRCUITS

Each resistor develops heat in accordance with the amount of current flowing through it. Regardless of the type of circuit involved, the power dissipated by R_1 is $P_1 = E_1I_1$. Likewise, $P_2 = E_2I_2$; $P_3 = E_3I_3$, and so on.

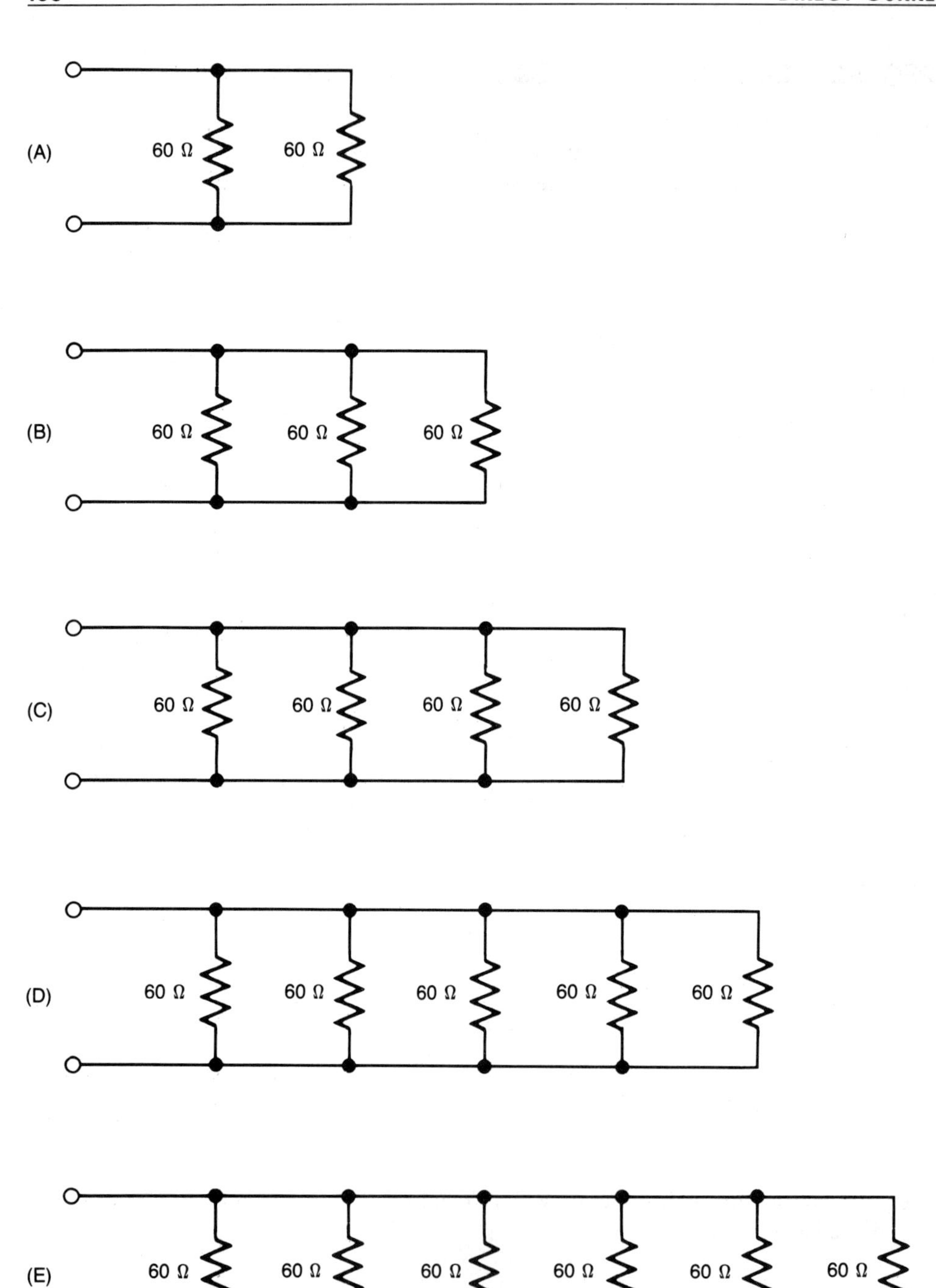
(A)
60 Ω
60 Ω
(B)
60 Ω
60 Ω
60 Ω
(C)
60 Ω
60 Ω
60 Ω
60 Ω
(D)
60 Ω
60 Ω
60 Ω
60 Ω
60 Ω
(E)
60 Ω
60 Ω
60 Ω
60 Ω
60 Ω
60 Ω

FIGURE 11–7

The total circuit power (P_T) is simply the sum of all the various power dissipations on the branch resistors. Mathematically stated

$$P_T = P_1 + P_2 + P_3 + \cdots + P_n$$

Of course, if the total current is known, it is even simpler to compute P_T directly by using the power equation $P_T = E_T I_T$ or $P_T = (I_T)^2 R_T$.

EXAMPLE 11–6

Given: The circuit shown in Figure 11–8.

Find: The total power consumption.

SOLUTION

Method #1

$$I_1 = \frac{E_1}{R_1} = \frac{240}{10} = 24 \text{ A}$$

$$I_2 = \frac{E_2}{R_2} = \frac{240}{20} = 12 \text{ A}$$

$$I_3 = \frac{E_3}{R_3} = \frac{240}{30} = 8 \text{ A}$$

$$I_4 = \frac{E_4}{R_4} = \frac{240}{40} = 6 \text{ A}$$

$$I_T = I_1 + I_2 + I_3 + I_4 = 50 \text{ A}$$

$$P_T = E_T I_T$$
$$P_T = 240 \times 50$$
$$P_T = 12{,}000 \text{ W}$$
$$P_T = 12 \text{ kW}$$

Method #2

$$P_1 = E_1 I_1$$
$$P_1 = 240 \times 24$$
$$P_1 = 5{,}760 \text{ W}$$

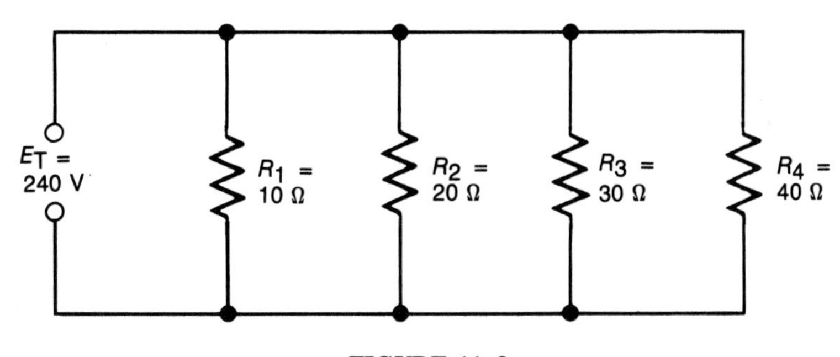

FIGURE 11–8

$P_2 = E_2 I_2$
$P_2 = 240 \times 12$
$P_2 = 2{,}880 \text{ W}$

$P_3 = E_3 I_3$
$P_3 = 240 \times 8$
$P_3 = 1{,}920 \text{ W}$

$P_4 = E_4 I_4$
$P_4 = 240 \times 6$
$P_4 = 1{,}440 \text{ W}$

$P_T = P_1 + P_2 + P_3 + P_4$
$P_T = 5{,}760 \text{ W} + 2{,}880 \text{ W} + 1{,}920 \text{ W} + 1{,}440 \text{ W}$
$P_T = 12{,}000 \text{ W}$
$P_T = 12 \text{ kW}$

And here is one more problem, solved by alternate methods, so that you may see the various options available to you.

EXAMPLE 11–7

Given: The circuit shown in Figure 11–9.

Find: The value of R_2.

SOLUTION

Ohm's law Method

Notice that no voltage has been specified. To use the Ohm's law method, we will have to assume our own voltage, say 210 volts.

$$I_T = \frac{E_T}{R_T}$$
$$I_T = \frac{210}{14}$$
$$I_T = \frac{30}{2}$$
$$I_T = 15 \text{ A}$$

$$I_1 = \frac{E_1}{R_1}$$
$$I_1 = \frac{210}{21}$$
$$I_1 = 10 \text{ A}$$

Since

$$I_T = I_1 + I_2$$

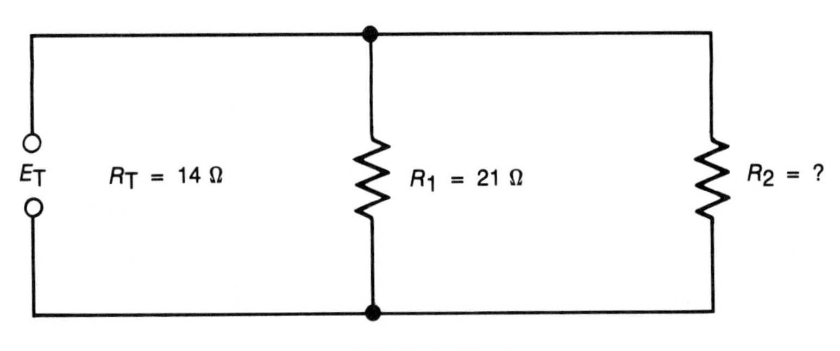

FIGURE 11–9

Then

$$I_2 = I_T - I_1$$
$$I_2 = 15 - 10$$
$$I_2 = 5 \text{ A}$$
$$R_2 = \frac{E_2}{I_2}$$
$$R_2 = \frac{210}{5}$$
$$R_2 = 42\ \Omega$$

Alternate Solution—The Reciprocal Formula

Since

$$\frac{1}{R_T} = \frac{1}{R_1} + \frac{1}{R_2}$$

Then

$$\frac{1}{R_2} = \frac{1}{R_T} - \frac{1}{R_1}$$
$$\frac{1}{R_2} = \frac{1}{14} - \frac{1}{21} \qquad \text{(L.C.D. = 42)}$$
$$\frac{1}{R_2} = \frac{3}{42} - \frac{2}{42}$$
$$\frac{1}{R_2} = \frac{1}{42}$$
$$R_T = 42\ \Omega$$

The Product Over Sum Formula

$$R_T = \frac{R_1 R_2}{R_1 + R_2}$$
$$14 = \frac{21\ R_2}{21 + R_2}$$

$$14(21 + R_2) = 21\ R_2$$
$$294 + 14\ R_2 = 21\ R_2$$
$$294 = 21\ R_2 - 14\ R_2$$
$$294 = 7\ R_2$$
$$R_2 = \frac{294}{7}$$
$$R_2 = 42\ \Omega$$

SUMMARY

- There is only one voltage in parallel circuits.
- In parallel circuits, each device has its own current, which is independent of other devices.
- In parallel circuits, the total current is equal to the sum of all the individual branch currents.

$$I_T = I_1 + I_2 + I_3 + \cdots + I_n$$

- In parallel circuits, the total circuit resistance is always smaller than any of the branch resistors.
- The total resistance of a parallel circuit can be calculated by the reciprocal formula.

$$\frac{1}{R_T} = \frac{1}{R_1} + \frac{1}{R_2} + \frac{1}{R_3} + \cdots + \frac{1}{R_n}$$

- Parallel circuits with only two resistors can be solved by the product over the sum equation.

$$R_T = \frac{R_1 R_2}{R_1 + R_2}$$

- Parallel circuits with all equal resistors have a total circuit resistance equal to the value of one resistor divided by the number of branches.

$$R_T = \frac{R_1}{n}$$

Achievement Review

1. Determine the total resistance of a 5-ohm resistor and a 15-ohm resistor connected in parallel.
2. Four 12-ohm resistors are connected in parallel. Calculate the total circuit resistance.
3. A 4-ohm, 12-ohm, and 16-ohm resistor are connected in parallel. Calculate the total circuit resistance by using the reciprocal formula.
4. The group of resistors mentioned in question 3 is connected in parallel across

a 120-volt DC supply. Calculate the current through each resistor. Find the total current and the total circuit resistance by applying Ohm's law.

5. Five lamps of equal resistance are connected in parallel across a 120-volt line. If the total current supplied measures 3 amperes, what is the resistance of *each* lamp?
6. The combined resistance (R_T) of two lamps in parallel is 35 ohms. If the resistance of one is 105 ohms, what is the resistance of the other?
7. Three equal resistors are connected in *series* and have a total resistance of 45 ohms. What would be their combined resistance if they were connected in *parallel?*
8. Find R_1 and R_T.

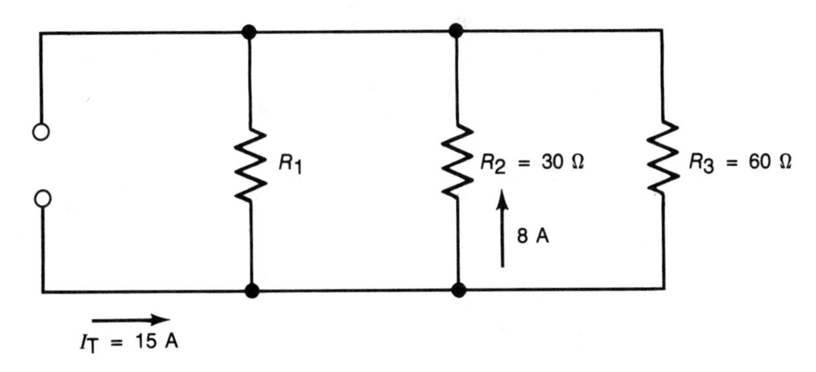

9. Find R_T.

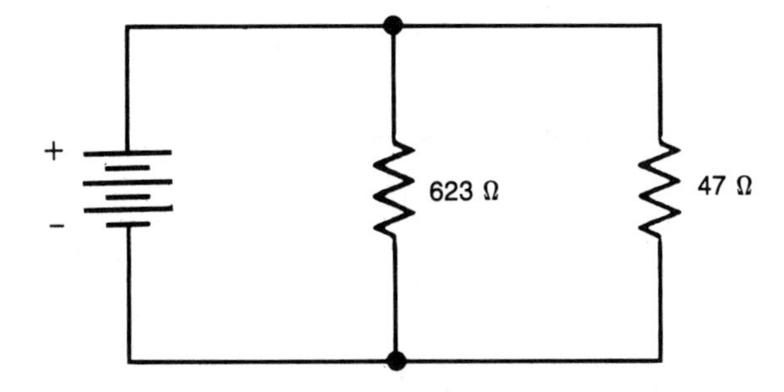

10. Find R_T.

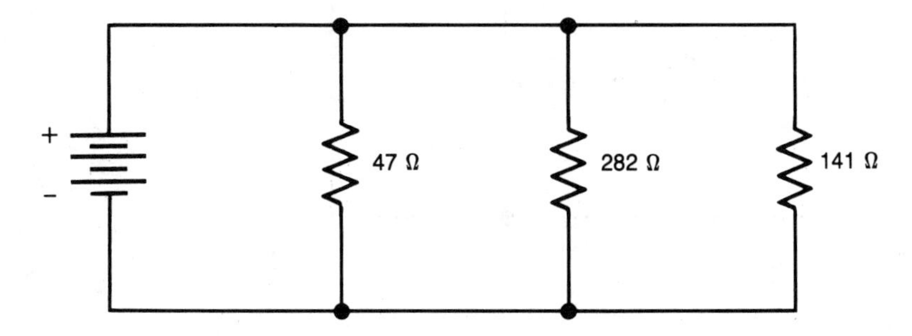

11. Find R_T.

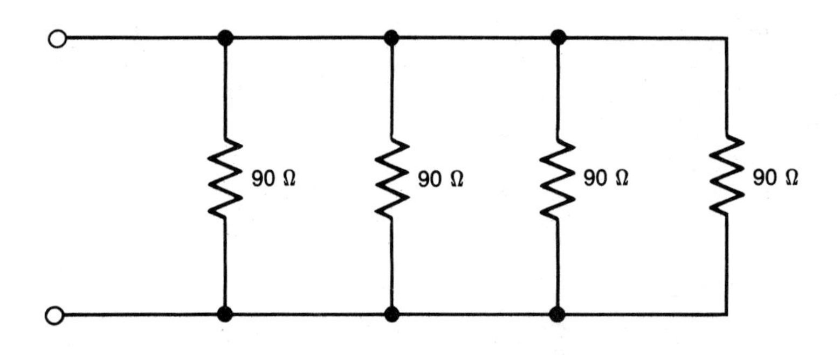

12. Find E_T.

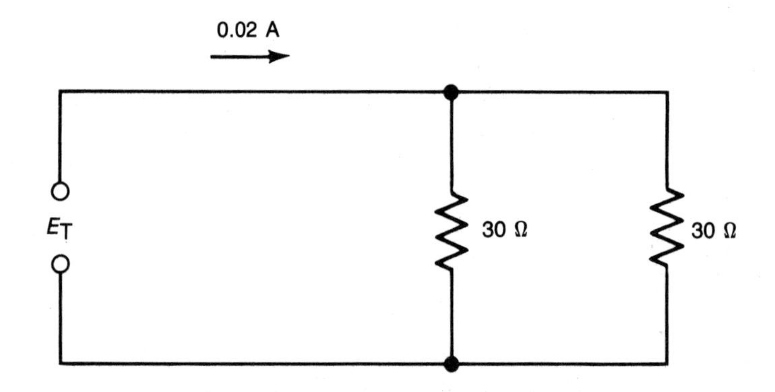

13. Find E_T.

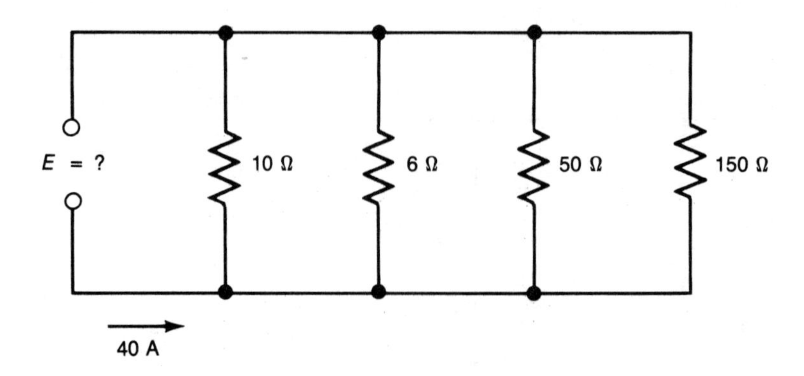

14. Find I_2 and E.

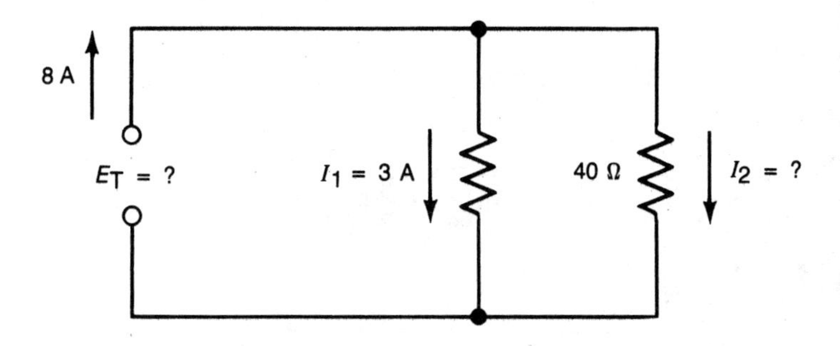

15. Find I_T.

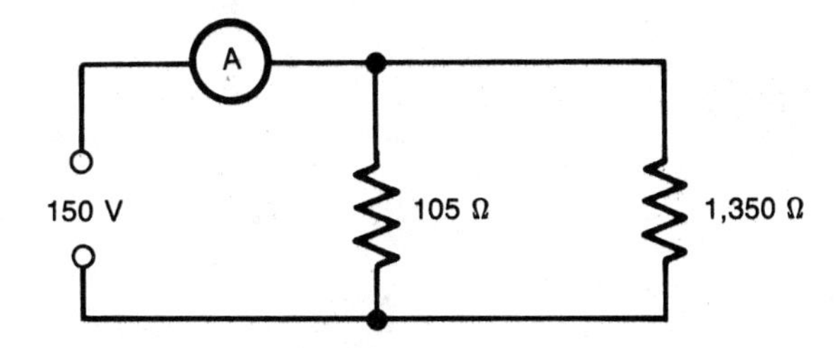

16. Solve for these quantities using the following schematic.

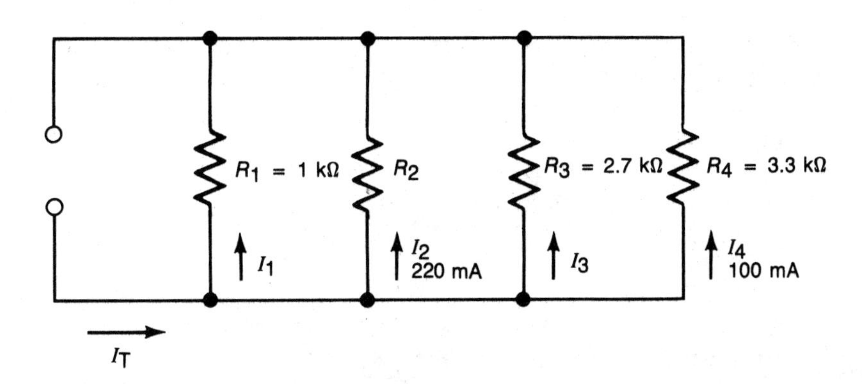

a. R_2
b. R_T
c. E_T
d. E_1
e. E_2
f. E_3
g. E_4
h. I_1
i. I_3
j. I_T
k. P_T
l. P_1
m. P_2
n. P_3
o. P_4

12 Series-Parallel Circuits and Loaded Voltage Dividers

Objectives

After studying this chapter, the student should be able to

- Define and explain the new technical terms introduced in this chapter

equivalent resistance	voltage regulation
bleeder resistor	loaded voltage divider
unloaded voltage divider	heavy load

- Identify the series-connected and parallel-connected components of a complex circuit
- Simplify complex circuits by redrawing them with resistor values equivalent to their series and parallel combinations
- Write and explain Kirchhoff's voltage and current laws
- Apply Kirchhoff's laws in the solution of complex circuits
- Design voltage divider circuits to meet specific load conditions

12–1 SIMPLIFYING SERIES-PARALLEL CIRCUITS

Pure series circuits and pure parallel circuits are seldom encountered in the practical applications of electricity and electronics. It is more common to deal with circuits that combine the aspects of both.

This chapter will introduce you to seemingly complicated circuit problems that may be solved by the same principles you have learned in the last two chapters. These principles can be summarized as follows:

> In series circuits, current is the same; voltages add. In parallel circuits, voltage is the same; currents add.

There is no one formula that can be applied to an entire circuit to obtain the desired answer or answers. Ohm's law must be applied first to one part of the circuit and then to another part. This procedure will lead from what we know to what we need to find.

EXAMPLE 12–1

Given: The circuit shown in Figure 12–1A.

Find: The total circuit resistance.

SOLUTION

a. Identify all parallel combinations and solve for their equivalent resistance values. Note the symbols $R_2 \parallel R_3$ and $R_4 \parallel R_5$ used to identify these values.

$$R_2 \parallel R_3 = \frac{20}{2}$$

$$R_2 \parallel R_3 = 10\ \Omega$$

$$R_4 \parallel R_5 = \frac{1}{12} + \frac{1}{6}$$

$$R_4 \parallel R_5 = \frac{12}{3}$$

$$R_4 \parallel R_5 = 4\ \Omega$$

Note: The symbol $\parallel$ means "the parallel combination of." In the computation here, $R_2 \parallel R_3$ is a symbolic way of saying "the equivalent resistance of the parallel combination of R_2 and R_3."

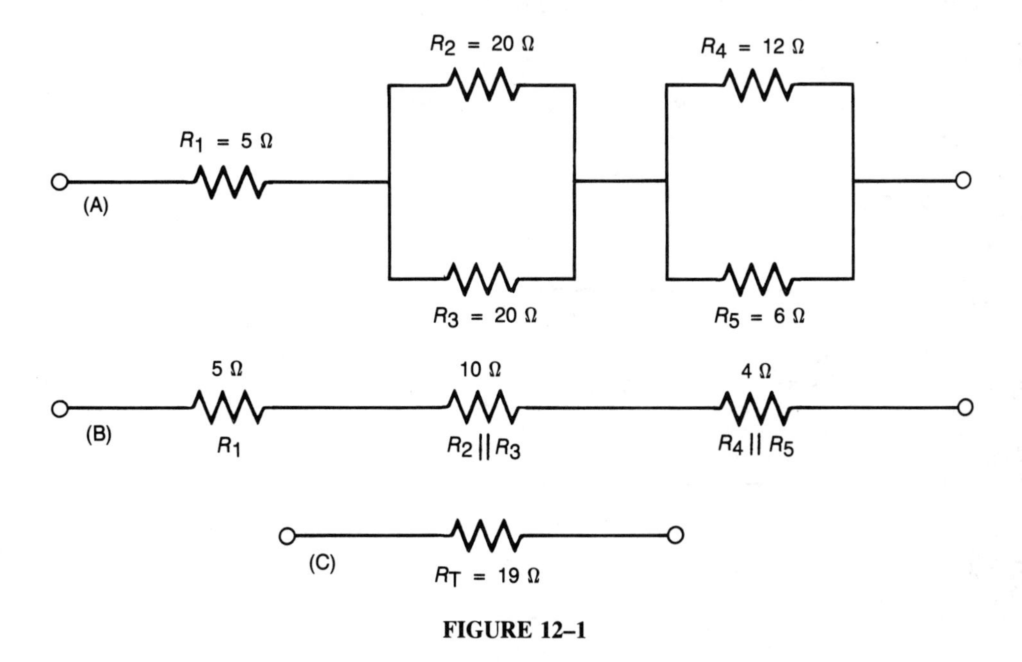

FIGURE 12–1

b. Redraw the circuit with the computed equivalent values, Figure 12–1B.

$$R_T = R_1 + R_2 \parallel R_3 + R_4 \parallel R_5$$
$$R_T = 5\ \Omega + 10\ \Omega + 4\ \Omega$$
$$R_T = 19\ \Omega$$

c. Reduce the circuit to a single resistance value, Figure 12–1C.

EXAMPLE 12–2

Given: The circuit shown in Figure 12–2A.

Find: The total resistance.

SOLUTION

Note: R_4 is *not* in parallel with R_2 as it may seem upon casual inspection. Instead, it is the series combination of R_3 and R_4 that is connected parallel to R_2. The result is a single parallel circuit with an equivalent resistance of 2.73 ohms.

$$R_3 + R_4 = 2.5\ \Omega + 7.5\ \Omega = 10\ \Omega$$
$$\frac{1}{R_T} = \frac{1}{R_1} + \frac{1}{R_2} + \frac{1}{R_3 + R_4}$$
$$\frac{1}{R_T} = \frac{1}{5} + \frac{1}{15} + \frac{1}{10}$$

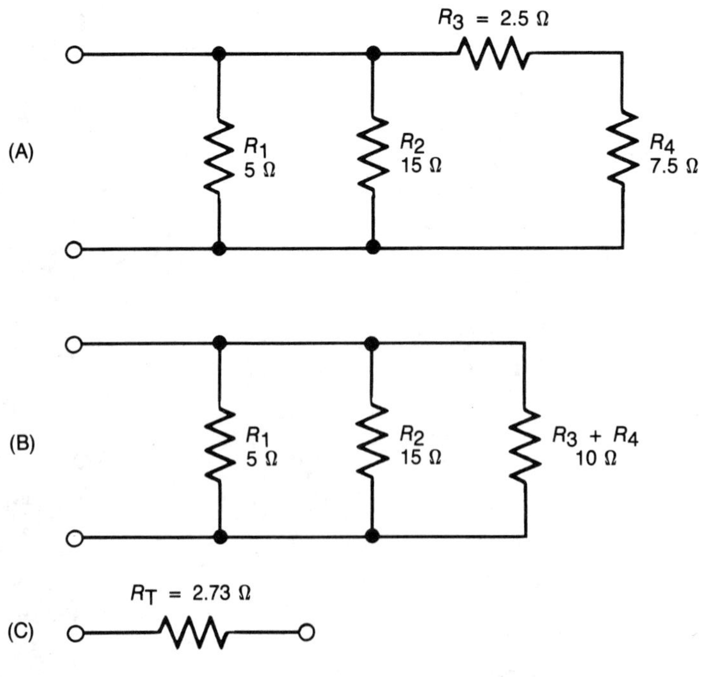

FIGURE 12–2

$$\frac{1}{R_T} = \frac{6 + 2 + 3}{30}$$

$$\frac{1}{R_T} = \frac{11}{30}$$

$$\frac{1}{R_T} = \frac{30}{11}$$

$$R_T = 2.73\ \Omega$$

EXAMPLE 12–3

Given: The circuit shown in Figure 12–3A.

Find: The total circuit resistance.

SOLUTION

Note: The parallel combinations have been identified for you, and the circuit has been redrawn accordingly, Figure 12–3B. Now it is your task to compute the equivalent values and to finish the problem.

The answer is 19.25 ohms.

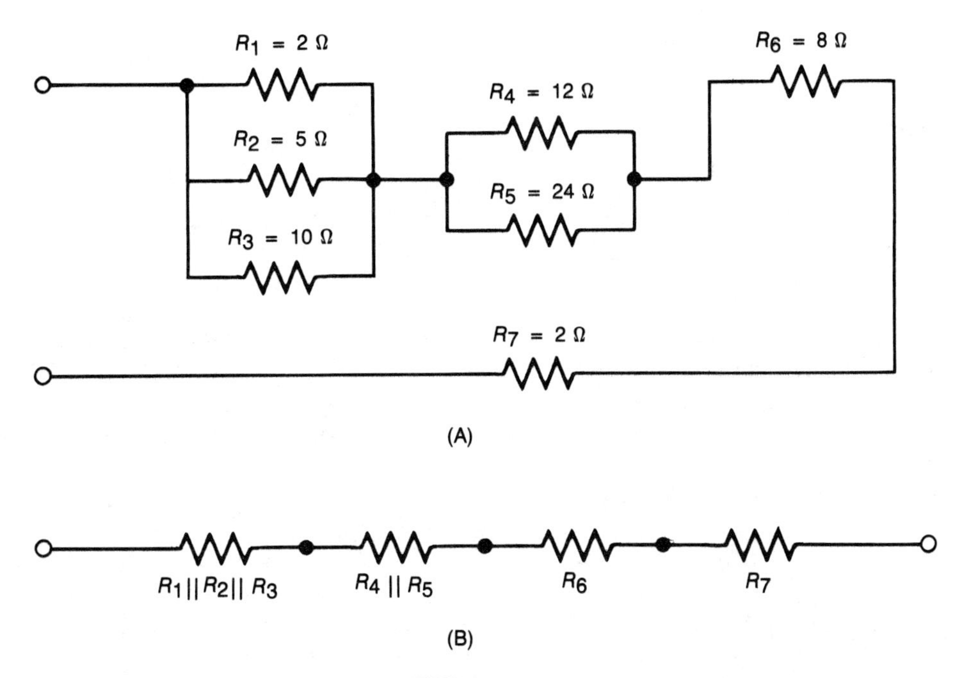

FIGURE 12–3

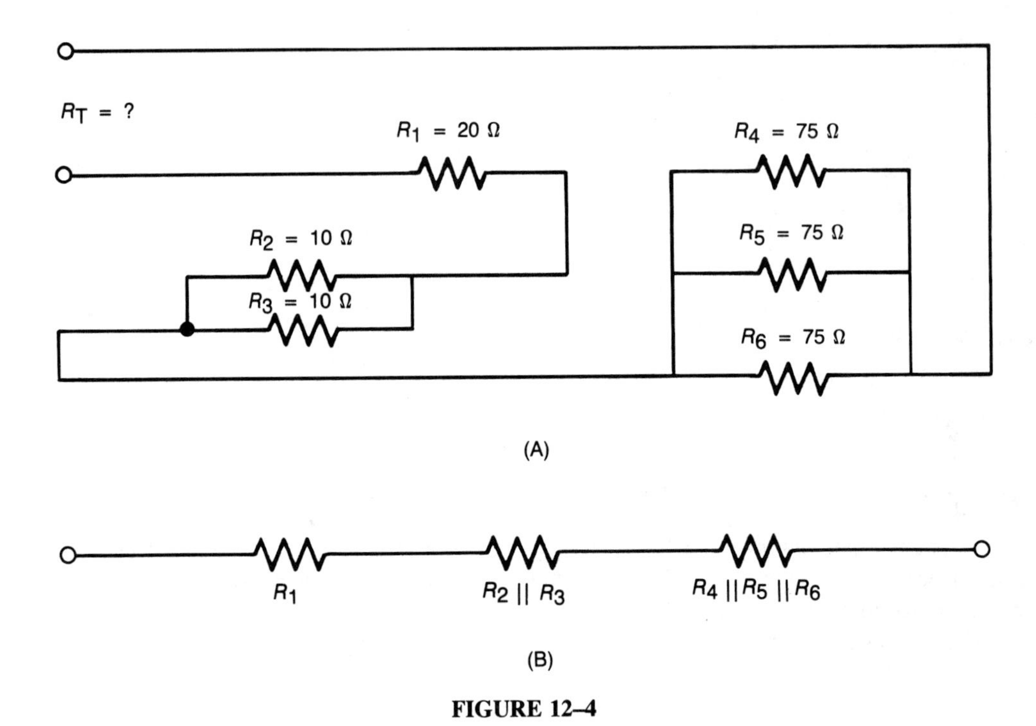

FIGURE 12–4

EXAMPLE 12–4

Given: The circuit shown in Figure 12–4.

Find: The total circuit resistance.

SOLUTION

Once again, the parallel combinations have been identified for you, and the circuit should be redrawn accordingly. Finish it! The answer is 50 ohms.

EXAMPLE 12–5

Given: The circuit shown in Figure 12–5.

Find:

a. The total circuit resistance R_T.
b. The current from A to B.
c. The current from C to D.
d. The total current supplied.
e. Each voltage drop across the resistors.

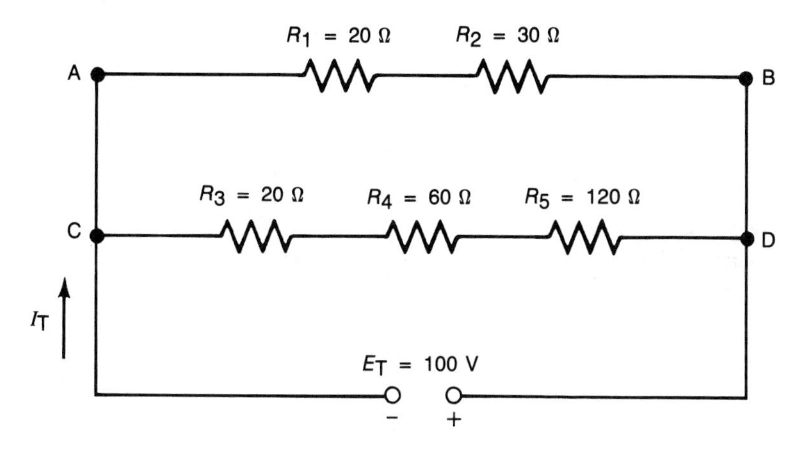

FIGURE 12–5

SOLUTION

a. Redraw the circuit, Figure 12–6.

$$\frac{1}{R_T} = \frac{1}{R_1 + R_2} + \frac{1}{R_3 + R_4 + R_5}$$

$$\frac{1}{R_T} = \frac{1}{50} + \frac{1}{200}$$

$$\frac{1}{R_T} = \frac{5}{200}$$

$$\frac{1}{R_T} = \frac{200}{5}$$

$$R_T = 40\ \Omega$$

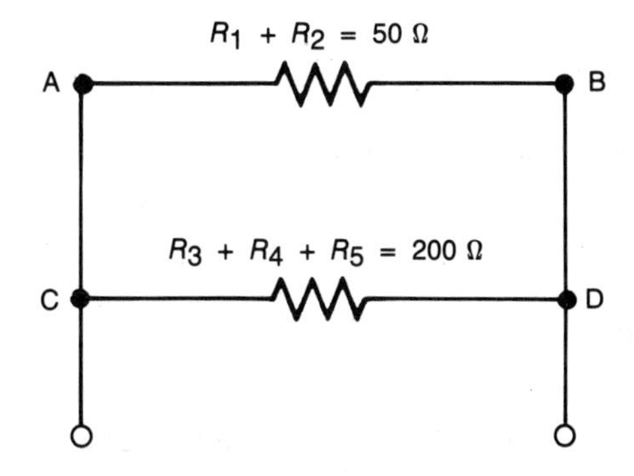

FIGURE 12–6

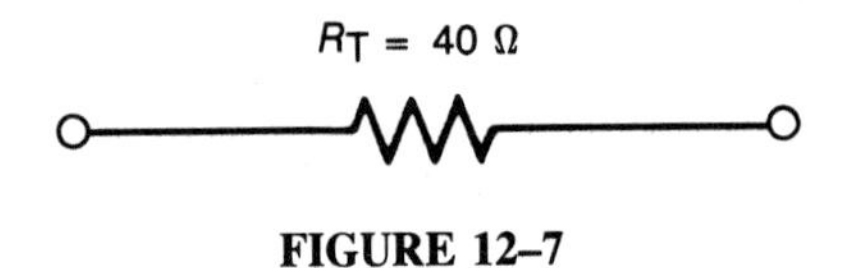

FIGURE 12–7

Redraw again, Figure 12–7.

b. *Note:* The line voltage of 100 volts is applied from A to B. Therefore

$$I_{AB} = \frac{E_{AB}}{R_{AB}}$$
$$I_{AB} = \frac{100}{50}$$
$$I_{AB} = 2\text{ A}$$

c. Likewise, the current from C to D is computed by Ohm's law.

$$I_{CD} = \frac{E_{CD}}{R_{CD}}$$
$$I_{CD} = \frac{100}{200}$$
$$I_{CD} = 0.5\text{ A}$$

d. The total current can be found by two different methods.

Method #1

$$I_T = \frac{E_T}{R_T}$$
$$I_T = \frac{100}{40}$$
$$I_T = 2.5\text{ A}$$

Method #2

$$I_T = I_{AB} + I_{CD}$$
$$I_T = 2 + 0.5$$
$$I_T = 2.5\text{ A}$$

e. To find each voltage drop, merely apply Ohm's law as follows:

$$E_1 = I_1R_1 = 2(20) = 40\text{ V}$$
$$E_2 = I_2R_2 = 2(30) = 60\text{ V}$$
$$E_3 = I_3R_3 = 0.5(20) = 10\text{ V}$$
$$E_4 = I_4R_4 = 0.5(60) = 30\text{ V}$$
$$E_5 = I_5R_5 = 0.5(120) = 60\text{ V}$$

Note: The voltage drop from A to B must equal the line voltage.

$$E_T = E_1 + E_2 = 40 + 60 = 100\text{ V}$$

Also, the voltage drop from C to D must equal the line voltage.

$$E_T = E_3 + E_4 + E_5$$
$$E_T = 10 + 30 + 60$$
$$E_T = 100 \text{ V}$$

12–2 KIRCHHOFF'S VOLTAGE LAW

The last sample problem, Example 12–5, will now be used to develop the ideas expressed by Kirchhoff's voltage law, which states

> Around any closed loop, the algebraic sum of the voltages is equal to zero.

Or, as some people prefer it

> Around any closed loop, the sum of the voltage drops is equal to the sum of the voltage rises.

Note that one key phrase is common to both statements, namely: "Around any closed loop . . ."

The problem of Example 12–5 has three closed loops as shown in Figure 12–8.

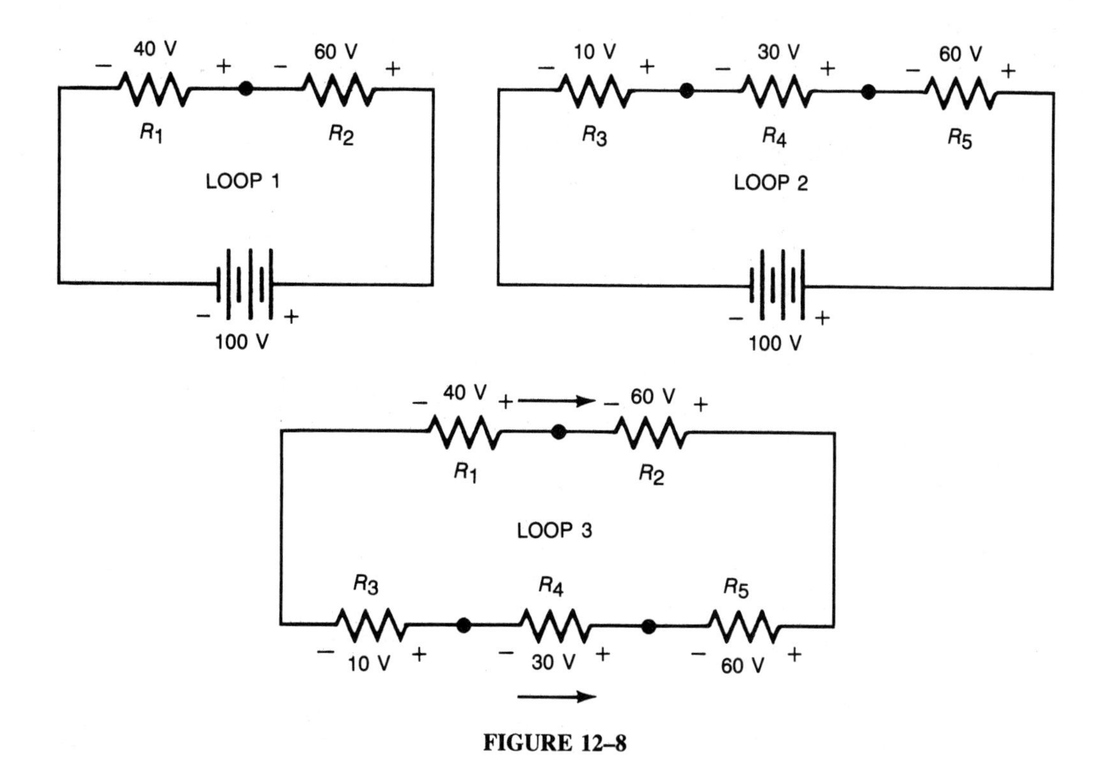

FIGURE 12–8

You should recall our convention of determining the polarity of a voltage drop. This was fully explained in Section 10–9, which you may have to review before you continue.

Tracing the loops in a clockwise direction, we obtain the following information:

for Loop 1: $-40\text{ V} -60\text{ V} +100\text{ V} = 0$
for Loop 2: $-10\text{ V} -30\text{ V} -60\text{ V} +100\text{ V} = 0$
for Loop 3: $-40\text{ V} -60\text{ V} +60\text{ V} +30\text{ V} +10\text{ V} = 0$

This, indeed, confirms Kirchhoff's law. Incidentally, the note at the conclusion of Example 12–5 is another statement confirming the truth of Kirchhoff's voltage law.

12–3 KIRCHHOFF'S CURRENT LAW

Consider a junction of three wires, called a *node,* as shown in Figure 12–9. Information is given concerning the amount and direction of the current in only two of the wires. With this information, however, it is possible to deduce the amount and direction of the unknown current in the third wire. This is what Kirchhoff's current law is all about. This law can be stated as follows:

At any node, the algebraic sum of the currents equals zero.

You have probably reasoned already that in Figure 12–9A, a current of 2 amperes is flowing from left to right, and in Figure 12–9B, a current of 8 amperes in the same direction. It appears that the total amount of current flowing *into* a junction must equal the current *coming out*. In short: What goes in must come out.

To apply Kirchhoff's law, we arbitrarily assign *positive* direction to currents flowing *into* the junction and *negative* direction to currents *coming out* of the junction. If X denotes the unknown quantity, it follows that for Figure 12–9A

$$+5 - 3 + X = 0$$
$$X = -5 + 3$$
$$X = -2\text{ A}$$

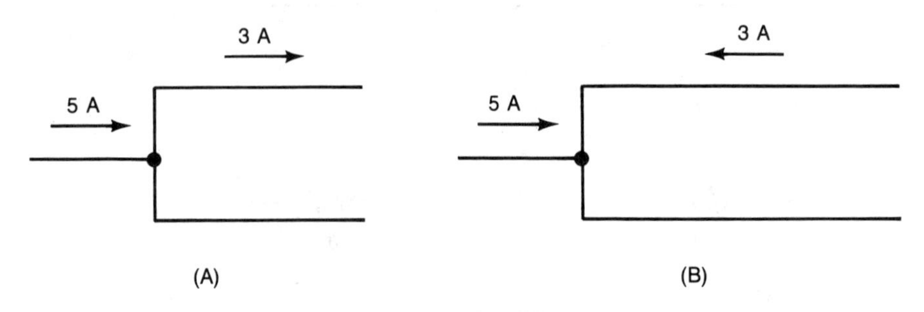

FIGURE 12–9

and for Figure 12–9B

$$+5 + 3 + X = 0$$
$$X = -5 - 3$$
$$X = -8 \text{ A}$$

Remember, the negative sign of the answer denotes the current flowing away from the junction.

The following example is designed to demonstrate the usefulness of Kirchhoff's voltage and current laws in the solution of complex circuits.

EXAMPLE 12–6

Given: The circuit shown in Figure 12–10.

Find: All voltage drops, all resistance values, and all branch currents that are not given.

SOLUTION

Note: Remember, our aim is to find all branch currents.

Applying the principles of Kirchhoff's current law, the following facts can be confirmed.

- The 8-ampere current entering node A divides in such manner that 3 amperes will flow upward (from A to B) through R_8.
- The power source provides a line current of 8 amperes to the circuit. The same 8-ampere current returns to the power source through R_1. Therefore, $I_1 = 8$ amperes.
- At point B the current I_8 combines with the current I_2 to constitute the 8-ampere line current leaving point B toward the left. This implies that I_2 must be equal to 5 amperes.

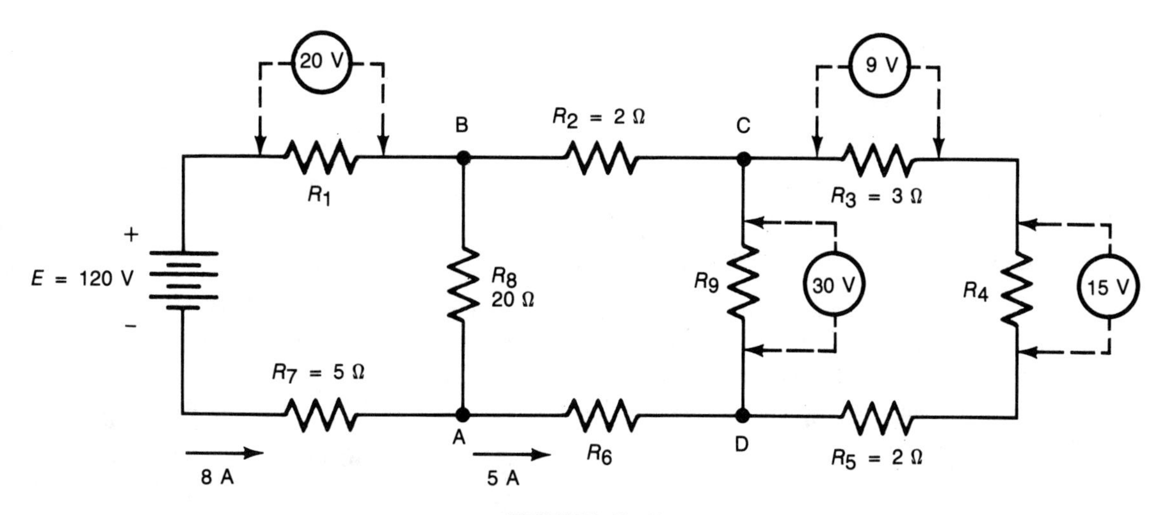

FIGURE 12–10

- I_3 can be computed by Ohm's law to be

$$\frac{9\ \text{V}}{3\ \Omega} = 3\ \text{A}$$

- Knowing that $I_2 = 5$ amperes and $I_3 = 3$ amperes, it follows that I_9, flowing from D to C equals 2 amperes.
- This completes all the computations for the different branch currents. It is suggested that you enter the new information into your diagram, Figure 12–10.

Using Ohm's law, we can now find the voltage drops of the first loop as follows:

$$E_7 = I_7R_7 = 8\ \text{A} \times 5\ \Omega = 40\ \text{V}$$
$$E_8 = I_8R_8 = 3\ \text{A} \times 20\ \Omega = 60\ \text{V}$$

Note: Our result confirms Kirchhoff's voltage law for the first loop.

$$E_7 + E_8 + E_1 = 120\ \text{V}$$

Next we solve for all values in loop 2.

$$E_2 = I_2R_2 = 5 \times 2 = 10\ \text{V}$$

By Kirchhoff's voltage law (tracing the loop counterclockwise)

$$E_6 + E_9 + E_2 + E_8 = 0$$
$$E_6 + (-30) + (-10) + 60 = 0$$
$$E_6 - 30 - 10 + 60 = 0$$
$$E_6 = -30 + 10 - 60$$
$$E_6 = -20\ \text{V}$$

Note: The negative sign of the answer signifies that the left terminal of R_6 is negative. Now Ohm's law can be used to solve for R_6 and R_9.

$$R_6 = \frac{E_6}{I_6} = \frac{20}{5} = 4\ \Omega$$
$$R_9 = \frac{E_9}{I_9} = \frac{30}{2} = 15\ \Omega$$

This completes all computations for loop 2.

12–4 LOADED VOLTAGE DIVIDERS

At this time students should have a complete understanding of simple voltage dividers, which were covered in the chapter on series circuits. A brief review of Sections 10–8 through 10–10 is highly recommended before proceeding with this section.

To understand the concept of loaded voltage dividers, consider the following hypothetical problem.

Let us assume you own a small portable radio that requires a 9-volt battery. All you have available is a 12-volt battery and a few resistors, so you decide to construct a voltage divider like the one shown in Figure 12–11. The values of the resistors are chosen in a ratio of 1 to 3; therefore, the supply voltage will be divided in a like ratio and you have your desired 9-volt supply, or so you think.

But watch what happens when you connect your radio across terminals A and B. Your radio, too, has a resistance, say 30 kilohms. When you connect the radio across the 60-kilohm resistor, you have a parallel circuit from A to B with an equivalent resistance of 20 kilohms. Now your voltage division is in a ratio of only 1 to 1, and your radio receives only 6 volts instead of the anticipated 9 volts. The radio has *loaded* the voltage divider. In other words, simple (unloaded) voltage dividers are series circuits, while loaded voltage dividers represent series-parallel circuits.

Note: The word *load* refers, of course, to any device that draws current from the voltage divider. That much is probably obvious to you. What may be new is the fact that

A *small resistance* is a *big load* and a *large resistance* represents a *small load.*

You see, the nominal size of a load is determined by the current it draws, not by its resistance value. Knowing this, you may appreciate the fact that the output from a voltage divider changes with varying load conditions. This fluctuation of voltage output is referred to as *voltage regulation.*

A high percentage of voltage regulation is undesirable and can be improved by minimizing the value of the so-called *bleeder resistor*. The bleeder is defined as that portion of the voltage divider that is connected parallel to the load.

The design of a voltage divider begins with choosing the size of the bleeder resistor so that it will draw at least 10% of the load current. An example will illustrate this.

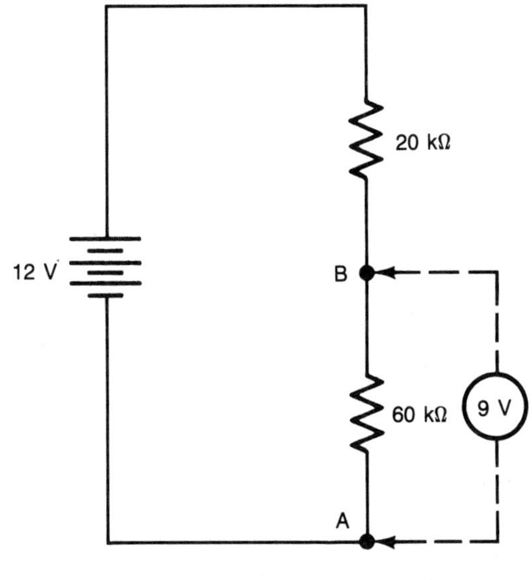

FIGURE 12–11

EXAMPLE 12–7

Given: A power source of 240 volts and a load that draws 60 milliamps at 150 volts, Figure 12–12.

Find: The proper resistance values of the voltage divider.

SOLUTION

R_2 is the bleeder and carries 10% of the load current, or

$$0.1 \times 60 \text{ mA} = 6 \text{ mA at } 150 \text{ V}$$

Therefore

$$R_2 = \frac{E_2}{I_2}$$
$$R_2 = \frac{150 \text{ V}}{6 \text{ mA}}$$
$$R_2 = 25 \text{ k}\Omega$$

By Kirchhoff's current law, R_1 will draw 66 mA, since

$$I_1 = I_2 + I_{\text{LOAD}}$$

Similarly, the application of Kirchhoff's voltage law results in

$$240 - 150 \text{ or } 90 \text{ V for } E_1$$
$$E_T = E_1 + E_2$$
$$E_1 = E_T - E_2$$

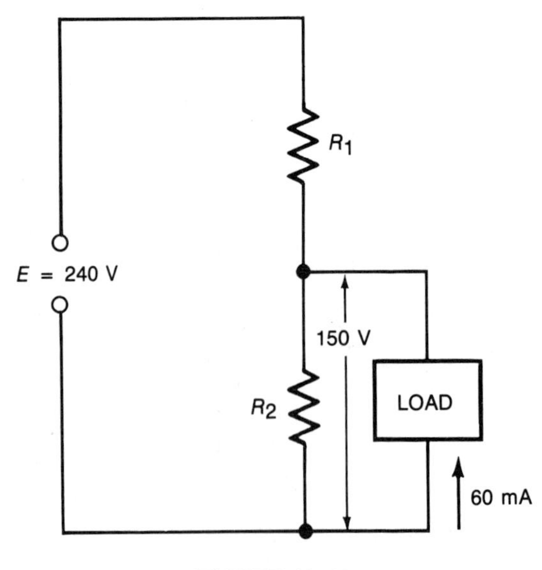

FIGURE 12–12

Therefore

$$R_1 = \frac{E_1}{I_1}$$

$$R_1 = \frac{90\ \text{V}}{66\ \text{mA}}$$

$$R_1 = 1.36\ \text{k}\Omega$$

EXAMPLE 12–8

Given: A voltage source of 240 volts and two loads with ratings as specified, Figure 12–13.

Load 1: 120 V at 0.03 A
Load 2: 200 V at 0.05 A

Find: The resistance values of the voltage divider shown.

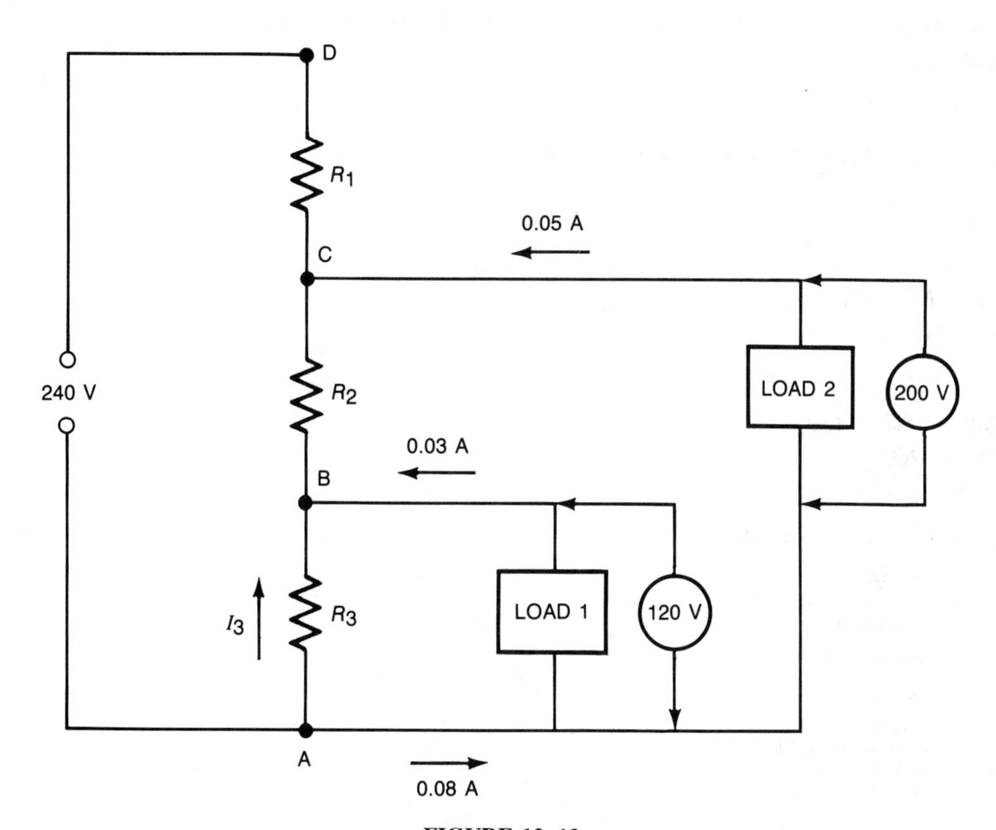

FIGURE 12–13

SOLUTION

Calculate the currents. The bleeder current (I_3) is 10% of the total load current,

$$I_{\text{LOAD TOTAL}} = 0.03 \text{ A} + 0.05 \text{ A} = 0.08 \text{ A, or } 80 \text{ mA}$$

which is

$$10\% \text{ of } 80 \text{ mA} = 8 \text{ mA}$$

Using Kirchhoff's current law, we find

$$I_2 = I_3 + I_{\text{LOAD 1}}$$
$$I_2 = 8 \text{ mA} + 30 \text{ mA}$$
$$I_2 = 38 \text{ mA}$$

and

$$I_1 = I_2 + I_{\text{LOAD 2}}$$
$$I_1 = 38 \text{ mA} + 50 \text{ mA}$$
$$I_1 = 88 \text{ mA}$$

Next, calculate the voltage drops. E_3 is specified by load 1 as being 120 volts. E_2, the voltage from B to C, is equal to the difference between the two load voltages specified, namely

$$200 - 120 = 80 \text{ V}$$

Kirchhoff's voltage law helps us to determine E_1. Since

$$E_T = E_1 + E_2 + E_3$$

it follows that

$$E_1 = E_T - E_2 - E_3$$

or

$$E_1 = E_T - (E_2 + E_3)$$
$$E_1 = 240 - (80 + 120)$$
$$E_1 = 240 - 200$$
$$E_1 = 40 \text{ V}$$

Applying Ohm's law, we calculate

$$R_1 = \frac{E_1}{I_1} + \frac{40 \text{ V}}{88 \text{ mA}} = 455 \ \Omega$$

$$R_2 = \frac{E_2}{I_2} + \frac{80 \text{ V}}{38 \text{ mA}} = 2{,}105 \ \Omega$$

$$R_3 = \frac{E_3}{I_3} + \frac{120 \text{ V}}{8 \text{ mA}} = 15 \text{ k}\Omega$$

SUMMARY

- Series-parallel circuits can be redrawn in a simplified form by substituting equivalent resistors for parallel and series combinations.
- Kirchhoff's voltage law states that the algebraic sum of voltages around any closed loop equals zero.
- Kirchhoff's current law states that the algebraic sum of currents at a node equals zero.
- Practical voltage dividers are series-parallel circuits whose resistance values are determined by the load(s).
- Heavy loads are characterized by a low resistance, and vice versa.
- Low-value bleeder resistors improve the voltage regulation of voltage dividers.

Achievement Review

1. Solve for these quantities using the following schematic.

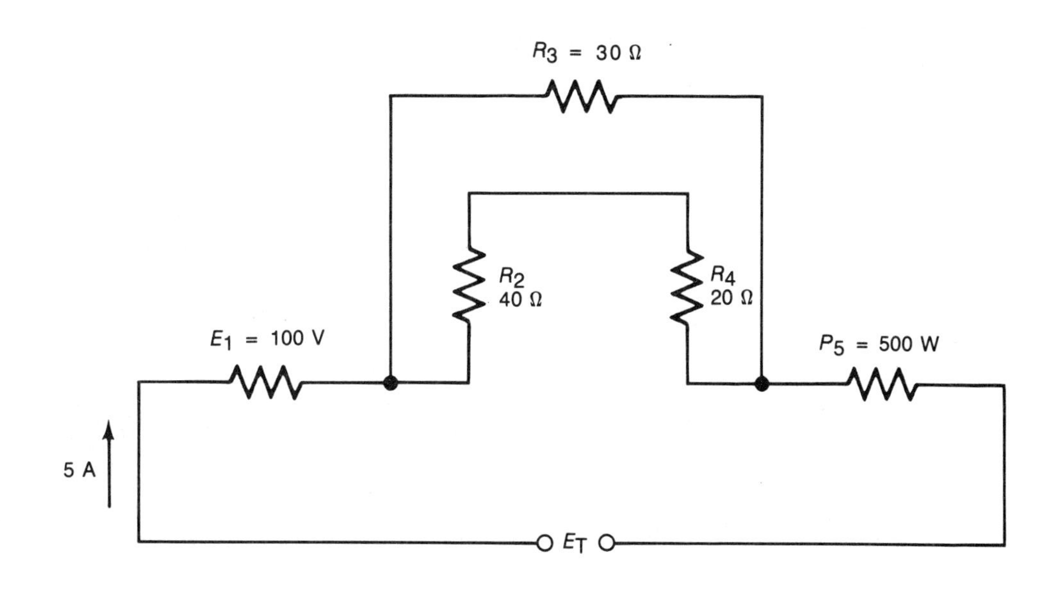

a. R_1
b. R_5
c. R_T
d. E_T
e. E_2
f. E_4
g. P_1
h. P_2
i. P_3
j. P_4
k. P_T

2. Solve for these quantities using the following schematic.

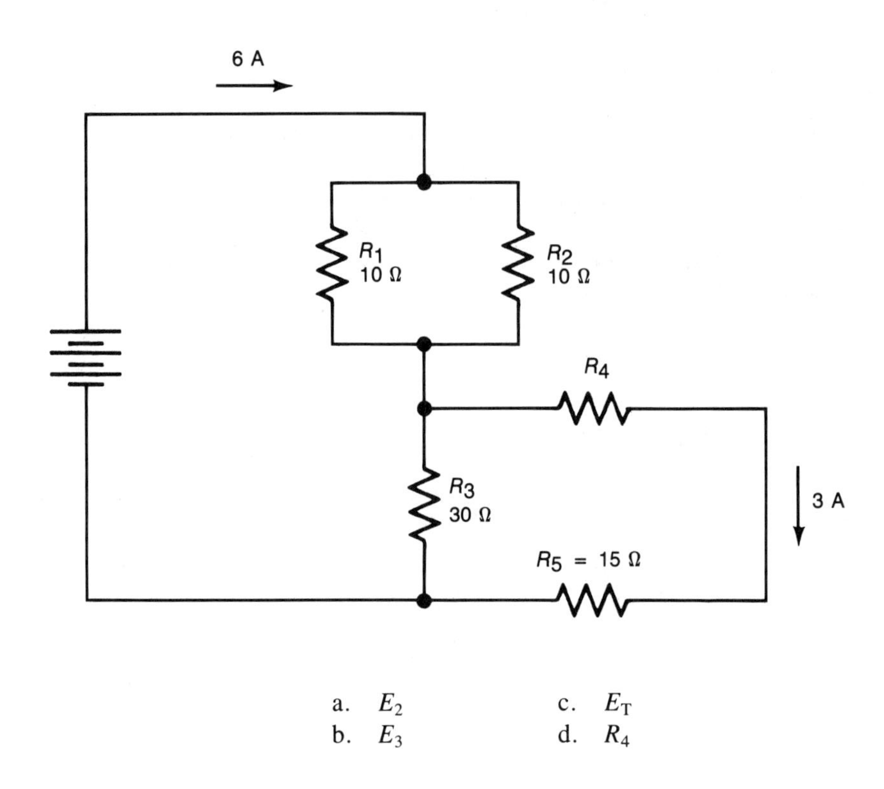

a. E_2
b. E_3
c. E_T
d. R_4

3. Given $R_1 = R_2$; $I_1 = 3$ mA; and $E_T = 12$ V, find R_T.

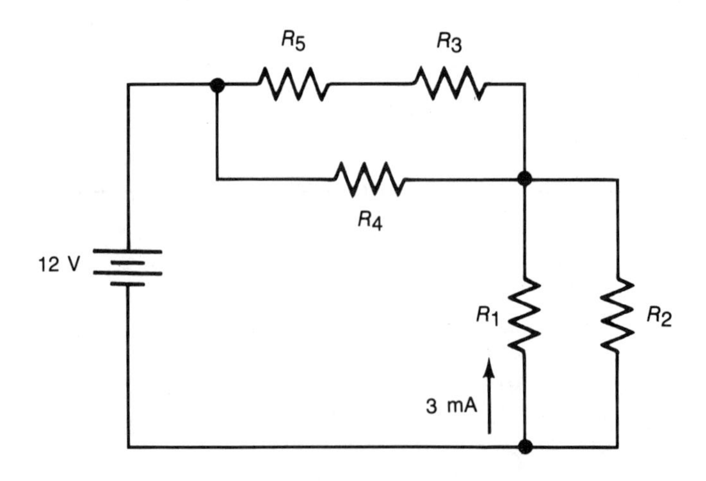

4. Solve for these quantities using the following schematic.

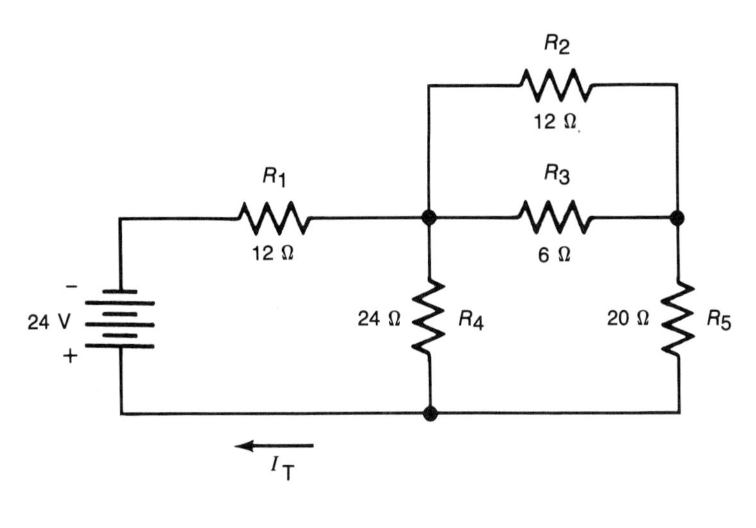

a. R_T	g. I_5	m. P_T
b. I_T	h. E_1	n. P_1
c. I_1	i. E_2	o. P_2
d. I_2	j. E_3	p. P_3
e. I_3	k. E_4	q. P_4
f. I_4	l. E_5	r. P_5

5. Solve for these quantities using the following schematic.

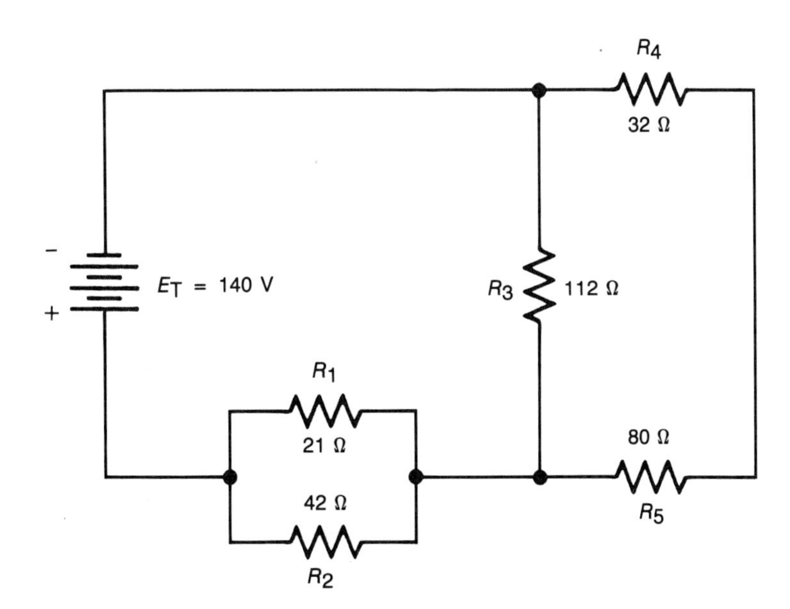

a. R_T
b. I_T
c. P_T
d. I_1
e. I_2
f. I_3
g. I_4
h. I_5
i. E_1
j. E_2
k. E_3
l. E_4
m. E_5
n. P_1
o. P_2
p. P_3
q. P_4
r. P_5

6. Given

$R_1 = 1\ k\Omega$
$R_2 = 1\ k\Omega$
$R_3 = 1.5\ k\Omega$
$E_T = 120\ V$

For each circuit, find the quantities as indicated in the answer charts. (T means total.)

a.

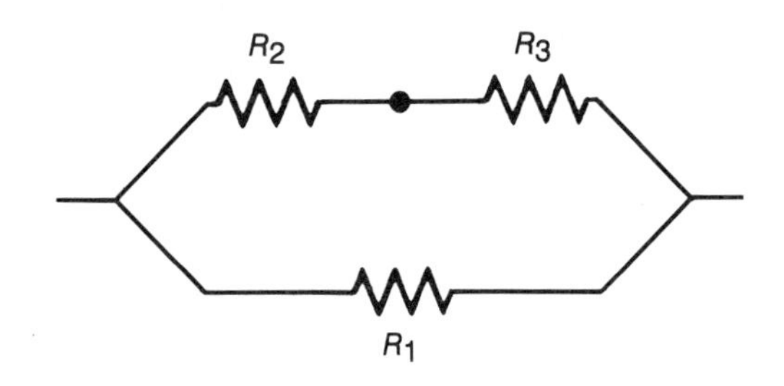

	T	R_1	R_2	R_3
E	120 V			
I				
R		1 k	1 k	1.5 k

b.

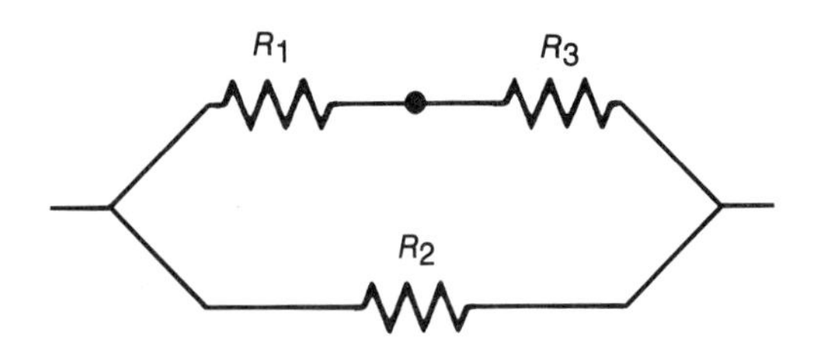

	T	R_1	R_2	R_3
E	120 V			
I				
R		1 k	1 k	1.5 k

c.

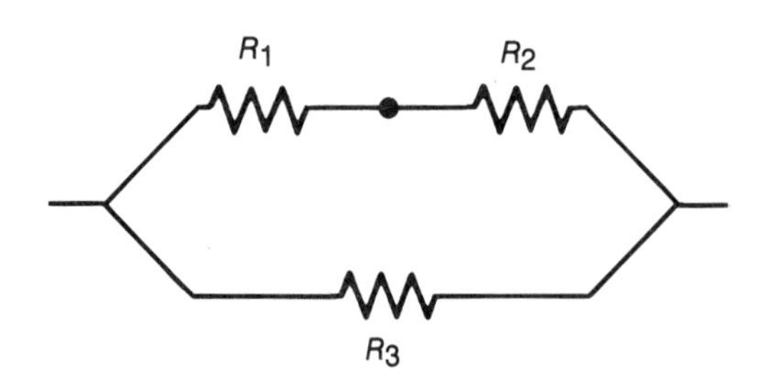

	T	R_1	R_2	R_3
E	120 V			
I				
R		1 k	1 k	1.5 k

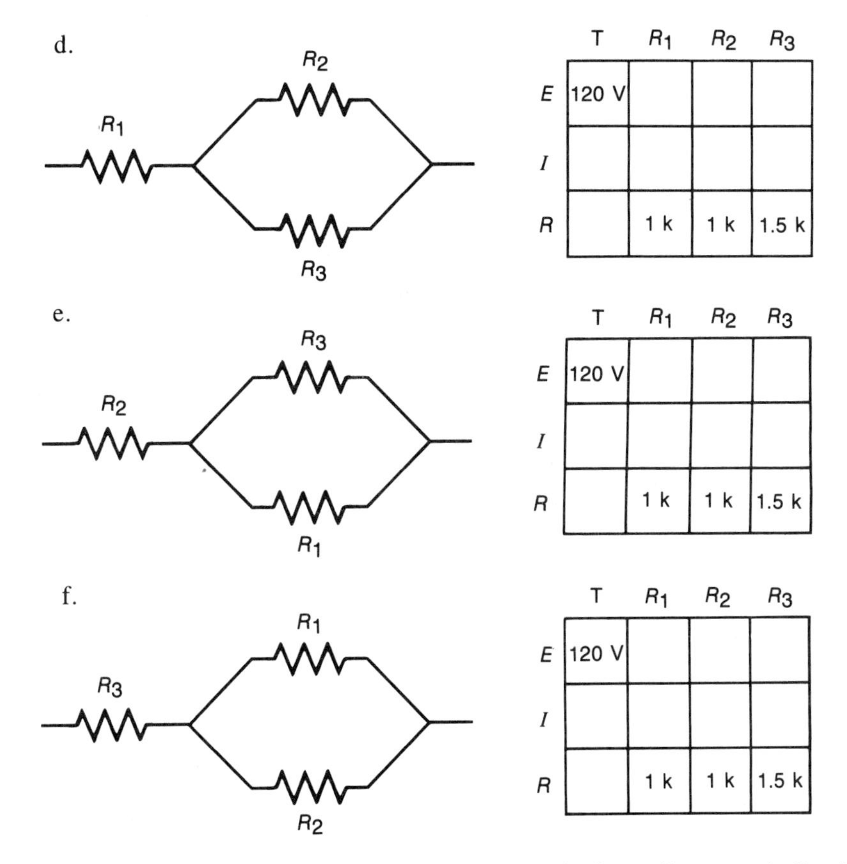

d.

	T	R_1	R_2	R_3
E	120 V			
I				
R		1 k	1 k	1.5 k

e.

	T	R_1	R_2	R_3
E	120 V			
I				
R		1 k	1 k	1.5 k

f.

	T	R_1	R_2	R_3
E	120 V			
I				
R		1 k	1 k	1.5 k

7. Redraw the circuit to show all resistors either horizontally or vertically. Then simplify the circuit.

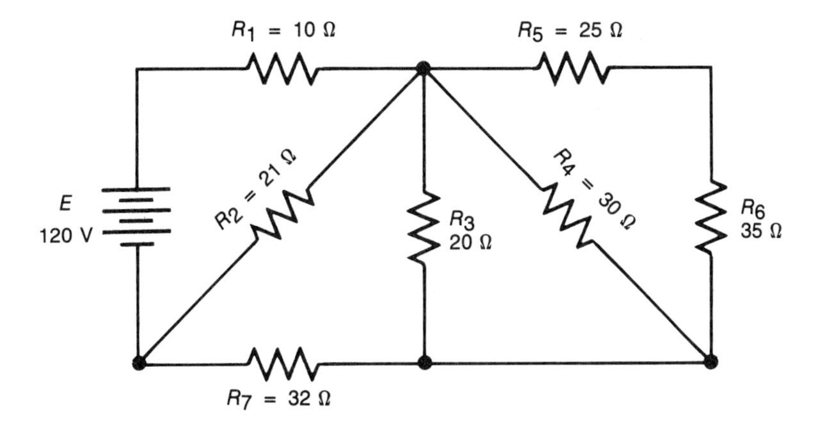

8. Solve the following quantities using the circuit in question 7.

a. R_T c. I_5
b. I_T d. P_4

9. Solve the following circuit to find all voltage drops and all the currents. Suggested sequence of operations:
   a. Find R_T and I_T.
   b. Apply Kirchhoff's and Ohm's laws to find unknown.
   c. Enter all your answers in the drawing. (It helps to use different colors.)
   d. Confirm your results by adding voltages around each loop.

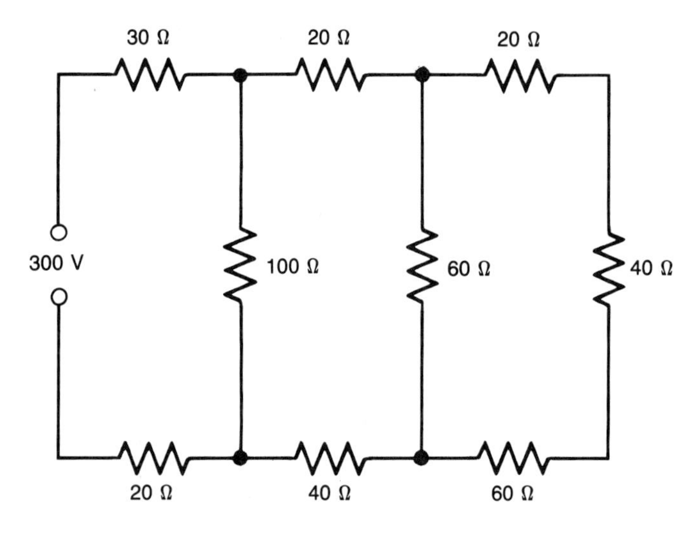

10. Find
    a. The total resistance of the circuit.
    b. The voltage drop across A-B.
    c. The current in the branch A-D-C.
    d. The potential difference between B and D.

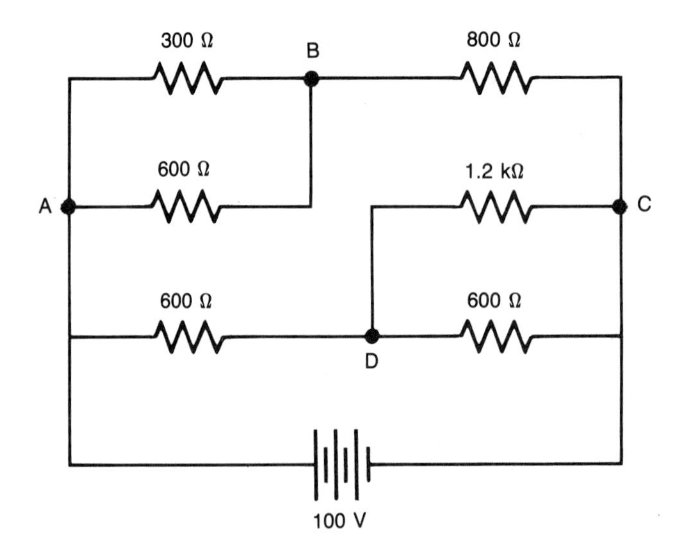

11. Solve these quantities using the following circuit.

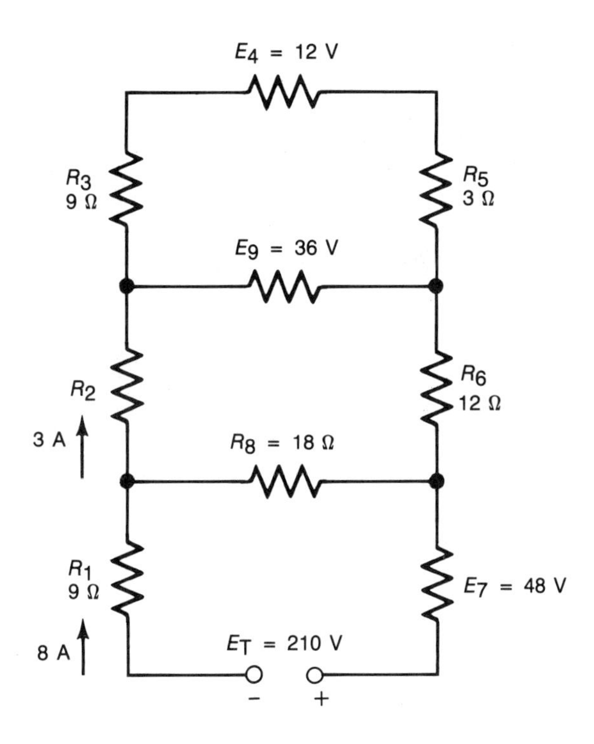

a. E_1
b. E_8
c. R_7
d. I_2
e. E_2
f. E_6
g. R_2
h. I_9
i. R_9
j. I_5
k. E_3
l. R_4

Confirm your results by adding the voltage drops around the following loops:

a. $E_1 + E_2 + E_3 + E_4 + E_5 + E_6 + E_7 =$ __________ V
b. $E_1 + E_8 + E_7 =$ __________ V
c. $E_1 + E_2 + E_9 + E_6 + E_7 =$ __________ V

12. Solve for these quantities using the following circuit.

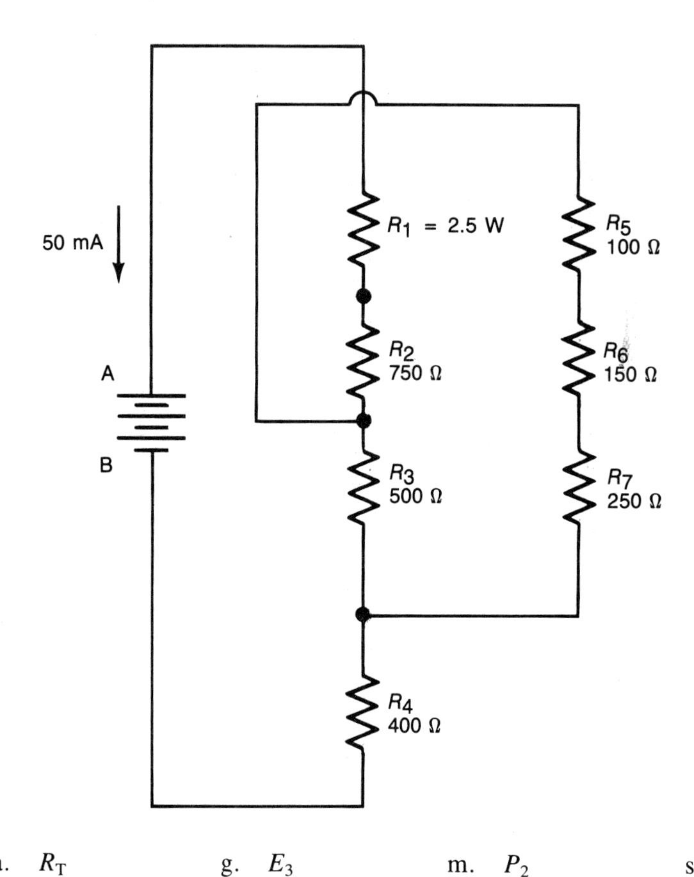

a. R_T
b. R_1
c. R_4
d. E_T
e. E_1
f. E_2
g. E_3
h. E_4
i. E_5
j. E_6
k. E_7
l. P_T
m. P_2
n. P_3
o. P_4
p. P_5
q. P_6
r. P_7
s. I_1
t. I_2
u. I_4
v. I_3
w. I_5
x. I_7

13 Conduction in Liquids and Gases

Objectives

After studying this chapter, the student should be able to

- Define and explain the new technical terms introduced in this chapter

positive ion	anode
negative ion	arcs
electrolyte	sparks
electrolysis	cathode ray
cathode	cathode-ray tube

- Describe how the ionization of liquids differs from that of gases
- Explain some useful processes that depend on conduction in liquids and gases

Many students of electricity associate the flow of electrical current with wires or other metallic conductors. Little thought is given to the important applications where conduction through liquids and gases is essential. An understanding of these principles is necessary if the student is to investigate processes such as electroplating, corrosion of metals, battery operation, and other chemical reactions that take place in the presence of an electric current through water solutions.

Likewise, similar principles apply to the operation of vapor lamps used to light highways, fluorescent lamps, and the neon lights commonly found in advertising signs.

13–1 THE IONIZATION PROCESS

In solid conductors, the transfer of electric charges is made by the movement of electrons. In liquids and gases, the transfer of electric charges depends on particles called *ions*. Ions were defined earlier in Chapter 1 as being atoms that have become electrically unbalanced by virtue of having lost or gained electrons.

To briefly review this concept, consider the model of a magnesium atom alongside a chlorine atom, Figure 13–1.

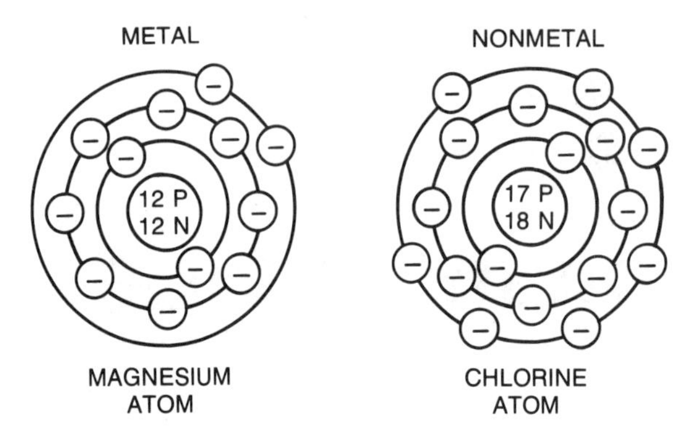

FIGURE 13–1

You should recall that magnesium is classified as a metal because it has only two valence electrons. By contrast, chlorine gas has seven valence electrons and is, therefore, classified as a nonconductor.

When a piece of magnesium is heated in the presence of chlorine gas, the two substances will chemically interact to form a new compound: a metallic salt called *magnesium chloride*. During this chemical process, each magnesium atom gives up its two valence electrons, which are then transferred to two neighboring atoms, Figure 13–2. These atoms are no longer known as atoms, since they now have been electrically charged. After an originally neutral atom becomes electrically charged, it is called an *ion*.

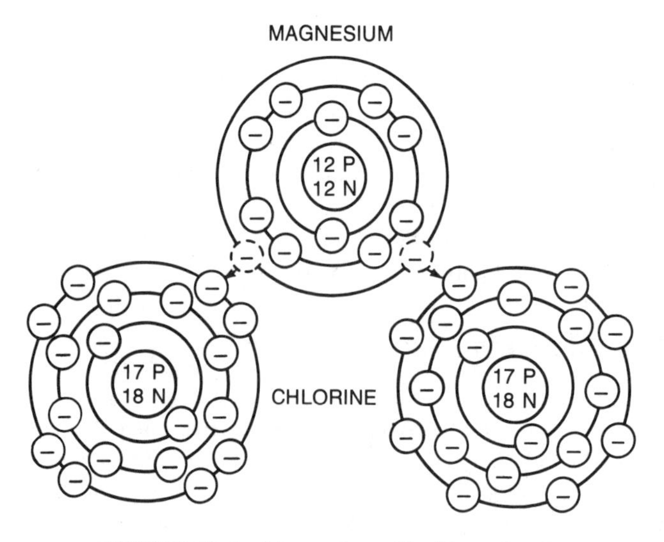

FIGURE 13–2 Magnesium chloride molecule

The magnesium atom has lost electrons (negative charges) and is now positively charged. Thus, it is called a *positive ion*. Likewise, the chlorine atoms have gained additional electrons and have been transformed into negative ions. Notice in Figure 13–2 that each ion has eight electrons in its outer shell.

This newly formed substance, magnesium chloride, is similar to table salt. The chemical name for table salt is *sodium chloride;* that is, the elements that make up table salt are a metal, sodium, and a nonmetal, chlorine.

The conducting ability of a solution of salt in water can be demonstrated with the equipment shown in Figure 13–3. First, fill the glass with kerosene, and then connect the circuit to the 120-volt line. Nothing happens to the lamp as a result of this step. Now add some table salt or magnesium chloride to the kerosene. The neutral molecules of the kerosene do not attract the undissolved charged ions of the salt. Again, there is no current through the lamp.

Fill another glass with pure distilled water (if available) and connect the circuit. Note that nothing happens. If water from the faucet is used in the glass, the lamp may light very dimly. When salt is added to the water and is stirred to dissolve it, the lamp brightens. What causes this increased conductivity when magnesium chloride or sodium chloride is dissolved in water? The dissolved substance does not retain its molecular structure but breaks up into its ion particles, which are free to move through the liquid.

Figure 13–4 shows a water solution consisting of negative chlorine ions and positive magnesium ions. When two electrically charged wires are submerged into the solution, a two-way movement of ions takes place. The negative chlorine ions are being attracted by the positive wire, and the positive magnesium ions are pulled toward the negative wire. When the ions reach their respective wires, the resulting action depends on the kinds of ions and the type of wire used.

In this manner, water solutions of acids, bases, and metallic salts can conduct large amounts of current. Such solutions are known as *electrolytes*.

The following points summarize the information presented so far:

- Metals and nonmetals unite to form compounds by transferring electrons from the metallic atoms to the nonmetallic atoms.
- As a result of the electron transfer, the compound consists of positively charged metallic ions and negatively charged nonmetallic ions.
- Some compounds dissolve in water and dissociate; that is, the ions become separated and move freely. Ions are much larger and heavier than electrons.

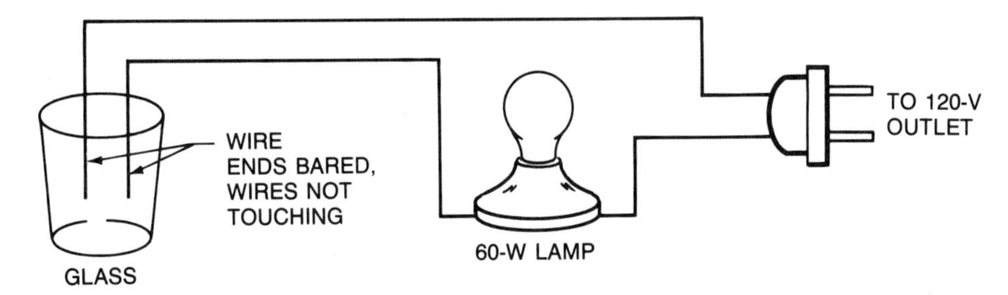

FIGURE 13–3 Conduction through liquid

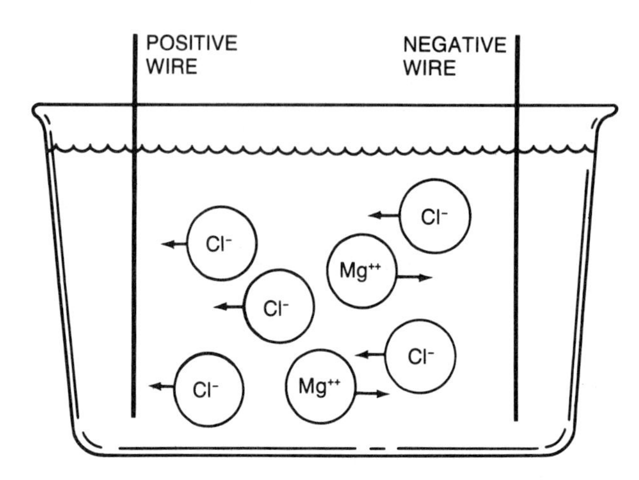

FIGURE 13–4 Moving ions in solution

- Conduction in metals consists of the movement of free (negative) electrons.
- Conduction in solutions, known as electrolytes, consists of the movement of free positive and negative ions in opposite directions.

13–2 USEFUL CHEMICAL COMPOUNDS

The ions covered so far are formed by the electrical charging (either positive or negative) of a single atom. More complicated ions, however, do exist. These ions consist of charged *groups* of atoms. The following four compounds, each of which is useful in electrical processes, are formed, in part, of these larger ions.

Sulfuric Acid. This compound is used as the electrolyte in automobile batteries. The chemical formula for sulfuric acid is H_2SO_4. In other words, one molecule of the acid contains two atoms of hydrogen, one atom of sulfur, and four atoms of oxygen. When mixed with water, each molecule can separate into the ions shown: H^+ H^+ SO_4^{--}. The + sign on the H indicates a hydrogen ion (a hydrogen atom that has lost its one electron). The SO_4^{--} is called a sulfate ion and consists of one sulfur atom, four oxygen atoms, and two extra electrons. The two electrons are necessary to hold the SO_4 group together. The electrons are part of an electron-sharing arrangement that keeps the electron rings of the sulfur and oxygen atoms tied together.

Ammonium Chloride. The electrolyte in a dry cell is a solution of ammonium chloride dissolved in water. Pure ammonium chloride is a white solid. The formula for ammonium chloride is NH_4Cl. Ammonium chloride separates into two ions when it dissolves: NH_4^+ and Cl^-. The symbol NH_4^+ means that one nitrogen atom and four hydrogen atoms are grouped together with the loss of one electron. The electron lost by the NH_4 group is taken by the chlorine atom to form a negatively charged chlorine ion. The NH_4^+ group is called an ammonium ion.

Cuprous Cyanide. The commercial process of electroplating copper on iron uses a poisonous solid called cuprous cyanide as the plating solution. The formula for cuprous cyanide is CuCN. In solution, the compound separates into the ions Cu^+ and CN^-. Cu is the symbol for copper. Cu^+ represents the cuprous ion (a copper atom that has lost its outermost electron). CN^- is one atom of carbon and one atom of nitrogen held together with the help of one electron taken from the copper atom. CN^- is called the cyanide ion.

Copper Sulfate. Some copper plating processes use a copper compound called copper sulfate, $CuSO_4$. Copper sulfate is a solid in the form of blue crystals. When copper sulfate dissolves in water, it forms Cu^{++} and SO_4^{--} ions. Under some conditions, a copper atom can lose two of its electrons to form an ion called a cupric ion. SO_4^{--} is the same sulfate ion that appears in sulfuric acid. An ion is an atom or group of atoms electrically charged.

13–3 ELECTROPLATING AND ELECTROLYSIS

To illustrate the electrical process that occurs in the plating of metals, let us examine the plating of an object with copper from a cyanide solution. The object to be plated, Figure 13–5, must be an electrical conductor. It is connected to the negative terminal of a battery or DC generator and is covered by the cuprous cyanide solution. The positive terminal of the supply is connected to a copper bar, which is also covered by the solution. When the circuit is complete, positively charged copper ions in the liquid move toward the object to be plated. When the copper ions touch the negatively charged object, the ions pick up electrons and become neutral atoms of copper. As the ions become atoms at the surface of the negatively charged object, they form a copper coating over the object. As long as the generator pushes electrons onto the object to be plated so that its negative charge is maintained, a copper coating of increasing thickness will be deposited on the object.

When the cyanide ions hit the positive copper bar, copper atoms on the surface of the bar lose electrons and become ions. In other words, the negative CN ion causes a copper atom to lose an electron and become a positive ion. The copper ion is attracted into the solution by the negative ions present in the solution. The electrons lost by the copper atoms drift toward the generator. Just as many copper ions are liberated from the copper bar as are deposited on the object to be plated; therefore, the solution stays at a constant strength.

In any electroplating process, the solution must contain the ions of the metal that is to form the coating. Since metal ions are all positively charged, the object to be plated is connected to the negative wire. The positive terminal normally is made of the same metal that is to form the coating. Plating solutions also contain ingredients other than the dissolved metal compound. These other ingredients are added to prevent corrosion of the object to be plated, to prevent poisonous fumes, and to aid in forming a smooth coating.

The purification of metals can be accomplished through a large-scale plating process. Almost all commercial copper is refined by plating in a copper sulfate solution.

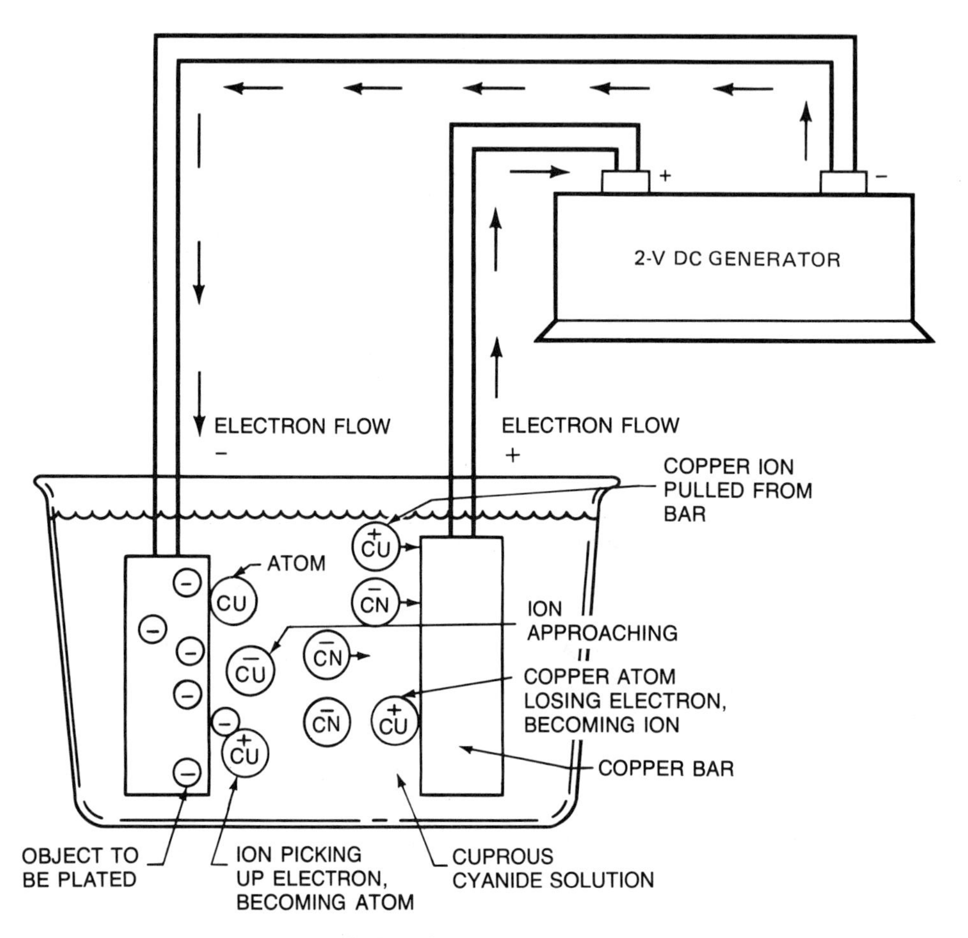

FIGURE 13–5 Copper plating

Pure copper is deposited on the negative plate. Impurities from the crude copper on the positive terminal either stay in the solution or never leave the copper bar at all. Impure copper has a high resistance. Copper wire manufacturers require electrolytically purified copper in the production of low-resistance conductor wire. Zinc and other metals are purified in a similar fashion.

The term *electrolysis* refers to the process of separating elements by the use of electrical energy. The commercial production of metallic and nonmetallic elements is often a matter of separating the element from others with which it is chemically combined. In 1885, aluminum was a rare and precious metal because it was difficult to separate from its abundant compound, aluminum oxide. Aluminum metal became inexpensive when a process was discovered for separating the aluminum from the oxygen. The extraction of aluminum is an electrical process in which electrons are removed from oxygen ions and are returned to aluminum ions. Magnesium metal is extracted from magnesium chloride (from sea water) by electrically separating the magnesium ions from the chlorine ions.

One process that has an important application in batteries is the electrolysis of water containing sulfuric acid. Hydrogen and oxygen gases are produced in this process; however, the commercial production of large amounts of these gases is achieved more economically by other processes.

Electrolysis of Sulfuric Acid in Water

Two new terms are shown in Figure 13–6: the *cathode* is the terminal or electrode where electrons enter the cell, and the *anode* is the electrode that carries electrons away from the cell.

Assume that the cathode and anode are made of materials that are not affected by the sulfuric acid solution or by the hydrogen or oxygen released in the process. When hydrogen ions touch the cathode, they pick up electrons from it and become neutral atoms. These atoms of hydrogen form hydrogen gas, which escapes from the solution as bubbles. When the sulfate ions approach the anode, they cause the water molecules to break up. The positive anode requires electrons, and the water molecules part with electrons more readily than the SO_4^{--} ions. Therefore, the removal of two electrons from the H_2O molecule leaves two H^+ ions and one oxygen atom. The oxygen atoms bubble away as gas at the anode, and the H^+ ions stay in the solution. For each pair of hydrogen ions formed at the anode, a pair of hydrogen ions is discharged as gas at the cathode. The amount of acid in the solution remains constant, and the water is gradually consumed.

If lead plates are used for the cathode and anode, then not all of the oxygen bubbles away. A portion of the oxygen combines with the lead anode to form lead dioxide. The hydrogen does not affect the cathode, which remains as pure lead.

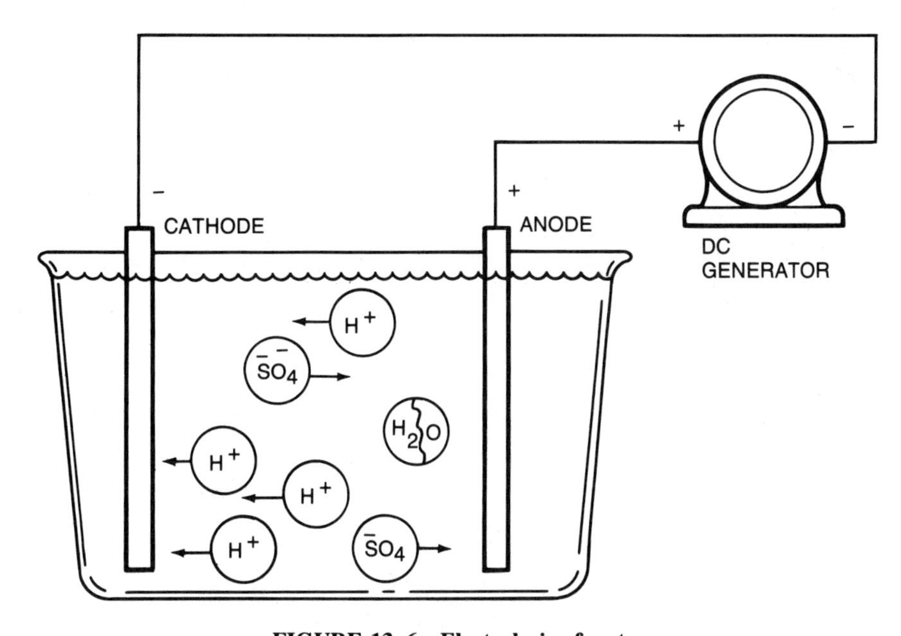

FIGURE 13–6 Electrolysis of water

Applications of plates of lead and lead dioxide in a solution of sulfuric acid relates to the working principles of storage batteries. More detailed information about this principle is offered in Section 14–5 of the following chapter.

Electrolytic Corrosion

Corrosion is the rusting or wearing away of a metal by the formation of a chemical compound of the metal. In some instances, the corrosion process is desirable. For example, in copper plating, the copper anode dissolves and in that way maintains the necessary copper ions in the plating solution. An example of an undesirable type of corrosion is shown when there is an electrical current through the earth that can follow buried water pipes for a portion of its path. At any points where the buried metal is positive (anodic) as compared to the earth, the attack on the pipe by negative ions converts the metal atoms into metal ions; in other words, iron pipe is converted into iron rust.

13–4 GASES AS INSULATORS

At atmospheric pressure, air and other gases are very close to being ideal insulators. When a gas is used as an insulator between two charged plates, Figure 13–7, the current shown by the ammeter is practically zero as long as a moderate voltage is applied. At a sufficiently high voltage, the gas suddenly becomes a conductor, and the current value is limited only by the circuit itself.

At less than atmospheric pressure, this breakdown in the insulating ability of a gas occurs at a lower voltage. However, if the gas pressure is increased, such as in compressed air, then the applied voltage necessary to start conduction in the gas must be increased a proportional amount. For example, a higher voltage is required to fire a spark plug under compression in an engine than is required to create the spark in the open air.

Sparks and Arcs

A *spark* is a noisy, irregular discharge; successive sparks follow separate paths. This statement refers to sparks that indicate electron paths. The sparks that are visible when a wire is brushed across the terminals of a car battery are similar to the sparks

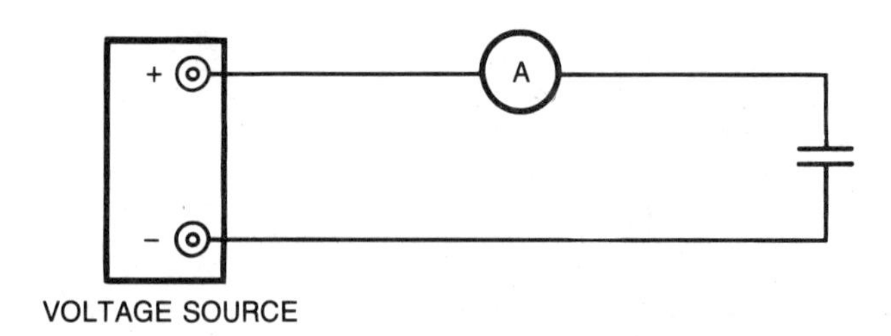

FIGURE 13–7 Gas insulation

from a grinding wheel; that is, they are tiny fragments of hot metal being sprayed in all directions.

An *arc* is quieter than a spark and is a continuous discharge. An arc, in air, consists of a conducting path of highly heated gas or metallic vapor. In this case, the air offers almost no resistance; therefore, an external resistance must be present in the circuit to limit the current. The potential drop across the arc itself is moderately low, in the order of 15 to 50 volts. An arc is accompanied by high-intensity light. Most of this light is due to gas or metallic vapor, but some light comes from the hot cathode. The cathode temperature of a carbon arc may reach 9,000°F.

13–5 GASEOUS CONDUCTION BY IONIZATION

Gas can be made to conduct a current if a moderately high voltage of 50 volts or more is applied to a gas at a low pressure inside a glass tube, Figure 13–8. (Note that the wires from the voltage source are sealed inside the tube.) In air, an arc can be started by touching and then separating a pair of conducting contacts, as in a welding arc or a carbon arc.

''How does the gas conduct?'' ''What happens in the gas so that it becomes a conductor?'' The process of *ionization* causes a gas to become a conductor. In this process, electrons are removed from gas atoms so they become positively charged ions. The free electrons are highly mobile. As a result, conduction in the gas is due mainly to electron flow (as in metals). Although mobile positive ions do exist (as in liquids), the nature of conduction in a gas differs greatly from conduction in either a solid or a liquid.

How Ionization Occurs

The most important process involved in ionization is *electron impact*. The ionization process can begin when an electron is freed from the negative wire. More likely sources of free electrons are radiation and heat (to be investigated shortly). Once an electron is set free, it gains speed in the enclosing glass tube or envelope as it is repelled by the negative (−) wire and attracted toward the positive (+) wire, Figure 13–9. If the electron hits an atom while it is still moving slowly, it merely bounces off the atom and again accelerates toward the positive terminal. When the electron is moving fast enough, any collision with an atom causes one or more electrons to be knocked free of the atom.

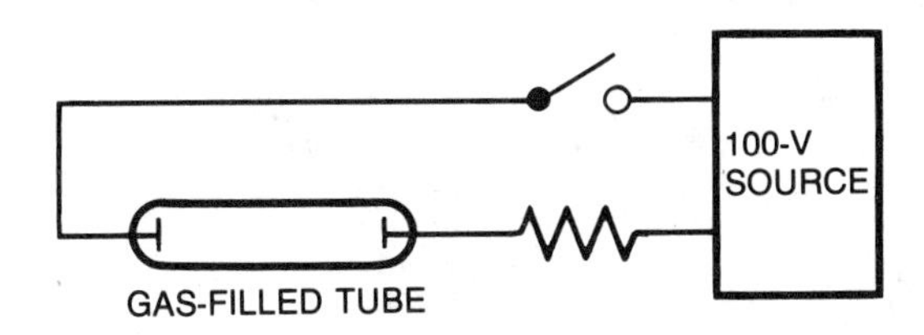

FIGURE 13–8 Gas conduction

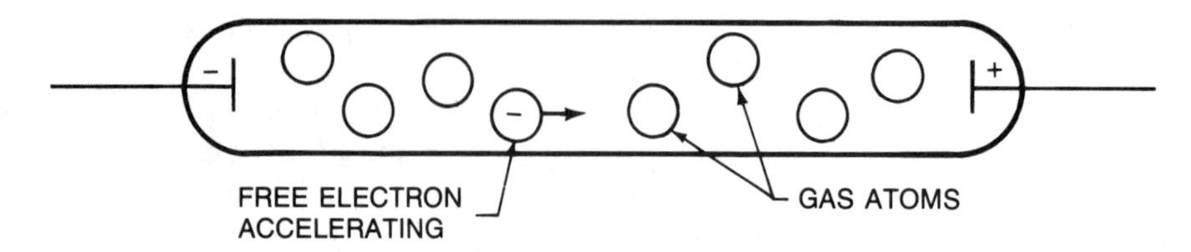

FIGURE 13–9 Ionization of gas in vacuum tube

The minimum amount of energy that an electron must possess to cause ionization (removal of electrons from atoms) is called the *ionization potential*. The amount of energy required depends on the kind of atom that is to provide the electrons. The ionization potential for sodium vapor is about 5 volts. For most gases, the ionization potential ranges between 10 and 25 volts (for example, mercury vapor requires 10.4 volts; neon, 21.5 volts; and helium, 24.5 volts). In other words, to ionize neon, an electron must fall through a potential difference of 21.5 volts.

The ionization process is shown in Figure 13–10. As one can see, the original colliding electron frees another electron. Now there are two free electrons that begin to accelerate toward the positive wire. These electrons collide with more atoms, and each electron frees an additional electron. Now there are four electrons. The subsequent series of collisions between electrons and atoms frees more electrons so that in a thousandth of a second, a million electrons may be released.

Let us compare this action with what happens to a heavily laden apple tree when one apple at the top starts to fall to the ground. The one apple will strike another apple, and the two apples will strike other apples, until finally, by the time the apple fall is completed, there is a bushel of apples on the ground.

Light is produced during gaseous conduction as a result of collisions that are not violent enough to free electrons. The collision of a moderately fast moving electron with an atom can transfer some of the energy of the electron to the electrons of the atom. This transfer of energy produces an *excited state* in the atom. In this state, the atom has excess energy and rids itself of this energy by emitting light. The color of the emitted light is characteristic of the energy state of the atom.

A second process that produces ionization in a gas is *radiation*. Rays striking the earth from space may ionize a few atoms and thus provide the first electrons required

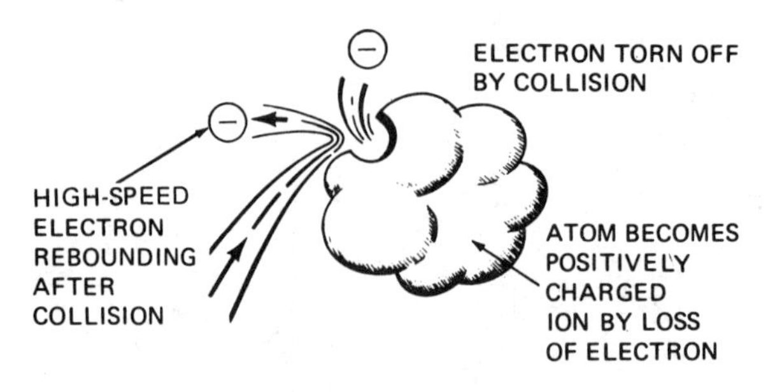

FIGURE 13–10 Ionization

to start the collision process. After conduction begins, electron disturbances in the atoms produce visible light and high-frequency (ultraviolet) radiation. This high-frequency radiation is absorbed by other atoms, thus providing them with enough energy to free electrons.

Still another phenomenon that produces ionization is *heat*. A hot gas is a better conductor and begins to conduct more readily than a cold gas. Heat is the movement of atoms and molecules. At higher temperatures, collisions between atoms can become violent enough to dislodge electrons. Many of the atoms may have a higher temperature than the average and thus can cause ionizing collisions. An ordinary flame contains many ions and is therefore a poor insulator. This statement may be proved by bringing a match flame near a charged electroscope.

Another source of electrons is the *collision of positive ions*. Although positive ions have lost one or more electrons, they still have electrons that can be set free. Therefore, when positive ions collide, electrons are released and the ions are even more positively charged. In addition, excited gas atoms can collide with other kinds of atoms, causing these atoms to release electrons. For example, conduction can be maintained more easily in neon gas if a trace of nitrogen gas is added to the neon. Excited neon atoms will collide with the nitrogen atoms. In the process, the nitrogen atoms are given enough energy to cause them to release electrons.

Why Low-Pressure Gas Conducts Better Than High-Pressure Compressed Gas

In gas at atmospheric pressure, a free electron collides with gas atoms so frequently that it never travels long enough to gain the speed required to ionize an atom by collision, Figure 13–11A. At lower pressures, gas atoms are farther apart; therefore, an electron has the opportunity to gain enough kinetic energy to make an ionizing collision, Figure 13–11B. As a result, designers of gas tubes must determine the *mean free path* of electrons to ensure that ionizing collisions will occur. The mean free path is the average distance traveled by a particle between collisions.

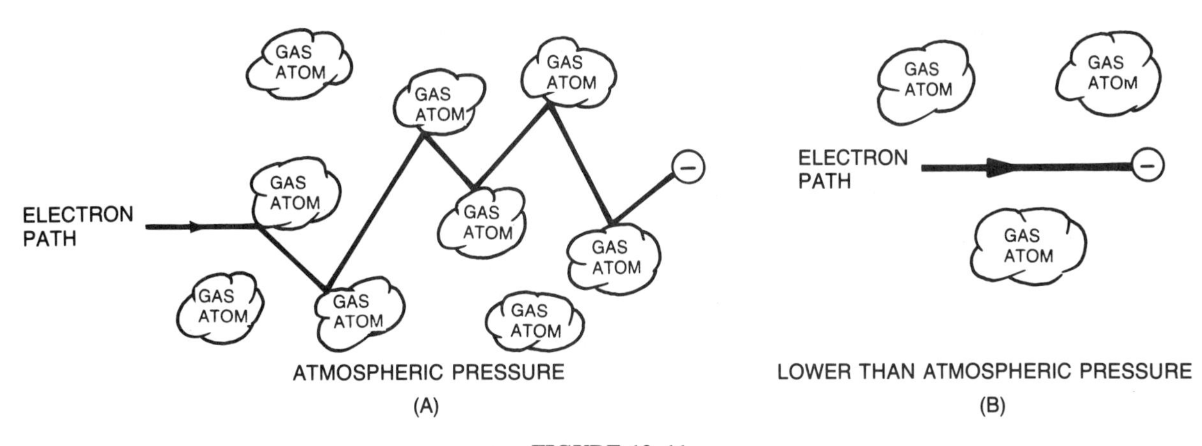

FIGURE 13–11

Electrons and Ions Formed by Collisions

The electrons that reach the positive terminal accomplish useful conduction. However, in a glass tube, many of the free electrons collect on the walls of the tube. Any positive ions that hit the walls are neutralized by the negative charges collected there.

On the average, approximately one atom out of every 1,000 atoms is ionized in a typical gas conduction tube. The glowing discharge in the tube is called *plasma* and consists of positive ions, electrons, and neutral atoms. The voltage drop in this region is not large; most of the voltage drop occurs near the electrodes. Scientists are still conducting experiments to define potential applications of the plasma phenomenon. For example, a number of years ago it was found that bursts of current (of a very high amperage) in hydrogen at low pressure produce plasmas that are contained by their own magnetic fields. The purpose of the experiments was to make the particles in the plasma reach speeds corresponding to temperatures of millions of degrees. It is at such temperatures that hydrogen nuclei fuse into helium. This hydrogen fusion process is the same as that occurring in the sun. Hydrogen fusion happens suddenly when an H-bomb explodes.

13–6 CONDUCTION AND IONS IN NATURE

An evening celestial display, which occasionally is visible in northern skies and is known as the Northern Lights (aurora borealis), is believed to be the result of conduction in the very thin upper part of the atmosphere. This display appears to be caused by electrons that are given off by disturbances in the sun and then travel through space until they reach the earth's atmosphere.

In addition to electrons, high-speed protons also reach earth from space. These protons strike air molecules and produce intense radiation that ionizes various layers of the atmosphere. These conducting layers have the ability to reflect radio signals; these reflections are often useful.

Lightning is a high-voltage spark. At times atmospheric conditions that are not violent enough to cause lightning produce a continuous discharge that can be seen on steeple tops or the masts of ships.

In the vicinity of pointed electrodes, a strong electric field will ionize the air. The discharge that occurs is visible if other illumination does not interfere. This type of discharge is called a *corona* and occurs frequently in high-voltage equipment if precautions are not taken to prevent it. Corona effects can be a serious source of power loss in high-voltage transmission lines.

13–7 CONDUCTION IN A VACUUM

If gas is pumped from a glass tube until the pressure in the tube is reduced to approximately 0.0001 mm of mercury (760 mm of mercury = atmospheric pressure), the brilliant glow of the conducting gas is no longer visible. The gas molecules are so far apart that few ions are formed. In this case, conduction is due almost entirely to electron flow. The glass tubing itself may glow because of electrons striking the glass.

If the gas pressure is reduced even more in a cold cathode tube, the discharge may stop entirely. If the cathode wire is then reheated by another current, conduction through the vacuum continues by means of the electrons that escape from the hot cathode.

Over a century ago, the people who were experimenting with conduction in gases and vacuums did not know of the existence of electrons. These experimenters gave the name *cathode ray* to the emissions from the cathodes of their tubes. In 1869, the German scientist Hittorf described the glow resulting when glass was struck by these rays. His experiments showed that the cathode rays travelled in straight lines. An English scientist, Crookes, found that the paths of these rays can be bent by a magnet. In addition, he found that these rays can be focused, that they heat the objects they strike, and that their speed depends on the applied voltage. In 1879, Crookes suggested that these rays might be "an ultra-gaseous state of matter." The French investigator Perrin showed in 1895 that the rays carried a negative charge and that positive ions were also formed.

Wilhelm Roentgen, while experimenting with cathode-ray tubes in 1895, found that when the cathode rays strike metal or glass, a new kind of radiation is produced. He determined that these invisible rays passed readily through air, paper, and wood; through thin metal better than thick; through flesh better than through bone; and through aluminum better than through lead. The rays caused fluorescence in some minerals, affected photographic plates, and were unaffected by a magnetic field. Roentgen called these rays *X rays*. Within a few months of his discovery, physicians were using X rays as an aid in setting broken bones.

In 1897, the English scientist J.J. Thomson completed a series of experiments in which he was able to measure the ratio of the weight of the negative particle to the electric charge, Figure 13–12. Modern cathode-ray tubes, such as those used in oscilloscopes and TV picture tubes, use the same principles of electron ray deflection that Thomson used in his experiments.

In oscilloscopes, Figure 13–13, the electron beam is accelerated from the cathode and is deflected horizontally at a known rate. The voltage (whose trace is to be observed) is used to deflect the beam vertically at the same time. The result is a time graph of the observed voltage.

In TV picture tubes, magnet coils are used to sweep the beam of electrons horizontally and vertically. At the same time, the intensity of the beam is changed to produce variations in the lightness and darkness of the picture. There continues to be some concern that harmful X rays are produced when the 20,000-volt electrons strike the face of the TV picture tube. This voltage generally is not great enough to produce the highly energetic and penetrating X rays. Any X rays that may be produced are absorbed in the glass and the few inches of air in front of the tube. A more serious source of X rays is the high-voltage rectifier tube in larger TV sets. In the past, some TV receivers radiated undesirable quantities of X rays through the bottom of the set because insufficient metal shielding was used around the high-voltage supply.

The American experimenter Edison might have discovered electrons. One day in 1883, one of his assistants accidentally connected a meter to a dead-end wire sealed

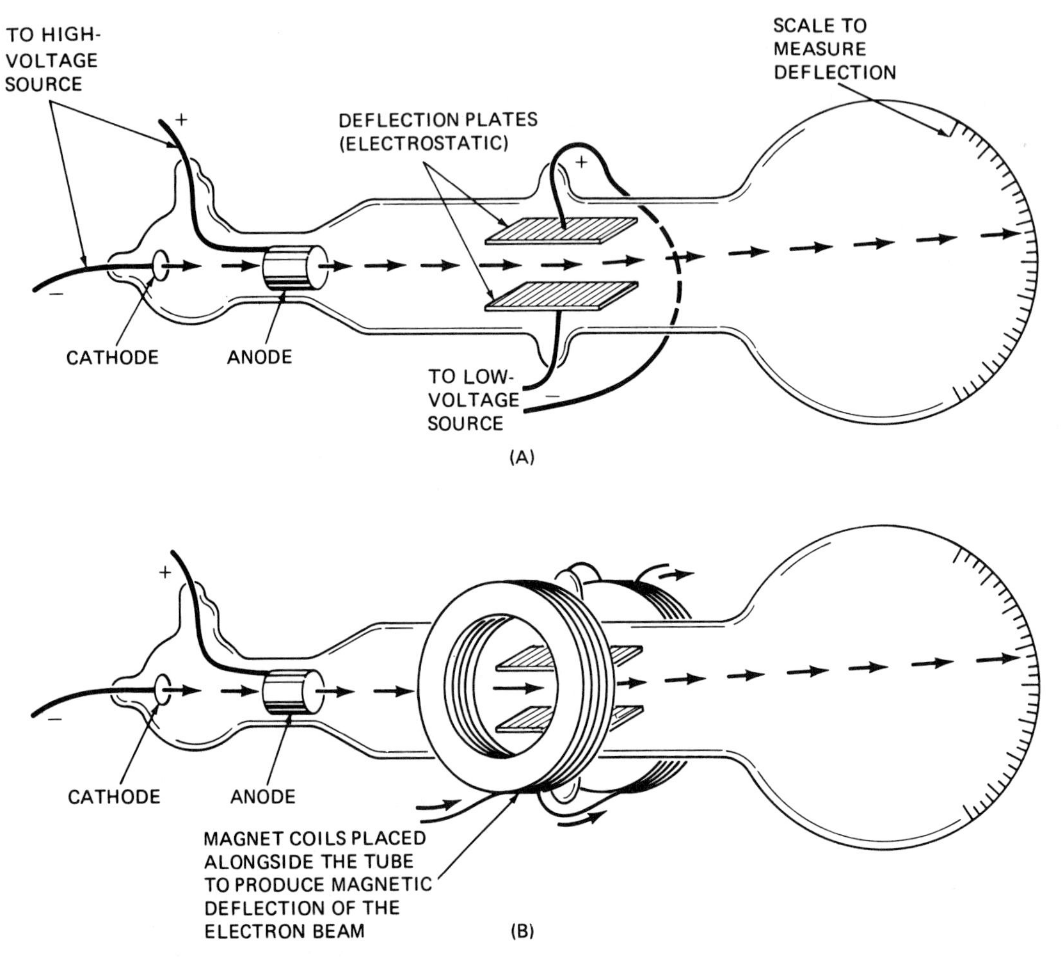

FIGURE 13–12

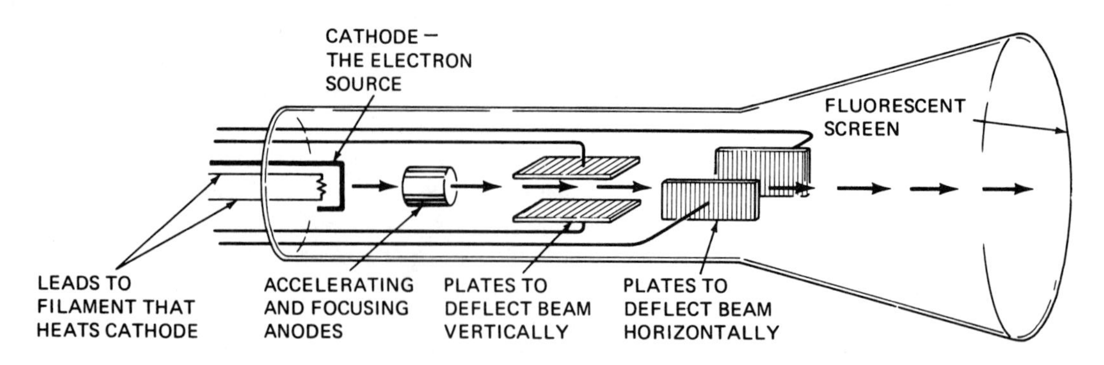

FIGURE 13–13 Oscilloscope cathode-ray tube

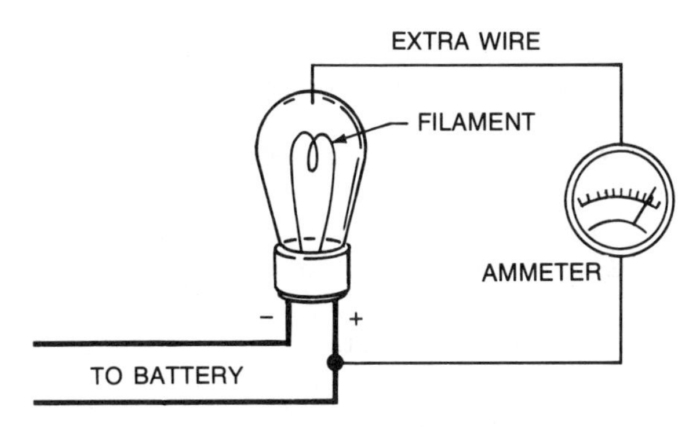

FIGURE 13–14 Edison's lamp

into one of Edison's experimental electric lamps. When the lamp filament was heated, a small current was indicated on the meter, Figure 13–14. The heated lamp filament was certainly emitting electrons. The free electrons in the lamp were attracted to the positive wire through the meter, thus giving rise to the current. This event was reported, but it was not until 20 years later that Fleming constructed a two-element valve to use as a rectifier and explained the action taking place in this device. The *elements* are the *filament,* which emits electrons, and the *plate,* which collects the electrons when it is positive. (*Valve* is the British term for vacuum tube.) In later years when Edison was asked why the third wire was sealed into his lamp bulb, he is reported to have said, "I have forgotten."

Fleming's concept of electron conduction in vacuum tubes has been applied to hundreds of devices designed to achieve specific tasks. Research conducted to improve glass-enclosed vacuum tubes led eventually to the discovery that special solid materials have the same characteristics as the simple vacuum tubes. As a result, the field of solid-state electronics was born.

SUMMARY

- Electrical conduction in liquids is the movement of positive and negative ions. There are no free electrons as in metals.
- Positive and negative ions are formed by a type of chemical combination in which one metallic element, or group, transfers electrons to a nonmetallic element or group.
- If a chemical compound can be dissolved in water, the charged ions of the compound become freely movable in the solution.
- In electroplating processes, the solution contains ions of the plating metal. The article to be plated is connected to the negative terminal of the current source. A bar of the plating metal is connected to the positive terminal.
- Metal removed from the plating solution and deposited on the object is replaced in the solution by metal dissolved from the positive bar.

- In liquid conduction, positive ions move in one direction and negative ions move in the other direction, resulting in a permanent separation of the parts of the compound. This decomposition process is called electrolysis.
- Gas is a good insulator until sufficient voltage is applied. The sudden ionization process changes the gas to a conductor.
- Ionization of a gas is the freeing of electrons from the gas molecules. The positively charged gas particles are the positive ions. The electrons are the movable negative particles.
- Conduction in a gas consists mainly of electron movement. As electrons collide with gas atoms, a new supply of electrons is continually set free from the atoms.
- Gases at low pressure conduct more readily than gases at high pressure.
- At pressures close to a vacuum, electrons from the cathode travel in straight paths. These electrons were called cathode rays. The discovery of electrons consisted of the measurement of their properties in cathode-ray tubes.

Achievement Review

1. How do metals differ from nonmetallic elements in the structure of their atoms?
2. What are ions?
3. Is dry salt a conductor?
4. Can a piece of wood be electroplated?
5. What happens if the wires leading to the generator in Figure 13–5 are reversed?
6. When a car battery is disconnected from the charging line by pulling the clip off the battery post, occasionally the top is blown off the battery. Why does this happen and how can it be avoided?
7. Water is regarded as an insulator. If water is an insulator, why is it that electrocution is possible by contact between a power line and wet earth?
8. Special problems in insulation had to be solved in the development of electrical control systems for high-altitude missiles. Why should any special problems exist?
9. Who discovered electrons?
10. What is the difference between *cathode rays* and *electrons?*
11. Name some useful examples of gaseous conduction.
12. Is it true that practically all of the useful discoveries in this field have already been made?

14
Batteries

Objectives

After studying this chapter, the student should be able to

- Define and explain the new technical terms introduced in this chapter

cell	battery
primary cell	fuel cell
secondary cell	hydrometer
specific gravity	polarization
amp-hour	

- Differentiate between the various kinds of commercially available cells
- Explain the advantages and shortcomings of various cells
- Implement sound maintenance practices for battery installations
- Install series and parallel cell combinations to meet specified voltage and current levels

14–1 CHEMICAL ENERGY: A SOURCE OF EMF

In the preceding chapter we learned about the ionization and conductivity of salt or acid solutions called electrolytes. If two electrodes of *different* metals are immersed into such electrolyte, a chemical reaction between these parts will produce a small emf. Such an arrangement of parts is known as a *cell*. Cells are the building blocks of batteries. In other words, batteries are combinations of two or more cells for the attainment of higher voltages or currents.

Cells are classified, according to the nature of their chemical activity, as being *primary* or *secondary*. A primary cell obtains its energy by consuming one of its electrodes and cannot be restored when its active materials have been depleted. It is discarded at the end of its useful life. The common flashlight battery is an example of a primary cell.

By contrast, a secondary cell may be repetitively recharged after it has run down. The reason for this is that the chemical process within the cell is reversible. The lead-acid cells within a car battery are secondary cells.

The voltage derived from a cell depends solely on the type of materials used in the construction of the cell.

Figure 14–1 lists a number of metals in the order of the ease with which electrons escape into a water solution. Metals at the top of the list lose electrons readily, and those at the bottom of the list lose electrons less readily. The *electromotive series* list is quite different from a list arranged according to conductivity.

Although it is theoretically possible to make batteries from many combinations of materials, there are actually only a few practical combinations. Many combinations promote undesirable chemical reactions that cause the active metal to corrode rapidly or to interfere with useful current production by building up resistance.

To gain some understanding of the process by which chemical interaction produces an emf, let us consider a simple, though impractical cell made by placing a strip of zinc and a strip of copper into a weak solution of water and hydrochloric acid, Figure 14–2. When a voltmeter is connected between the strips, an emf of approximately 1 volt is indicated.

From where does this energy come? The energy is due to the electrons of the zinc atoms and the hydrogen ions (H^+) of the acid. All acids consist of positively charged H^+ ions plus negative ions. The H^+ ions attract electrons. Each zinc atom has two electrons in its outer orbit that have sufficient energy to leave the zinc atom when they are subjected to the attraction of the H^+ ions.

The transfer of electrons from atoms of zinc to H^+ ions takes place when a piece of zinc is placed into an acid (whether or not another metal is present). When H^+ ions take electrons from the zinc, they become neutral H atoms. Once the atoms are no longer positively charged, they are not attracted to the negative ions in the water. The

Sodium
Calcium
Magnesium
Aluminum
Zinc
Chromium
Cadmium
Iron
Cobalt
Nickel
Tin
Lead
Hydrogen
Copper
Mercury
Silver
Platinum
Gold

FIGURE 14–1 Electromotive series of elements

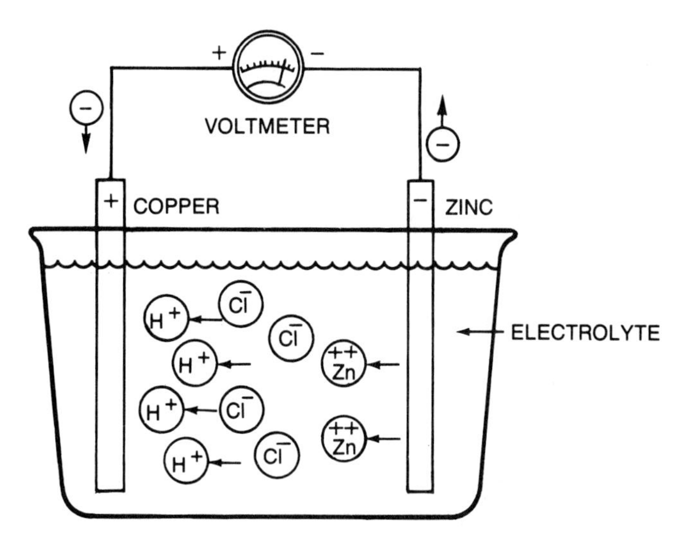

FIGURE 14–2 Single voltaic cell

H atoms then pair to form H_2 molecules. These molecules are ordinary hydrogen gas that is visible as it bubbles away. The circle labelled Zn^{++} in Figure 14–3 represents a zinc ion (a zinc atom that has lost two electrons). Zinc ions (Zn^{++}) are attracted from the metal into the solution by the negative ions of the acid.

Electron transfer at the surface of the zinc bar does not result in a useful electric current. To obtain a useful current, a wire is connected between the zinc strip and the copper strip in the acid. The copper in the acid provides another source of electrons that

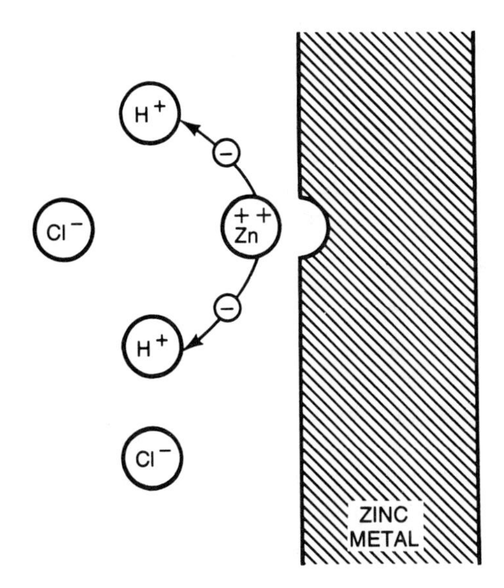

FIGURE 14–3 Zinc ion leaving metal

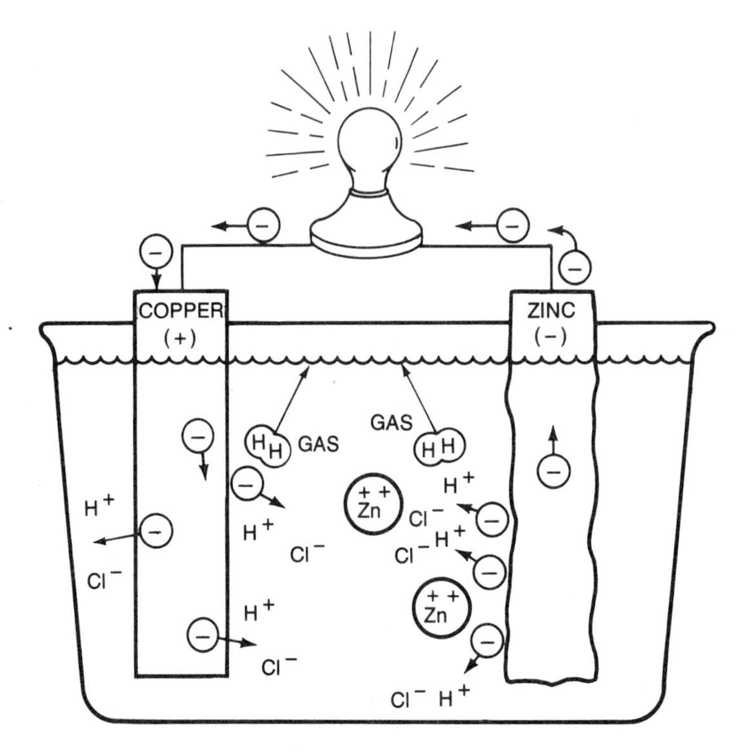

FIGURE 14–4 Using electric current from a simple cell

can be attracted by the H^+ ions, Figure 14–4. The electrons taken by the H^+ ions at the copper surface are simultaneously replaced by electrons at the wire connection supplied by the zinc. The copper serves as a conductor and carries electrons away from the zinc and to the H^+ ions.

Figure 14–4 shows a simple battery cell. However, this type of cell is impractical because of the rapid, noncurrent-producing transfer of electrons at the surface of the zinc. A small strip of zinc gives away all of its energetic electrons and is distributed in the solution as zinc ions after just a few minutes of operation.

The Carbon-Zinc Cell (Leclanché Cell)

Leclanché improved the basic battery cell by using a solution of water and ammonium chloride instead of hydrochloric acid. Ammonium chloride consists of NH_4^+ and Cl^- ions. The ammonium ions, NH_4^+, do not cause the zinc to be consumed as rapidly as the H^+ ions of the HCl solution. Instead of a copper strip, Leclanché used a carbon rod surrounded by packed manganese dioxide. Manganese dioxide (MnO_2) is a solid that does not dissolve in water. The manganese atoms in manganese dioxide have a strong positive charge and thus provide a strong attraction for electrons.

When a wire is connected between the zinc and the carbon rod, electrons flow from the zinc through the wire and carbon rod and are attracted to the positive manganese. At the same time, zinc ions are attracted into the ammonium chloride solution.

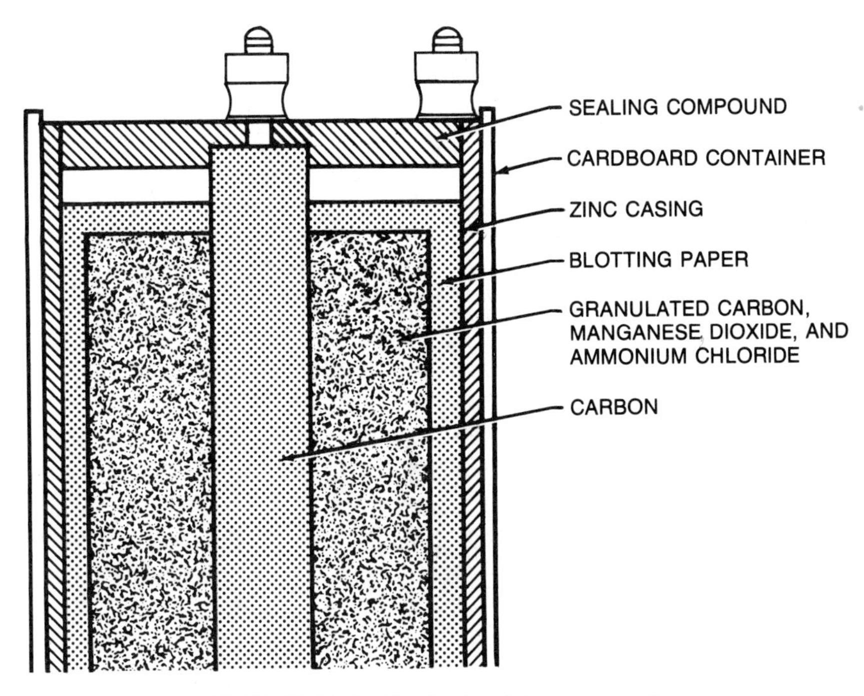

FIGURE 14–5 The Leclanché primary cell

The combination of the carbon powder mixed with the manganese dioxide and the carbon rod acts as an inert conductor to carry electrons to the manganese dioxide in the cell. (The Mn^{++++} ions gain two electrons each to become Mn^{++} ions. However, these ions do not have an attraction for electrons that is strong enough for the electrons to be of any further use in the cell.) The typical *dry cell* used in flashlights and portable radios is a sealed Leclanché cell, Figure 14–5. The inside of the cell is not dry but rather is a moist paste (when the cell is new). Zinc is often used as the outer container of a cell in which the container is the negative electrode. Dry cells cannot be recharged.

If a cell is to produce a large current in amperes, both the positive and negative terminals must have a large area of metal plate in contact with the electrolyte solution. This large area of contact makes it easy for electron transfers to occur. In other words, the resistance of the cell is low. A small penlight dry cell has just as much emf as the larger #6 dry cell, but the penlight cell cannot produce the same amount of current, because it has more internal resistance (this statement can be verified by Ohm's law).

14–2 ANODES AND CATHODES

The terminal by which electrons enter a device is called the *cathode;* the terminal where electrons leave the device is called the *anode*. The energy user in Figure 14–6 can be any device (a lamp; an electroplating bath) in which electrons are forced onto the cathode by some other energy source. Electrons in the device are repelled from the

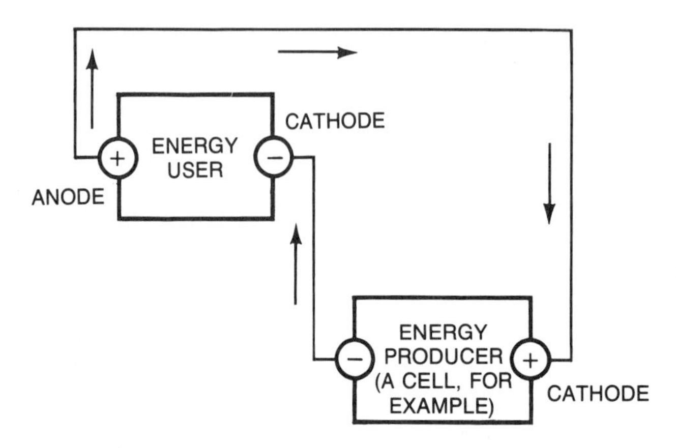

FIGURE 14–6 Relationship of anode and cathode when producing and using energy

cathode toward the anode, where they are attracted by an external energy source. The cathode of the energy user is negative (containing excess electrons) and the anode is positive. Electrons flow through the device from the cathode (negative) to the anode (positive).

For the energy producer in Figure 14–6, the anode is the metal that is rich in electrons (such as zinc). The anode supplies electrons to the external energy user. Electrons enter the cell at the cathode, since they are attracted to this location by a relatively positive electrode. The anode is negative; the cathode is positive. Thus, the energy in the cell pushes electrons out at the anode (−) and attracts them at the cathode (+).

14–3 PRIMARY CELLS

Each cell listed in Figure 14–7 is called a *primary cell*. In this type of cell, electron transfer is not readily reversible; that is, the dissolved zinc is not easily returned to its original metallic form. The dry cell cannot be reclaimed. A dry cell is the most commonly used primary cell. (Cells that can be recharged, such as those used in car batteries, are called *secondary cells*.)

Alkaline dry cells provide a greater current than the standard dry cells of equal size. Alkaline cells also provide a moderate current for a much longer time and can provide more power at low temperatures. These cells are useful in portable photoflash equipment, motion picture cameras, robot model planes, and for other applications where a better performance is worth the extra cost. A type of alkaline dry cell is rechargeable.

A *mercury cell* has a still higher energy potential for its size and weight when compared with the cells covered previously. In addition, the cost of this cell is greater. A mercury cell will maintain its 1.35-volt emf steadily for a long time. The mercury cells are used in hand-held communication sets, hearing aids, and portable electronic equipment and appliances. This type of cell usually consists of a negative center terminal that is connected to a zinc cylinder or pellet. The positive case is connected to a

	Negative	Positive	Electrolyte Solution	Emf
Leclanché (dry cell)	Zinc	Carbon, manganese dioxide	Ammonium chloride	1.5
Alkaline (dry cell)	Zinc	Manganese dioxide	Potassium hydroxide	1.5
Mercury cell (R-M cell)	Zinc	Mercuric oxide	Potassium hydroxide & potassium zincate	1.35
Silver-zinc	Zinc	Silver oxide	Potassium hydroxide or sodium hydroxide	1.5
Zinc-air cell	Zinc	Oxygen	Potassium hydroxide	1.4
Edison-Lalande	Zinc	Copper oxide	Sodium hydroxide	0.8

FIGURE 14–7 Composition of primary cells

mixture of mercuric oxide and graphite. To avoid mercury contamination of soil and water, ordinary methods of trash disposal are not suitable for discarded mercury cells or other devices containing mercury.

The three types of dry cells just covered (Leclanché, alkaline, and mercury) are the most widely used dry cells. The *silver-zinc cell,* which is similar in construction to the mercury cell, maintains its emf throughout its life. Although the silver-zinc cell is expensive, it has a high ratio of energy to weight. This factor makes it useful in hearing aids, electric watches, and spacecraft. Some silver cells are rechargeable.

In the *zinc-air cell,* the negative terminal leads to a porous zinc anode, which is soaked with the electrolyte solution. The cathode is a thin conductive plastic arrangement. The cell has enough wet resistance to prevent slow loss of moisture, yet it is porous enough to permit the entry of oxygen from the air. Finely divided platinum (or a silver alloy) on the cathode promotes an oxygen and water reaction that removes electrons from the cathode to form OH^- ions. Electrons are supplied to the load circuit by zinc atoms as they ionize. Some zinc-air cells are mechanically rechargeable; that is, the oxidized zinc anode (consisting mainly of potassium zincate) is removed, the cell is refilled with water, and a new porous zinc plate containing potassium hydroxide is inserted in the cell. Portable military communications equipment commonly uses zinc-air cells.

Many other primary cell combinations have been investigated, including *magnesium-air, iron-air, lithium-nickel,* and a 2-volt magnesium cell with *magnesium bromide* as the electrolyte and a cathode made of manganese dioxide.

For special applications, there are primary cells that become active only when water or electrolyte is added. The *Lalande cell,* using zinc and copper or copper oxide, is activated by adding a solution of sodium hydroxide. Another type of cell activated by water uses a magnesium anode with a copper chloride or silver chloride cathode.

14–4 FUEL CELLS

All of the cells described to this point base their operation on the tendency of energetic electrons to transfer to a material that lacks electrons, such as a positively charged ion. During this process, the anode material itself is used up in the sense that it is converted to a useless low-energy form. The fuel cell also makes use of this same type of electron transfer but with the important difference that the material giving electrons and the material taking electrons are not contained within the cell. The solid electrodes of the cell are not consumed in the process.

The term *fuel* makes one think of something burning. During the burning (combustion) of any fuel, coal, oil, or gas, atoms of the fuel give electrons to oxygen in the air. The energy of the electrons immediately appears as heat. It is correct to think of the gradual consumption of the zinc in a flashlight battery as a slow oxidation (a low burning process). Research over a number of years has produced very efficient fuel cells. The purpose of such cells is to control the electrons of inexpensive fuels and make them perform useful work as they leave the fuel atoms to join atoms of oxygen or some other electron taker. Fuel cells require high-purity fuel and inexpensive but reliable catalytic surfaces on which the essential reactions take place. Fuel cells also require auxiliary equipment such as gas containers and pressure controls. Research is under way to develop efficient, compact fuel cells for use in electric automobiles and tractors and as a source of power for a residence.

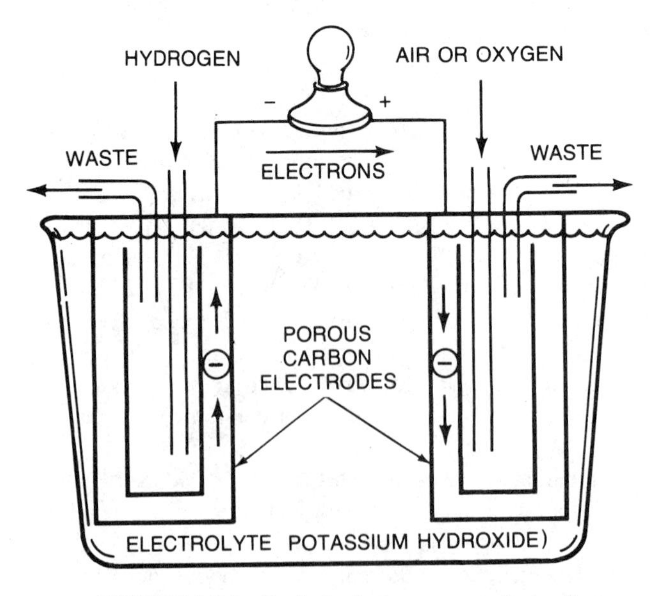

FIGURE 14–8 Basic hydrogen-oxygen fuel cell

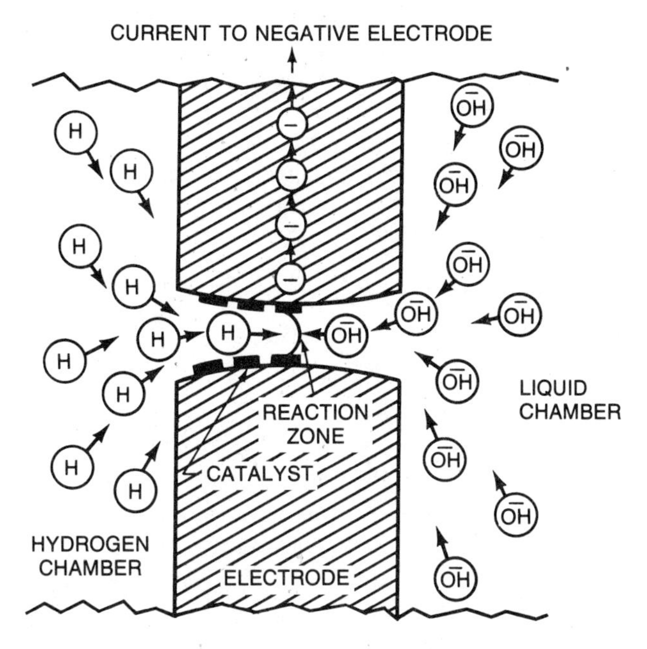

FIGURE 14–9 Reaction in pores of electrode between hydrogen and hydroxide produces electron flow.

The hydrogen-oxygen fuel cell, Figure 14–8, has hollow, porous carbon electrodes immersed in a potassium hydroxide solution. Hydrogen gas is pumped into one electrode and oxygen gas is pumped into the other. The porous carbon also contains certain metals or metal oxides called *catalysts*. (A *catalyst* promotes a chemical reaction.) In this case, the catalyst aids hydrogen molecules (which are pairs of hydrogen atoms) to separate into single atoms that can then combine with the negatively charged hydroxide ions (OH^-) of the electrolyte. The combination of H and OH^- forms a molecule of water, H_2O, with one electron left over. These surplus electrons are attracted to the electrode supplied by oxygen, Figure 14–9, where they take part in the reaction of oxygen + H_2O + electrons to form hydroxide ions, OH^-. Thus, hydroxide ions are reformed at the cathode at the same rate as they are used up at the anode. The important resultant change is the conversion of hydrogen and oxygen to water.

14–5 SECONDARY CELLS

In one family of cells, the electrochemical action is reversible; that is, if the discharged cell is connected to another electron source so that electrons can be put back into the negative terminal of the cell, the cell is recharged to its original emf value. Cells of this type are called *storage cells* or *secondary cells*. Figure 14–10 lists five types of secondary cells in order of increasing cost.

Type	Nominal Voltage	Anode	Cathode	Electrolyte	Advantages	Uses
Lead-acid	2.2 V	Lead	Lead dioxide	Dilute sulfuric acid	Cheapest	Autos, submarines, aircraft
Edison (nickel-iron)	1.4 V	Iron	Nickel oxides	Potassium hydroxide solution	Long life without careful maintenance	Electrical traction, auxiliaries, marine, lighting
Nickel-cadmium	1.2 V	Cadmium	Nickel hydroxide	Potassium hydroxide solution		Engine starting, controls, and communications
Silver-zinc	1.5 V	Zinc	Silver oxide	Potassium hydroxide solution	High energy-to-weight ratio	Military, satellites, communications
Silver-cadmium	1.1 V	Cadmium	Silver oxide	Potassium hydroxide solution		

FIGURE 14–10 Storage cells (secondary cells)

The Lead Storage Battery

Each lead cell produces 2 volts. Automobile batteries rated at 12 volts contain six cells connected in series, Figure 14–11.

The action of a storage cell is shown in Figure 14–12. The negative plate consists of metallic lead. When the cell is producing current, lead atoms on the surface of the plate lose two electrons each to become Pb^{++} ions. These Pb^{++} ions do not dissolve into the electrolyte but remain on the plate and attract SO_4^{--} ions from the sulfuric acid solution. As a result, an invisible thin layer of $PbSO_4$ is formed on the negative lead plate.

The positive plate consists of lead dioxide, PbO_2. Each lead particle in the plate lacks four electrons (these electrons were given to the oxygen when the compound PbO_2 was formed). Each Pb^{++++} ion takes two electrons from the external circuit to become a Pb^{++} ion.

The energy for the electron transfer results from the tendency of neutral lead atoms to give two electrons each to Pb^{++++} ions, so that both Pb^{++++} ions and Pb atoms become Pb^{++} ions.

When each Pb^{++++} ion of the lead dioxide takes up two electrons, the ion can no longer hold the oxygen. The oxygen, therefore, goes into the acid solution and combines with the hydrogen ions of the acid to form water molecules. The Pb^{++} re-

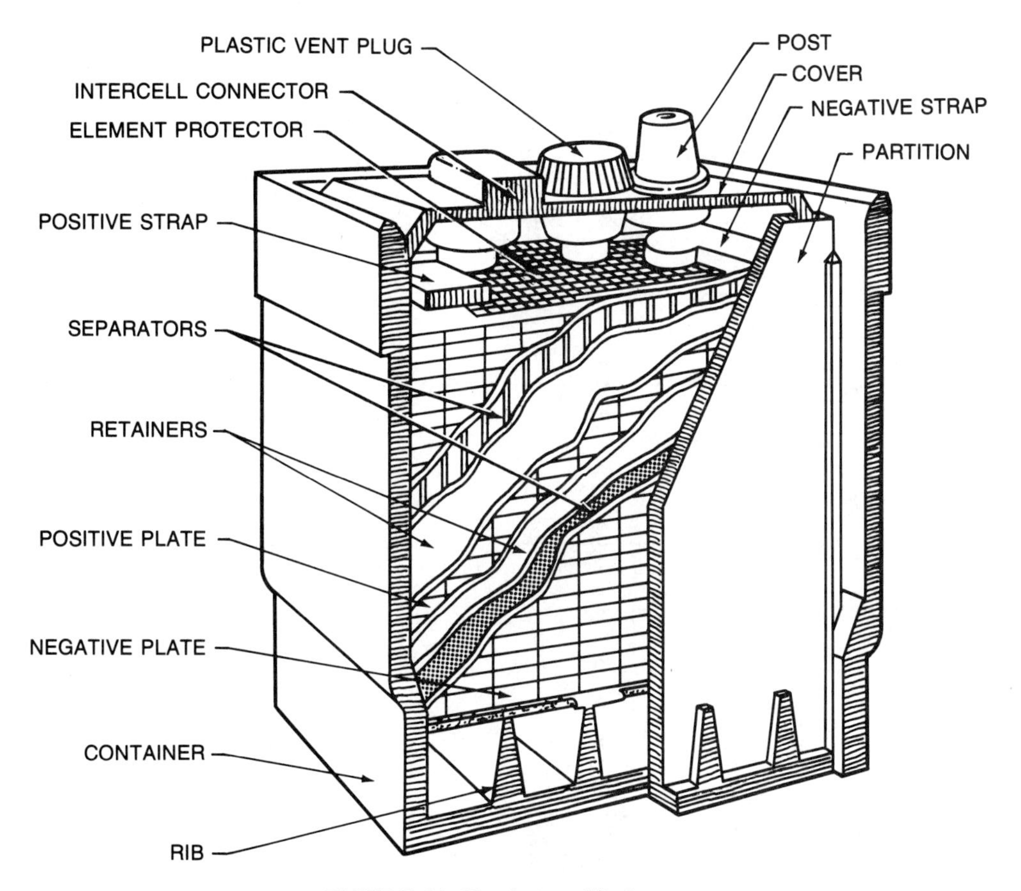

FIGURE 14–11 Automobile battery

mains on the plate and picks up SO_4^{--} from the sulfuric acid to form lead sulfate.

The chemical action in the lead storage battery can occur only where the plates are in contact with the sulfuric acid solution. If a large current is required, the plates are constructed so that the surface area in contact with the electrolyte solution is large. In a cell, the plates are arranged as shown in Figure 14–13. As indicated in Figure 14–12, the negative plate is made of lead sponge and the positive plate of porous lead dioxide. The porosity of these plates makes it possible for a large surface area of material to be wet by the electrolyte. Separators of wood, glass fibers, or similar porous material keep the plates from touching each other. To provide mechanical strength, both the negative and positive plates consist of an open framework of an alloy of lead-antimony. The active material is pressed into this framework. The electrolyte is dilute sulfuric acid having a specific gravity of 1.28.

Batteries can be shipped wet (filled with the electrolyte) or dry (with the electrolyte packaged in a separate container). The relative packing and shipping costs determine which shipment method is used.

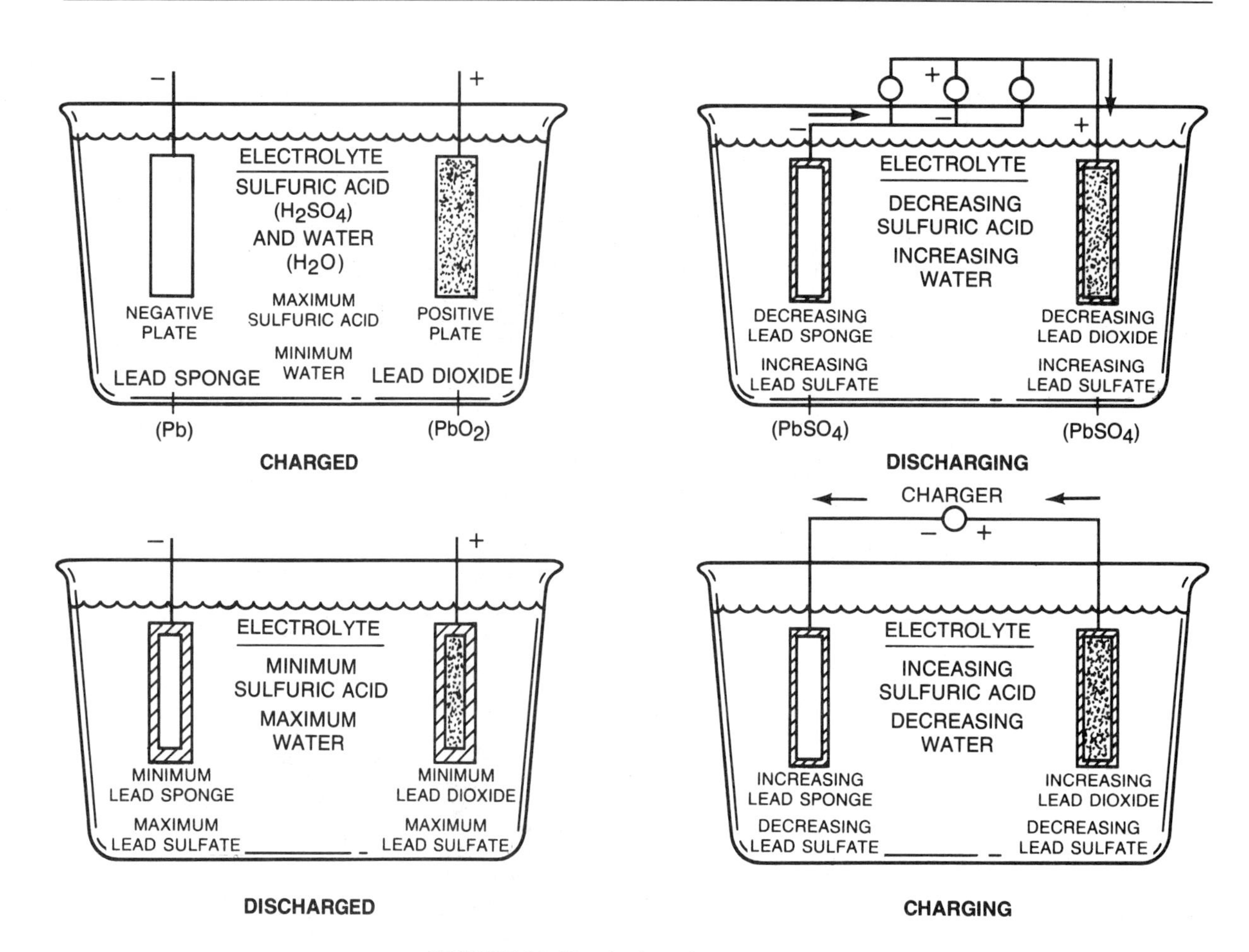

FIGURE 14–12 Action of a storage cell

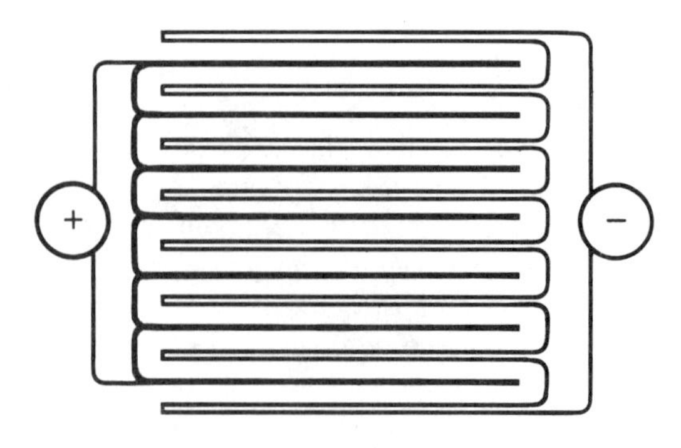

FIGURE 14–13 Top view of plate arrangement in a storage cell

The Ampere-Hour Rating

An ampere-hour is the amount of charge delivered by 1 ampere in 1 hour (1 ampere-hour = 3,600 coulombs). The ampere-hour rating of a battery is determined from its measured ability to produce current for 20 hours at 80°F; therefore, a battery that produces 6 amperes steadily for 20 hours at 80°F has a rating of 120 ampere-hours. If the discharge rate of the battery is 1 or 2 amperes instead of 6 amperes, a 120-ampere-hour battery can produce more than 120 ampere-hours. The battery cannot produce 120 amperes for 1 hour; the actual value of ampere-hours produced depends on the current.

Battery Charging

A rectifier or DC generator is required to charge a battery. The battery is charged by forcing a current through it in a direction opposite to that of normal battery operation. In the lead cell, this reversal of current reverses the chemical changes that take place in the cell when *it* furnishes energy. (Recall that in primary cells, the chemical changes cannot be reversed.)

The charging process in the lead cell is a series of chemical reactions. Recall that a current through a solution of sulfuric acid in water produces hydrogen at the negative plate and oxygen at the positive plate. If the plates are already covered with a thin layer of lead sulfate (from the discharging process), then the H^+ ions forced toward the negative plate combine with the SO_4^{--} ions to reform sulfuric acid. The electrons forced onto the negative plate by the generator combine with the Pb^{++} ions to form lead atoms, Figure 14–14A.

At the same time, water decomposes at the positive plate. The hydrogen set free by this process combines with the SO_4^{--} ions on the plate to form more sulfuric acid. The oxygen resulting from the breakdown of the water combines with the lead to form lead dioxide. The Pb^{++} ions of the discharged positive plate are converted to Pb^{++++} ions as the generator removes electrons, Figure 14–14B.

The following equation summarizes the charge/discharge process of the battery.

$$\text{(Negative)(Positive)(Liquid)} \xrightarrow[\text{discharging}]{} \text{(Negative)(Positive)(Liquid)}$$

$$Pb + PbO_2 + 2H_2SO_4 \xleftarrow[\text{charging}]{} PbSO_4 + PbSO_4 + 2\,H_2O$$

In a charging circuit, such as the one shown in Figure 14–14, the generator voltage must (1) equalize and overcome the battery emf and (2) have a large enough value to produce a current through the resistance of the circuit.

A charging rate that is too high damages a battery by causing it to overheat. It can also cause gas bubble formation inside the spongy plate material, which forces active material to break away from the plate structure. The safest charging procedure is a 10-ampere rate, or less, requiring about 24 hours. A battery can be charged on a constant voltage circuit in six or eight hours, starting with a 30-ampere or 40-ampere rate, which tapers down as the battery charge builds up. With this method, however, the battery should be checked to see that it is not overheating. The accepted limit is usually 110°F.

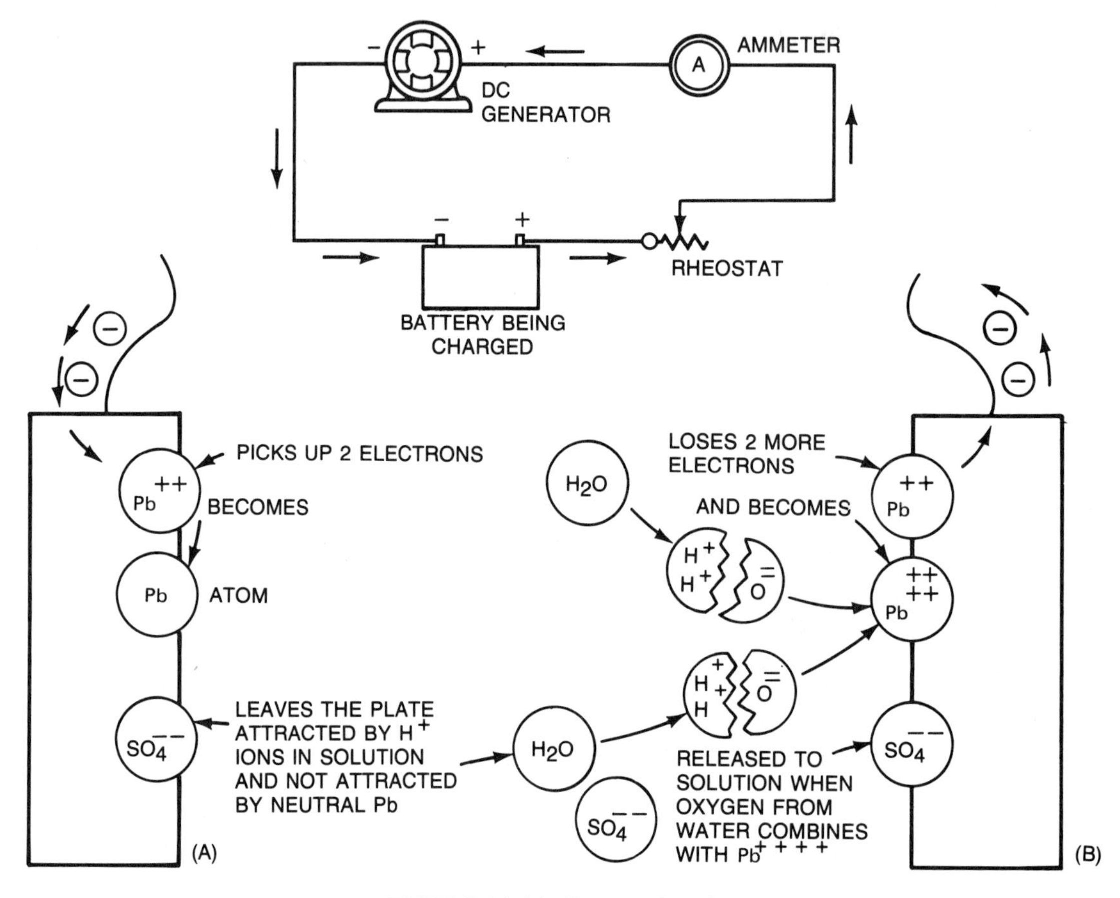

FIGURE 14–14 Battery charging

Battery Testing

A *hydrometer* measures the *specific gravity* of the electrolyte in each cell. A charged battery has enough sulfuric acid in the electrolyte so that its specific gravity is 1.25 to 1.28 (water = 1.00). As the battery discharges, SO_4 ions of the acid are tied up on the plates, and the electrolyte becomes more like plain water, its specific gravity approaching 1.1. Since occasionally a cell may not operate properly even when it has sufficient acid, a better method of battery testing measures the voltage of each cell when it is producing current. Voltage of a good cell will remain at 2 volts even though the cell is producing 20 or 30 amperes. Voltage of a dead cell will be 2 volts when the cell is not producing current but will drop off when it is producing current.

Hydrometer calibrations normally omit the decimal point. For example, the reading for a fully charged cell will be from 1250 to 1280, which is understood to mean the specific gravity is equal to 1.25 to 1.28. The indication for a completely discharged cell will be about 1100 (meaning specific gravity equals 1.1).

Battery Care

Particular care should be taken to avoid getting dirt into a cell. When the cap is removed, it should be set in a clean place, if it has to be set down at all. Dirt, especially a few flakes of iron rust, can spoil a cell permanently. Probably a great many automobile batteries go bad because of the accidental entry of dirt into a cell.

A battery should not be allowed to remain in a discharged condition. If a battery is completely run down, it should be charged within a few hours at a slow rate. If the discharged battery is allowed to stand discharged, the lead sulfate apparently hardens or crystallizes and is difficult to restore to lead and lead dioxide. Also, the watery solution in a discharged battery can freeze in winter, cracking the battery.

The liquid should be maintained at a level that covers the plates. Distilled water is preferable, but faucet water is better than none at all. (Melted frost from the refrigerator is distilled water.) Water is lost from a battery mainly by evaporation or slightly by hydrogen and oxygen forming during charging. Acid is lost only by cracking the case or tipping it over, so acid seldom is needed.

There is no point in adding various amazing powders and liquids to a battery to improve its performance, life, and complexion. For many, many years it has been known that Epsom salt ($MgSO_4$) or sodium sulfate (Na_2SO_4) can be added to a lead storage cell electrolyte in small amounts without damage to the battery but without doing it any good either. If acid has been lost from the cell, these materials provide ions that can be useful. But a teaspoonful of sulfuric acid (worth $0.02, like the Epsom salt) is more desirable in such a battery. When new and important discoveries appear, more reliable remedies will be announced in news articles and electrical engineering journals rather than in exclamatory and sensational advertising.

Nickel-Cadmium and Nickel-Iron (Edison) Storage Batteries

These truly long-life batteries are unaffected by the mechanical or electrical misuse that spoils a lead battery. But, they cost several times as much as lead batteries of similar size. Nickel and cadmium are much more costly than lead. In addition, expensive manufacturing processes are required for these batteries.

The *nickel-iron (Edison)* cell was originally developed in 1899 for electric motor-driven vehicles. Research was aimed at producing a battery having small weight and volume for a given ampere-hour rating. One that could withstand repeated cycling, (frequent charging and complete discharging) was especially desired.

The Edison cell is structurally stronger and lighter in weight than lead cells of the same current rating. The negative plates consist of a nickeled steel grid containing powdered iron, with some FeO and $Fe(OH)_2$. The iron is the source of the electrons, which are attracted through the external circuit toward nickel ions, Ni^{++} and Ni^{+++}, on the positive plate. The positive plates are nickel tubes containing a mixture of nickel oxides and hydroxides, with flakes of pure nickel for increased conductivity. The electrolyte solution is 21% KOH (potassium hydroxide, caustic potash), which is chemically a base rather than an acid. The Edison cell is therefore called an alkaline cell.

The Edison cell has two disadvantages: the initial cost is high, and the maximum current is limited by high resistance, especially when the cell is cold. It is not suitable for starting gasoline or diesel engines, because its internal resistance puts a limit on its current output. These disadvantages limit its use. However, the Edison cell has an advantage that makes it useful for specific purposes. It is not damaged by remaining in a discharged condition; therefore, it is useful in some portable lighting equipment and in a few marine installations, where it neither receives nor needs the attention that lead cells require. It is also appropriate for running traction equipment, such as mine locomotives and forklift trucks.

The *nickel-cadmium* (ni-cad) battery (Junger & Berg, Sweden, 1898) was developed not for frequent cycling but rather for general purposes. It enables the user to draw as many amperes as possible from a battery of given ampere-hour rating, without an excessive decline in voltage.

Both sets of plates are mechanically alike. The active materials are held in finely perforated, thin, flat steel pockets that are locked into a steel frame. The active material put in the positive plate is nickel hydroxide mixed with graphite to improve conductivity. Cadmium oxide is put into the negative plates. The electrolyte is potassium hydroxide (KOH) of about 1.2 specific gravity. When the cell is charged, the electrons forced onto the negative plate combine with Cd^{++} ions of the cadmium oxide (CdO), converting them to uncharged cadmium metal atoms. On the positive plate, electrons are removed from Ni^{++} ions of Ni $(OH)_2$, converting them into more strongly positvely charged Ni^{+++} ions. (The compound changes to Ni $(OH)_3$.) Therefore, the active materials in the charged cell are

On the Negative Plate	On the Positive Plate
Cadmium atoms, which provide electrons	Ni^{+++} ions in nickel hydroxide, which accept electrons

The chemical reaction may be written as

$$Cd + 2\,Ni(OH)_3 \underset{\text{charge}}{\overset{\text{discharge}}{\rightleftarrows}} CdO + 2\,Ni(OH)_2 + H_2O$$

The H_2O that is formed stays tied up in combination with the Ni $(OH)_2$ in solid form on the plate and does not dilute the electrolyte; therefore, the density of the electrolyte remains constant, and a hydrometer does not indicate the amount of charge.

The nickel-cadmium battery is similar in its chemical composition to the Edison cell. Both contain alkaline electrolytes. Its electrical characteristics are similar to those of the lead cell; therefore, its applications are competitive with those of the lead cell. Like the Edison cell, the nickel-cadmium battery is chiefly used in industrial service. Its ruggedness, long life, and low-maintenance cost outweigh the high original cost. (Nickel-cadmium batteries cost nearly twice as much as do lead acid batteries of comparable energy ratings.)

Nickel-cadmium cells are marketed in the popular sizes of zinc-carbon cells (sizes AA, A, C, and D) for replacement in toys and consumer appliances. The buyer should consider the following potential disadvantages:

1. For any given number of cells, the voltage output is only 80% that of dry cells.
2. Unused nickel-cadmium cells tend to lose a significant amount of their emf when standing idle for several days. This effect is aggravated by higher temperatures.
3. Ni-cad batteries tend to remember repeated demands for low output and sometimes persist in delivering such low levels, even though they are designed to meet the needs of increased output.

On the positive side, it should be mentioned that nickel-cadmium batteries are good for nearly 2,000 recharge cycles and can be left on a trickle charge for indefinite periods of time.

Commercially, the nickel-cadmium battery is used in railroad signals systems; fire alarm systems; relay and switchgear operation; missile controls; aircraft engines; and diesel engines in locomotives, buses, and oil well pumps. Figure 14–15 shows a large, industrial-type cell. Nickel-cadmium cells in smaller, sealed, cylindrical forms, with no problems of gas or spillage, are used in a great variety of communication equipment and portable appliances.

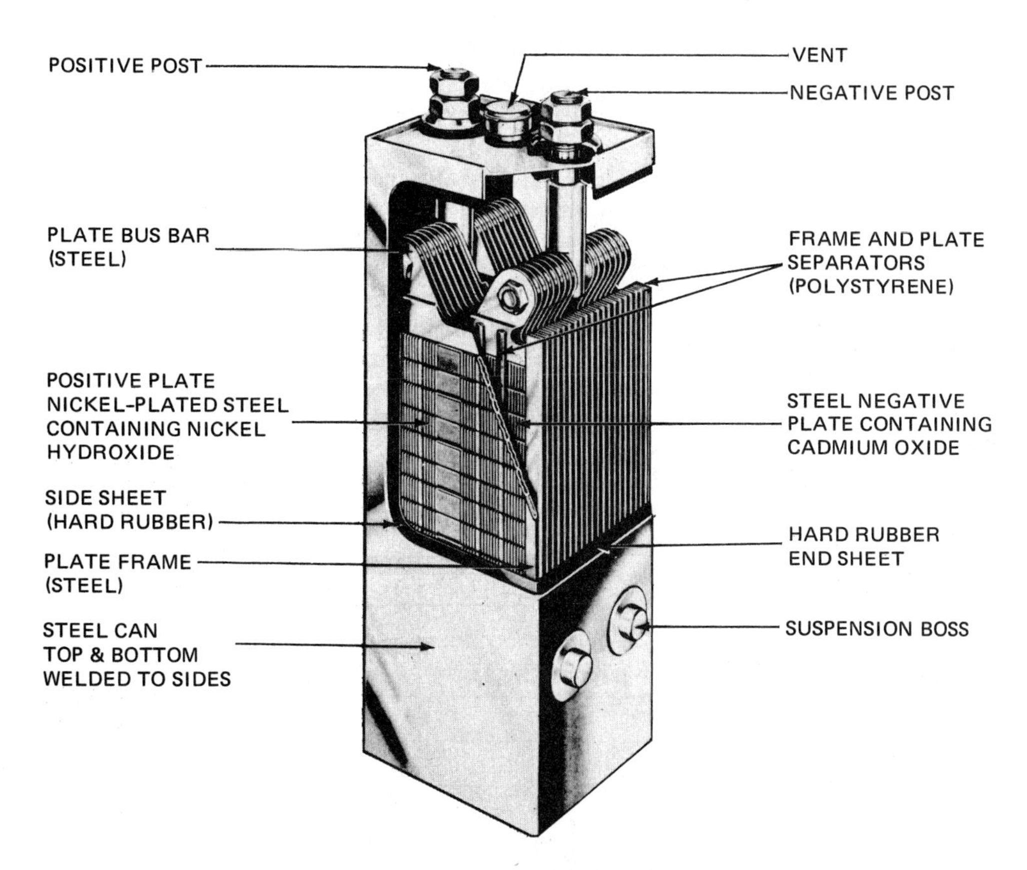

FIGURE 14–15 Nickel-cadmium battery

14–6 MISCELLANEOUS ASPECTS OF BATTERIES

Internal Resistance (R_i) of Cells

Cells themselves have an internal resistance that enters into a circuit calculation. When electrons flow from the negative plate to the positive plate, there is a movement of ions (current) in the electrolyte in the cell. This movement in the cell, like any current, does not occur with perfect ease; there is some resistance in the internal material of the cell.

This internal resistance of the cell (or battery) is generally shown in series with the load, Figure 14–16. This means that every time current flows in the circuit, a voltage drop (*IR* drop) will occur across the internal resistance. This voltage drop subtracts from the battery's emf and is lost to the load.

Let us assume that a 12-volt car battery has an internal resistance of 0.2 ohms and is connected to a load that draws 5 amperes. The internal voltage drop of the battery will be $= I \times R = 5 \times 0.2 = 1$ volt. In other words, the load receives only 11 volts, Figure 14–17A.

If the current is doubled to 10 amps, the corresponding voltage drop will be 2 volts and consequently the load "sees" only 10 volts, Figure 14–17B. Likewise, if the load current would increase to 20 amperes, the load voltage would be only 8 volts, Figure 14–17C.

The internal resistance of a cell increases with age, which in turn diminishes the voltage delivered to the load. Such a run-down battery may still yield a good reading when checked with a voltmeter. This happens because high-quality voltmeters have a

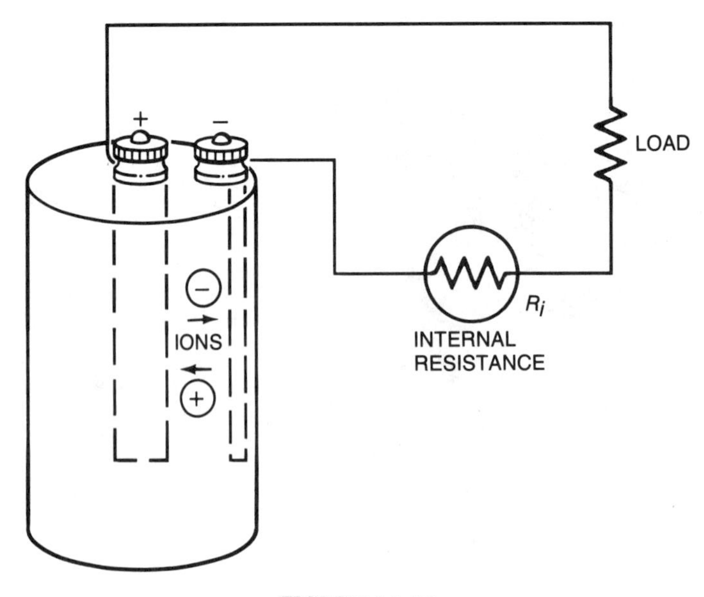

FIGURE 14–16

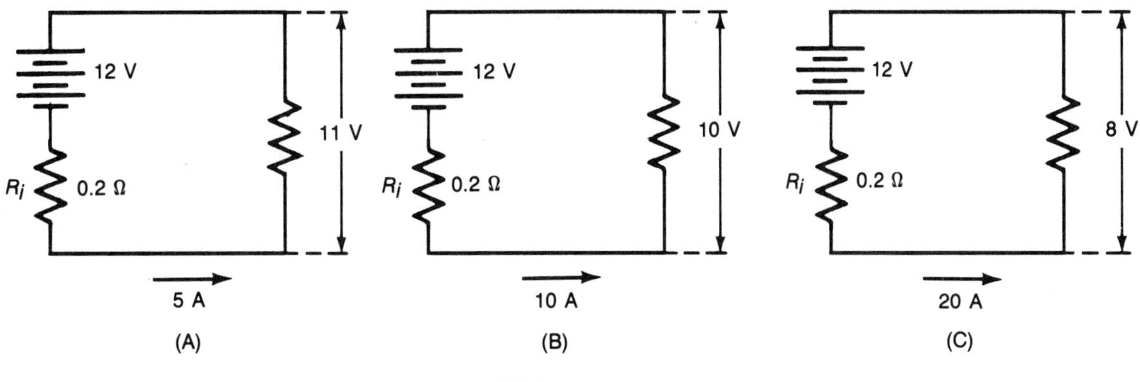

FIGURE 14–17

very high resistance and draw virtually no current from the source. In other words, checking a battery's condition with a voltmeter can be meaningless unless the test is performed when the battery is loaded down, Figure 14–18.

Maximum Current from a Cell

A cell produces its greatest current, uselessly, when it is short-circuited. Assume we connect a wire of practically 0 ohms resistance across the terminals of a 1.5-volt, 0.035-ohm dry cell. The amount of current is limited only by the internal resistance of the cell: $I = 1.5 \div 0.035 = 42.8$ amperes. The terminal voltage is now 0, because all of the cell emf is used inside the cell. If this condition exists for more than a few seconds, the cell overheats, gases form in it, the electrolyte starts boiling out the top of the cell, and the cell is destroyed.

If a dry cell is used to produce a moderate current for 10 or 15 minutes, there can be a noticeable drop in terminal voltage and current by the end of this time. This drop is caused by the temporary increase in internal resistance due to the formation of a very small amount of hydrogen around the positive plate. When the cell is allowed to stand on open-circuit, this hydrogen is reconverted to H_2O, and the cell is restored to its original low internal resistance.

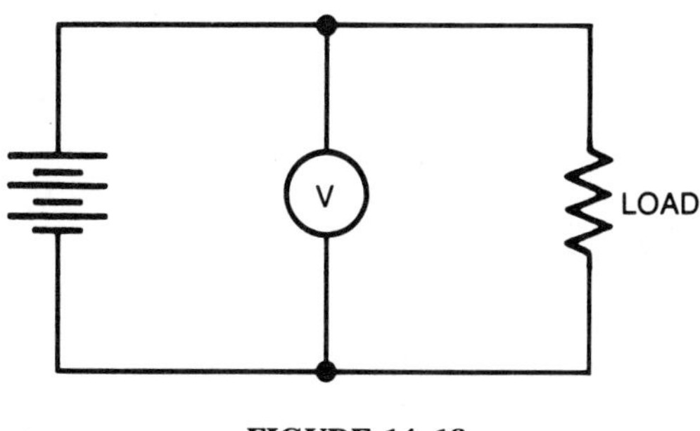

FIGURE 14–18

Dry cells generally fail because they really become internally dry. An unused dry cell on a warm shelf for two years may lose its moisture by evaporation, despite the manufacturer's attempt to seal the top. A cell in use loses its moisture when a hole is finally dissolved in the zinc case. Drying out causes a great increase in internal resistance. A cell with 1 or 2 ohms internal resistance is of no value. It may produce enough current to make a voltmeter read 1.5 volts but not enough current to light a flashlight bulb brightly.

Cells Connected in Series

The voltage output from *series-connected* power sources can be additive or subtractive, depending on the polarity of the power sources. This has been fully explained in Section 10–11. The students are encouraged to review this material, if necessary.

Cells Connected in Parallel

When multiple power sources are connected in parallel, there is no increase in emf. Why, then, would anybody want to connect batteries in parallel?

1. The current capacity of such an arrangement is increased. In other words, three cells can deliver three times as much current as one cell can.
2. The lifespan of such a parallel arrangement is increased. This is an advantage in that it is more convenient to replace a group of three cells every 30 days than to replace one cell every 10 days.
3. In addition, the internal resistance of the parallel group is less than that of a single cell; therefore, the terminal voltage will be closer to the battery emf.

When batteries or other power sources are connected in parallel, it is important to observe the *proper polarity* (positive to positive and negative to negative) and to assure that all batteries have *equal* voltages.

Cells Connected Series-Parallel

If the current and voltage requirements of a load are higher than those of a single cell, a series-parallel arrangement of multiple cells is in order. An example will make this clear.

EXAMPLE 14–1

Given: A load rated at 6 volts and 2.4 amperes; available cells are rated at 1.5 volts and 800 milliamperes.

Find: A suitable arrangement of series-parallel connected cells to satisfy the load requirement.

SOLUTION

1. Connect four 1.5-volt cells in series to satisfy the voltage requirement of 6 volts. Such an arrangement of cells, known as a *series-connected bank,* Figure 14–19A, is capable of delivering no more current than one single cell; in this case, 0.8 ampere.

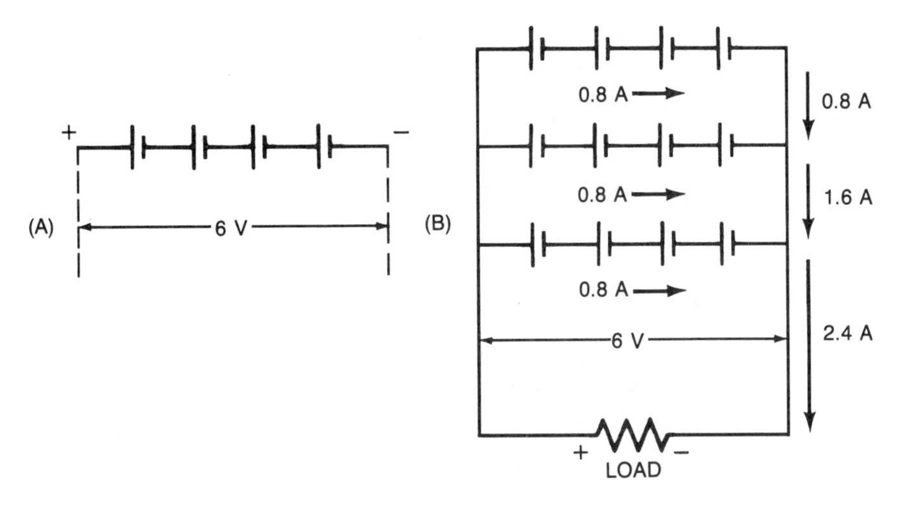

FIGURE 14–19

2. Connect additional banks of series-connected cells to increase the current capacity of the bank. In this case, 2.4 divided by 0.8 = three banks, Figure 14–19B.

Note: The current capacity of a battery is equal to that of one cell multiplied by the number of parallel banks.

Advantages and Disadvantages of Batteries

Advantages of batteries are that they are portable, reliable, and self-contained. Submarines, automobiles, aircraft, and small boys with flashlights are four examples of devices that need to be independent of powerline operation. Powerline service can be interrupted by storms. In such situations, batteries can provide emergency lighting and communication service.

The great disadvantage of batteries is that the energy they produce is *expensive,* so batteries are used only where the convenience outweighs the cost. For example, how much energy is there in an automobile battery? A 6-volt car battery can produce 6 amperes for 20 hours (120 ampere-hours). Watt-hours of energy is volts × amperes × hours = 6 × 6 × 20 = 720 watt-hours, which is about 0.75 kilowatt-hour. Power corporations charge much less for 1 kilowatt-hour of energy. An ordinary flashlight cell (size D) contains a little over 0.5 ounce of zinc. According to the chemistry book, this amount of zinc is 1.5×10^{23} atoms. Each atom supplies 2 electrons, so if all of the zinc is dissolved usefully, 3×10^{23} electrons are produced. One coulomb is 6×10^{18} electrons; 3×10^{23} electrons is equal to 50,000 coulombs.

$$\frac{3 \times 10^{23}}{6 \times 10^{18}} = \frac{3 \times 10^{5}}{6} = \frac{300{,}000}{6} = 50{,}000$$

It takes 3,600 coulombs to equal an ampere-hour, so the flashlight cell, even if 100% efficient, can produce only 14 ampere-hours. Fourteen ampere-hours at 1.5 volts equals

21 watt-hours. Therefore, 50 flashlight cells would be needed to produce 1 kilowatt-hour of energy.

Similar calculations can be made to show even more drastic examples of the cost of energy provided by cells. For example, if one considers the price paid for those tiny silver-oxide cells used to power electronic watches and hearing aids, the cost of energy obtained from such cells might amount to more than $3,000 per kilowatt-hour.

Safety Precautions with Batteries

1. The hydrogen and oxygen released by a battery being charged form an explosive gas mixture that can be ignited by sparks or open flames.
2. If acid needs to be diluted with distilled water (for use as an electrolyte), *never pour water into acid*. Such action may cause splattering and cause serious acid burns. Instead, slowly pour the acid into the water.
3. When handling acids, wear safety goggles, rubber gloves, rubber aprons, or similar protective garb.

SUMMARY

- Batteries consist of cells that deliver an emf from chemical energy.
- Primary cells cannot be recharged when run down.
- Secondary cells are rechargeable.
- The emf obtainable from a cell depends on its chemical makeup.
- The current rating, or amp-hour capacity, of a cell depends on the size of its electrodes.
- Cells have an internal resistance that causes them to drop some of their emf. This *IR* drop is the difference between their open-circuit and closed-circuit voltage.
- Voltage readings from batteries should be taken under loaded condition.
- Cells connected in series can deliver more voltage.
- Cells connected in parallel can deliver more ampere-hours.

Achievement Review

1. Define the words *primary cell* and *secondary cell*.
2. What is meant by the word *electrolyte?*
3. Explain the term *specific gravity*.
4. What kind of instrument is used for testing the specific gravity?
5. Someone reports the specific gravity of a battery to be eleven hundred. What does that say about the battery?
6. What is meant by the term *ampere-hour?*
7. Name two common types of secondary cells.
8. A battery is generally considered to be a *source*. Can it ever be regarded as a *load?* Explain your answer.
9. Why do most battery chargers contain a rectifier?
10. Is it advisable to store batteries in a discharged condition? Why or why not?

11. What kind of substance is recommended to clean and neutralize corrosion on the terminals of a battery?
12. Why does a discharged battery freeze more easily than a fully charged battery?
13. Why should lighted matches never be used for illumination when inspecting the electrolyte of a battery?
14. Is the voltage output of a battery dependent on
    a. The size of the plates?
    b. The number of the plates?
    Explain.
15. Draw four cells connected
    a. For highest voltage output
    b. For maximum life expectancy and high current output
16. Two 24-volt batteries in series are being charged by a 60-volt generator. Each battery has 0.02 ohm internal resistance. Calculate how much additional resistance is needed in the circuit to limit the current to 6 amperes.
17. Make a drawing to show how seven dry cells (1.5 V each) can be connected to yield 6 volts.
18. Now show seven wet cells (2 V each) to yield 10 volts.
19. In the following schematics, you see different circuit arrangements of batteries composed of 1.5-volt dry cells.
    a. Check each one of the circuits to see if it is properly or incorrectly connected.
    b. Also, estimate the voltage output of each combination.

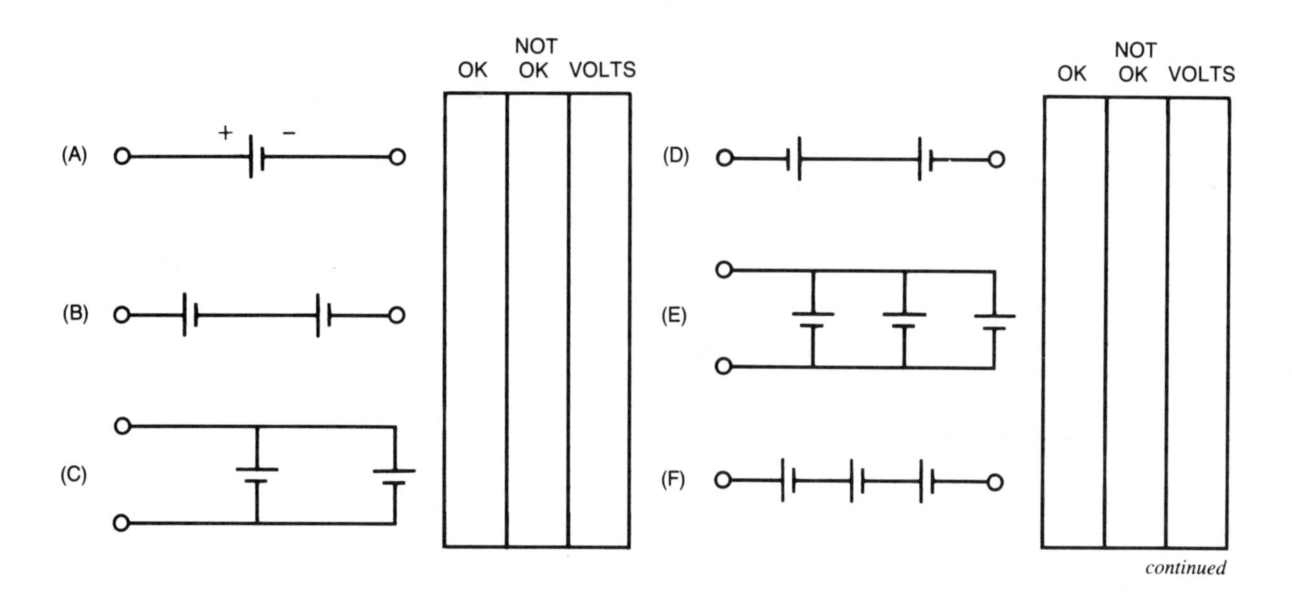

continued

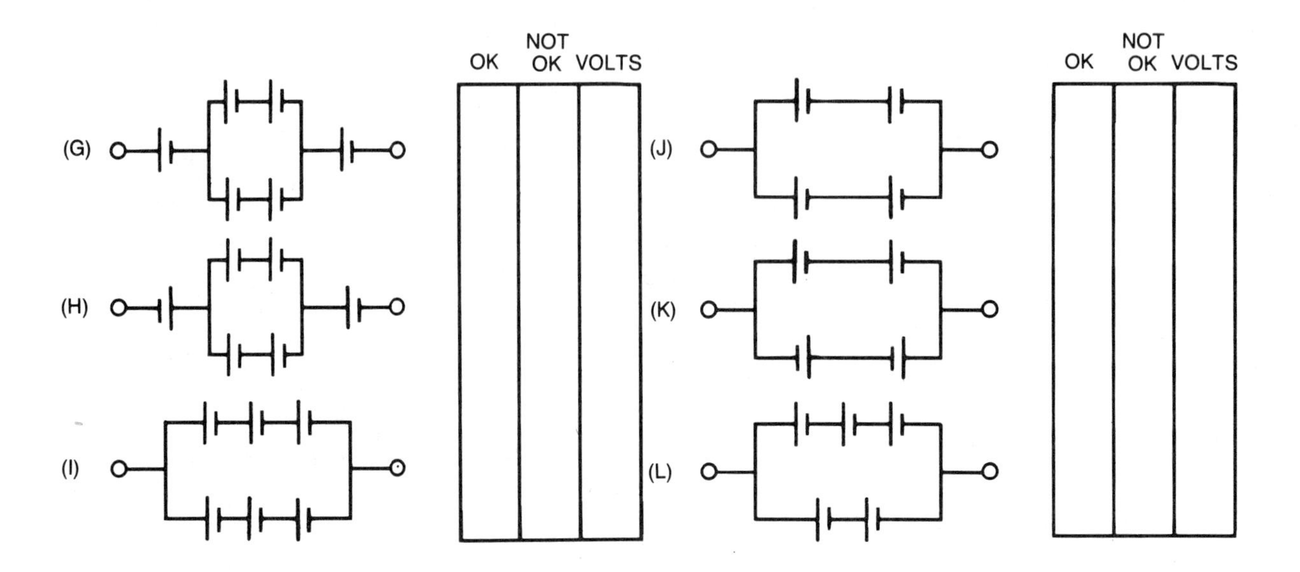

20. The sketches below represent various possible ways of connecting two dry cells and a lamp (as viewed from the top). The + and − are the terminals of the cells; the resistance represents the lamp. For each: State the voltage at the lamp (0, 1.5, or 3 V). Classify the circuits as *good* or *poor*.

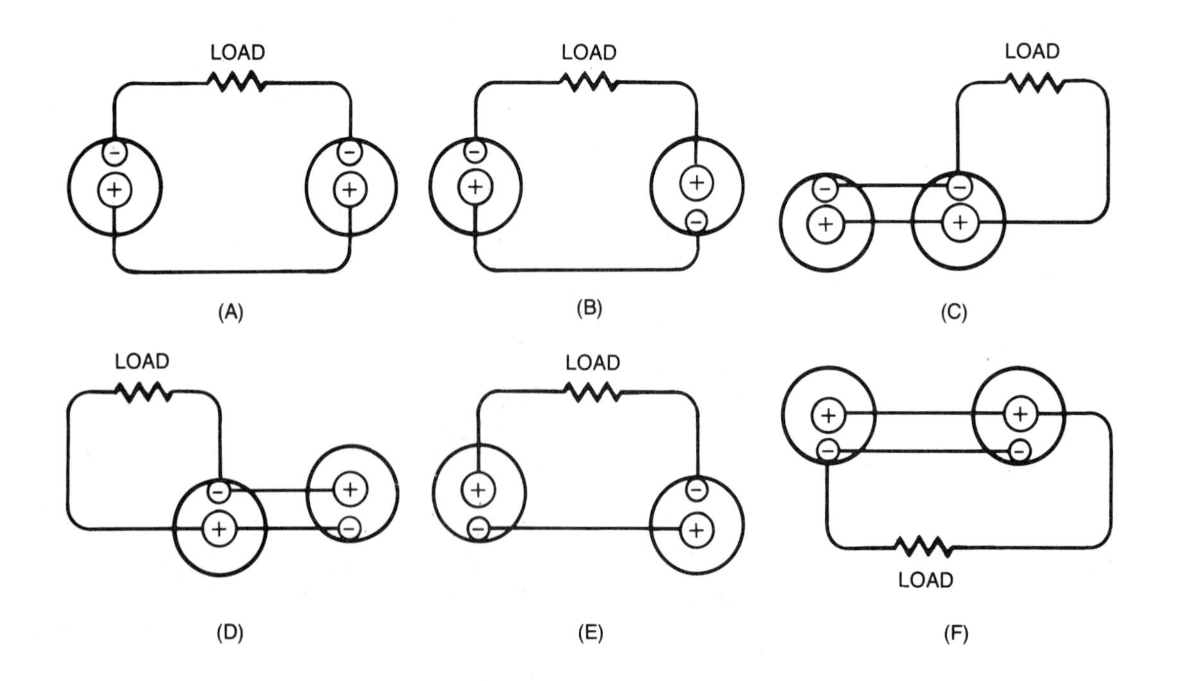

21. Eight lead storage cells are arranged as in the drawing below (top view). What is the emf of this battery?

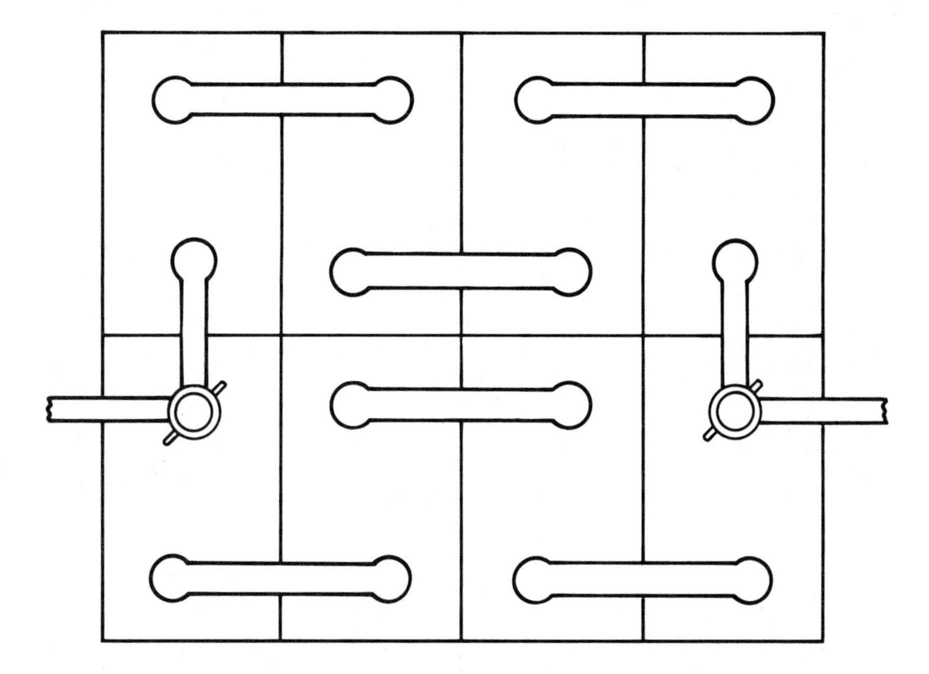

15
Magnetism and Electromagnetism

Objectives

After studying this chapter, the student should be able to

- Define and explain the new technical terms introduced in this chapter

magnetic domains	ferromagnetic
flux	paramagnetic
flux density	diamagnetic
sintering	alnico
coercive force	solenoid
powder metallurgy	magnetomotive force
saturation	residual magnetism
permeability	magnetizing intensity
maxwell	permalloy
tesla	weber
gilbert	gauss
retentivity	reluctance
hysteresis	eddy current
ampere-turns	Oersteds

- Correctly employ the graphic representation associated with electromagnetism
- Predict magnetic polarity and/or direction of current flow by use of left-hand rules
- State the law of magnetic attraction and repulsion
- Explain the factors determining the strength of an electromagnet

15–1 ELECTRICITY AND MAGNETISM

One of the most familiar and most frequently used effects of electric current is its ability to produce the force we call *magnetism*. This force is responsible for the operation of motors, generators, electrical measuring instruments, communication equipment, transformers, and a great variety of electrical control devices.

All magnetism is essentially electromagnetic in origin. Electromagnetism results from the energy of motion of electrons. In fact, every time a current flows through a wire, there are magnetic forces at work. Electric current and its associated magnetic forces are inseparable.

Because of this close relationship between electricity and magnetism, and due to certain similarities, some students tend to confuse one with the other. Some of these pitfalls will be pointed out to you as you progress with this chapter.

To begin our study of magnetism, we will investigate some of the earliest-known properties of magnetism and then explore how these properties can be explained by the action of electrons.

15–2 SIMPLE MAGNETS

A *magnet* is a piece of material that attracts a number of other materials such as iron, steel, nickel, cobalt, and a few minerals and alloys. Magnets do not attract copper, aluminum, wood, or paper. In fact, magnets have no effect on most substances. Magnetic attraction is quite unlike electrical attraction, which affects all materials.

The force of the magnet is strongest at two areas on the magnet called the *poles*. If a magnet is supported in the center by a string or is on a pivot, one of its poles turns toward the north and the other pole turns toward the south. Thus the end of the magnet pointing to the north is called the north pole, and the other pole is called the south pole. The needle of a compass is just a lightweight magnet (strip of magnetized steel) mounted on a pivot.

If a compass or magnet is brought near another compass or magnet, the north end of one compass repels the north end of the other compass and attracts its south end, Figure 15–1. Similarly, the south pole of one magnet (or compass) will repel the south pole of another magnet and attract the north pole. This effect is summarized in the

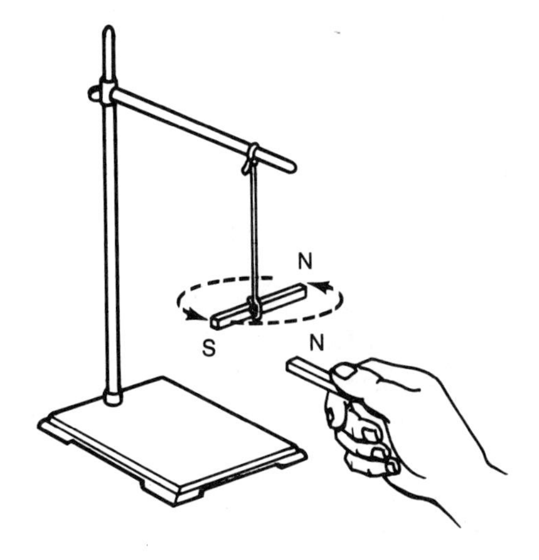

FIGURE 15–1 Like poles repel; unlike poles attract.

magnetic attraction and repulsion law: Like poles repel and unlike poles attract. (Even though this law appears to be similar to the electrical attraction and repulsion law, remember that magnets and electrical charges are different.)

The term *poles* means points where opposite properties exist, such as in the positive and negative poles of a battery or the north and south geographic poles of the earth. The poles of a magnet could have been assigned names other than north and south. In fact, it would be less confusing if the poles had been given a pair of opposite names such as Black and White, or Right and Left. The geographic poles of the earth are the ends of the axis on which the earth turns; they are *not* areas of magnetic attraction. The earth does have magnetic poles, however. There is a place in northern Canada that has the same kind of magnetic force as the south pole of a steel magnet; similarly, there is a place in the Antarctic that has the same kind of magnetic force as the north pole of a steel magnet.

15–3 THE MAGNETIC FIELD

You should recall from our discussion of electrical charges that the attraction and repulsion of electrical charges was explained by the existence of an invisible field of force between the charges. The pattern of an electrostatic field was shown in Section 3–8, Figure 3–13.

Similarly, the force existing in the space around a magnet is shown by the pattern resulting when iron filings are sprinkled on a card placed over a magnet, Figure 15–2. Compare the similarities of these phenomena, but keep in mind that we are dealing with two entirely different forces. Magnetism is not the same force as the attraction and repulsion forces caused by static electrical charges.

These lines of force, often referred to as *flux lines,* have specific characteristics attributed to them.

- Flux lines are directional. They are said to exit from the north pole and enter into the south pole, forming a closed loop through the magnet.
- Flux lines do not cross each other.

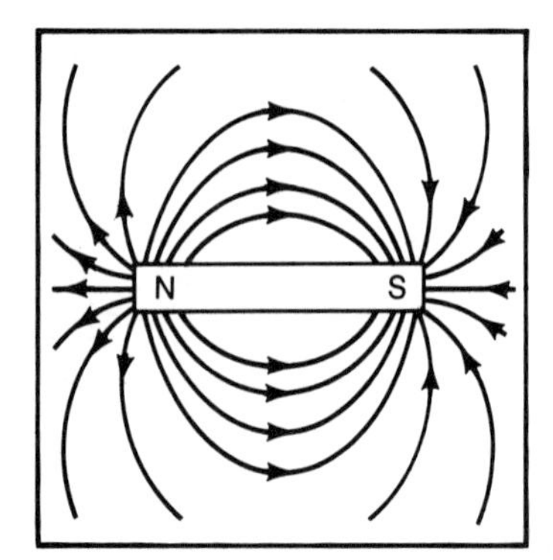

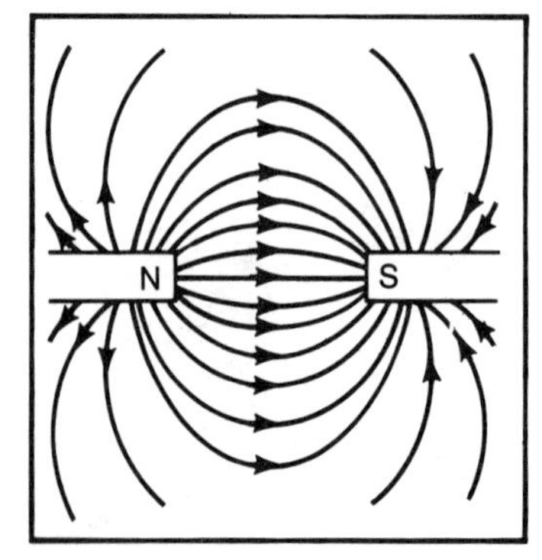

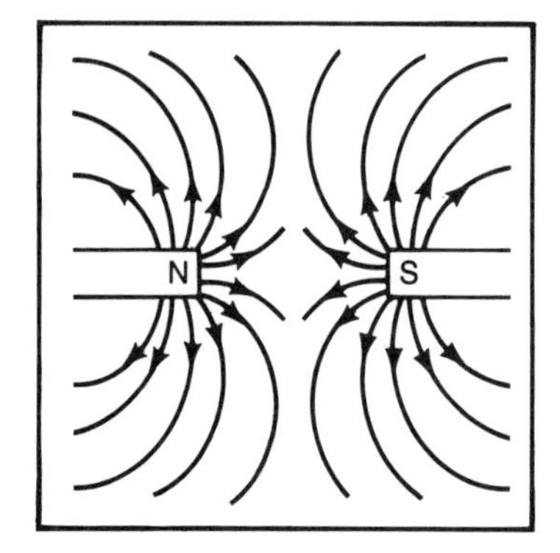

FIGURE 15–2 Magnetic lines of force

- Flux lines act like stretched rubber bands; they tend to contract.
- The *flux density,* or concentration of flux lines at a point, determines the amount of magnetic force. The greater the concentration of flux lines, the stronger the magnetic field. Flux lines are most densely concentrated at the poles.
- Flux lines facing the same direction attract each other, but flux lines facing opposite directions repel each other.
- The concentration of flux lines, and therefore the strength of a given magnet, is limited. When a magnet achieves maximum flux density, it is said to be *saturated.*

15–4 FERROMAGNETIC MATERIALS AND THE MAGNETIZING PROCESS

Iron, nickel, cobalt, and some oxides and alloys are called ferromagnetic materials. A magnet is a piece of ferromagnetic material that has magnetic poles developed in it by placing it inside a current-carrying coil of wire or by placing it near another magnet.

Early experimenters found that heating or hammering a magnet causes the magnet to lose some of its strength. Both of these processes disturb the atoms of the metal. Furthermore, it was found that if an ordinary steel bar magnet (or any magnet) is cut into fragments, each fragment has a north pole and a south pole, Figure 15–3. If we can continue to cut this material into smaller pieces, eventually we will reach the smallest possible fragment of iron. This fragment is an atom. Thus, scientists stated that all atoms of magnetic materials are themselves permanent magnets.

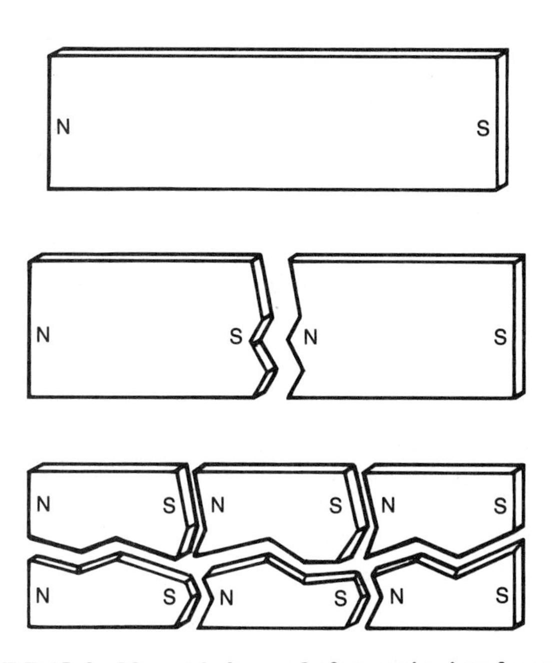

FIGURE 15–3 Magnet before and after cutting into fragments

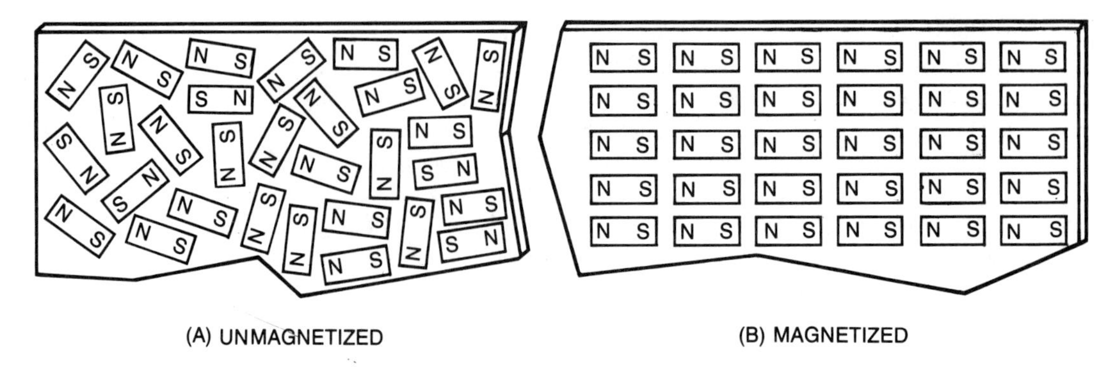

FIGURE 15–4

In an unmagnetized piece of iron, the atoms of iron are arranged in a disorganized fashion; that is, the north and south poles of these atom-sized permanent magnets point in all directions, Figure 15–4A. When the iron is magnetized, the atoms are rotated and aligned so that the north pole of each atom faces in the same direction, Figure 15–4B.

If a magnet is cut without disturbing the atom arrangement, the atomic south poles are exposed on one side of the break and the north poles are exposed on the other side. Before the magnet is cut, these poles exert their attractive forces on each other so that there is no force reaching out into space around them.

Some of the previous conclusions about magnets have changed slightly over the years because of the discovery of a degree of order in an unmagnetized piece of iron. Within a crystal grain of iron, several thousand atoms form a group called a *magnetic domain*. Within one domain, the atoms are lined up with the north poles all facing in one direction. This group of atoms acts like a minute permanent magnet.

15–5 MAGNETIC MATERIALS AND THE ATOMIC THEORY

Why do atoms of a magnetic material behave like iron magnets? The answer to this question is the result of a long series of complex scientific investigations of the behavior of electrons in atoms. All electrons are constantly spinning on their own axes within an atom. This spin is the reason that each electron is a tiny permanent magnet, Figure 15–5. In most atoms, electrons spinning in opposite directions form pairs. In other words, their north and south poles are so close together that their magnetic effects cancel out, as far as any distant effect is concerned. (Compare this situation with two permanent bar magnets placed together with their north and south poles adjacent to each other.)

An atom of iron contains 26 electrons. Twenty-two of these electrons are paired. Each electron of a pair spins in a direction opposite to that of the other electron so that the external magnetic effect is cancelled. In the next-to-the-outermost ring of electrons, four electrons are uncancelled. These four electrons, because they are spinning in the same direction, are responsible for the magnetic character of the atoms of iron.

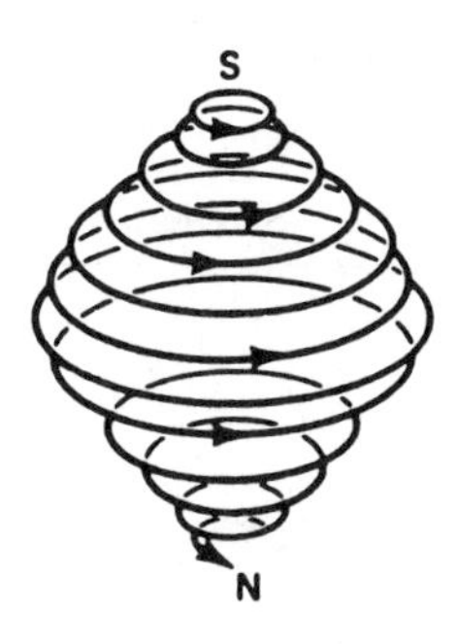

FIGURE 15–5 A spinning electron

There is still a great deal more to discover about electrons in atoms. Electron spin directions in an atom are affected by temperature and by the presence of other atoms. AT 1,420°F, iron loses its magnetism due to a rearrangement of electron spin patterns. Strongly magnetic alloys and compounds have been made from elements that are either weakly magnetic or not magnetic at all in their uncombined form.

15–6 PERMANENT MAGNETS

A rough classification divides magnets into two groups: permanent or temporary. A permanent magnet will keep its strength after the magnetizing force is removed; that is, the magnet maintains its orderly atomic arrangement. Before 1920, high-carbon tool steels (cobalt, molybdenum, and chrome-tungsten) were the only useful permanent magnet materials. These materials were and still are used in engine magnetos, telephone ringers and receivers, electrical measuring instruments, and compass needles. The discovery that weakly magnetic and nonmagnetic metals can be alloyed to make strong magnets promptly led to the commercial development of dozens of new magnetic materials that have better magnetic properties than the steel materials.

Alnico 5 is presently the most frequently used permanent magnet material. It can be found in loudspeakers, toys, and door latches. Alnico 5 is also used in magnetic separators; rotors for small generators; magnetic chucks and holding devices; and in motors used in aircraft, automobiles, computers, and small appliances. Permanent magnets are replacing small electromagnets in many simple applications.

In Figure 15–6, the term *coercive force* refers to the difficulty of magnetizing or demagnetizing a magnet. The coercive force of permanent-magnet steels is about 200. Alnico 5 makes a stronger magnetic field than do the magnets made of the steel materials and is about three times as difficult to demagnetize.

The properties of magnetic materials depend on their composition and are influenced by the methods used to manufacture and treat the materials. Alloys can be cast from molten metal or formed by sintering. (*Sintering* is the high-temperature heating of a compressed fine powder mixture.) Rolling and heat treatment cause changes in the grain structure and magnetic properties of alloys. The compounds barium ferrite and strontium ferrite are formed by sintering and are called ceramic magnets. These materials have the properties of stone rather than those of metal. Ferrites are useful in several

Composition of Several Permanent Magnet Materials	Approximate Relative Strength*	Coercive Force (Approx.)
Alnico 5 (8% A1, 13–14 1/2% Ni, 24% Co, 3% Cu, 50–52% Fe)	13	620
Alnico 8 (7% A1, 15% Ni, 35% Co, 4% Cu, 5% Ti, 34% Fe)	8.5	1,400–1,500
Platinum Cobalt (77% Pt, 23% Co)	6.5	1,500
Alnico 2 (10% A1, 19% Ni, 13% Co, 3% Cu, 55% Fe)	7–7.5	550
Ferrite Ceramics:		
$BaO \cdot 6Fe_2O_3$	2–4	1,500–2,000
$SrO \cdot 6Fe_2O_3$	3–4	2,500
*Number of lines per cm^2 (in thousands)		

FIGURE 15–6 Permanent magnet materials

different forms: They can be used in the bulk solid form, they can be powdered and mixed with plastic or rubber, or they can be mixed with liquid to make a product such as the magnetic printing ink used for the numbers at the bottom of bank checks. The brown material on magnetic recording tapes is a magnetic iron oxide.

The strength of a given permanent magnet is limited. When all of the atoms are facing in the same direction, the magnet achieves its maximum strength. This state is called *saturation*.

15–7 ELECTROMAGNETISM OF A STRAIGHT WIRE

The first experimentation leading to the conclusion that magnetism was in some way connected with electrical behavior occurred in 1819. Hans Oersted, a physics professor in Denmark, noted that a magnetic compass needle was affected by a wire that was connected to a battery.

Assume that a wire connected to a battery is inserted through the black dot in the center of Figure 15–7 and is held perpendicular to the paper. The wire is connected to a battery so that electrons come from behind the page toward the reader. Compasses placed on the paper near the wire will point as shown. The north ends of the compasses point in the direction shown by the clockwise arrows around the wire.

In other words, Oersted discovered that a current-carrying conductor produces a circular magnetic field about itself. The circular pattern becomes more evident when iron filings are sprinkled on a plane through which the wire is stuck, Figure 15–8.

It can be shown that the density of these circular flux lines is strongest near the wire and weakens with distance from the wire. More importantly, the direction of the circular flux depends on the direction of the current flow.

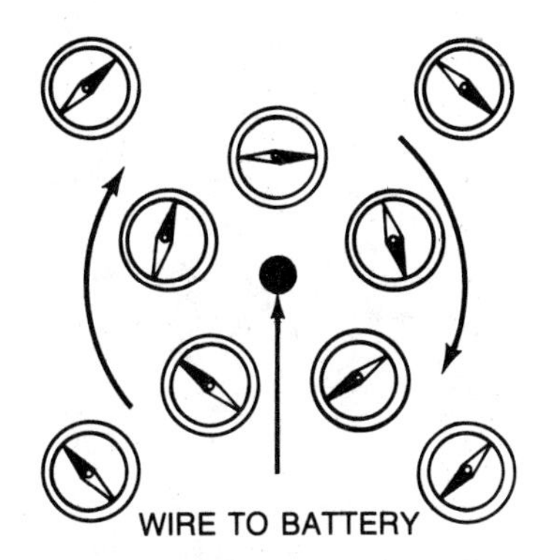

FIGURE 15–7

If the direction of electron flow is known, the direction of the magnetic field can be found as shown in Figure 15–8. If a wire is grasped with the left hand so that the thumb points in the direction of the electron current, the fingers will encircle the wire in the same direction as the magnetic lines of force. (The direction of the field is the direction in which the north pole of the compass is pointing.)

Conversely, if it is necessary to determine the direction of the electron current, the field direction can be found with a compass. Then, if the wire is grasped so that the fingers point around the wire in the direction indicated by the north pole of the compass, the thumb gives the direction of the electron flow.

The pattern of the magnetic field is shown in Figure 15–9. The dot in the center of the left-hand wire indicates that the arrow showing current direction is pointing toward the observer. The X in the right-hand wire indicates that the current arrow is pointing away from the observer.

If two current-carrying conductors are placed in close proximity to one another, a force of attraction or repulsion may be observed between the two wires.

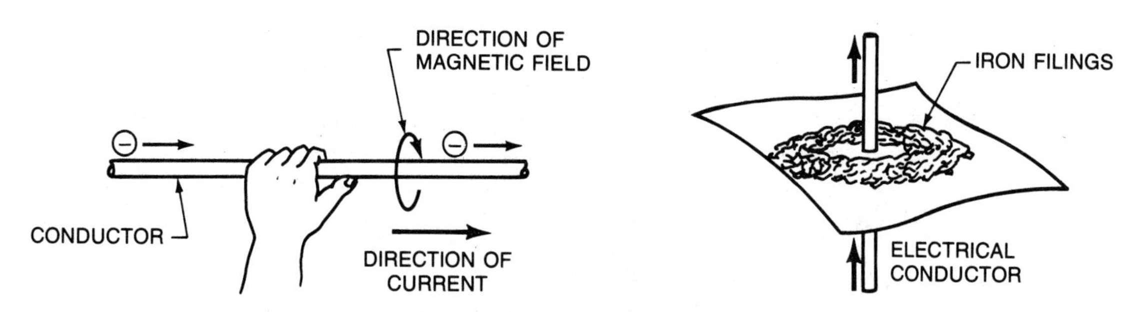

FIGURE 15–8 Left-hand rule for magnetic field around a wire

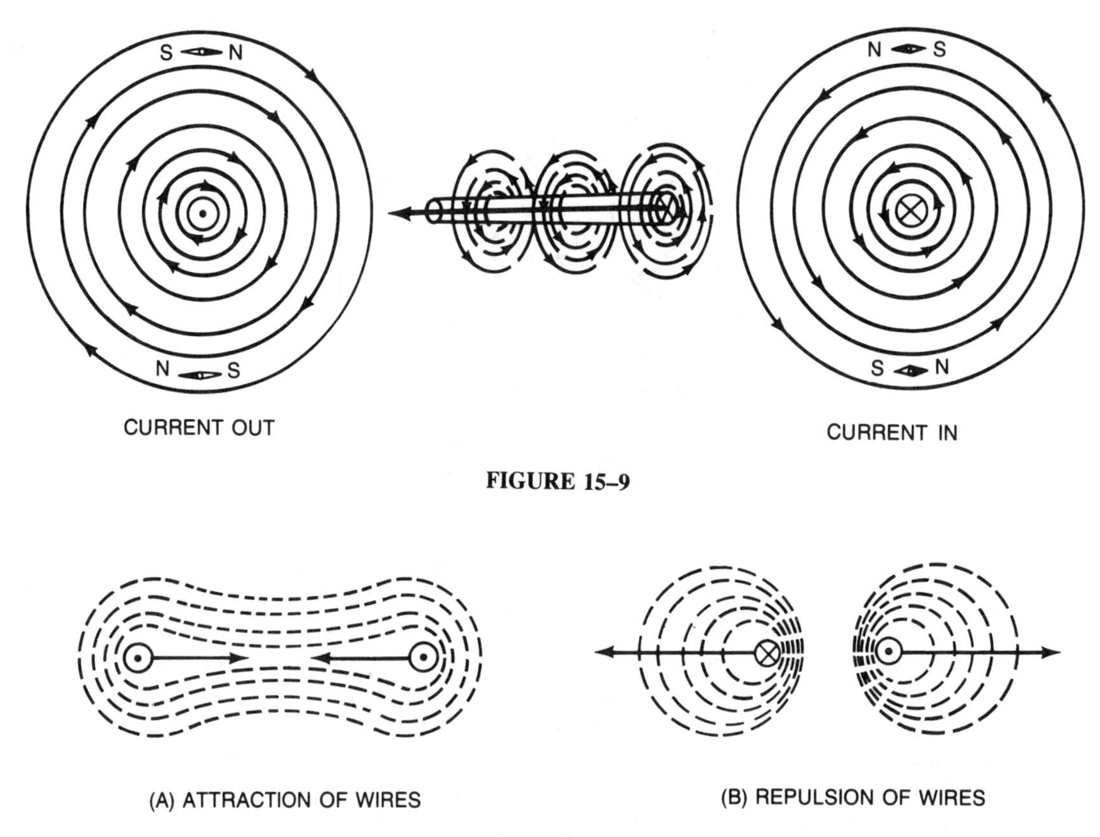

FIGURE 15–9

FIGURE 15–10

In 1822, Andre Ampere reported: "I observed that when I passed a current of electricity in both of these wires at once, they attracted each other when the two currents were in the same direction, and repelled each other when the currents were in opposite directions . . ."

Figure 15–10 illustrates this principle. Figure 15–10A shows how currents flowing in the same direction cause the magnetic fields about the two wires to join and reinforce each other. Since flux lines are said to contract like stretched rubber bands, the conductors will move toward one another.

Compare this with Figure 15–10B, where currents in opposite directions cause opposing magnetic fields. Recall from Section 15–3 that flux lines in opposite directions repel each other; thus, the two wires will be forced apart.

15–8 ELECTROMAGNETISM OF A COIL

When a wire is wound into the form of a coil, Figure 15–11, each turn of wire is surrounded by its own circular magnetic field. These little whirls of magnetic force combine to produce one large field that surrounds the entire coil.

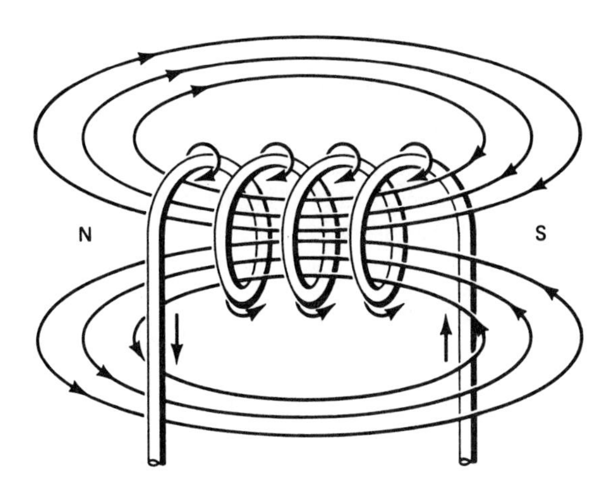

FIGURE 15–11 Magnetic polarity of a coil

The magnetic coil shown in Figure 15–12 is termed a *solenoid*. The figure indicates a way of remembering the relationship between the current direction and the field direction for a coil.

The ends of the coil are, in effect, magnetic poles (whether or not there is an iron core in the coil). Therefore, if the coil is grasped with the left hand so that the fingers point in the same direction as the electron current in the wires, the thumb points toward the north end of the coil. If the current direction is unknown but the field direction of the coil is known or can be found with a compass, the current direction can be found by the use of this left-hand coil rule.

Carefully compare the two drawings presented in Figure 15–12. Can you tell why the two electromagnets show different magnetic polarity? Note that the first coil is wound in a clockwise direction. The other coil is wound counterclockwise. Consequently, the left hand must be switched around so that the fingertips line up with the current.

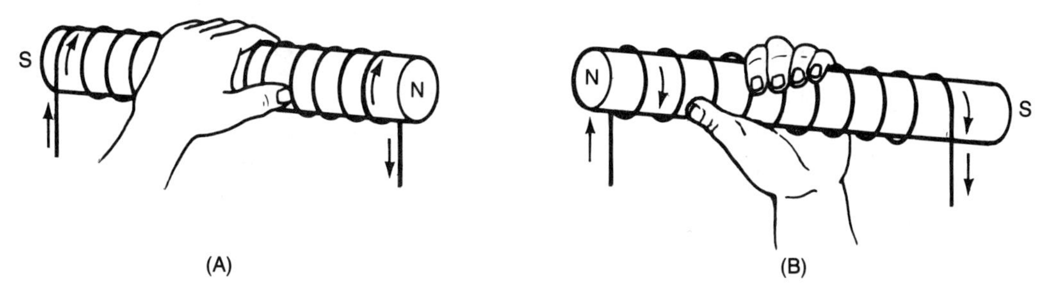

FIGURE 15–12 Left-hand rule for a coil

15–9 THE MAGNETIC CORE IN THE COIL

The illustrations of Figure 15–12 show the coils wrapped around a core. The material of such a core greatly determines the magnetizing ability of a solenoid. If the core is made of a ferromagnetic substance, the *magnetomotive force* (mmf) is greatly enhanced by the core's ability to concentrate the lines of flux. The magnetomotive force (mmf) of a coil is described and measured in either of two ways.

1. The magnetizing ability of a coil, or magnetic strength, is represented by a certain number of lines of force in each square inch of sectional area of the coil. The number of lines of force per square inch is called the *flux density*. *Flux* means the total number of lines of force.
2. The magnetizing ability can also be represented by the number of *ampere-turns* of the coil. This quantity is obtained by multiplying the current (in amperes) by the number of turns of wire in the coil. A current of 2 amperes in a coil of 20 turns provides the same magnetic effect as a current of 4 amperes in a coil of 10 turns, or 0.5 ampere in an 80-turn coil. A current of 2 amperes in a 100-turn coil has five times as much magnetizing force as 2 amperes in a 20-turn coil.

The presence of nonmagnetic materials in the DC magnet coil has no appreciable effect on its magnetic force. The insertion of magnetic material inside the coil results in a great increase in the total force. Assume that we have a long coil of wire with enough current in it to produce a magnetic field whose strength is indicated by 10 lines of force, Figure 15–13A. When a bar of magnetic material is inserted inside the coil, the material

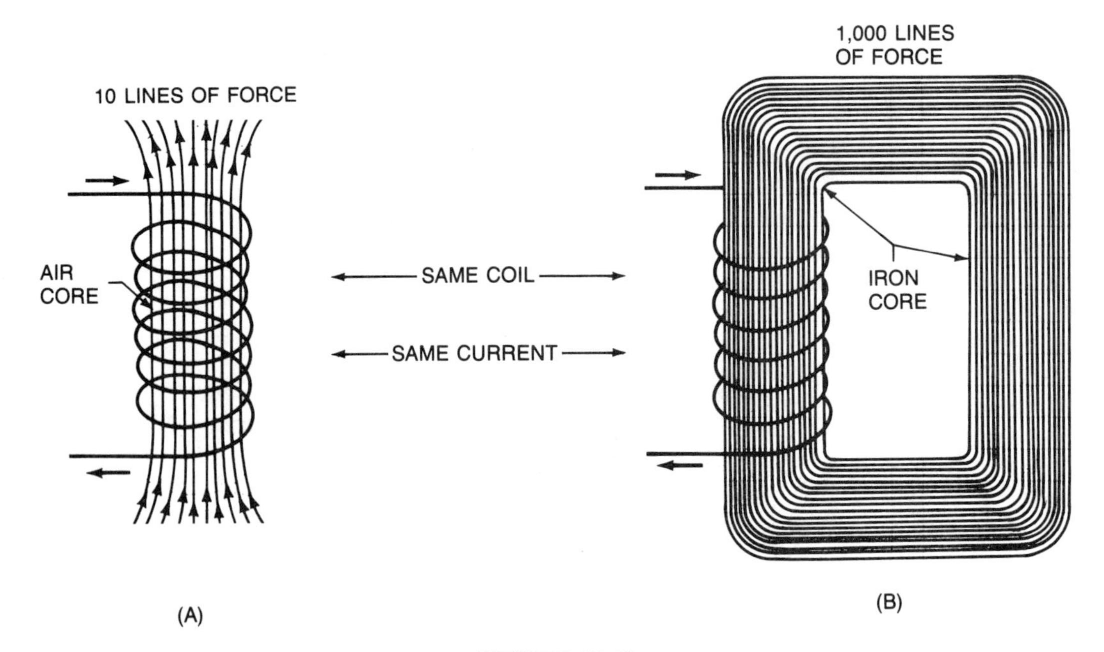

FIGURE 15–13

is magnetized and there are now 1,000 lines of force, Figure 15–13B. The insertion of the bar of material increases the magnetic field 100 times.

The ability of a magnetic material to increase the field strength is called *permeability*. In other words, *permeability* is the number of times that the flux density is increased by the addition of the magnetic material. The permeability of the iron core in Figure 15–13B is 100.

Magnetic flux density also increases when the current in the coil increases. The manufacturers of magnetic materials provide graphs showing this relationship for each of their products, Figure 15–14. The expression *magnetizing intensity* on the graph means the ampere-turns of the coil divided by the total length of the magnetic path, in inches. (Magnetic units are more fully explained in Section 15–10.)

Note that the increase of flux density has a limit at the saturation point (point S). Beyond point S the increase of current is wasteful, because no significant increase in magnetic flux can be achieved.

Two other properties of special interest can also be indicated on a graph of this type. The dashed line shows the magnetic behavior of the material as the current in the magnetizing coil is reduced. The height of point R above 0 represents *residual magnetism*. This is the amount of magnetism remaining in the core after the magnetizing force (the current in the coil) is removed. In good temporary magnets, residual magnetism is very low.

The relative strength of permanent magnet materials, Figure 15–6, is given in thousands of lines per square centimeter. For example, Alnico 8 has a residual magnetism of 8,500 lines per square centimeter (or 55,000 lines per square inch).

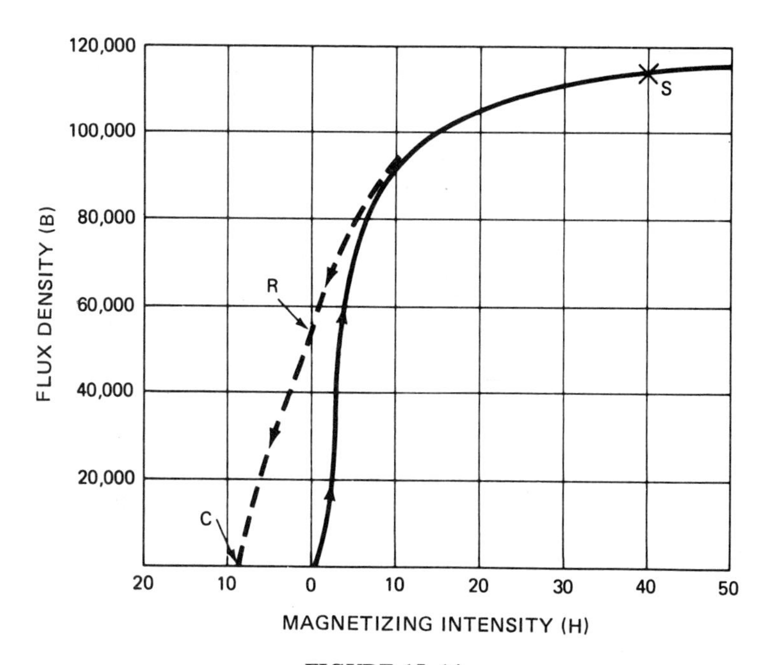

FIGURE 15–14

Returning to Figure 15–14, the distance on the horizontal scale from 0 to point C is a measure of *coercive force*. The measurement to the left on the scale indicates the amount of current in the reverse direction that must be put through the coil to remove the residual magnetism (to reduce the magnetism in the core to the zero level). If the coercive force is large, the magnet is difficult to demagnetize. A large coercive force is a desirable property for permanent magnets. However, the best materials for temporary magnets have a coercive force that is very close to zero.

The relationship between B and H, shown in the graph of Figure 15–14, can be expanded to explain a magnetic phenomenon known as *hysteresis*.

Let us assume that the current of the coil has been reduced from its point of saturation (point S in Figure 15–14) to its point of residual magnetism (point R) when the current is equal to 0 amperes. A subsequent reversal of current will not only remove the residual magnetism but reverse the magnetic field to saturation in its opposite direction, point S′ in Figure 15–15.

The graph shown in Figure 15–15 shows the kind of loop generated when alternating current (AC) is applied to the coil. This is known as a *hysteresis loop*. The size of its area under the curve is directly related to the energy losses suffered by the coil. Such losses, given off in the form of heat, are the consequence of the molecular friction caused by the continuous reversal of the magnetic domains under the influence of alternating current.

Solenoids are temporary magnets; therefore, solenoids require core materials that readily lose their magnetism when the current to the coil is removed. In a soft alloy such as silicon-iron (2% to 4% Si), the atoms slide and rotate easily, making the mag-

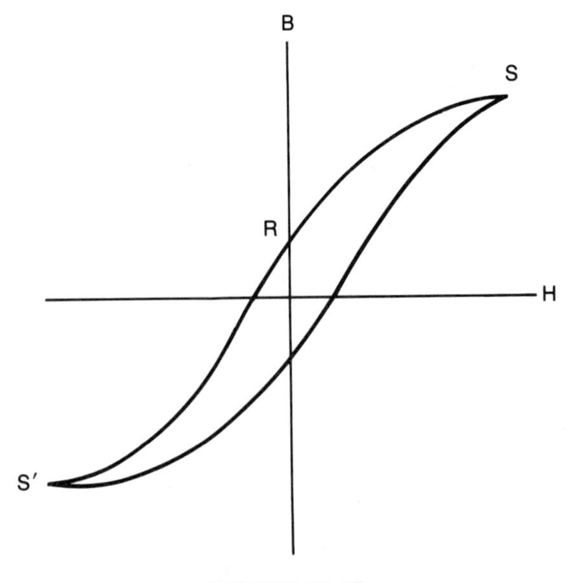

FIGURE 15–15

netic organization and disorganization of atoms very easy. Silicon-iron is widely used in the production of transformers, motors, generators, relays, and in other equipment.

Many alloys that are easier to magnetize and demagnetize than iron or silicon-iron have been developed. Some of the more useful are

Permalloy (78% nickel, 21% iron)
4–79 Permalloy (4% molybdenum, 79% nickel, 17% iron)
Mu-metal (75% nickel, 2% chromium, 5% copper, 18% iron)
Supermalloy (79% nickel, 5% molybdenum, 15% iron)
1040 Alloy (72% nickel, 14% copper, 3% molybdenum, 11% iron)

The specialized field of *powder metallurgy* (the process of forming alloys by mixing powdered metals and subjecting them to high pressure) produces many alloys that are useful for both temporary and permanent magnets. One alloy produced by the process of powder metallurgy, supermendur (2% vanadium, 49% iron, 49% cobalt), has the lowest coercive force of all the iron-cobalt alloys.

When iron oxide is mixed with oxides of manganese, cobalt, nickel, copper, or zinc and then is pressed and fired, the resulting temporary magnetic materials are called *ferrites*. These materials have good magnetic properties and high electrical resistance and thus are preferred for use in electronic equipment.

15–10 MAGNETIC QUANTITIES

A discussion of magnetic units can be somewhat confusing because three different systems of measurement have been commonly used in the analysis of magnetic circuits. Besides the English measuring system, we have to consider the metric system with two different measuring standards: the mks and the cgs system. These abbreviations simply refer to the basic units of length, mass, and time, where m stands for meter, c stands for centimeter (0.01 m), g stands for gram, k is for kilogram (1,000 grams), and the s stands for time in seconds.

This fact is merely mentioned in passing so that the student may appreciate the reason for the different names, symbols, and units of measurement.

It is the purpose of this text to convey the important concepts of magnetism without emphasizing the units and formulas necessary for a mathematical analysis of magnetic circuits.

Magnetic Flux

One line of magnetic flux is generally called a *maxwell*. One hundred million (10^8) lines of flux are called a *weber* (wb). The Greek letter ϕ (phi) denotes magnetic flux.

Flux Density

The number of flux lines per unit area is called the flux density, denoted by the letter B. This relationship is mathematically stated by the formula

$$B = \frac{\phi}{A}$$

The flux density is expressed either in teslas, gausses, or lines per square inch.

tesla = webers per m^2 (mks)
gauss = maxwell per cm^2 (cgs)

Magnetomotive Force (mmf)

The term mmf was explained in Section 15–9 as being equivalent to the product of the current (in amperes) and the number of turns.

mmf = ampere-turns (A-turns) in both the mks and the English systems
mmf is measured in gilberts (in the cgs system)
1 gilbert = 1.26 ampere-turns

Note: The student should realize that in addition to the number of ampere-turns, the flux density depends on the design of the core—for example, its length, cross-sectional area, and type of material (permeability).

Magnetic Intensity (H)

Magnetic intensity is defined as mmf per unit length. In other words, a given number of ampere-turns is more intense on a short core than on a long core. The letter H denotes the magnetizing intensity. The basic mathematical statement is

$$H = \frac{\text{mmf}}{l}$$

Magnetizing intensity can be expressed in ampere-turns per meter (mks system), ampere-turns per inch (English system), or gilberts per centimeter (cgs system). In the cgs system,

$$H = \frac{1.26 \text{ A-turns}}{\text{cm}} = \text{gilberts per cm}$$

Note: Gilberts per centimeter are called *Oersteds*.

Permeability (μ)

The word *permeability,* as has been mentioned earlier, refers to the ability of a substance to conduct and concentrate lines of magnetic flux. In this respect, permeability, denoted by the Greek letter μ (mu), is to magnetism what the word conductivity is to electricity.

Permeability is a pure number (without units) indicating how much better a material can establish magnetic flux within itself as compared to the permeability of a vacuum (or air).

For practical purposes, the permeability of nonmagnetic materials, such as wood, aluminum, and plastic, is the same as the permeability of a vacuum (or air). (By contrast, the ferromagnetic substances and alloys have permeability numbers ranging in the thousands.)

But, strictly speaking, even some of those nonmagnetic substances can become slightly magnetized under the influence of strong magnetic fields. Such substances are classified as being either *paramagnetic* or *diamagnetic*.

Paramagnetic substances have permeability ratings *slightly greater* than that of air. Aluminum is an example of a paramagnetic substance.

Diamagnetic substances have a permeability rating *slightly less* than that of air. Diamagnetic materials are characterized by the fact that their magnetization is in opposite direction to that of the external, magnetizing force. (Copper is an example.)

Reluctance

Reluctance is to magnetism what resistance is to electricity. In other words, *reluctance* opposes the passage of flux lines.

You should recall that the resistance of a conductor can be computed by the equation

$$R = \frac{Kl}{A}$$

Similarly,

$$\mathcal{R} = \frac{l}{\mu A}$$

This means, like resistance, reluctance varies directly with length and inversely with the cross-sectional area. The unit of reluctance is the rel (English) or the ampere-turn per weber (mks).

Ohm's Law for Magnetic Circuits

The similarities of electric circuits to that of magnetic circuits have been demonstrated. Comparing the two circuits in Figure 15–16, we acknowledge that each circuit has a force (emf and mmf) between two distinctly different poles (+ and −, as well as N and S). This force must overcome a unit of opposition (resistance or reluctance) to set up a closed path for current (or flux) with a specific direction (negative to positive or north to south). Thus, it is possible to write an Ohm's law equation for each circuit as shown in the drawing.

$$\text{Flux} = \frac{\text{Magnetomotive force}}{\text{Reluctance}}$$

$$\text{Ohm's law: Current} = \frac{\text{Electric force}}{\text{Resistance}}$$

The term *magnetic circuit* can be used to mean the path of the lines of force through a magnetic device, even though there is no motion along these lines. All magnetic devices have a magnetic circuit. Beginning at any point, the magnetic circuit can be traced by following the lines of force through the iron and the air and returning to the starting point.

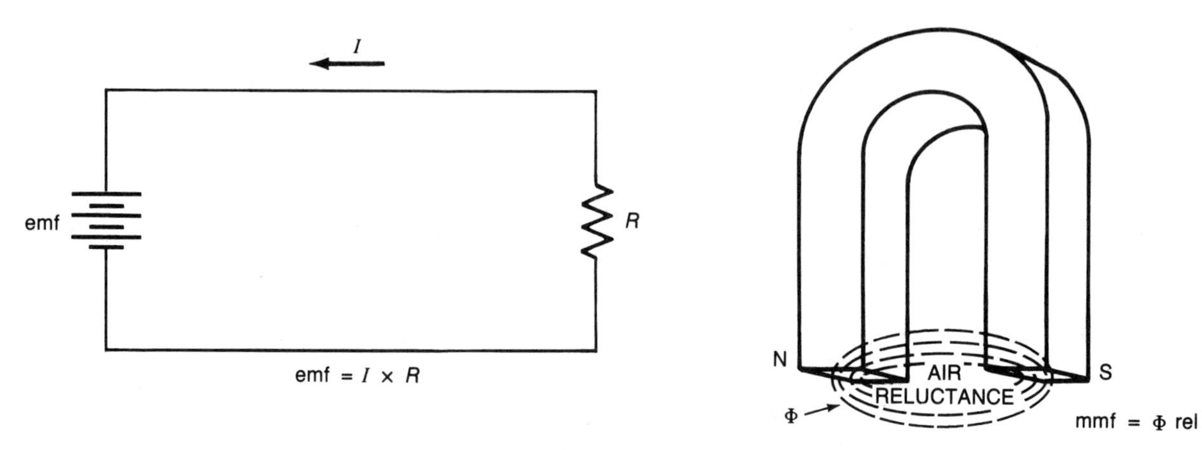

FIGURE 15–16

The student should recall that it is easy to magnetize iron and is difficult to produce lines of force in air; therefore, the path of the lines (magnetic circuit) through air or other nonmagnetic material in all magnetic devices should be as short as possible. Figure 15–17 shows two possible ways of supporting a pivoted iron bar so that it can be attracted to an electromagnet. If the number of ampere-turns is the same in each coil,

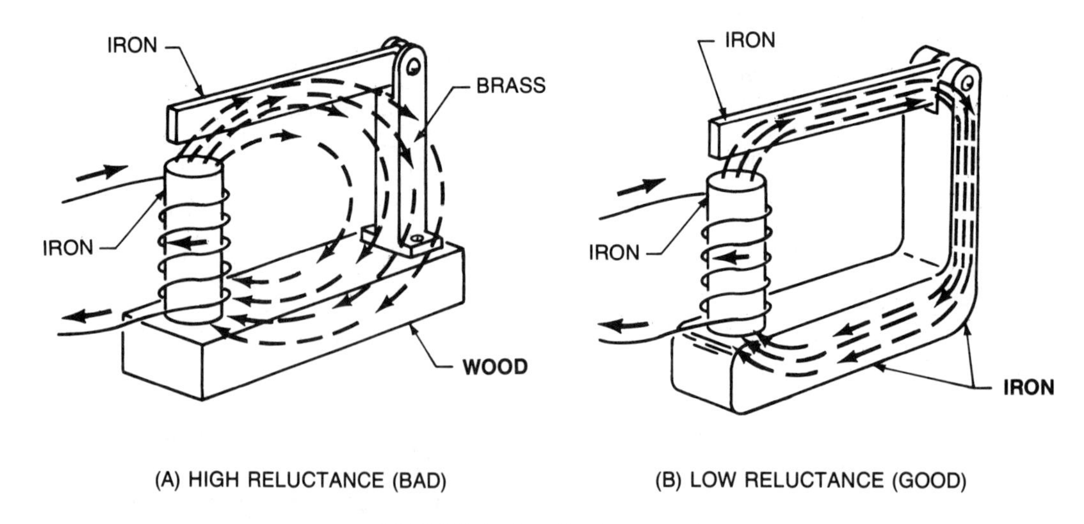

FIGURE 15–17 Electromagnets

then the force pulling on the iron bar in arrangement A is only a small fraction of the force exerted on the bar in arrangement B. The reluctance in the wood mount in A is so large that a greater number of ampere-turns is required for A to equal the amount of flux produced by B, which has an iron mount.

Core Design

Many DC electromagnets are solid bars or rods. However, the magnets for DC vibrators, motors, and generators and for AC equipment are an assembly of thin sheets of iron called *laminations*. For this type of magnet, the magnetic fields are in motion; that is, the lines of force whip back and forth as the current is turned on or off or is reversed.

Whenever lines of force pass through any kind of metal, they tend to generate a current in the metal. For example, the dotted line in Figure 15–18A indicates the current that is generated inside the iron core of the electromagnet each time the amount or direction of the current in the coil is changed. The generated current is called an *eddy current*. An eddy current is considered to be a nuisance, since it takes energy from the coil circuit and heats the iron core.

The magnet core in Figure 15–18B consists of a stack of thin sheets of iron. Only a small amount of current can circulate within the laminations of this type of core because of the poor electrical contact between the sheets. The small amount of electrical contact between the sheets is due to a coating of lacquer on each sheet and the buildup of iron oxide on the sheets.

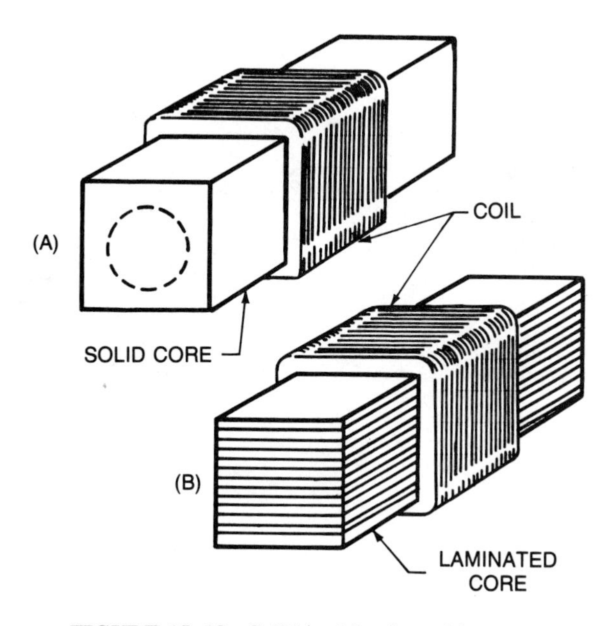

FIGURE 15–18 Solid and laminated iron cores

If eddy currents can be prevented, the loss of energy and the production of heat can be reduced. Iron containing a small amount of silicon is used for the core laminations, because this combination possesses high permeability and high electrical resistance. A high *electrical* resistance in a core material hinders the formation of eddy currents in the core material.

Demagnetizers

Occasionally, it may be necessary to demagnetize an object. This is accomplished by placing the object into a varying magnetic field powered by a source of alternating current. This procedure will upset the orderly alignment of the atoms within the object to be magnetized.

To illustrate the procedure, we are using a convenient demagnetizer fashioned from the stator winding of a small discarded motor. This winding is then connected to an AC power line through a current-limiting resistor, Figure 15–19.

In Figure 15–19, at position 1, a piece of steel is being magnetized; that is, the upper end of the steel piece is changing from north to south and back again at the rate of 60 times per second on an AC line. The piece of steel is now moved away from the coil to position 2. The steel is still being magnetized 60 times per second but not as strongly as at position 1. At position 3, the magnetizing effect is weaker and at position

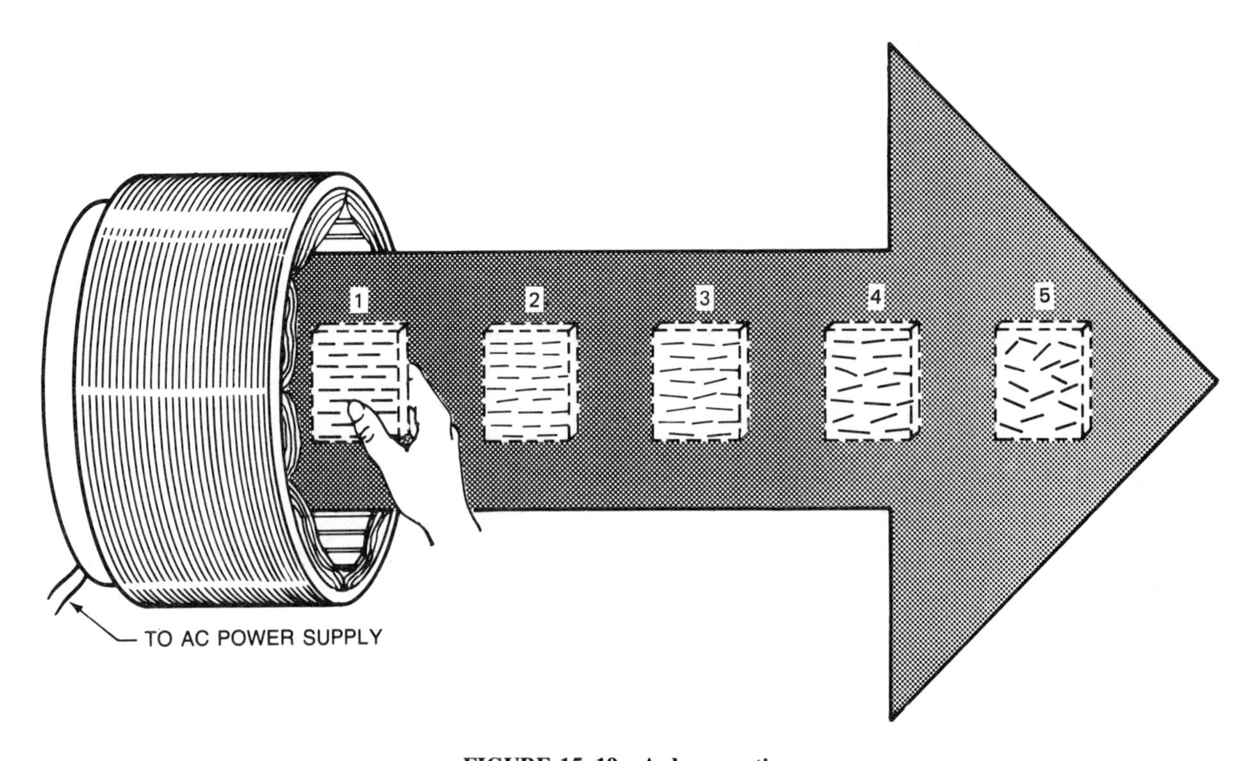

FIGURE 15–19 A demagnetizer

4, weaker still, the upper end of the steel piece alternating between a very weak north and a very weak south. By the time the steel piece reaches position 5, the magnetizing force of the coil is so weak that the steel atoms are left in general disarrangement and the steel piece is no longer magnetized.

Residual Magnetism

After an electromagnet has been de-energized (power turned off), a ferromagnetic core is not completely demagnetized. The core has retained some of the magnetism. This is known as *residual magnetism*. The amount of residual magnetism depends on the *retentivity* of the steel used. Generally speaking, hard steels have a high degree of retentivity as opposed to soft iron, which has low retentivity.

In some instances, residual magnetism may be considered to be a nuisance factor. In some applications, however, the residual magnetism is a desirable and necessary characteristic of the appliance.

SUMMARY

- All magnetism is due to electron motion, either the movement of electrons as they pass through a coil or the spinning motion of electrons in atoms.
- Like poles repel, unlike poles attract.
- The strength of a magnetic field is represented by the density of the lines of force. The direction of the field means the direction in which the north pole of a compass points.
- Most materials are nonmagnetic. The spinning of the electrons in the atoms of iron, nickel, and some alloys and oxides makes them magnetic.
- Atoms of iron are little permanent magnets. The magnetizing of a piece of iron is a matter of arranging these atoms so that like poles face in the same direction.
- The left-hand rule for a single wire states that the thumb is in the direction of electron flow and the fingers are in the direction of the magnetic field.
- The left-hand rule for a coil states that the thumb is at the north pole and the fingers point in the direction of the electron current.
- Parallel currents in the same direction attract; parallel currents in opposite directions repel.
- The magnetic strength of a coil can be measured in ampere-turns (amperes × turns).
- Permeability is the ability of a material to become magnetized. Residual magnetism is the flux density that is retained in the material after the magnetizing force is removed. Coercive force is a magnetizing force applied in the opposite direction to demagnetize the material.
- A magnetic field contains useful energy.
- Moving electrons exert force on magnets; moving magnets exert force on electrons.
- Maintaining the strength of a permanent magnet depends on
  1. The type of alloy
  2. Avoidance of excessive heat, shock, and AC magnetic fields
  3. The presence of an iron keeper between the poles

- It is easy to remagnetize a permanent magnet.
- The magnetizing force of a coil depends on the number of ampere-turns per inch of the magnetic circuit. Equal magnetizing forces can be produced by large current and few turns or by small current and many turns.
- The complete path of lines of force through a magnet, the iron that the magnet attracts, the air, and so on are called the *magnetic circuit*.
- The unwillingness of a material to be magnetized is called *reluctance*.
- $\text{Total number of lines of force} = \dfrac{\text{Magnetizing force in ampere-turns/inch}}{\text{Reluctance of the magnetic circuit}}$
- The use of laminated cores prevents energy loss due to eddy currents.
- Silicon steel is preferred to ordinary steel for laminated magnet cores because of its improved permeability and low eddy current energy loss.

Achievement Review

1. Naturally magnetic metals are called ferromagnetic materials. There are only three of those. Name them.
2. Explain the ferro- in *ferromagnetic*.
3. How can a ferromagnetic metal be magnetized? Explain two methods.
4. The theory of magnetic domains explains the difference between magnetic and nonmagnetic substances. Draw a neat sketch and write a few brief sentences explaining this theory.
5. State the magnetic laws of attraction and repulsion.
6. Review your knowledge of electrostatic fields. How do those forces compare to the forces of magnetism? Explain.
7. Explain three ways to demagnetize a magnet.
8. Draw a neat sketch of a magnetic field around a bar magnet.
9. Flux lines have a specific direction. They are assumed to leave the ________ pole and enter into the ________ pole.
10. Magnetic lines of force will not cross one another. True or false?
11. What is reluctance? Explain.
12. Magnetic lines of flux can pass through all materials, even those that have no magnetic properties. True or false?
13. In Figure 15–9 you see pictures of concentric circles that resemble a rifle target of sorts. Explain
    a. What these drawings represent
    b. The significance of the innermost circle (containing a dot or cross mark)
    c. The significance of the dot and the cross mark
14. What happens when a coil or wire is connected to a source of direct current?
15. What happens to a piece of steel when it is placed in a coil that is connected to a source of direct current?
16. Name at least ten electrical devices that use electromagnets.
17. What effect does the steel core have on the magnetic field of an electromagnet?
18. There are two left-hand rules. How do they differ from one another?

19. Explain, in your own words, the left-hand rule for straight conductors.
20. Explain, in your own words, the left-hand rule for coils.
21. Name three factors that determine the strength of an electromagnet.
22. What happens to an electromagnet when the current through the coil is reversed?
23. Explain the word *solenoid*.
24. For what are solenoids used? Explain. Name some practical applications.
25. Solenoid cores are generally constructed from a soft iron rather than hard steel. Why is soft iron preferred? (Explain this in terms of magnetic domains as explained in Figure 15–4.)
26. What effect, if any, does the polarity of a solenoid have on its electromagnetic strength?
27. How would the electromagnetic power of a solenoid be affected if one added
    a. More insulation to the wire.
    b. More resistance to the solenoid circuit.

16 Applications of Electromagnetism

Objectives

After studying this chapter, the student should be able to

- Define and explain the new technical terms introduced in this chapter

solenoid	relay
generator action	motor effect
relay ladder logic	micron
recording head	playback head
normally open contact	normally closed contact

- Describe the uses and applications of solenoids
- Draw a functional circuit diagram involving a control relay
- Explain how electromagnetism can produce vibrating motion
- Differentiate between the motor effect and generator action of conductors within a magnetic field
- Name and describe some industrial applications of electromagnetism

16–1 SOLENOIDS FOR LATERAL MOTION

One of the major applications of electrical energy is the production of mechanical energy. The motion desired may be in a straight line (lateral motion) or rotating (as in motors). Either way, the mechanical energy is produced by the attractive or repulsive forces of electromagnetism.

Solenoids are electromagnetic coils that use a movable plunger to translate the electrical energy into straight-line motion. The moving plunger, activated by the magnetism of the coil, can be used to operate valves, set brakes, or position an object. Solenoids are extensively used for control of hydraulic and pneumatic circuits.

16–2 THE ELECTROMAGNETIC RELAY

A relay can be described as a solenoid with switching contacts attached to its movable plunger. In other words, a *relay* is an electromagnetic switch, often with multiple switching contacts, that may open or close when the relay coil is energized.

For the better part of this century, millions of relays have been used in the following applications:

1. *Remote control* in locations that may be inaccessible or hazardous to the operator
2. *Automated industrial processes,* where the relay automatically responds to monitoring devices that can sense environmental changes, such as temperature, light, sound, or position of a machine
3. *Controls* for very *strong currents* or *high voltages,* with a relatively small voltage or current source

Notice in Figure 16–1 that the relays have two distinctly separate circuits, namely

1. *The control circuit,* which has a weak current flowing through the coil to energize relay M in order to attract armature A and close contact C. This, in turn, completes
2. *The power circuit,* in which a substantially stronger current is delivered to the controlled device.

Consider, for example, the circuit shown in Figure 16–2. This circuit will illustrate the use of a starting relay (sometimes called a *starting solenoid*) in your automobile.

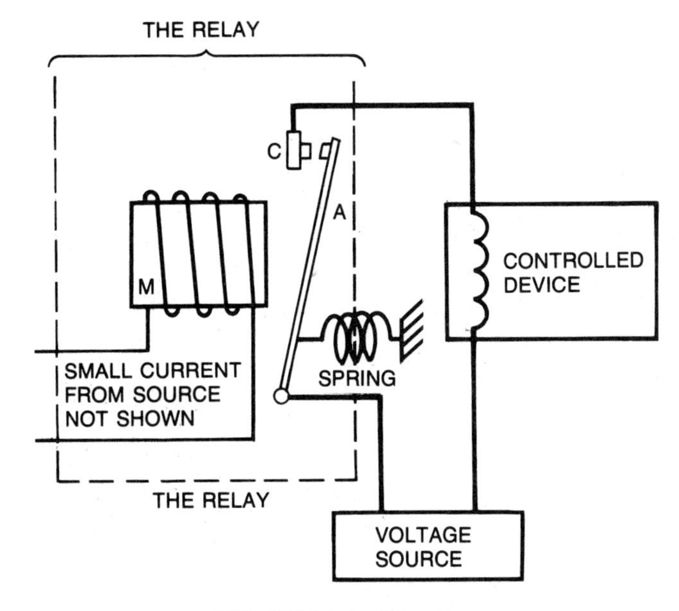

FIGURE 16–1 The relay

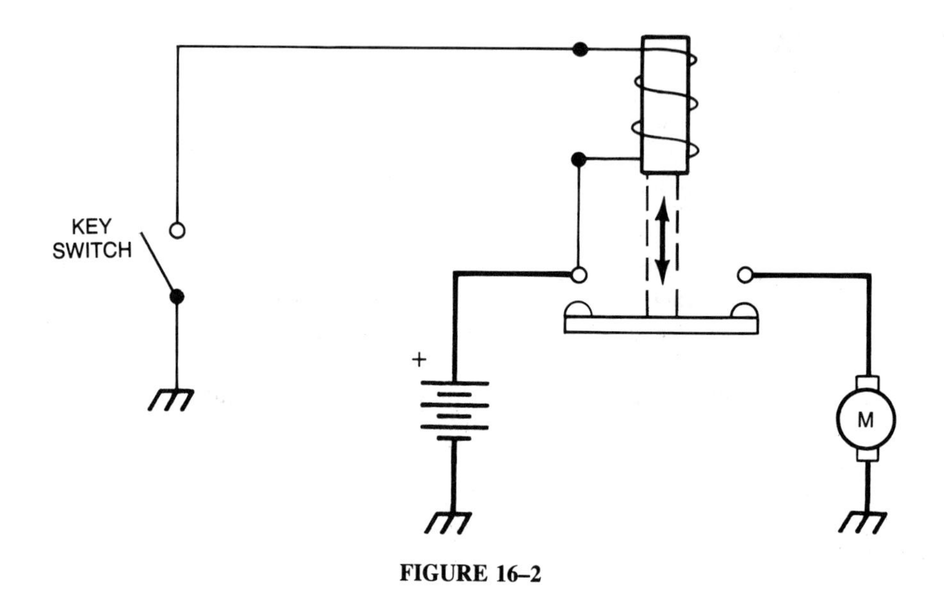

FIGURE 16–2

Notice that the starter motor is connected to the battery with a heavy-duty cable. (Starter motors often draw more than 200 amperes and thus require heavy cables.)

By contrast, the control circuit is shown with thin lines, representing a relatively small wire. (Only a weak current is going to flow through the coil.) It is important to see that the two circuits are electrically isolated from each other.

Furthermore, since an automobile is made from steel, it is customary to use its body as an electrical conductor, thereby eliminating almost 50% of the wiring. The battery, therefore, has its negative pole attached to the chassis. This is called a negative ground. Likewise, the other components of the circuit are also connected to the ground.

Relays have been used so extensively in industry that electrical blueprints came to be referred to as *relay ladder diagrams* or *relay ladder logic*. This concept of relay ladder logic plays an important role in the solid-state devices known as *programmable controllers,* which are rapidly replacing electromechanical devices for the control and operation of industrial machinery.

It is conceivable that electronic devices may some day supplant electromagnetic relays. For the time being, however, the multitude of relays still in use demand that electricians and technicians be familiar with their use and applications.

16–3 MAGNETIC VIBRATORS AND BELL

A relay circuit can be modified to produce vibratory motion, which may be utilized for different applications. The electric bell (or buzzer) demonstrates this principle well.

The flat spring and iron armature shown in Figure 16–3 comprise a movable assembly that pivots at the left end of the flat spring. When the bell is not in use, the free (unattached) end of the spring touches the stationary contact. When an external switch (push button) is closed, the bell is connected to a battery. The resulting current path is

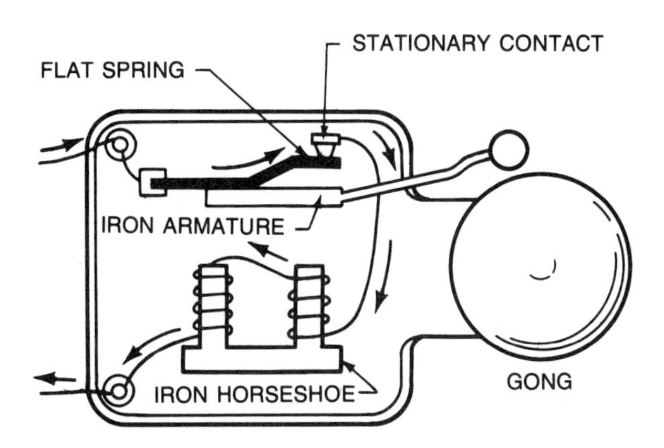

FIGURE 16–3 The electric bell

shown by the arrows in the figure. When the iron horseshoe is magnetized by the current, it attracts the armature, and the spring is pulled away from the stationary contact, breaking the circuit. When the spring leaves the contact, the current in the circuit stops and the magnet loses its magnetism. Since the magnet can no longer hold the armature, the elastic spring moves the armature and spring away from the magnet until the spring touches the contact again. Then the entire process can be repeated. Removal of the gong converts the bell to a buzzer.

Magnetic vibrators have been used in the past to rapidly switch the current on or off in a circuit to produce specific results. For instance, earlier models of car radios utilized vibrators to chop up the battery's DC and produce current pulsations suitable for use with transformers. (Transformers normally do not work on DC.) Another example of magnetic vibrators is the example of a spark coil in the old Model-T automobile. Ignition coils, also called induction coils, produce a high voltage when a direct current through them is rapidly switched on or off. Most of these functions of magnetic vibrators are now accomplished more efficiently by the use of electronic devices.

16–4 ELECTROMAGNETISM FOR ROTATIONAL MOTION

Electrical, rotating machines encompass both motors and generators. All of these machines operate on the principle of electromagnetism.

Motors and generators are more completely covered in Chapters 19 and 21. At this time we merely introduce you to the concepts of generator action and motor effect.

The term *generator action* refers to the phenomenon that an electric current can be generated simply by moving a wire through a magnetic field. The wire is moved across the magnetic field so as to cut lines of magnetic flux. This important principle, upon which all electrical generators work, is more fully explained in Chapter 18, where it is illustrated in Figures 18–1 and 18–2.

The term *motor effect,* or motor action, is used to describe the phenomenon that a current-carrying wire within a magnetic field will move. The reason for this motion is based on the fact that the current flowing through the wire produces its own magnetic

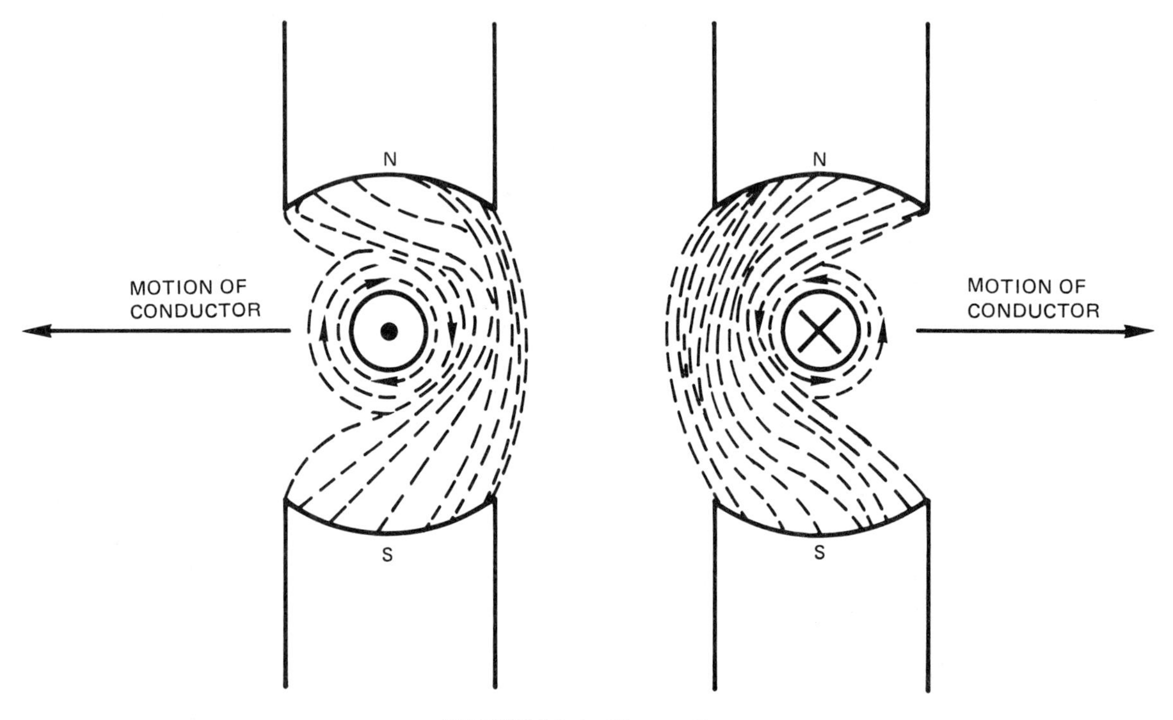

FIGURE 16–4 Motor effect

field around the wire. This magnetic field will interact with the flux lines of the two poles between which the wire is situated. The direction of motion is entirely predictable and depends on the direction of the current and the orientation of the magnetic field. The sketch in Figure 16–4 illustrates this motor effect. More will be said about this in Chapter 20.

16–5 OTHER APPLICATIONS OF THE MOTOR EFFECT

The term motor effect does not apply just to motors. Many other magnetic devices operate on the principle that a current-carrying wire will move within a magnetic field; for instance, (1) electrical measuring instruments, (2) loudspeakers, and (3) TV picture tubes.

Many electrical *measuring instruments,* or meters, depend on the interaction of magnetic fields to move a tiny coil of wire, thereby causing a deflection of the meter movement. The amount of deflection depends on the strength of the magnetic field produced by the flowing current. The following chapter is entirely devoted to this principle. The reader is encouraged to look ahead at the introductory illustrations in Chapter 17.

Loudspeakers also operate on the motor effect by causing a tiny coil (known as the voice coil on a speaker) to vibrate within a magnetic field.

Mechanical vibration produces sound. Vibrations in the range from 20 to 18,000 vibrations per second can be heard by the human ear. As the frequency of vibration increases, the pitch increases, and as the amount of back and forth movement of the mechanical vibration increases, the sound produced by the vibration increases in loudness.

Figure 16–5 shows a coil of wire wound on a paper sleeve that is suspended so that it can move freely near the pole of a permanent magnet. If an alternating current is applied to the coil of wire, the coil is alternately attracted to the permanent magnet, Figure 16–5A, and repelled by it, Figure 16–5B. The coil vibrates (moves back and forth) at the same frequency as the frequency of the electron vibration of the alternating current.

Many radio loudspeakers are constructed as in Figure 16–6. To obtain a uniform magnetic field in which the moving coil can vibrate, one pole of the magnet is located just inside the moving coil. The second pole is constructed so that it surrounds the moving coil. The moving coil is attached to a cone made of composition paper. The vibration of the cone produces sound when an alternating current from an amplifier is applied to the movable voice coil.

Early telephone receivers, Figure 16–7, used a stationary coil consisting of many turns of fine wire wrapped around the poles of a permanent horseshoe magnet. A receiver of this type operates on a much smaller current than is required by a loudspeaker; therefore, the coil must have a large number of wire turns. The alternating current in

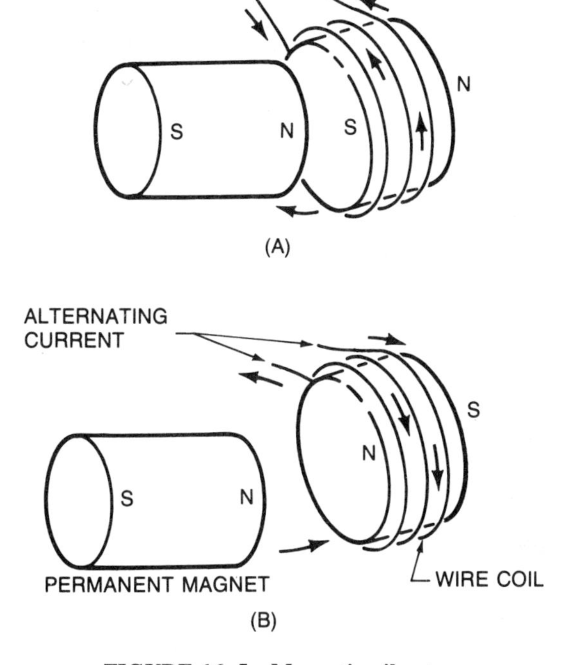

FIGURE 16–5 Magnetic vibrator

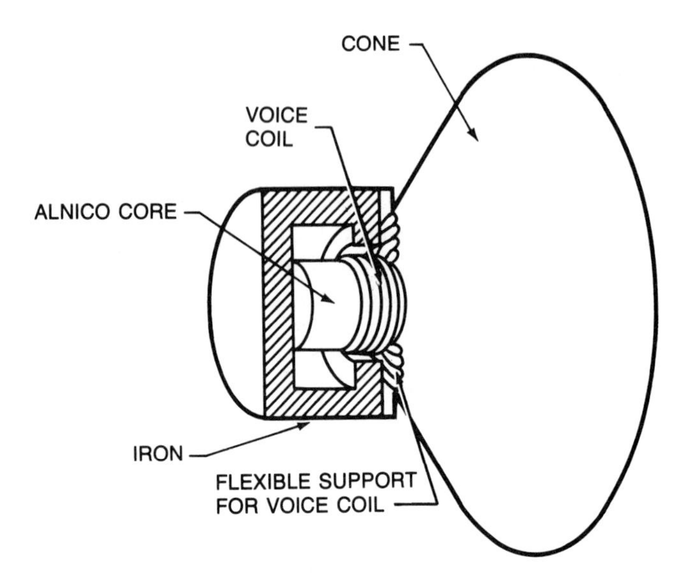

FIGURE 16–6 Loudspeaker

the coils strengthens and weakens the pull of the magnet. These variations in the strength of the magnet cause the flexible iron disk (diaphragm) to vibrate.

Television picture tubes also operate on the motor effect when they develop the picture on the screen. You may recall, from our discussion in Chapter 13, that the electron beam traversing a cathode-ray tube can be deflected by the field of an electromagnet. It really makes no difference whether the electrons travel through a metal con-

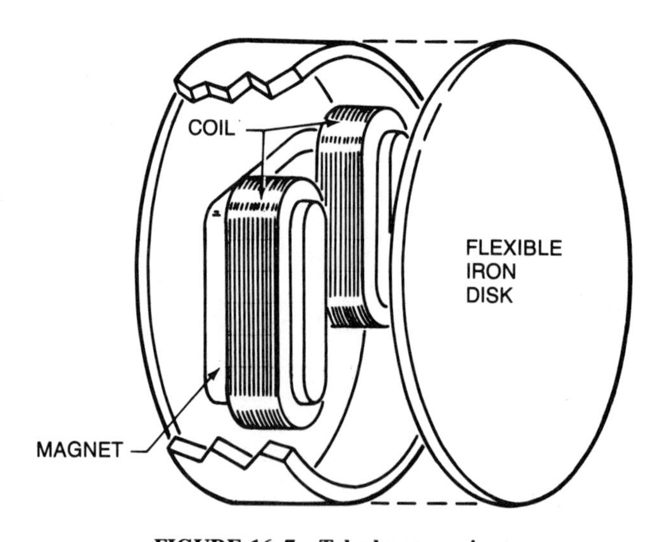

FIGURE 16–7 Telephone receiver

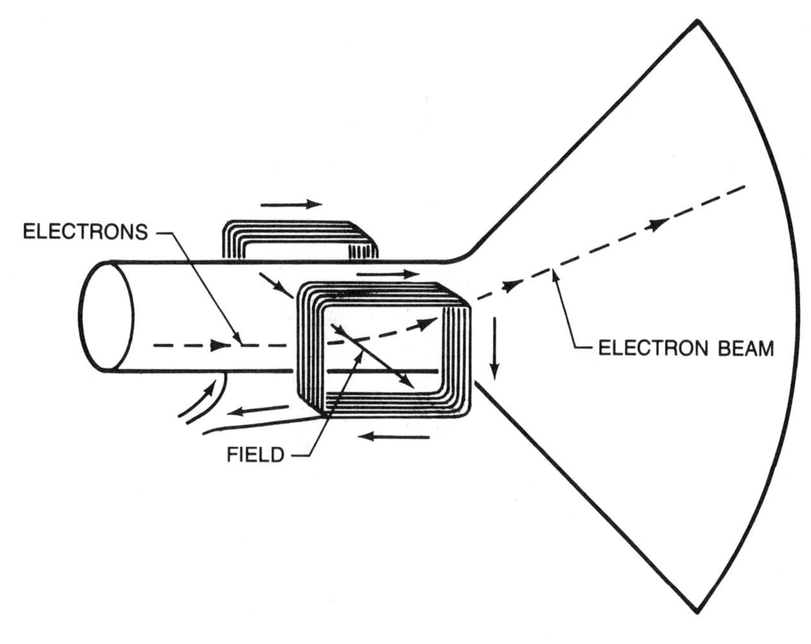

FIGURE 16–8

ductor or move as part of a cathode ray through a gas or vacuum; the effect is the same. In either case, the deflection is caused by the interacting magnetic fields.

The picture on the fluorescent coating on the face of a TV picture tube is caused by an electron beam that sweeps the screen horizontally at 15,750 times per second and vertically at 60 times per second. These motions of the beams are accomplished by magnetic fields.

Figure 16–8 shows the vertical deflection coils. The current in this pair of coils controls the vertical position of the electron beam. A similar pair of coils, one above and one below the neck of the tube, controls the horizontal movement of the electron stream. A coil encircling the neck of the tube focuses the electron beam, again using a magnetic field to control electron movement.

16–6 ELECTROMAGNETISM AT WORK

We have surveyed the use of magnetism from the perspective of mechanical motion, either lateral or rotational. But there are numerous other applications where the only motion involved is that of a changing magnetic field, as encountered with alternating currents.

Your future studies of electronics will reveal that electromagnetism enters into almost every aspect of electronic communication and industrial processes.

So it was with inventions of nearly 100 years ago: Joseph Henry's *telegraph,* Alexander Graham Bell's *telephone,* and Thomas Edison's *voice recorder.* All have one element in common: namely, electromagnetism.

And so it is with the sophisticated electronic devices of modern times. From *sound* and *video equipment* to *computers,* and from the *broadcasting stations* to *radar installations,* one feature is held in common: namely, electromagnetism.

Consider, for example, the magnetic tape recorders we enjoy for home entertainment. Audio and video recorders alike operate on the principle of storing electronic signals by producing variations in the strength of a magnetic field and storing these signals by magnetizing the red oxide particles deposited along the length of the tape.

As stated earlier, sound vibrations can be converted into corresponding electrical signals (by a microphone, for instance), which are then amplified and converted to electromagnetic variations in the *recording head.* As the tape is fed across the recording head, the needle-shaped oxide particles, which are about 1 micron long (1 *micron* = 0.000001 inch), are rearranged in conformity with the magnetic variations.

In audio recorders, the tape head is generally stationary and the pattern of magnetization is longitudinal along the length of the tape, Figure 16–9A. Many video recorders employ rotating recording heads, producing an oblique recording pattern on a helically guided tape, Figure 16–9B. Some earlier, commercial-type recorders have successfully employed four tape heads positioned 90° apart on a rotating disk. This results in a transverse recording pattern on the magnetic tape, Figure 16–9C.

Thus, the tape remembers; and when it is pulled across the *playback head,* the stored-up magnetism induces voltage variations in the electromagnetic coil of the playback head. The varying voltage signals so produced contain all the elements of speech or music, which then can be processed to activate the loudspeaker.

This is merely one example to demonstrate the widespread use of electromagnetism in modern electronics. Your future studies in this subject will introduce you to many more such applications.

Our discussion of electromagnetism would not be complete without mentioning one of the first applications of magnetic pull in lifting magnets, which are widely used for the transfer of scrap steel.

The lifting magnet shown in Figure 16–10A is constructed so that the coil is nearly surrounded by iron. One pole of the magnet is formed on the core inside the coil, and the other pole is formed on the shell that surrounds the coil, Figure 16–10B. This type of circular horseshoe magnet produces a strongly concentrated magnetic field.

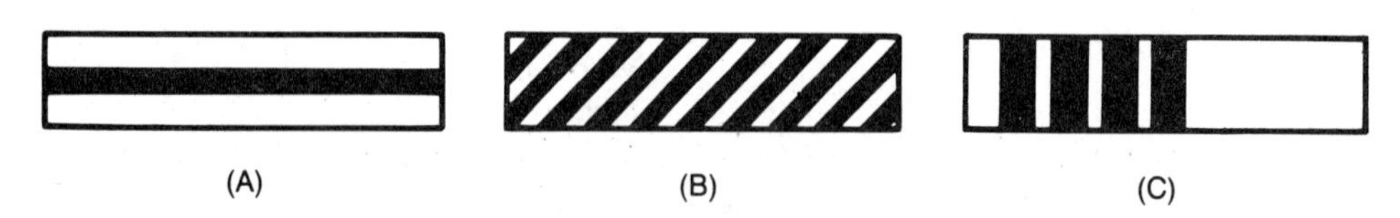

FIGURE 16–9

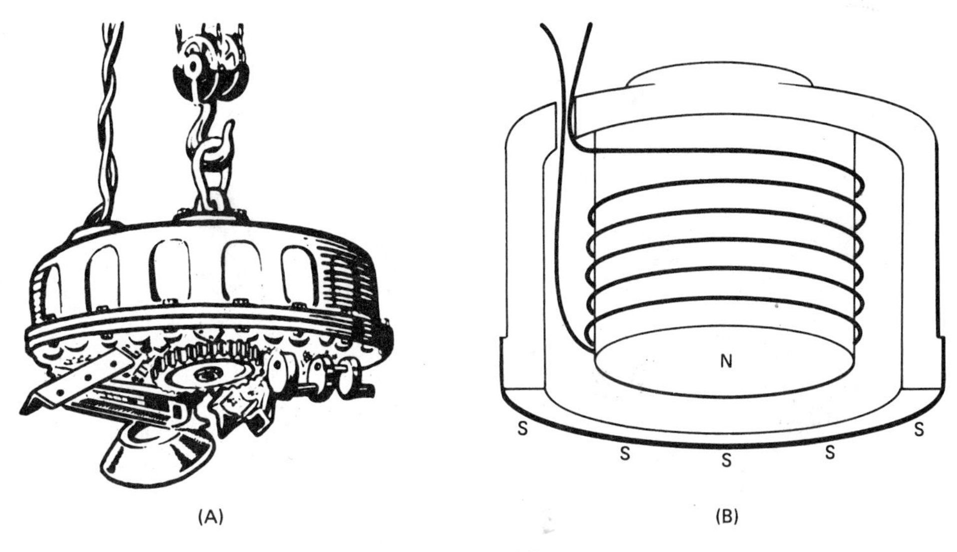

FIGURE 16–10 Lifting magnets

SUMMARY

- Solenoids are electromagnets with a movable plunger, designed to change electrical energy into straight-line motion.
- Relays are electromagnetic switches that can be used for remote control, automation, or for control of high voltages and currents.
- Relays have two distinct circuits that are electrically isolated from each other.
- The concept of relay ladder logic carries over into modern applications of solid-state control.
- Electrical, rotating machinery operates on magnetic concepts known as generator action and motor effect.
- The concept of motor effect is used in the explanation of operating electrical meters, loudspeakers, and TV picture tubes.
- Electromagnetism finds extensive applications in electronics for communication and industrial processes.

Achievement Review

THE ELECTRIC BELL

1. Finish the drawing that follows question 3 by connecting the parts of the bell to the push button and the battery. Be sure to notice the letters N and S in the drawing, indicating magnetic polarity.

2. Draw tiny arrowheads on the wires of the coil in the drawing following question 3 to show direction of the current, proving the magnetic polarity by the left-hand rule.
3. Sketch with fine, dashed lines the path of the magnetic flux.

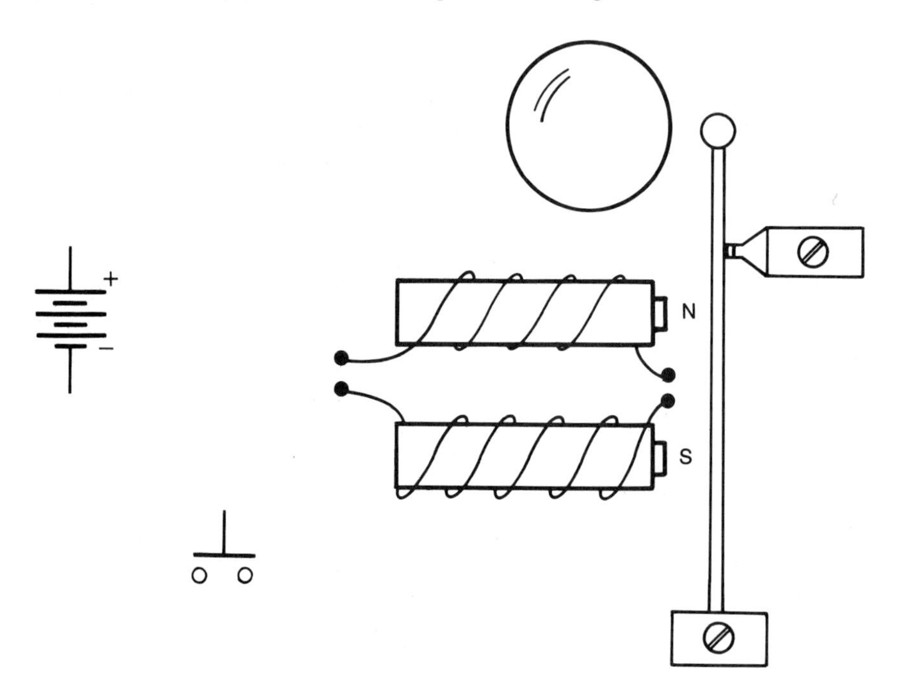

4. Write a brief but complete explanation of the theory behind the bell. Explain how it works.

ELECTRICAL DOOR CHIMES

1. Finish the drawing at the top of page 261 by connecting the solenoids in the chime to the pushbuttons and to the step-down transformer.
2. Assuming that the top wire of the voltage supply is positive (as indicated), trace the current through the solenoids by drawing tiny arrows in the drawing at the top of page 261. Using the left-hand rule for coils, determine the north and south poles on the solenoids.
3. Write a brief but complete explanation of the theory behind the door chimes. Explain how it works.

THE RELAY

Shown at the bottom of page 261 is a relay with two sets of switching contacts. The abbreviation N.C. stands for normally closed and means that the contacts are in a

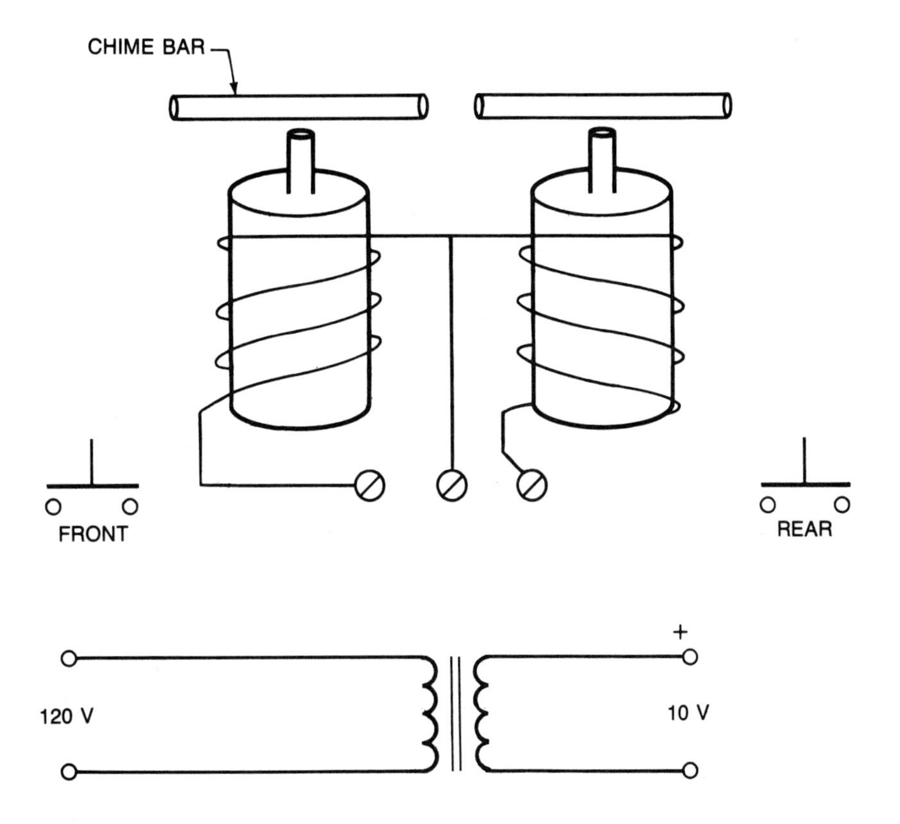

closed position as long as the coil is de-energized. Similarly, the N.O. contact remains open as long as there is no current flowing through the coil.

Finish the drawing by connecting all parts in such a manner that lamp A is burning all the time but turns off when the push button is depressed. Lamp B will turn on at the same time lamp A is extinguished.

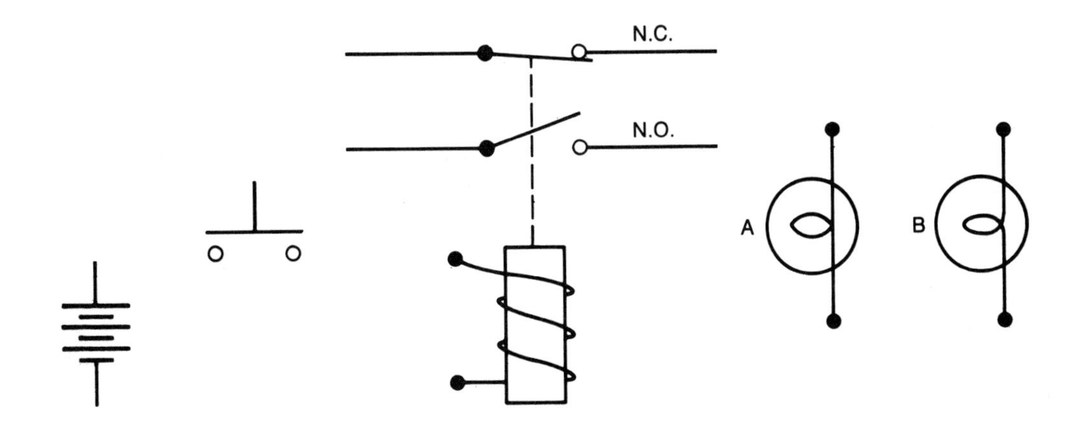

17 Electrical Measuring Instruments

Objectives

After studying this chapter, the student should be able to

- Define and explain the new technical terms introduced in this chapter

analog meters	multiplier
digital instruments	ohms per volt rating
d'Arsonval meter	megohmmeter
galvanometer	Wheatstone bridge
sensitivity	shunt resistor

- Describe and explain the operating principles of voltmeters, ammeters, ohmmeters, wattmeters, megohmmeters, and Wheatstone bridges
- Calculate the value of shunt resistors to extend the range of an ammeter
- Calculate the value of multiplier resistors to extend the range of a voltmeter
- Compute an unknown resistor from the settings of a Wheatstone bridge

Instruments used for measuring electrical quantities are generally classified as analog or digital meters. Unlike the digital meter with its precise numerical readout, the analog meter has a moving pointer that slides over a number scale from which the reading must be interpolated by the observer.

Such meters have served technology well for nearly a century, and thousands upon thousands of analog instruments are still in service. Therefore, we shall focus our attention on such instruments, even though it appears that at some time in the future they may become obsolete.

17–1 d'ARSONVAL METERS

The instruments most commonly used to perform measurements are ammeters, voltmeters, and ohmmeters. These instruments are all similar in construction and are modifications of a basic instrument called a *galvanometer*. Galvanometers are often known as d'Arsonval meters or as *permanent-magnet meter movements*. The action of galvanometers and that of most measuring instruments depends on the magnetic effects of a small current.

Two forms of permanent magnets for use in DC meters are shown in Figure 17–1. The steel horseshoe magnet in Figure 17–1A has been used in instruments for many years. Also in common use is an Alnico magnet in the form of a rectangular slug, Figure 17–1B. The flux for this type of magnet divides into the two sides of the soft iron rings surrounding the slug. These rings also act as shields to protect the magnet assembly from outside magnetic disturbances. In both forms of magnet construction, a stationary cylindrical iron core is located between the poles. Due to this core, there is an evenly distributed, uniformly strong magnetic field in the space where the moving coil operates. This uniform field makes it possible to space numbers evenly on the meter scale. To obtain a support that is nearly free of friction, the moving pointer can be supported either by jewelled pivots (similar to those that support the balance wheel of a watch) or by a taut springy wire or band, Figure 17–2.

What makes the pointer move? The pointer is fastened to a coil that becomes an electromagnet when there is a current through it. In Figure 17–3, it can be seen that a current moving in a clockwise direction in the coil causes the coil to act as a magnet, having a north pole on the near side and a south pole on the far side.

Due to magnetic attraction and repulsion forces, the coil tries to turn between the poles of the magnet so that unlike poles will be as close together as possible. The amount of the turning force depends on the strength of the permanent magnet and on

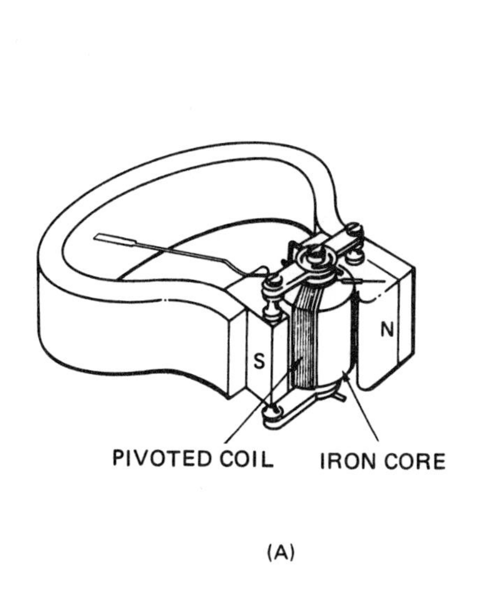

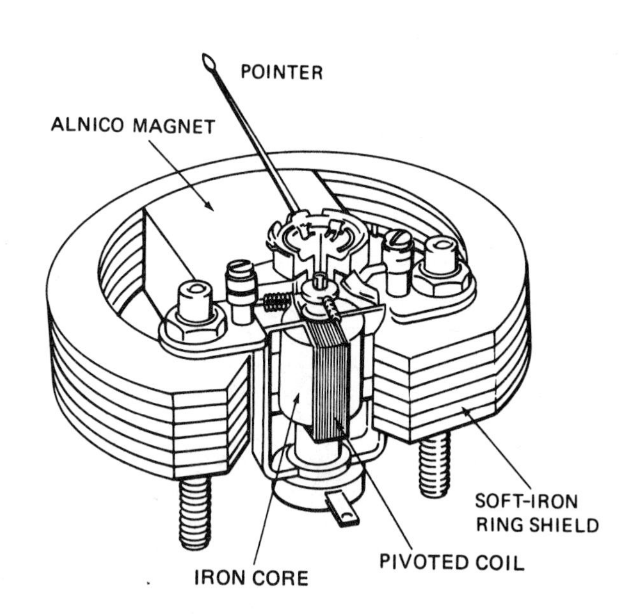

FIGURE 17–1

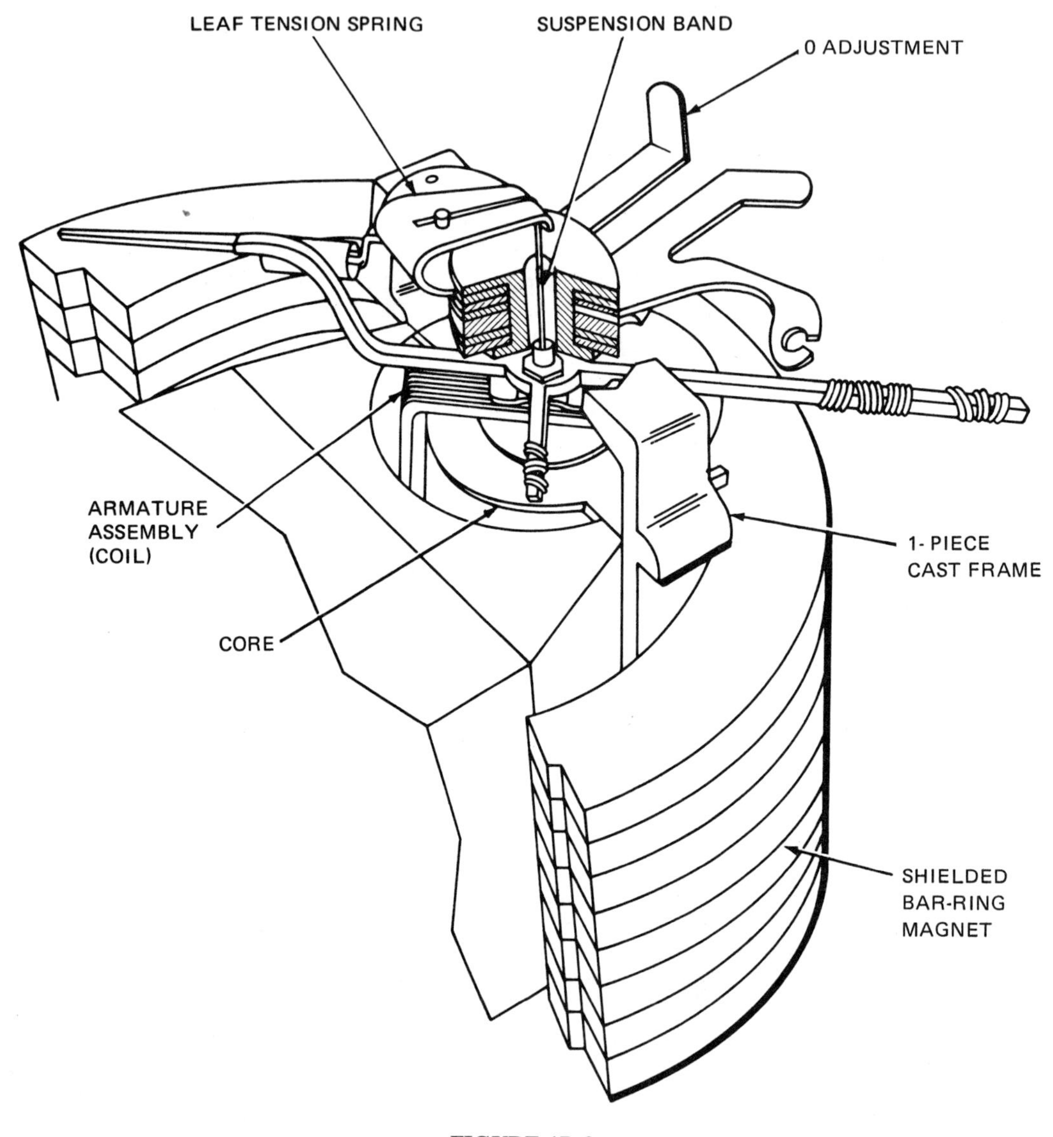

FIGURE 17–2

the number of ampere-turns of the movable coil. The springy coil support offers mechanical resistance to the motion of the coil. If the current in the moving coil is increased, the magnetic effect of the coil is stronger and the coil turns even more. As a result, the pointer indicates the increased current on the scale, Figure 17–4. When the current stops, the spring or twisted band returns the coil and pointer to the 0 mark. The coil assembly is constructed so that one end of the spring or band is fastened to the

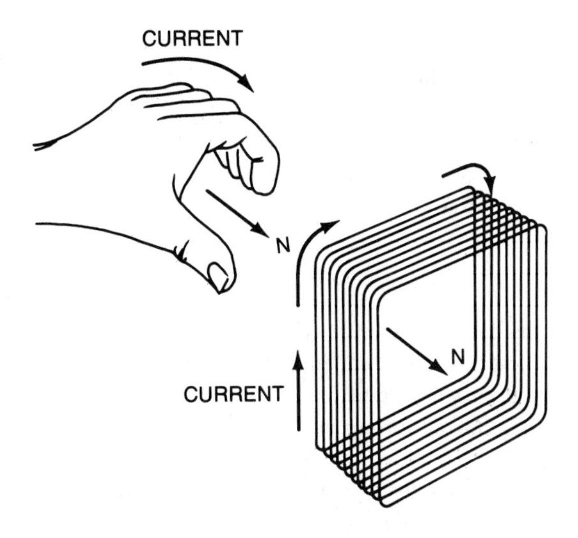

FIGURE 17–3 Left-hand rule

moving coil and the other end is stationary. As a result, the spring can serve as a conductor to connect the movable coil to the stationary wiring in the meter.

The meter shown in Figure 17–4, as well as other meters based on this type of construction, operates only on direct current. If an alternating current is applied to this meter, the magnetic poles of the coil reverse rapidly. Since the coil is too large to swing back and forth at the AC frequency of 60 times per second, the coil does not turn at all. A meter that is meant to operate on 1 ampere DC is not damaged by 1 ampere AC, but the meter reads 0 on AC.

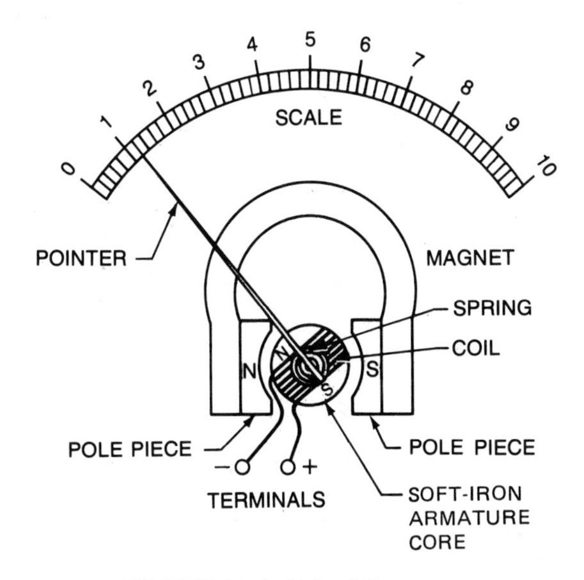

FIGURE 17–4 DC milliammeter

A simple galvanometer, by itself, has a very limited use. If a galvanometer is to be used as a current indicator, only a very small current is allowed in the fine wire of the moving coil. Since this coil has a low resistance, it is possible to apply only a very small voltage to the moving coil. The most useful galvanometers are scaled as milliammeters or microammeters. (These meters will indicate how many thousandths or millionths of an ampere pass through the meter.)

17–2 DC AMMETERS

To measure large currents with the galvanometer, a known large fraction of a large current is bypassed through a parallel low resistance called a *shunt,* Figure 17–5. In this arrangement, only a small fraction of the total current passes through the moving coil. The scale is marked to indicate the total current through the entire ammeter (galvanometer plus shunt circuit).

In order to calculate the value of a shunt resistor, it is necessary to know two other points of information about the meter to be modified.

1. We need to know how much current it takes to drive the pointer from its 0 position to the end of the scale. This full-scale deflection current is known as the *sensitivity* of the meter.
2. In addition, we must know the internal resistance of the meter movement.

EXAMPLE 17–1

Given: A d'Arsonval meter movement with a sensitivity of 1 milliamp and internal resistance of 50 ohms.

Find: The value of a shunt resistor necessary to convert this meter into an ammeter with a 5-amp range.

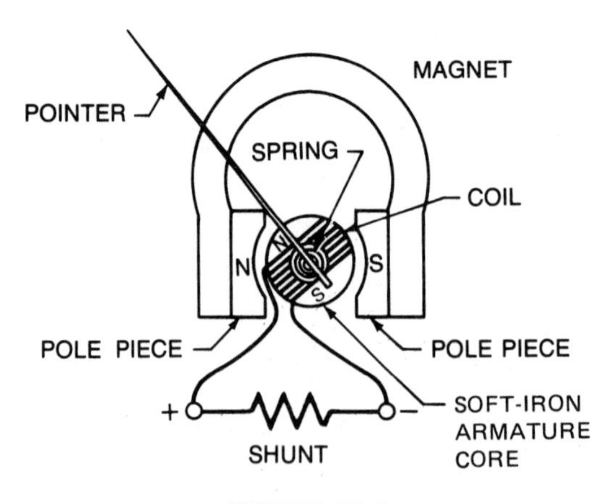

FIGURE 17–5

SOLUTION

Look at Figure 17–6. Note that the shunt is a parallel resistance. When there is a 5-ampere current through the meter, only 0.001 ampere passes through the moving coil. The balance of the current, 4.999 amperes, must go through the shunt. Ohm's law can be used to find the resistance in ohms of the shunt if the potential difference between A and B is known. The voltage between A and B can be found, since it is known that there is a current of 0.001 ampere in the 50-ohm coil.

$$E = IR$$
$$E = 0.001 \times 50$$
$$E = 0.050 \text{ V}$$

This voltage is the same for the two parallel parts of the circuit. Therefore, this value can be used to find the resistance of the shunt.

$$E = IR$$
$$\frac{E}{I} = R$$
$$R = \frac{0.050}{4.999}$$
$$R = 0.01\Omega$$

A 5-ampere ammeter is formed by combining a resistance of 0.01 ohm in parallel with the 1-milliamp meter.

Note: This low value of shunt defines the low resistance of an ammeter. Ammeters must have a very low resistance so that the insertion of the meter into a circuit does not reduce the circuit current to be measured.

An experimeter planning to make this meter conversion need not look for a 0.01-ohm resistor in a supply catalog. Copper wire can be used to make a shunt. The wire

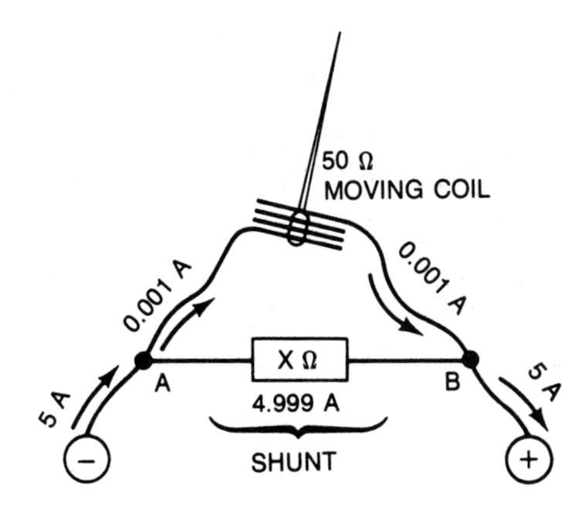

FIGURE 17–6

is soldered or firmly attached to the meter movement to avoid the introduction of resistance due to a poor contact. The following procedure is used to add a shunt made from copper wire to a meter.

1. Decide on a reasonable length of wire (3 inches, for example).
2. Find the resistance of 1,000 feet of this wire (3 inches = 0.01 ohm, 1 foot = 0.04 ohm, and 1,000 feet = 40 ohms).
3. Refer to Figure A–2 in the Appendix to find the copper wire size that has approximately the same value in ohms per 1,000 feet as the value determined in step 2. (No. 26-gauge wire has a resistance value close to 40 ohms per 1,000 feet.)
4. Use the available wire size that has the nearest resistance value, and calculate the required length. (Either reverse the previous calculations or use the methods shown in Section 7–4.

Meter manufacturers generally make meter shunts of a material called manganin (a copper-nickel-manganese alloy) rather than copper. The advantage of using manganin is that the resistance of this material does not change appreciably with temperature changes. Furthermore, since the resistivity of manganin is greater than that of copper, a sturdy assembly that takes up a small amount of space can be made using only a short strip of this material.

17–3 MULTIRANGE AMMETERS

The diagram in Figure 17–7 represents the preferred arrangement of shunts for an ammeter with two scales. The circles marked 2 and 10 represent either binding posts or selector switch contacts. A set of possible values of shunt resistance is shown.

- When the 2-ampere contact is used, the shunt consists of R_1 and R_2 in series.
- When the 10-ampere contact is used, R_2 acts as the shunt; R_1 is in series with the moving coil.

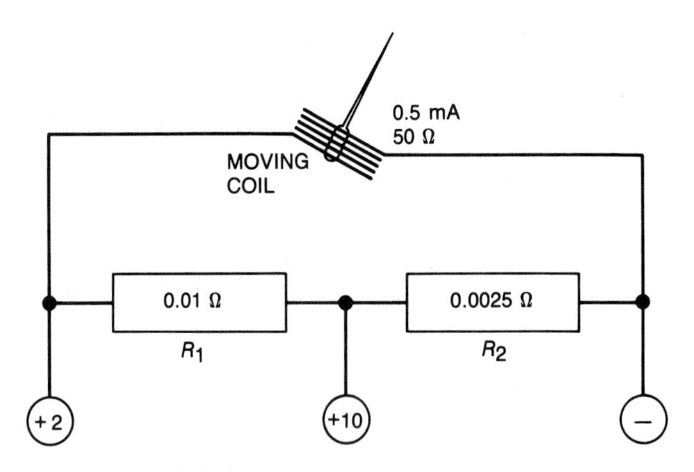

FIGURE 17–7 Two-scale ammeter

This arrangement is called an *Ayrton shunt*. A three-scale ammeter contains a three-section shunt.

17–4 VOLTMETERS

A voltmeter is obtained by connecting a high resistance in series with the galvanometer, Figure 17–8. Such voltage-dropping, series-connected resistors are known as *multipliers,* because they multiply the usable range of the basic meter movement.

Unlike an ammeter, a voltmeter must be connected directly across (parallel to) the source of energy. In addition, the voltmeter should be connected in parallel with any device supplied by the measured voltage.

There are two reasons for making the voltmeter a high-resistance instrument.

- Only a tiny current can be permitted through the moving coil.
- The addition of the voltmeter in the circuit should not reduce the electrical pressure being measured.

If this statement is not very meaningful, review our discussion of loaded voltage dividers (Section 12–4).

The conversion of a galvanometer (milliammeter or microammeter) to a voltmeter is a simple process both in the calculation of the required resistance and in the construction of the meter, Figure 17–9. It is necessary to calculate the value of the series *resistor* that limits the *current* to the galvanometer's full-scale capability when the desired *voltage* is applied. Once again, Ohm's law is used to determine the necessary value.

EXAMPLE 17–2

Given: A galvanometer with 200-microampere sensitivity.

Find: The value of the multiplier needed to construct a voltmeter with a 200-volt range, Figure 17–9.

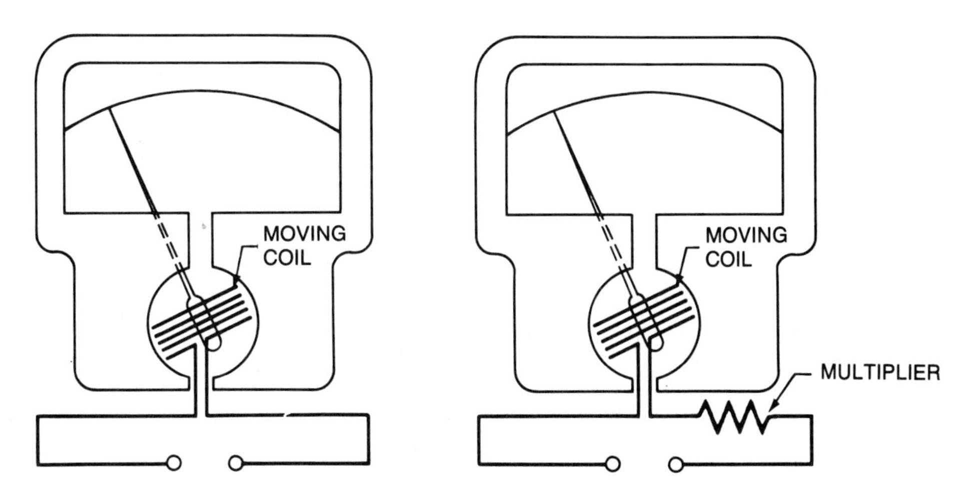

FIGURE 17–8 A resistor converts the galvanometer to a voltmeter.

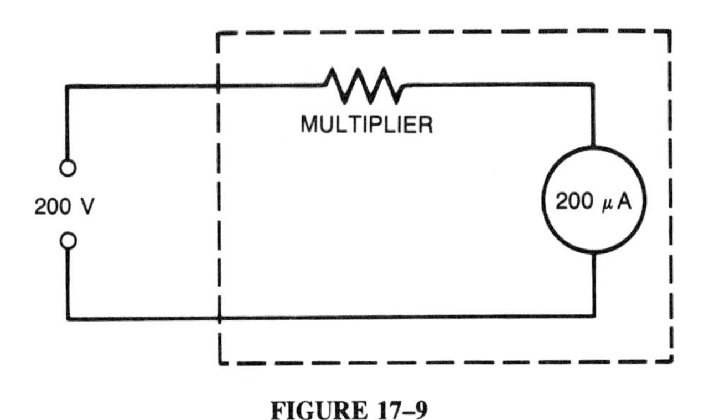

FIGURE 17–9

SOLUTION

$R = E/I$

$R = \dfrac{200}{0.0002}$

$R = 1{,}000{,}000\ \Omega$

$R = 1\ \text{M}\Omega$

This value is the total resistance of the voltmeter (the moving coil plus the series resistor). In general, the resistance of the moving coil is so small (in the order of 50 to 100 ohms) that it is disregarded. Therefore, a 1-megohm resistor (1 Mohm or 1 MΩ) connected in series with the galvanometer coil results in a 200-volt voltmeter.

Anyone who is concerned about the inaccuracy introduced by disregarding the coil resistance (50 to 100 ohms) should consider the following questions. If a resistor marked with a value of 999,900 ohms is issued for a job, how can the technician know if it is 999,900 ohms or 1,000,000 ohms? How accurate is the microammeter movement assumed for the problem? How accurately can a voltmeter be read? In general, lower priced meters have a 2% accuracy and higher priced ones, a 1% accuracy.

Multirange voltmeters contain several resistors. The meter range to be used is determined by the choice of binding posts or by the selector switch setting.

Figure 17-10 shows calculated values for the series resistors of a multirange meter. The actual values can vary from the stated values by 1% or 2%. With the selector switch at the 2.5-volt position, $R = 2.5/0.001 = 2{,}500$ ohms total (30 ohms in the meter plus 2,470 ohms in the series resistor).

At the 25-V position: $R = 25/0.001 = 25{,}000\ \Omega$

At the 250-V position: $R = 250{,}000\ \Omega$

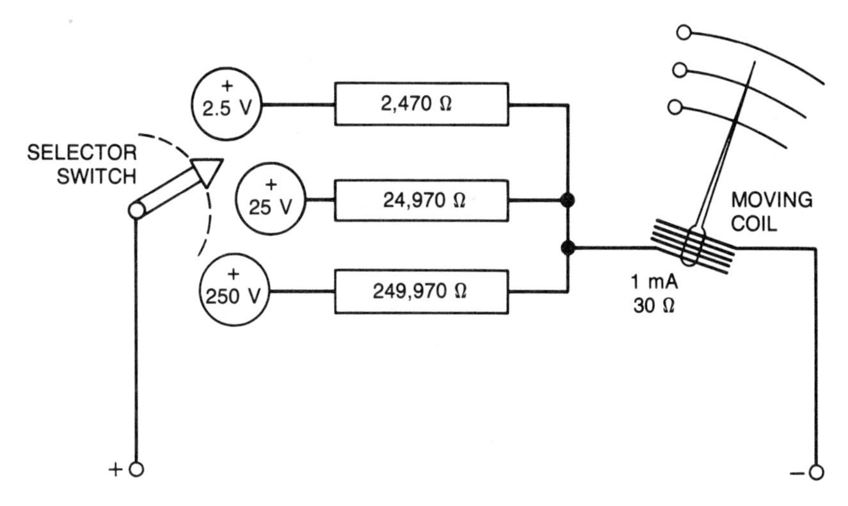

FIGURE 17–10 Multirange voltmeter

Sensitivity of Voltmeters

The *sensitivity* of a voltmeter is stated in *ohms per volt* (Ω/V). In Example 17–2, the 200-volt meter has a resistance of 1,000,000 ohms; therefore, the sensitivity of this meter in ohms per volt is

$$\frac{1{,}000{,}000}{200} = 5{,}000\ \Omega/\text{V}$$

The multirange voltmeter in Figure 17–10 has a sensitivity of 1,000 ohms per volt on all scales. A large value of sensitivity is desirable, since a high-resistance voltmeter uses a very small current to operate the meter movement. A meter rated at 1,000 ohms per volt will take a current of one milliampere for a full-scale reading.

A voltmeter with a sensitivity of 20,000 ohms per volt operates on 50 microamperes (μA) at full scale.

17–5 OHMMETERS

An ohmmeter contains a battery, series resistors, and a galvanometer (microammeter) movement, Figure 17–11. The battery ranges from 1.5 to 45 volts. The same type of coil assembly is used in the ohmmeter as is used in the types of meters covered previously. An increase in the current in the meter causes the pointer to move to the right. The meter is scaled so that it indicates the amount of resistance in ohms when the meter is placed between the tips of the external test leads.

To use an ohmmeter, the tips of the test leads are first held together (short circuited) and the rheostat (variable resistor) adjusted so that the meter pointer moves to the right-hand end of the scale and points to the 0 ohms mark. The meter now indicates a condition that is already known, that there is no resistance between the test leads. The

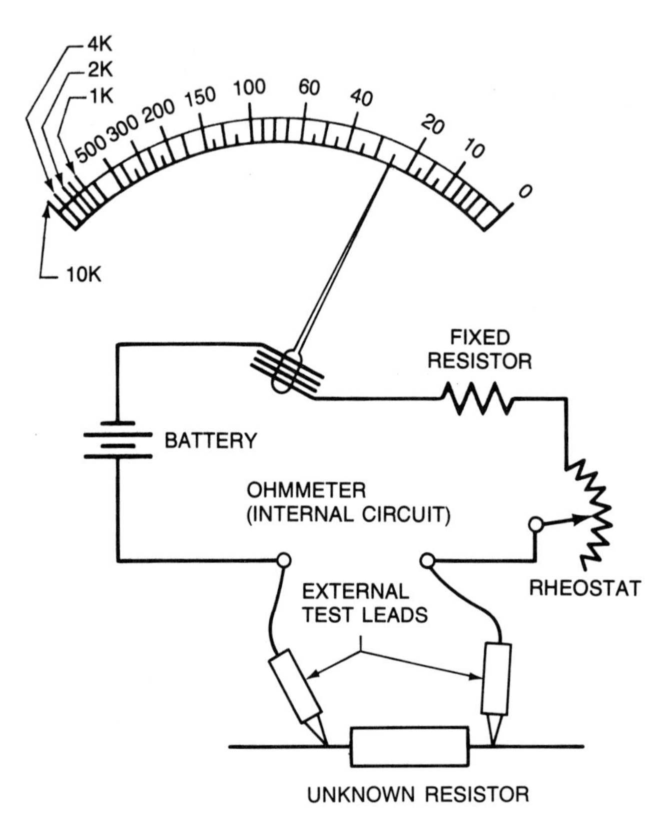

FIGURE 17–11 Ohmmeter circuit

rheostat adjustment is made to compensate for changes in the resistance of the battery as it ages.

When the test leads are separated, there is no current in the circuit, and the pointer drops back to the left end of the scale to the indication of infinite resistance. (The presence of several inches of air between the test leads means that there is high resistance between them.)

When the test leads are touched to the ends of a resistor of unknown value, the resistance is read directly from the ohms scale. An ohmmeter normally has several ranges, where different combinations of series resistance and battery voltage are used for the individual range.

The ohmmeter shown in Figure 17–11 is called a *series ohmmeter*. When it is installed in a case containing multiple-contact switches, voltmeter resistors, and ammeter shunts, the resulting assembly is called a *multimeter* or *volt-ohmmeter*. Volt-ohmmeters are widely used in testing electronic equipment.

To measure very low resistance values, a *shunt ohmmeter* is used, Figure 17–12. The scale reads from left to right because the high resistance permits more current through the meter. Zero resistance in the test lead circuit permits most of the current to bypass the meter.

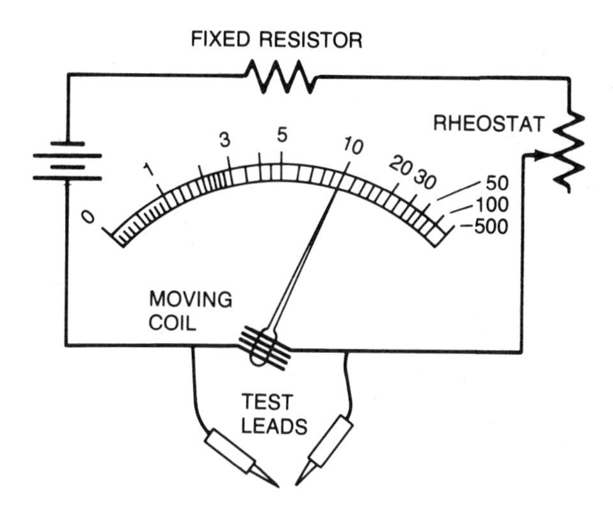

FIGURE 17–12 Shunt ohmmeter

17–6 MEGOHMMETERS

An instrument called a *megohmmeter* is used for insulation testing and similar high-resistance tests. The megohmmeter contains a high-voltage generator that supplies current through the series resistors and the unknown resistance *(R)* to the two-coil assembly that operates the pointer, Figure 17–13.

Note in the figure that permanent magnets supply the field for the DC generator and the field for the moving coil assembly. The potential coil is connected in series with R_2 across the generator output. The current coil is connected in series with the unknown resistance. The current in the coil depends on the value of the unknown resistance. The potential coil and the current coil are fastened together and can rotate only as a single unit.

Since there is no spring in the coil and pointer assembly, the pointer can take any position on the scale when the meter is not in use. If there is no external connection across the ground and line terminals, when the generator is operated the current in the potential coil causes a magnetic force that rotates the coil assembly counterclockwise, moving the pointer to the infinity (∞) end of the scale (open circuit point). If the ground and line terminals are shorted (if the unknown *R* has a value of 0 ohms), there is almost no current in the potential coil. Thus, the strong field of the current coil will rotate the assembly clockwise and move the pointer to the 0 end of the scale.

When the value of the unknown *R* is neither very high nor very low, the currents in the two coils produce opposing torques. As a result, the coil and pointer assembly comes to rest at the position near the middle, where these torques balance each other. A low value of external resistance permits the current coil to turn the pointer assembly closer to the 0 end of the scale. As the assembly is moved closer to the 0 side, the potential coil is pushed far enough into the north pole field to prevent further turning.

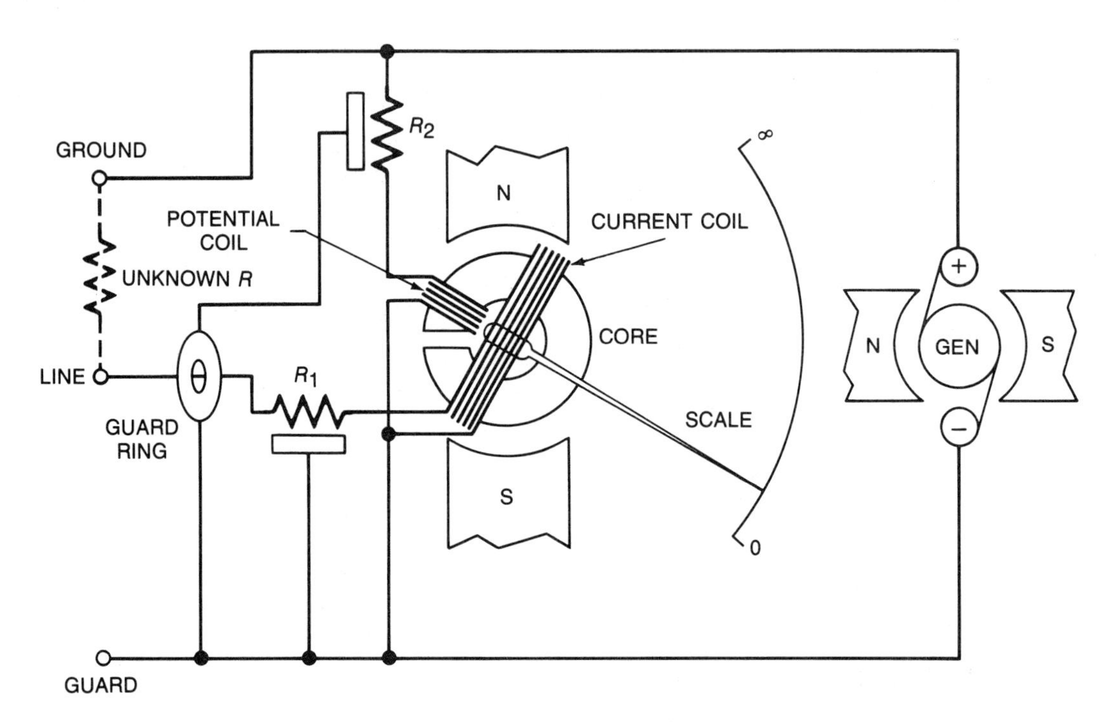

FIGURE 17–13 Megohmmeter

In the presence of a high external resistance, the current coil has less effect and the potential coil moves the pointer closer to the ∞ end of the scale. The scale is marked to show external resistance in megohms.

In addition to the megohmmeter, various electronic instruments operated from a 120-volt AC line can measure high resistance.

17–7 WATTMETERS

As stated in Chapter 9, watts = volts × amperes in DC circuits. To measure the wattage of a circuit, a meter must have two coils; one coil is affected by the voltage and one is affected by the current. The voltage coil is the moving coil and is connected across the current line so that the magnetic strength of the coil is proportional to the line voltage. The combination of the moving coil and its series resistor in Figure 17–14 is similar to a voltmeter, described previously. However, instead of having a permanent magnet to provide the magnetic field for the moving coil, the wattmeter has current coils to provide the magnetic field. The magnetic strength of these coils is proportional to the current through them (the current supplied to the device being tested).

The amount of movement of the coil and pointer depends on the strength of both coils. If there is a voltage but no current, then there is no magnetic field to turn the moving coil; therefore, the pointer reads 0. If the magnetic strength of either coil is increased, the turning force increases; that is, the turning force depends on the product

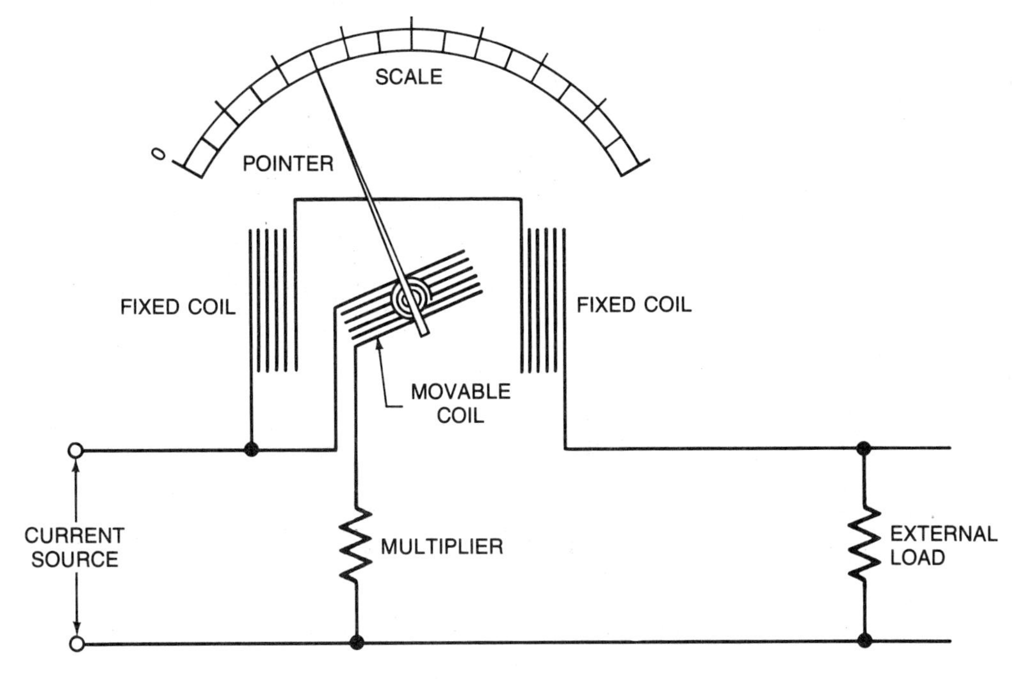

FIGURE 17–14 Wattmeter

of the magnetic strengths of the two coils, just as the force between any two magnets depends on the product of their magnetic strengths (Chapter 16).

With this coil arrangement, the pointer reading depends on the product of the voltage on one coil and the current in the other coil. Thus, the meter scale is calibrated in watts. The coils used have air cores, not iron cores. A wattmeter can operate on AC as well as DC, because the magnetic polarity of *both* coils reverses when the current reverses, and the turning force remains in the same direction.

Wattmeters are more necessary for AC measurements than for DC measurements. In DC circuits, watts are always equal to volts × amperes, and wattmeters are not required. In AC circuits, there are occasions when watts are not equal to volts × amperes, and wattmeters are needed to indicate the power consumption in the circuit.

17–8 WHEATSTONE BRIDGES

The *Wheatstone bridge* is a circuit frequently used to make accurate measurements of resistance. The Wheatstone bridge readings are considerably more precise than readings obtained with an ohmmeter.

A Wheatstone bridge basically consists of a galvanometer placed within a diamond-shaped arrangement of four resistors. One of these four resistors is the unknown quantity to be measured (R_X in Figure 17–15). The other three resistors are variable resistors and can be adjusted by the operator to balance the bridge. The bridge is balanced when no

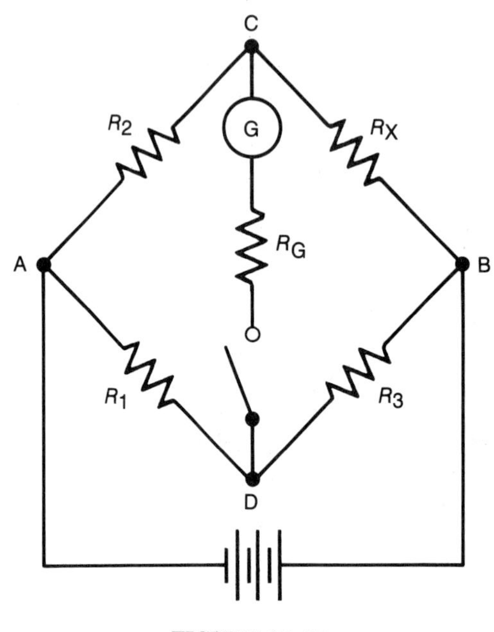

FIGURE 17–15

potential difference exists between points C and D. This condition is indicated when no current flows through the galvanometer and its reading is equal to 0.

Note that a resistance (R_G) has been placed in series with the galvanometer. The purpose of this resistor is to protect the sensitive galvanometer movement from overcurrent during the initial adjustment period. When an approximate balance is reached, this resistor is shorted out in order to obtain maximum accuracy.

When the bridge is balanced, all resistors are in a fixed ratio to one another. It can be shown mathematically that at the time of balance, *the cross products of the opposite sides are equal*. In our example of Figure 17–15

$$R_1 R_X = R_2 R_3$$

The values of the adjustable resistors can be accurately read from a finely divided scale on the bridge; thus, we are in a position to accurately determine the unknown resistance R_X. The following example will make this clear.

EXAMPLE 17–3

Given: The *balanced* bridge circuit of Figure 17–16.

Find: The unknown resistor R_X.

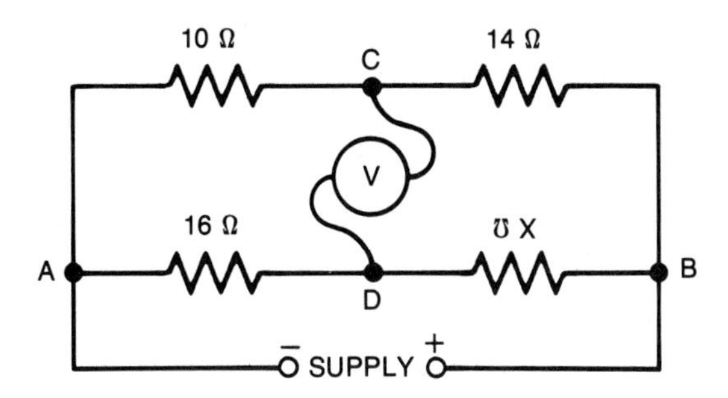

FIGURE 17–16 Wheatstone bridge

SOLUTION

Since the bridge is balanced, the cross-products of the opposite resistors are equal. Thus

$$10\,X = 14 \times 16$$

$$X = \frac{224}{10}$$

$$X = 22.4\ \Omega$$

SUMMARY

- A DC galvanometer consists of a small coil of wire and a pointer connected in an assembly, moved by magnetic action in the field of a permanent magnet; small spiral hairsprings return the pointer back to the 0 mark on the meter scale.
- Galvanometers are also known as d'Arsonval meters or as permanent magnet meters.
- An ammeter is a low-resistance meter and consists of a galvanometer and shunt resistance of low value in parallel with the galvanometer.
- An ammeter is connected in series with the device in which current is to be measured.
- A voltmeter is a high-resistance meter and consists of a galvanometer plus a resistor in series with the galvanometer.
- A voltmeter can be connected directly to a voltage source and must be connected across (in parallel with) the device in which voltage is to be measured.
- An ohmmeter consists of a galvanometer, dry cells, and series resistors. It measures the resistance between the test leads and the instrument. An ohmmeter must be used in a *dead* circuit; this instrument must NOT be used on resistors that have current in them from some other source.
- A wattmeter contains a voltage coil across the line and a current coil in series with the line. A wattmeter will indicate either DC or AC watts.
- There are more electrons on the negative terminal of a device (such as a meter, resistor, or battery) than there are on its positive terminal. (Remember this point

when you must determine either the polarity of a device or the electron current direction. Electrons move from the negative toward the positive terminal.)

- Electrical measurements are based on the accuracy of mathematical formulas and the reliability of precision instruments.
- Wheatstone bridges are laboratory-type, precision instruments for the measurement of unknown resistors.
- When a bridge circuit is balanced, the cross-products of the opposite sides are equal.

Achievement Review

1. State the purpose of each of the following: the hairsprings in a galvanometer, a shunt in an ammeter, and a series resistor in a voltmeter.
2. Under what conditions is each of the following devices used: a Wheatstone bridge, a megohmmeter?
3. Diagram the internal circuit of a voltmeter, an ammeter, an ohmmeter, and a wattmeter.
4. The user of an ohmmeter finds that when the test leads are shorted, the adjustment does not cause the pointer to move to the 0 mark. Instead, the pointer stops at about $R = 5$. Why?
5. Calculate the resistance of the shunt required to convert a 100-microampere meter with a 40-ohm moving coil to a 10-milliampere meter.
6. Calculate the series resistor required to convert the 100-microampere meter in question 5 to a voltmeter with a full scale of 100 millivolts.
7. Calculate the series resistor required to convert the 100-microampere meter in question 5 to a voltmeter scaled to 100 volts.
8. A pair of #28-gauge copper wires in a telephone cable are accidentally short circuited. A Wheatstone bridge is connected to the accessible ends of the wires. The values for the resistances are as follows: $R_1 = 100$, $R_2 = 327.8$, $R_3 = 100$. Calculate the distance from the accessible end of the cable to the point where the pair of wires is shorted. (The temperature is 70°F.)

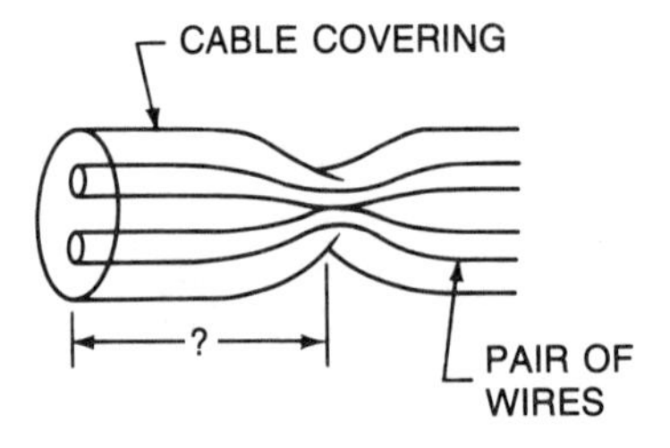

9. a. Using the data from the illustration in Figure 17–7, calculate the voltage across the moving coil and the voltage across the shunt when 2 amperes pass through the meter (use terminals 2 and −).
   b. Using the 10 and − terminals, find the voltage across the moving coil and the voltage across the shunt when 10 amperes pass through the meter.

10. Two resistors, A and B, are connected in parallel. This combination is placed in series with a third resistor, C. This entire group is connected across a 120-volt, DC supply. The resistance of A is 20 ohms. The current in B is 3 amperes, and the current in C is 5 amperes.
    a. Determine the voltage across A, the resistance of B, and the resistance of C.
    b. If resistor A is accidentally open-circuited, find the new voltages across resistor A and resistor C.
11. A 0- to 150-volt voltmeter has a resistance of 2,000 ohms per volt. It is desired to change this voltmeter to a 0 to 600-volt instrument by the addition of an external multiplier. What is the resistance, in ohms, of this external multiplier?
12. A 0- to 150-volt DC voltmeter has a resistance of 100 ohms per volt.
    a. What is the instrument resistance?
    b. What is the instrument full-scale current?
    c. Extend the range of the voltmeter to 750 volts by adding an external multiplier. What is the resistance of this external multiplier?
    d. What is the power dissipation of the external multiplier when the voltmeter is used to measure 750 volts?
13. A d'Arsonval movement has a full-scale deflection at 25 milliamperes, and the coil has a resistance of 2 ohms.
    a. What is the resistance of the multiplier required to convert this instrument into a voltmeter with a full-scale deflection at 300 volts?
    b. What is the resistance of the shunt required to convert this instrument into an ammeter with a full-scale deflection at 25 amperes?
14. The power to a 25-watt lamp is being measured with a voltmeter and an ammeter. The voltmeter has a resistance of 14,160 ohms. The meter is connected directly across the lamp terminals. When the ammeter reads 0.206 ampere, the voltmeter reads 119 volts.
    a. What is the true power taken by the lamp?
    b. What percentage of error is introduced if the instrument power is neglected?
15. A DC instrument has a resistance of 2.5 ohms. It gives a full-scale deflection when carrying 20 milliamperes.
    a. What is the resistance of the shunt required to give the instrument a full-scale deflection when the current is 10 amperes?
    b. What resistance is connected in series with the instrument movement so that a full-scale deflection occurs when the instrument is connected across 150 volts?
16. The resistance of a 0- to 50-millivoltmeter is 10 ohms. This meter is connected with an external shunt in a circuit in which the current is 100 amperes.
    a. Draw a diagram showing the method of connecting the instrument and the shunt in the circuit.
    b. What instrument current causes full-scale deflection?
    c. Determine the resistance of the shunt that is used with the instrument to cause a full-scale deflection.

18 Electromagnetic Induction

Objectives

After studying this chapter, the student should be able to

- Define and explain the new technical terms introduced in this chapter

induction	slip rings
commutator	Lenz's law
sine wave	oscilloscope
cycle	frequency
prime mover	

- Explain the principle of electromagnetic induction
- Predict the direction of induced emf by Fleming's left-hand rule
- Discuss the meaning of Lenz's law
- Describe the induction process within rotating generators

The principal source of electrical energy for all industry is the *generator,* a device for changing the mechanical energy of motion into electrical energy. This energy conversion takes place by the action of magnetic forces, thus the name *electromagnetic.*

An *induced voltage* or *induced current* is one that is produced by the action of magnetic forces rather than by chemical action or other methods.

18–1 MOVING COILS—STATIONARY FIELDS

A simple way of demonstrating the *induction* process is shown in Figure 18–1. A piece of copper wire is connected to the terminals of a sensitive meter and moved downward through a magnetic field, the wire cutting across the lines of force. While the wire is moving, a voltage, or emf, is produced, tending to drive electrons from A toward B. This emf *induced* by the movement of the wire across the field produces a current if a complete circuit exists. When the magnet and wire are kept stationary, no

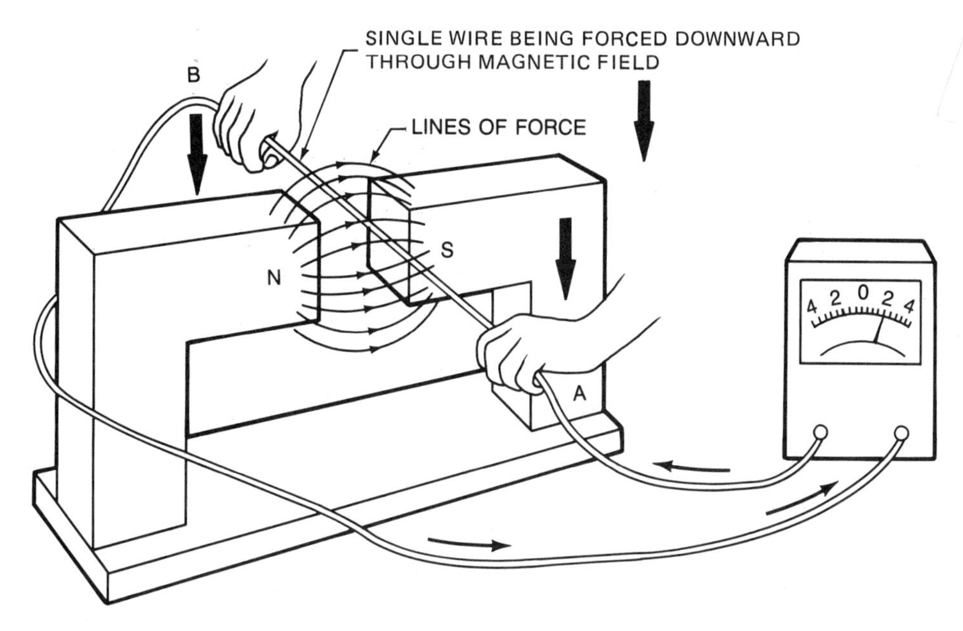

FIGURE 18–1

emf is produced; motion is necessary. A strong horseshoe magnet (preferably Alnico) should be used. The meter can be a millivoltmeter, milliammeter, microammeter, or galvanometer, preferably a zero-center type.

When the wire is moved upward through the magnetic field, Figure 18–2, the meter needle is deflected in the direction opposite to its previous motion. This change shows that the induced emf and induced current have been reversed in direction. If the wire is repeatedly pumped up and down, the pointer on the meter will fluctuate from left to right, indicating the generation of alternating current (AC).

When the wire is moved endwise through the field, as from A to B and back again, no emf is produced. If the wire is moved in a direction parallel to the lines of force, as from S toward N or N toward S, no induced emf is generated. The wire has to move so that it cuts across the lines of magnetic force. This cutting is a quick way of describing the motion that must occur if any voltage is to be produced. (No one need be concerned about any damage to these imaginary lines during the cutting process, for the field is just as strong after the wire has passed through as it was before.)

The induction process is greatly enhanced if the wire is shaped into a coil. Remember, to generate a useful voltage, we can move a coil through a magnetic field.

18–2 MOVING FIELDS—STATIONARY COILS

Up to this point, we have discussed generation of voltage by moving wire so that it cuts across a magnetic field. Actually, it is often just as practical to produce emf by moving the magnetic field so that it cuts across stationary wires.

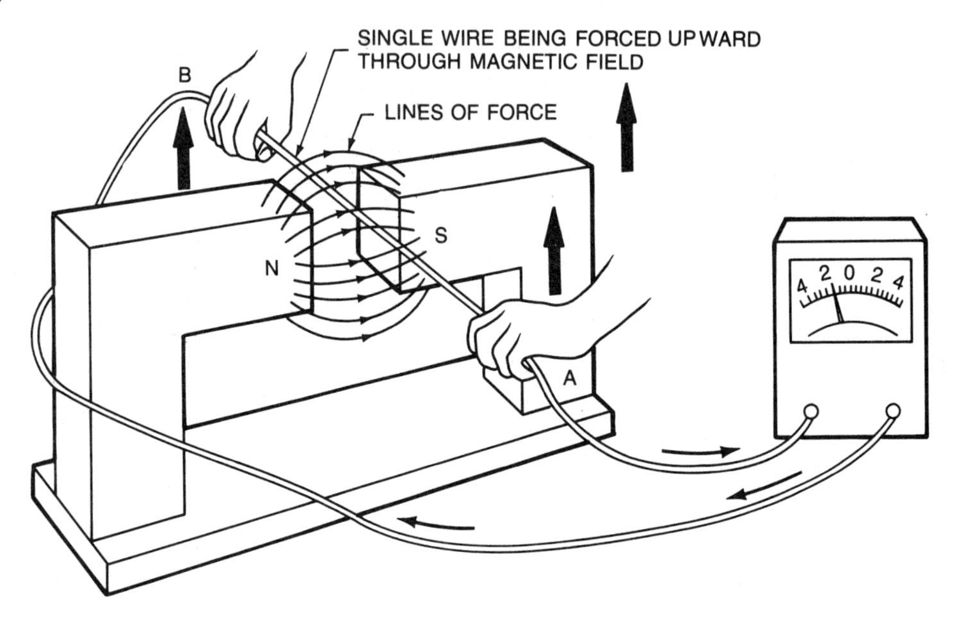

FIGURE 18–2

An emf is produced in a stationary coil, Figure 18–3, when a bar magnet is withdrawn from or inserted into the coil. The lines of force, moving with the magnet, cut across the wires of the coil. By moving different magnets of varying strengths, one can see that a greater voltage is produced by a stronger magnet. By trying coils that differ in number of turns, one can see that more turns produce more emf.

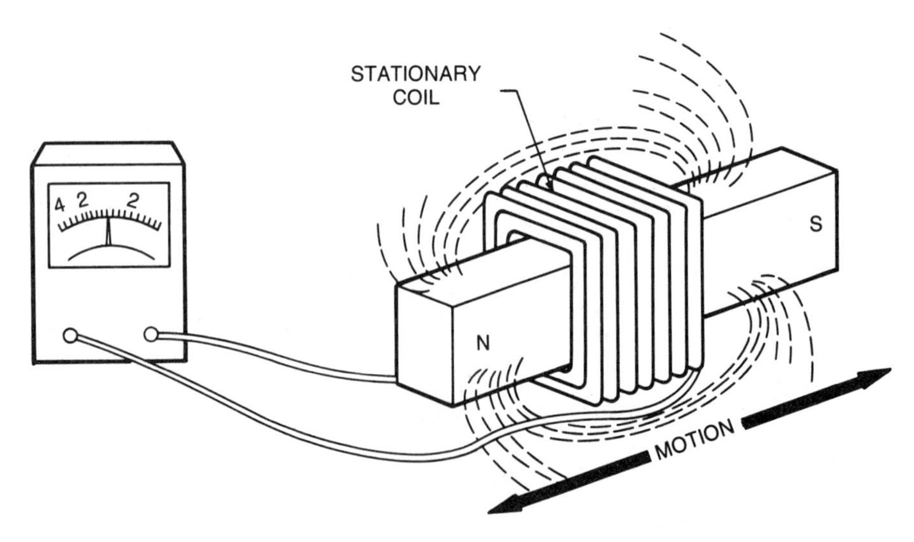

FIGURE 18–3

This last effect can be explained in either of two ways, both amounting to the same thing.

- A coil of ten turns has ten loops that are in series with each other. If 1 volt is produced in each turn, then a total of 10 volts is produced in 10 turns.
- The amount of voltage produced depends on the amount of cutting of lines of force. One line of force cutting across ten wires produces the same voltage as ten lines cutting across one wire. The total amount of cutting across is the same.
- Moving the magnet slowly produces a small emf; faster motion produces more emf.

The amount of emf induced is proportional to the product of three factors that determine the rate of cutting.

- The number of lines of force
- The number of turns of the wire
- The speed of the motion of wires through field or field through wires

When lines of force are cut by wire coil, the relationship of voltage measurements to magnetic field measurements is this.

> To produce 1 volt, lines of force must be cut by wire at the rate of 100,000,000 lines per second.

Recalling that 100,000,000 lines of flux are called a *Weber,* we can state that

> Cutting 1 Wb/sec results in the generation of 1 V.

For instance, when 50,000 lines of force are cut by a coil of 2,000 turns in 1 second, the total cutting is $50{,}000 \times 2{,}000 = 100{,}000{,}000$ lines per second, and 1 volt is induced.

EXAMPLE 18–1

Given: 300,000 lines of flux cut by a coil of 5,000 turns in 2 seconds.

Find: The value of the induced emf.

SOLUTION

The total number of lines cut in 2 seconds is equal to

$$300{,}000 \times 5{,}000 = 15 \times 10^8$$

The rate per second is equal to

$$\frac{15 \times 10^8}{2 \text{ sec}} = 7.5 \times \frac{10^8}{\text{sec}}$$

It follows that

$$7.5 \times \frac{\cancel{10^8}}{\cancel{\text{sec}}} \times \frac{1 \text{ V}}{\cancel{10^8}/\cancel{\text{sec}}} = 7.5 \text{ V}$$

All generators operate on this principle of cutting flux lines with the relative motion of magnetic fields and coils of wire. Remember, cutting 100,000,000 lines of flux per second generates 1 volt.

18–3 LEFT-HAND RULE FOR GENERATORS

The relation of direction of motion of the wire in a field to direction of induced emf can be determined by Fleming's *left-hand rule*. With the thumb, forefinger, and middle finger of the *left* hand each placed at right angles to the other two fingers, Figure 18–4, the *f*orefinger (or *f*irst finger) gives the direction of the *f*ield, the thu*m*b gives the direction of *m*otion of the wire, and the *c*enter finger gives the direction of the induced *c*urrent. This rule does not explain anything; it is merely one of the ways of determining one of these directions when the other two are known.

One way of remembering this rule is to associate the first letters of the fingers used with the first letters of the indicated quantity.

*TH*umb stands for *TH*rust
*F*orefinger stands for *F*lux
*C*enter finger stands for *C*urrent

If you think you understand this concept well, look back at Figures 18–1 and 18–2. Look carefully at all the arrowheads indicating the various directions of field, current, and conductor motion, and see if you can verify the correctness of the information presented in these drawings.

Now let us see if you can apply this left-hand rule to find the direction of current flow in the coil of Figure 18–5 when the magnet is withdrawn from the coil.

Figure 18–5 shows the withdrawal of the magnet from the coil. The magnetic field of the N-pole is moving to the right, cutting across the wires of the coil. If one wishes

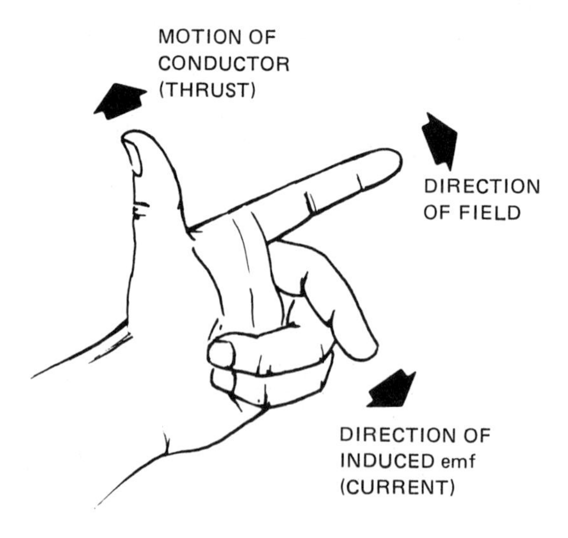

FIGURE 18–4 Left-hand rule

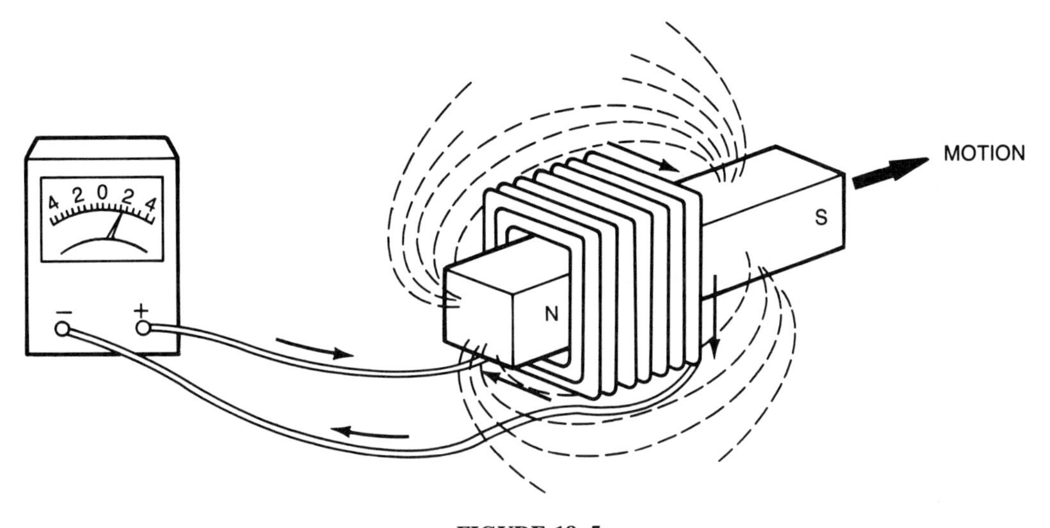

FIGURE 18–5

to determine the direction of induced emf by using the left-hand rule, it must be remembered that the hand rule is based on the relative motion of the *wire*. Pulling the magnet to the right is equivalent to moving the coil of wire to the left. To use the hand rule, the thumb must point to the left, representing the relative motion of the wire through the field.

To find the direction of emf over the top of the coil, the thumb points to the left, the forefinger (for field) upward, and the center finger then gives the current direction toward the observer at the top of the coil. At the bottom of the coil, the thumb still points to the left, the field is downward, and the center finger points away from the observer, giving the current direction around the coil as shown by the arrows on the wire in Figure 18–5.

18–4 LENZ'S LAW

There is a more fundamental way of determining the current direction. As pointed out before, electrical energy is produced by mechanical energy; the hand that removes the magnet from the coil must do some work. The magnet does not push out of the coil by itself; the hand must pull it out.

The coil itself, however, makes it difficult to move the magnet. It creates magnetic poles of its own that oppose the motion of the hand. Figure 18–6 shows the magnet a little more removed from the coil than in Figure 18–5. The induced current in the coil is in such a direction as to develop poles in the coil as shown. (Recall the left-hand rule for a coil.) The attraction of these opposite poles pull the magnet toward the coil.

If the motion of the magnet is reversed, that is, pushed into the coil, the induced current reverses also, developing poles on the coil that repel the approaching magnet.

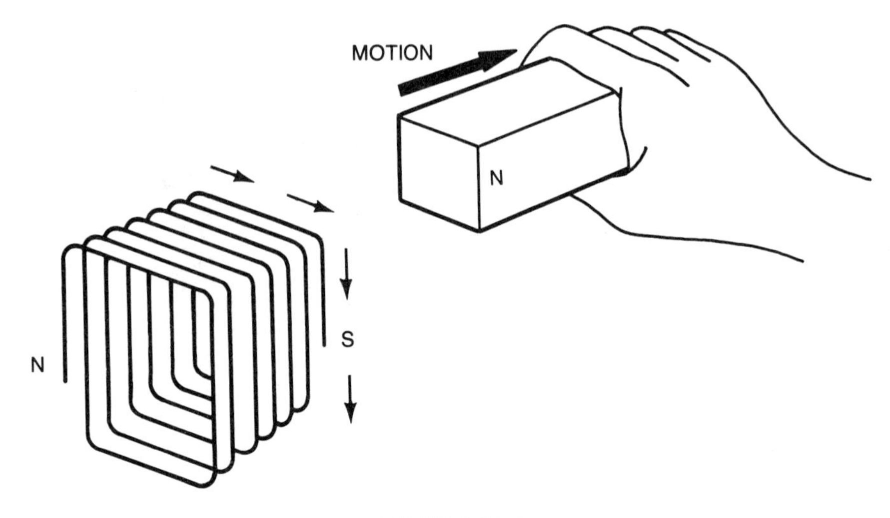

FIGURE 18–6

This general idea was recognized years ago by Heinrich Lenz and is summarized in Lenz's law.

An induced voltage or current opposes the motion that causes it.

Lenz's law is also useful in determining induced current direction in more complex machinery. Note that in Figures 18–3 and 18–5 the coil is surrounded by the magnetic field of the bar magnet. Removing the magnet removes this magnetic field. The induced current in the coil tries to maintain a field of the same strength and direction as the field that is being removed, Figure 18–7. (Apply the hand rule to the coil in Figure 18–6 to determine direction and shape of the field of the coil due to current in it.)

There is another way of looking at Lenz's law. Originally, the coil in Figure 18–8 had no magnetic field in it. As the approaching magnet's lines of force enter the coil, the coil develops a field of its own that tends to restore conditions in the coil to the original zero-field condition, that is, tends to cancel out the oncoming field. Therefore, there is an alternate way of stating Lenz's law.

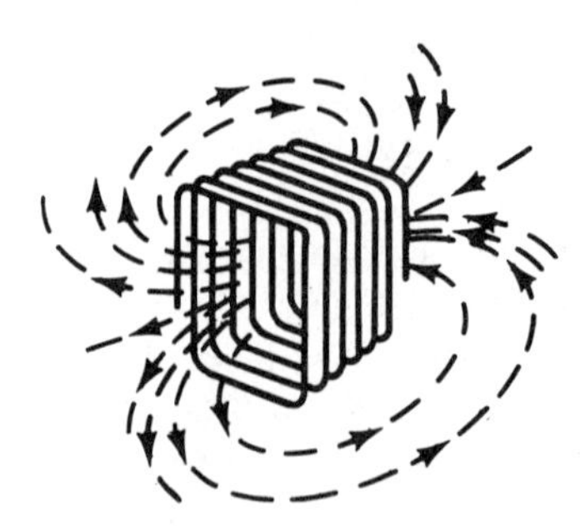

FIGURE 18–7

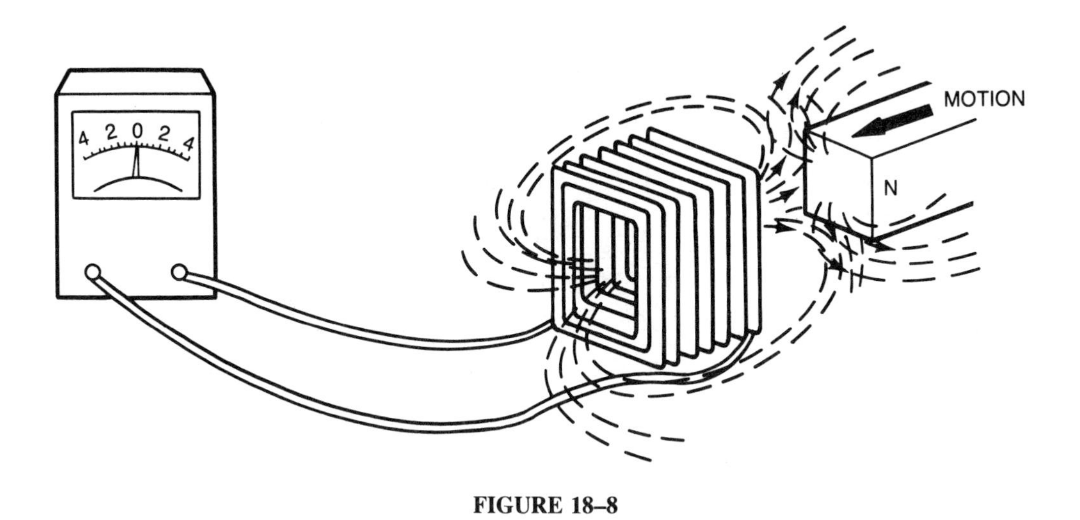

FIGURE 18–8

An induced voltage or current opposes a change of magnetic field.

In the preceding discussion, the terms induced voltage and induced current seem to have been used interchangeably. One should understand that the relative motion of wire and magnetic field always induces a voltage, or emf. If there is a closed circuit, this induced emf causes a current.

18–5 INDUCTION IN ROTATING MACHINES

Commercial generators produce electrical energy either by rotating coils of wire in a stationary magnetic field or by rotating a magnet inside a stationary coil of wire. Let us explore the rotation of a coil within a magnetic field. To analyze this action, we are going to reduce the coil to one single loop and examine the result of its rotation for one revolution. The single loop shown in Figure 18–9 represents the concept of such a simple generator.

Mechanical energy must be expended to rotate this coil within the magnetic field and cut the lines of flux. In practical applications, such energy is provided by turbines or engines called *prime movers*.

As the coil is rotated, an emf is induced, which appears between the two ends of the loop. Two metal bands, called *slip rings,* are attached to these ends to facilitate the transfer of this voltage to an external load. Carbon brushes riding on these smooth metal rings conduct the generated electricity to the circuit, where it can be utilized.

The voltage generated in this manner continually changes in magnitude and direction. Such emf is known as alternating voltage and can be made visible on an instrument known as the *oscilloscope*. The pattern displayed is called a *sine wave,* which shows the variations of the ever-changing voltage as the coil is rotated for one complete revolution.

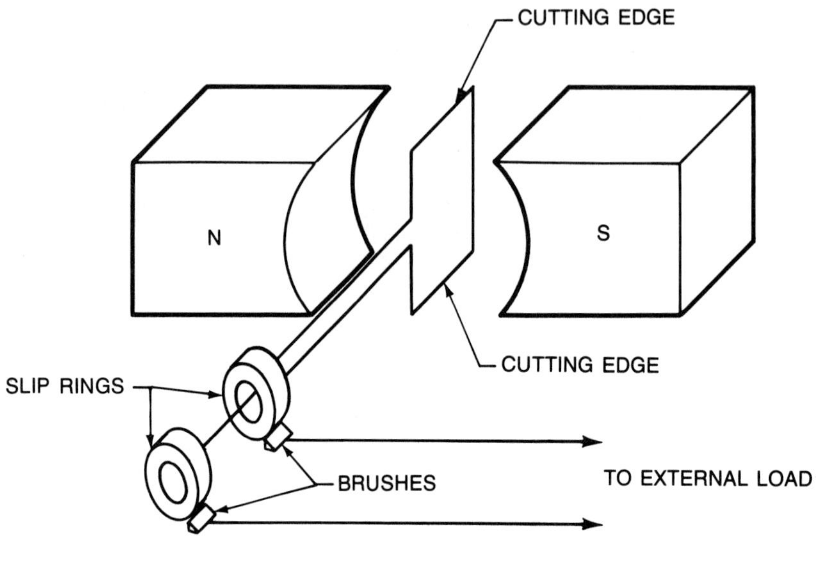

FIGURE 18–9

Figure 18–10 shows such a sine wave, representing the voltage output generated during one revolution. Note the voltage scale to the left of the curve showing both positive and negative values. The sine wave shows how these values change with respect to the angle of rotation through which the coil has travelled. (One circle of rotation = 360 degrees.)

In tracing this curve, we find moments when the voltage output is 0. This happens every 180 degrees, when the cutting edge of the coil moves parallel to the lines of flux,

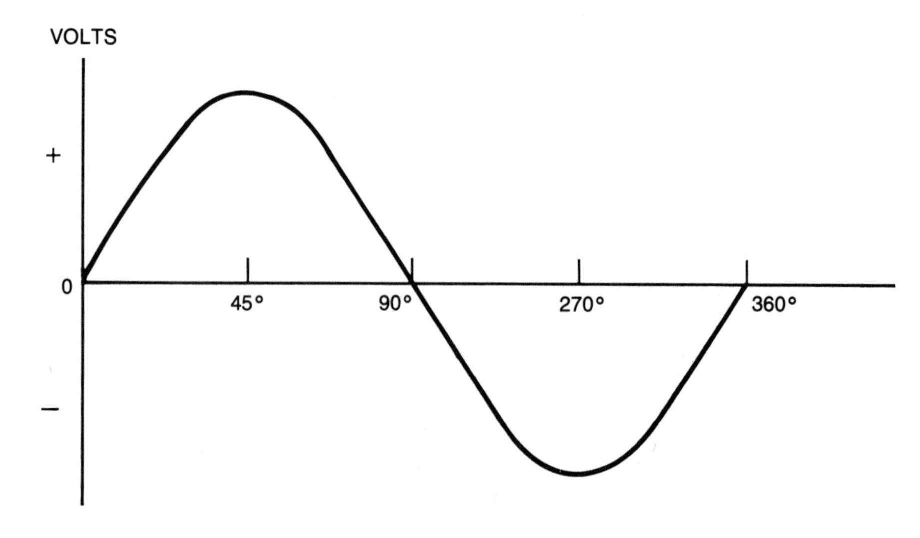

FIGURE 18–10

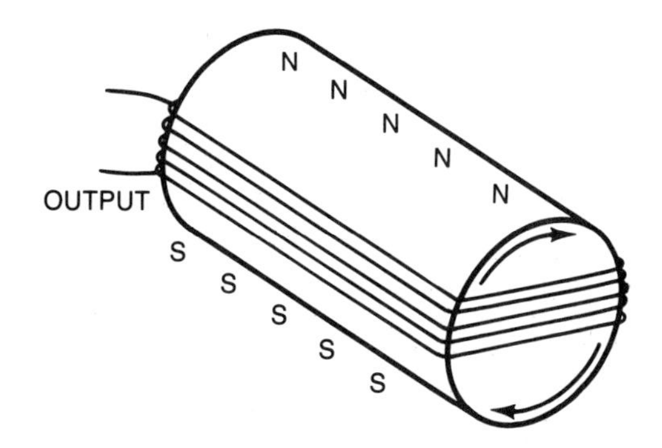

FIGURE 18–11 Rotating magnet in stationary coil

Figure 18–9. By contrast, when the cutting edges move perpendicular to the flux lines (beneath the poles), the rate of cutting flux lines—and therefore the voltage output—is at its maximum. This is shown at the 90-degree and 270-degree points of the sine wave.

Recall that not all generators are designed to rotate coils within a magnetic field but, instead, rotate a magnet within a stationary field.

The largest generators built contain stationary coils in which useful current is induced by a rotating magnet. Figure 18–11 shows one stationary coil. The cylindrical object shown within the coil represents a rotating magnet that is magnetized along a diameter.

Let us assume that this cylindrical magnet is rotated within the coil and, as the magnetic poles revolve, its flux lines are cut by the stationary conductors of the coil. Such rotation produces an alternating current, just as the repeated motion of the wire in Figure 18–1 produces AC.

To stress an important point once more, we can state that one such revolution produces one cycle of alternating voltage output in the form of a sine wave. Commercial generators in the United States are standardized to produce 60 cycles per second. This standard is known as the *frequency* of the power supply.

As you progress in your study of electricity, you will learn more about AC. What you need to know at this time is that the induction process in rotating machines produces AC. If direct current (DC) is needed, it can be obtained from AC by using rectifiers.

DC generators achieve rectification by using a *commutator* instead of slip rings. This is one of the facts to be presented in the next chapter.

SUMMARY

- The voltage produced by generators is called induced emf. This emf is produced either by the motion of wires across a magnetic field or by the motion of a magnetic field across wires.
- This induction process in a generator converts mechanical energy into electrical energy.

- The amount of induced emf depends on the strength of the magnetic field, the number of turns of wire in the device, and the speed of motion.
- To produce 1 volt, lines of force must be cut by the wire at the rate of 100,000,000 per second.
- Lenz's law: An induced voltage or current opposes the motion that causes it. An induced voltage or current opposes a change of magnetic field.
- Alternating current is generated in stationary coils when a magnet is rotated inside the coil.
- Alternating current is generated in a coil when the coil is rotated in a stationary magnetic field. By means of a commutator, this current is fed into the outside circuit as direct (one-way) current.

Achievement Review

1. A wire is moved through a field, as in the sketch below. What is the direction of the induced emf?

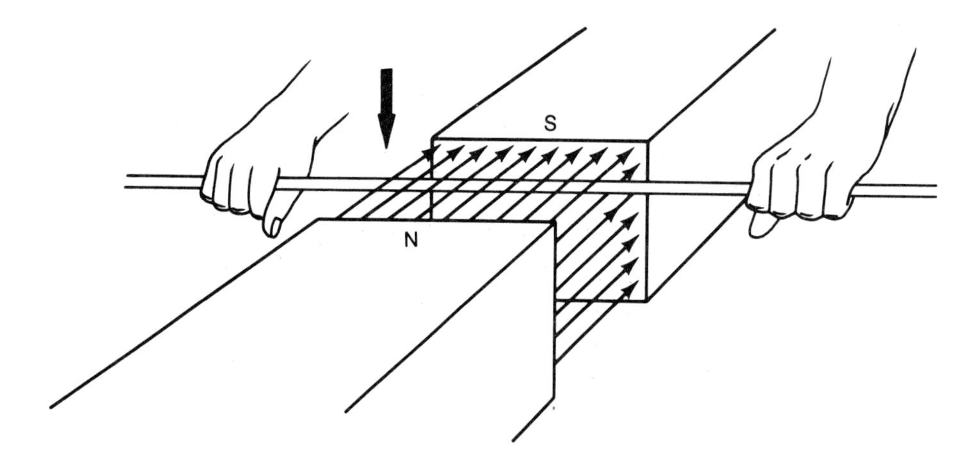

2. In a generator, from where does the electrical energy come?
3. The electric automobile of 1912 was powered only by storage batteries and an electric motor. The necessity of frequent battery charging helped make these cars obsolete. Could a generator, belt-driven from the wheels of a trailer, charge the batteries to make longer trips possible?

4. State three factors on which the amount of induced voltage depends.
5. In a certain generator, like that shown in Figure 18–11, the loop of wire encloses 10,000 lines of force. What is the total number of lines of force cut during one complete rotation of the magnet?
6. If the magnet of the generator in Figure 18–11 is rotated at a speed of 2,400 rpm, how many lines of force are cut each second? How much voltage is produced?
7. In the generator in the sketch below, determine which brush is positive and which is negative.

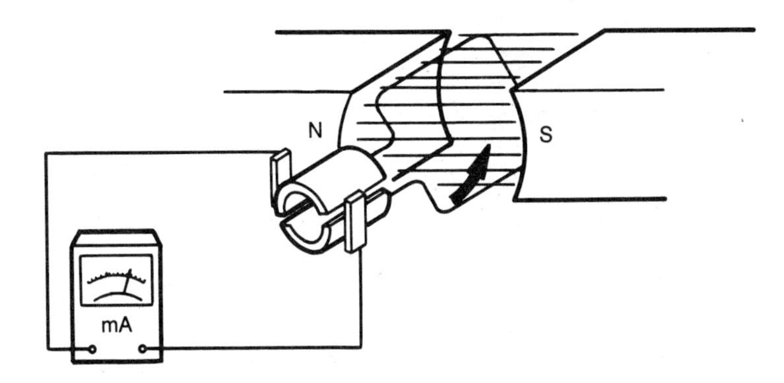

8. State Lenz's law. State the left-hand generator rule. Which one of these is the more important fundamental principle?
9. Figure 18–5 shows a magnet being pulled out from a coil to the right. If the magnet is pulled out to the left, is the induced current in the same direction as shown? Explain your answer.
10. A copper disk is supported between magnet poles so it can be rotated, as shown in the sketch. What happens in the disk as it rotates?

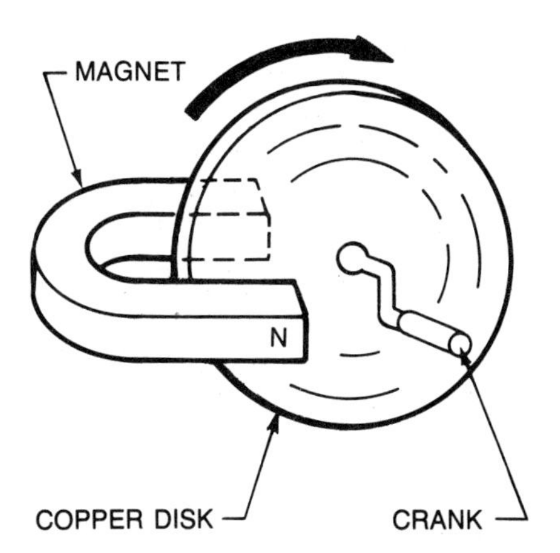

11. Does the strength of the magnet in the previous sketch have any effect on the amount of force needed to turn the disk?

12. A coin spinning on the end of a thread is lowered into a magnetic field. Explain what happens and why it does happen.
13. An aluminum cup is hung upside down on a pivot over a rotating magnet, as shown in the sketch below. Explain what happens to the cup.

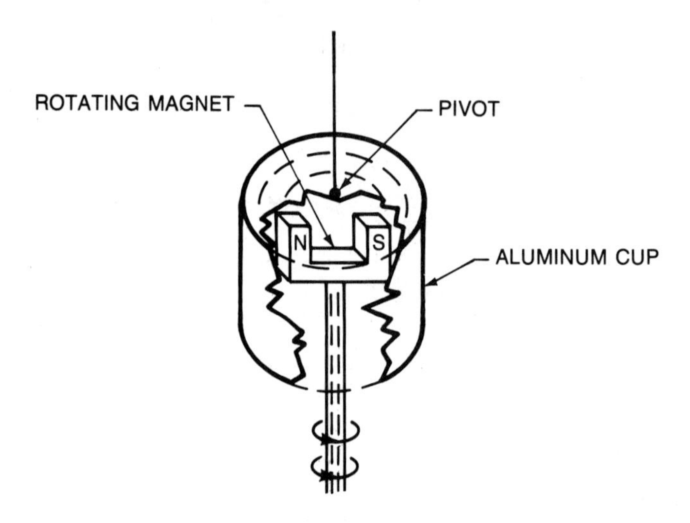

19 DC Generators

Objectives

After studying this chapter, the student should be able to

- Define and explain the technical terms introduced in this chapter

brushes	magnetohydrodynamic
field coils	armature
drum winding	lap winding
frogleg winding	wave winding
eddy current	mica
form winding	lamination
separately excited	magneto
armature reaction	self-excited
series generator	shunt generator
compound generator	interpole
compensating winding	voltage regulation
overcompounded	undercompounded
flat compounded	differentially compounded
stray losses	temperature rise
copper losses	

- Name all components of a DC generator
- Explain the function of various generator components
- Describe the theory of operation of different types of DC generators
- Evaluate the operating characteristics of different types of DC generators
- List advantages and disadvantages of different types of generators
- Recognize different kinds of generators by their construction details and hookup
- Draw accurate schematic diagrams of different generator configurations
- Connect generator windings in accordance with established practices
- Correctly connect rheostats for the purpose of controlling voltage output
- Perform correct mathematical calculations to predict electrical quantities related to generators

19–1 PRINCIPLES OF DC GENERATORS

Basically, direct current is generated in rotating coils surrounded by a stationary field magnet, Figure 19–1. First, observe in detail what happens when one loop of wire is rotated in the field between the N and S poles. In order to produce direct (one-way) current in the outside circuit served by the generator, the ends of the loop are fastened to semicircular metal strips, insulated from each other, that rotate with the loop. These two half-circle segments form the part of this generator that is called the *commutator,* Figure 19–2.

Two stationary blocks of conducting graphite, called *brushes,* are held in contact with the commutator. These stationary brushes carry the generated DC to the outside circuit.

The purpose of this arrangement is to achieve a sort of mechanical rectification to

FIGURE 19–1 Two-pole field magnet

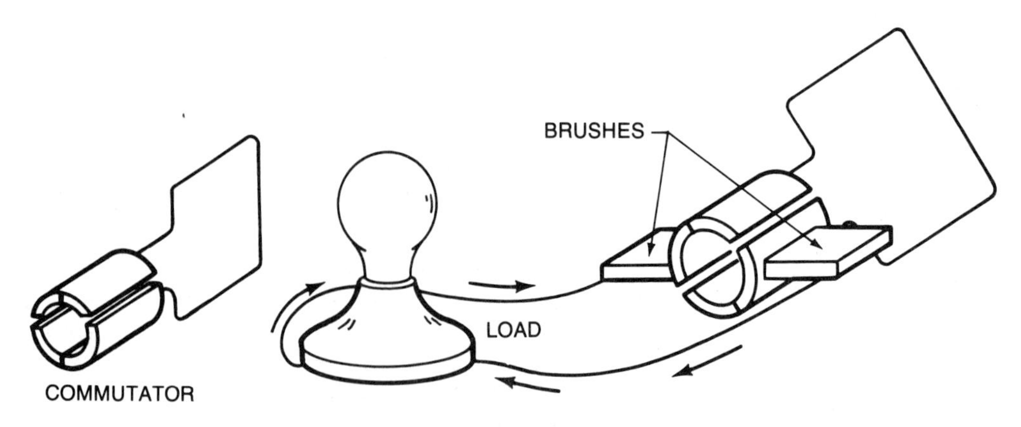

FIGURE 19–2

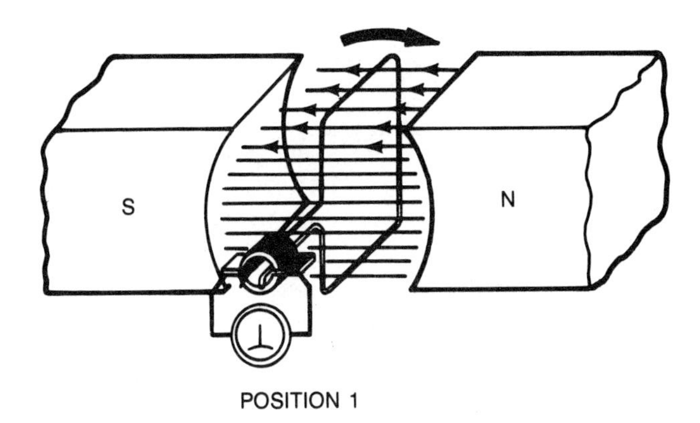

FIGURE 19–3 Rotating loop between poles of stationary magnet

assure that current always flows in the same direction through the load. To achieve this task, the brushes must be placed on opposite sides of the commutator in such a manner that the rotating commutator segments pass from one brush to the next at the moment the induced emf drops to zero. Then, the brushes will maintain a constant polarity, even though the current in the loop continues to reverse itself periodically.

The following series of diagrams show successive positions of the steadily rotating loop. N and S represent poles of a horseshoe magnet.

At the instant represented by position 1, Figure 19–3, no emf is produced, because the wires are moving parallel to the field and are not cutting the lines of force. (The black half of the loop is at the top in the vertical position.)

As the loop moves from position 1 to position 2, lines of force are cut at an increasing rate, even though the rotation rate is steady. A, B, C, and D in Figure 19–4 show the relative number of lines cut during the *equal* time intervals between positions 1 and 2.

At position 2, Figure 19–5, the sides of the loop are cutting lines of force at the maximum rate. The induced current *in the loop* (found by either left-hand rule or by Lenz's law) is a flow of electrons directed *toward* the brush on the right and *away* from the brush on the left. This forcing of electrons from the rotating coil toward the right-

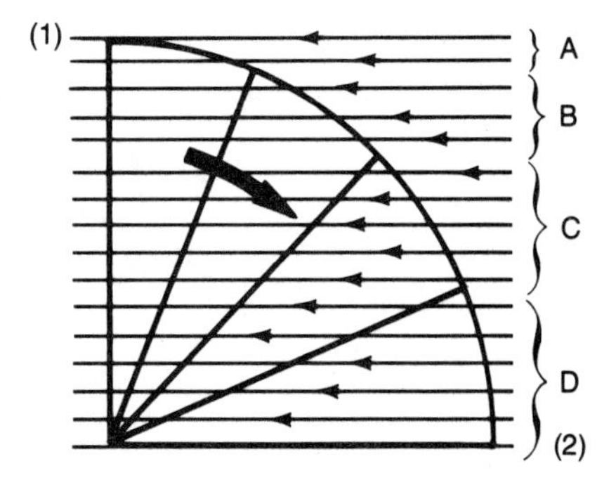

FIGURE 19–4 Lines of force cut during equal time intervals A, B, C, and D.

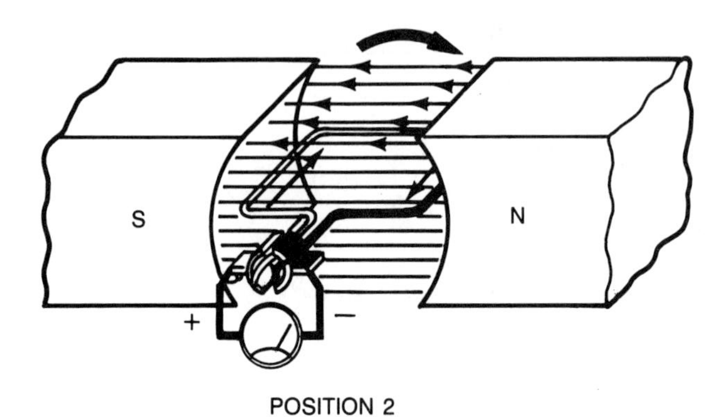

POSITION 2

FIGURE 19–5

hand brush gives this brush a negative charge. The removal of electrons from the left-hand brush gives it a positive charge. In the stationary wiring of the external circuit, electrons flow from the negative brush to the positive brush.

Between positions 2 and 3, Figure 19–6, current continues to be induced in the same direction but decreases in amount as the wire approaches position 3. At position 3, the two segments of the commutator are short-circuited but with no ill effect, because at this instant no emf is being generated. At this position no lines of force are being cut.

Between positions 3 and 4, Figure 19–7, increasing voltage and current are produced, just as between positions 1 and 2. But, notice that the black side of the loop, previously moving downward through the field, is now moving upward; thus, the current direction in the black side is reversed. (Compare diagrams for positions 2 and 4.) However, the black side of the loop is now taking electrons from the left brush, forcing

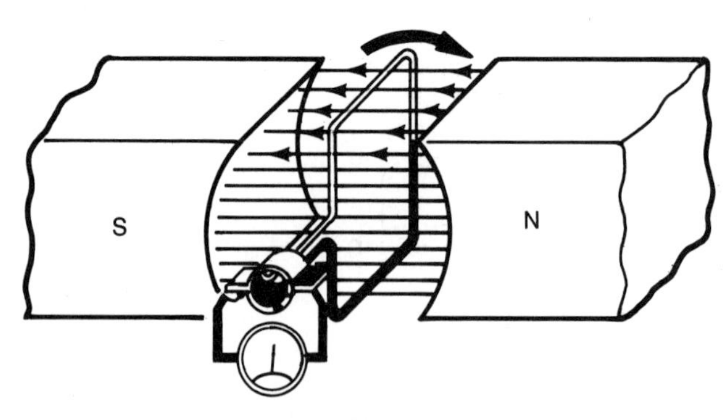

POSITION 3

FIGURE 19–6

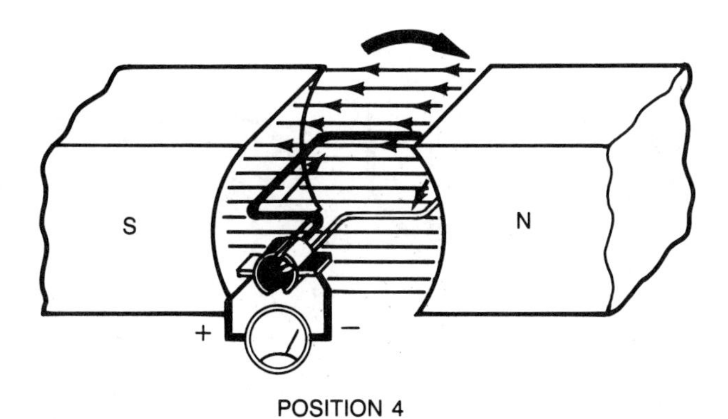

POSITION 4

FIGURE 19–7

them around through the white side of the loop toward the right-hand brush; thus, the charges on each brush remain the same as before.

In the rotating loop itself, the generated current *alternates* in direction. Watch the black half of the commutator as it rotates. In position 2, the black segment is negative and supplies electrons to the right-hand brush. In position 4, the black segment is positive; but at the instant it becomes positive, it pivots away from the right-hand brush and contacts the left brush again. Similarly, the white segment of the commutator changes position as the current direction in the loop changes. As a result, the right-hand brush is always supplied with electrons and the left-hand brush is always losing electrons into the rotating loop. Thus, the electron flow through the outside load is maintained in one direction.

The graph in Figure 19–8 shows how the voltage (or current) changes in amount during one revolution of the loop. This type of current is called pulsating DC in contrast to the smooth and steady DC obtained from a battery.

- Because one loop of wire produces such a small voltage, a coil of many turns is used on a real generator. A coil of 100 turns makes 100 times as much voltage as a single-turn coil at the same rpm and magnetic field strength. If the rotating coil is wound on an iron core, the added iron creates a stronger magnetic field where the coil operates. Adding the iron core between the poles of the same stationary (field) magnet increases the number of lines of force by several times, resulting in increased emf generated, Figure 19–9.
- One disadvantage of the simple generator still needs correcting. This coil of wire, rotating in a strong field, produces an emf that rises and falls to zero twice during each rotation, Figure 19–8. Mechanically, this generator is like a one-cylinder gas engine. Electrically, current that varies in this fashion causes difficulties in the generator and in the circuit that it feeds.

To produce a smoother and steadier emf and current, more coils are placed on this rotating iron core. They are connected to a several-segment commutator in such a fashion that the voltage produced by the generator is the sum of several individual coil

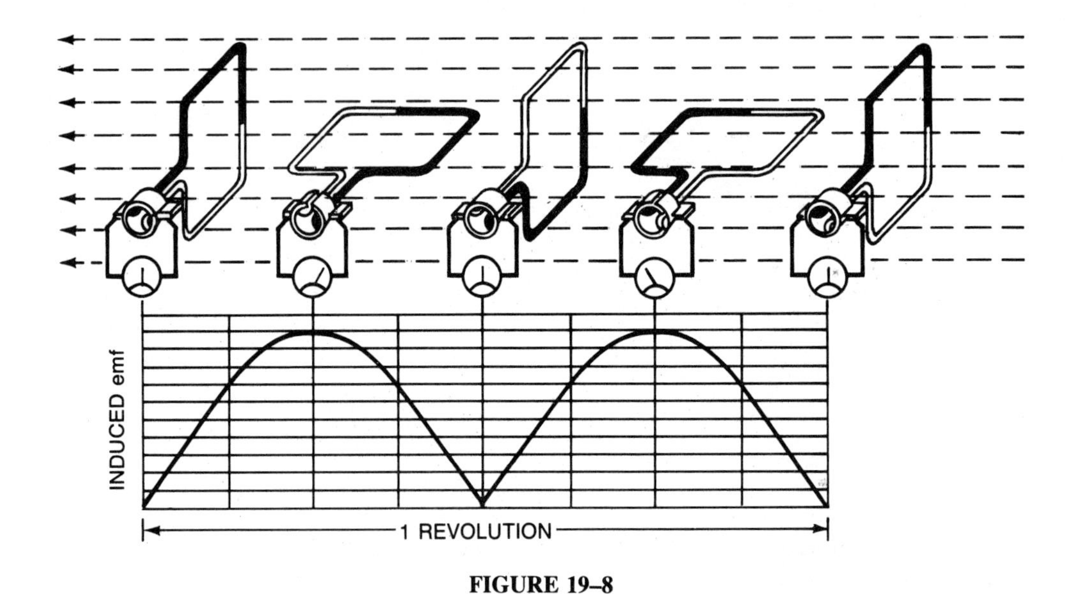

FIGURE 19–8

voltages. At certain instants, the voltage generated in one coil still drops to zero, but other coils in series are then producing voltages high enough so that the total output voltage is fairly constant.

Let us summarize the foregoing statements.

> A practical generator requires more than one loop. Additional loops will reduce the fluctuations of the pulsating DC output and, at the same time, increase the voltage output.

Figure 19–10 illustrates a two-loop generator. Note that the added loop, placed 90 degrees apart, requires two additional commutator segments. (There are always twice

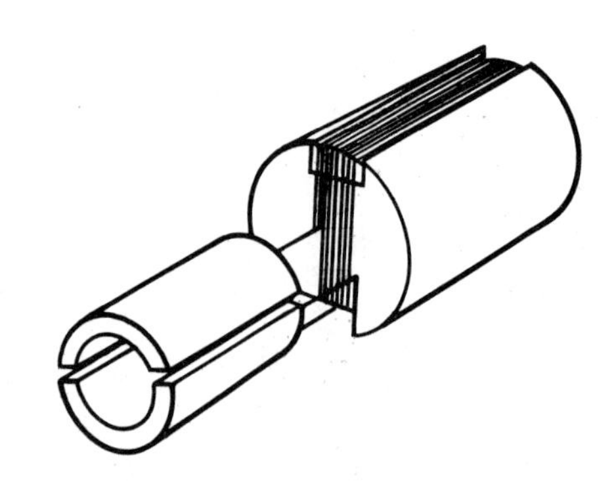

FIGURE 19–9 Rotating coil wound on iron core

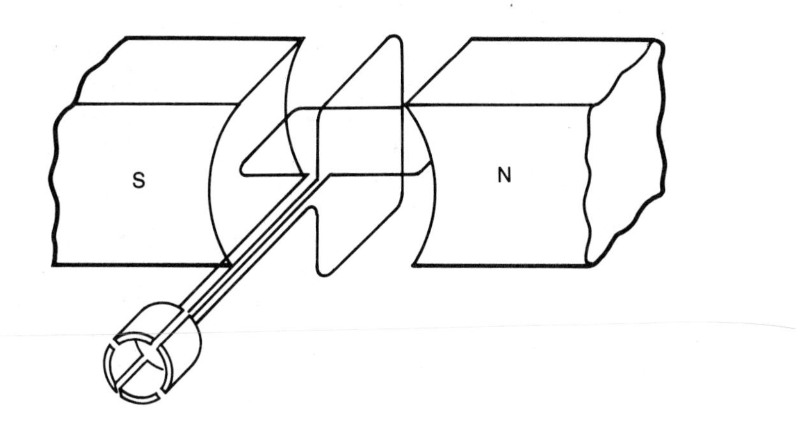

FIGURE 19–10

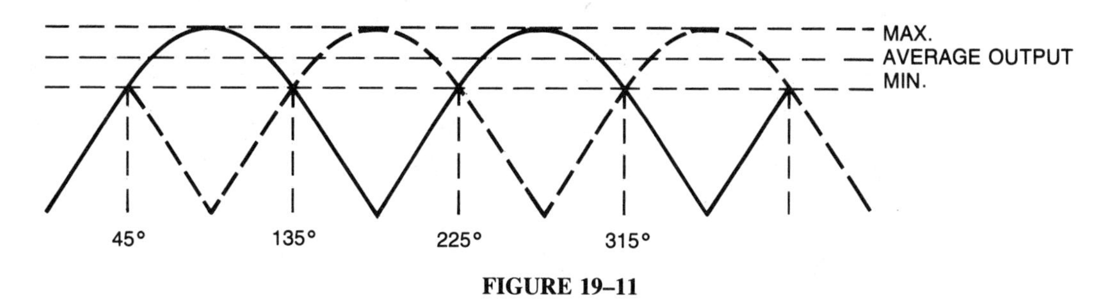

FIGURE 19–11

as many commutator segments as there are loops.) The voltages delivered from these two loops are equal but 90 degrees apart, Figure 19–11.

Note where the two waveforms intersect (at 45°, 135°, 225°, and 315°). These are the points where the brushes switch from one commutator segment to the next. The significance of these points can be appreciated when we realize that the voltage no longer drops down to zero but rises again to its maximum value. In fact, the switchover points represent the lower limit of the fluctuating voltage output. Careful inspection of Figure 19–11 reveals that the average voltage output lies halfway between this minimum and the maximum.

19–2 ARMATURE DESIGN

This rotating assembly of coils, placed in slots in an iron core and connected to commutator segments, is called the *armature* of the generator, Figure 19–12.

Small DC armatures are easy to obtain for study and disassembly. There are millions in use in automobiles and aircraft. Armatures from DC motors and from universal (AC-DC) motors, used in vacuum cleaners and hand drills, are similar in construction.

The core of such armatures is generally constructed from soft silicon-steel laminations. The laminations, punched from thin sheet steel, are insulated from one another by a thin coating of enamel to prevent the circulation of induced eddy currents within

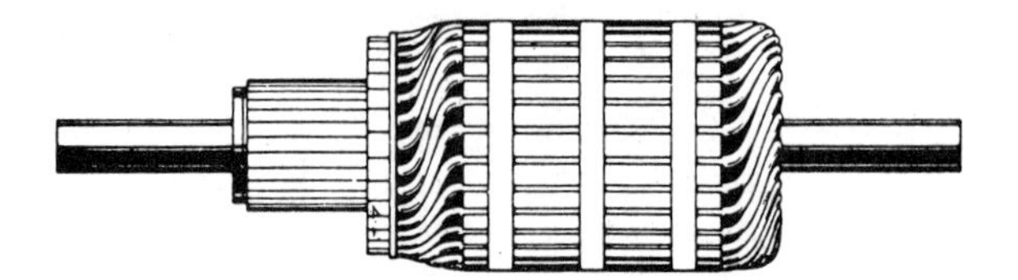

FIGURE 19–12 Generator armature

the core. In addition, hysteresis losses are minimized by the use of silicon-enriched steel alloys for the construction of rotating machines.

The armature winding may be placed into the slots of the core, Figure 19–13. A different technique, known as *form winding,* utilizes preformed coils, which are preshaped, insulated, and fitted to be dropped, ready-made, into their assigned armature slots. The coils are then held in place by small, insulating wedges.

The coil leads are then brought back to the commutator, where they are terminated by soldering. Examine the construction of a commutator, if available, to see that all segments are insulated from one another with a heatproof insulating substance known as *mica*.

Figure 19–13 shows the wiring of an armature suitable for use in a simple generator that has one pair of field-magnet poles. At first glance, this sketch seems complicated,

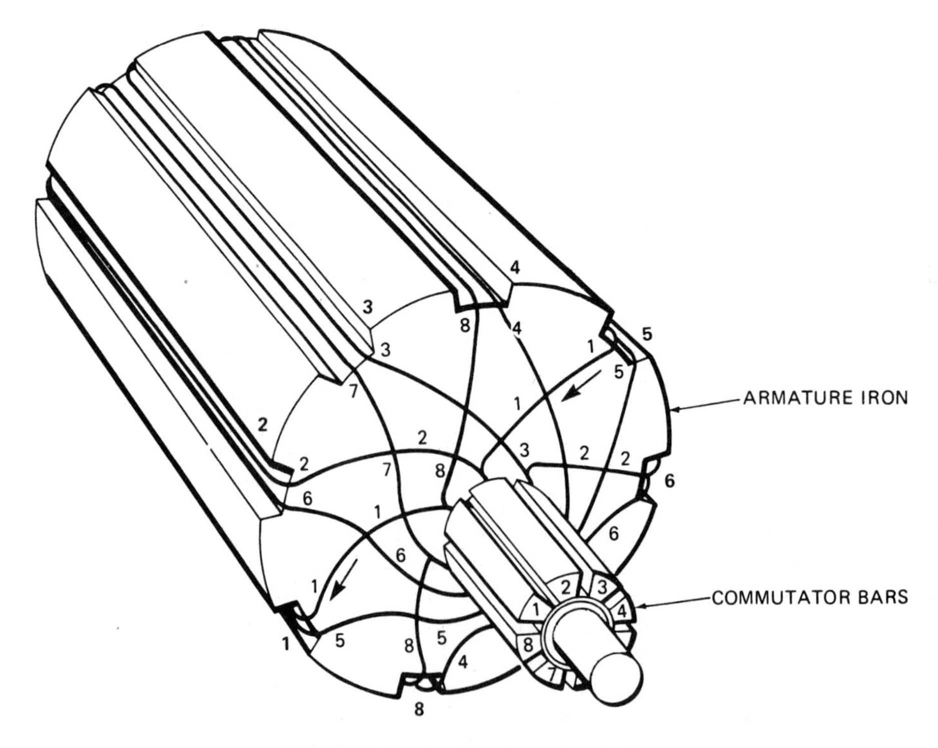

FIGURE 19–13 Simplex lap winding

but like everything else, it becomes simple after one takes a few minutes to trace the circuit and understand it.

To trace the wiring, start on commutator bar 1. Wire 1 leads to the left through slot 1, is brought over to slot 5 at the back end of the armature, and through slot 5 to the front end, where it connects to bar 2. This traces one loop of coil 1. In actual practice, wire 1, starting from bar 1, is wound through slots 1 and 5 many times, forming the coil, with the end of the coil connecting to bar 2.

Similarly, trace coil 2. Starting on bar 2, wire 2 leads to slot 2, across the back of the armature to slot 6 (and around through slots 2 and 6 many times), with the end of the coil connecting to bar 3. Coil 3 uses slots 3 and 7; its ends connect to bars 3 and 4. Coil 4 uses slots 4 and 8, connecting to bars 4 and 5. Coil 5 uses slots 5 and 1, connecting to bars 5 and 6. Coil 6 uses slots 2 and 6, along with coil 2, but coil 6 connects to bars 6 and 7. Coil 7 uses slots 3 and 7, and connects to bars 7 and 8. The last coil, 8, starts on bar 8, through slot 8, around the other end of the armature and through slot 4, after which it connects to bar 1.

Types of Armature Windings

There are several ways, many of them obsolete, of arranging wire and iron to make an armature. The so-called *drum winding* is the only one now in extensive use. Drum windings can be arranged in several ways.

- Low-voltage, moderately high current generators have a *lap winding*. Figure 19–13 shows a simplex lap winding. Two simplex windings on the same armature form a duplex winding. Some generators have more than 1 pair of field poles, which changes the armature coil layout because each coil must be placed so that when one side of the coil is passing an N-pole, the other side is passing an S-pole. Lap-wound generators have as many pairs of brushes as pairs of poles.
- High-voltage, low-current machines use a somewhat different pattern of wires on the armature drum called a *wave winding*.
- The largest DC generators are built with a combination lap and wave winding, which is called a *frogleg winding*.

Regardless of the type of winding chosen, the coil is always wound so that its opposite sides pass beneath opposite magnetic poles simultaneously.

Each of these three types can be subdivided into many varieties, depending on voltage and current requirements, number of commutator segments, and number of field poles. A complete discussion of them fills entire textbooks on armature wiring. A text should be sought by anyone who needs further details on armature wiring.

Lap Windings Analyzed

To see how the drum-wound armature behaves when rotated between the poles of its field magnet, return to Figure 19–13. Imagine the N-pole of a large magnet placed at the left of the armature and the S-pole at the right. Thus, the armature is in a magnetic field directed from left to right. Assume the armature is rotated counterclockwise; the wires in slots 1, 2, and 3 are moving downward through the field. Applying the left-

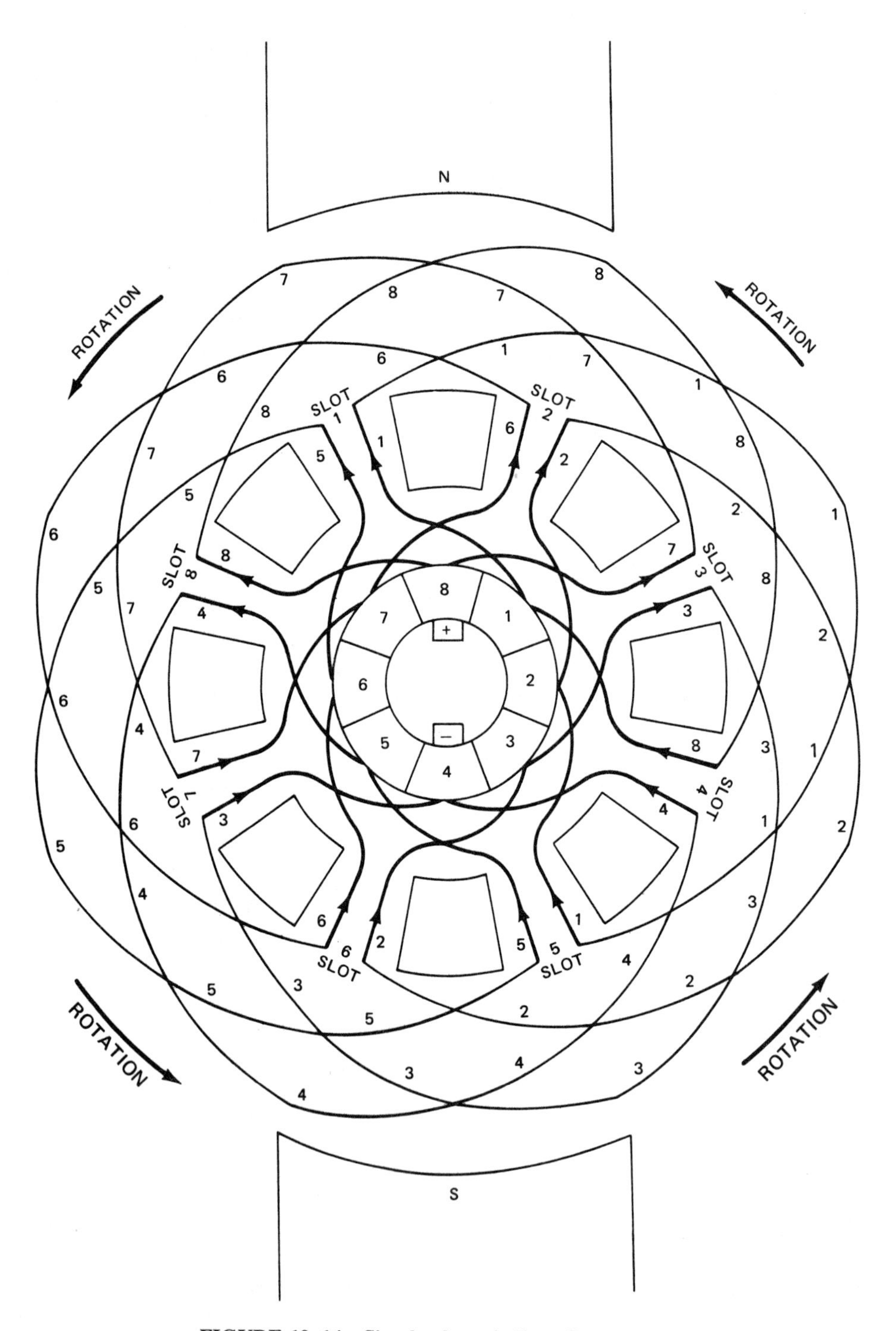

FIGURE 19–14 Simplex lap winding, flattened view

hand rule to these wires (forefinger . . . field, toward the right; thumb . . . motion, downward; center finger . . . current, away from the observer), we find the direction of emf in these wires is away from the commutator, toward the back of the armature. In slots 5, 6, and 7, the wires have an induced emf toward the commutator. (Use three-finger left-hand rule again.)

To obtain a more complete view of all of the wires and what occurs in them, we use a different type of diagram. Let us flatten out the armature of Figure 19–13 in Figure 19–14.

The front (commutator) end stays in place, and the back and wiring are spread and stretched so that they form the outside circumference of this flattened armature, Figure 19–14. Near the center are the numbered ends of the eight commutator segments, as in Figure 19–13. Around the commutator are the commutator end wiring connections, just as in Figure 19–13. The eight near-rectangles are the eight cylindrically curved faces of the iron armature. The outermost lines are the expanded crossovers of the coils at the back, not visible in Figure 19–13.

A diagram like this is useful for tracing the current path for each coil. (The use of colored pencils is very helpful.) This technique should reveal that each coil overlaps the preceding one; hence the name *lap winding*. It can also be seen that the ends of each coil are connected to adjacent commutator segments, which is typical of lap windings.

Careful analysis of the tracing process results in a schematic diagram that is even more descriptive. Figure 19–15 is the end product of our tracing process and reveals two parallel sets of coils.

This demonstrates another characteristic of lap windings, having as many parallel pathways as they have poles. Remember, this is a two-pole machine. If we were dealing with a four-pole machine, there would be four parallel branches.

One set of brushes is provided for any two parallel branches, and the instantaneous voltages, generated by the coils between them, add up to the same amount. At the instant pictured in Figure 19–14, for example, coils 1, 2, 5, and 6 might be generating 9.2 volts each and coils 3, 4, 7, and 8 generating 3.8 volts each.

In Figure 19–15 notice that coils 8, 1, 2, and 3 are in series. Their individual voltages add: $3.8 + 9.2 + 9.2 + 3.8 = 26$ volts. Coils 7, 6, 5, and 4 in series also produce 26 volts. With the two sets of coils in parallel, the total instantaneous voltage is still 26.

The current (amperes) delivered by this generator is the sum of two equal currents in the two parallel sets of coils. If the generator delivers 10 amperes to the external circuit, then the current in each coil is 5 amperes.

Figure 19–15 may be puzzling to someone who has learned that electrons flow from negative to positive. They do, in the *external* circuit. *Inside* the generator, however, electrons whipped by lines of force are driven from the positive terminal toward the negative terminal. This motion is the cause of the positive and negative charges developed at the terminals.

Lap-wound armatures require as many brushes as there are field poles and have as many parallel paths through the armature as there are brushes. This quantity of brushes and paths is advantageous in low-voltage, high-current armatures. The total current output can be large without excessively large wire on the armature. For example, a six-

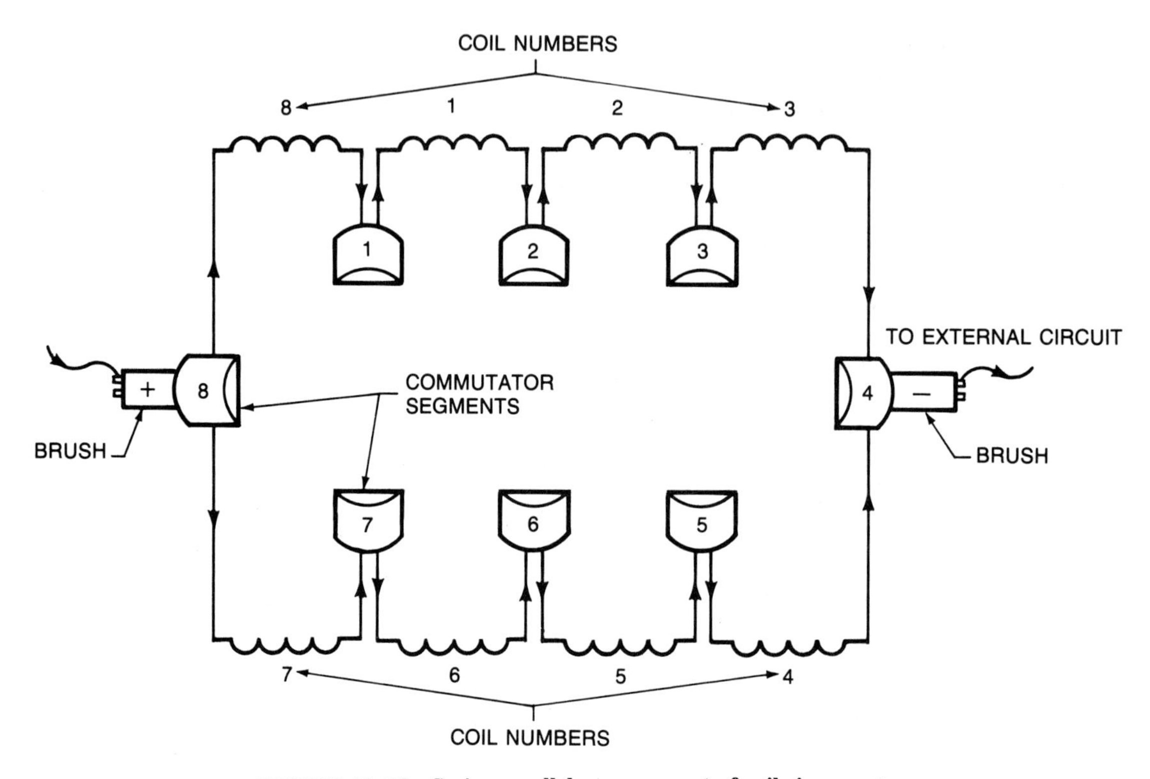

FIGURE 19–15 Series-parallel arrangement of coils in armature

field-pole lap-wound generator has six brushes and six paths through the armature. To produce 180 amperes, each coil in the armature need carry only 30 amperes.

Wave windings have only two parallel paths, regardless of the number of field poles.

19–3 GENERATOR FIELD STRUCTURES

DC generators can be classified by the method used for providing the magnetic field. This classification can be tabulated like this.

1. Permanent magnet generators
2. Separately excited generators
3. Self-excited generators
   a. shunt generators
   b. series generators
   c. compound generators

Permanent magnet generators are reserved for a few low-power applications where control of field strength is not needed. Such constant-field generators are useful in control devices or circuits. In such applications, use is made of the permanent magnet generator's characteristic to deliver a voltage output proportional to its speed.

Permanent magnet generators are also known as *magnetos* and find applications with the electrical systems of motorcycles, small tractors, and the like.

The *field structure* of a permanent magnet generator is similar in design to that shown in Figure 19–1. Generally, though, electromagnets are used instead of permanent magnets. In this case, the circular frame, or yoke, is fitted with laminated iron pole pieces to accommodate the field winding, Figures 19–16 and 19–17.

Look at these drawings and note how the magnetic poles are developed in accordance with the left-hand rule for coils and the direction of the electron current. Furthermore, note that the *magnetic* circuit is completed by the iron yoke, or frame, which carries and concentrates the lines of magnetic flux. Note the positive and negative polarity marking on the wires that supply the field coils. It is the electric polarity of the power source that determines the direction of the current flow and, therefore, the orientation of the magnetic poles.

This is shown by the schematic diagram in Figure 19–18. If these field coil wires are attached to a storage battery or to a rectifier, the generator is said to be *externally excited,* or *separately excited.*

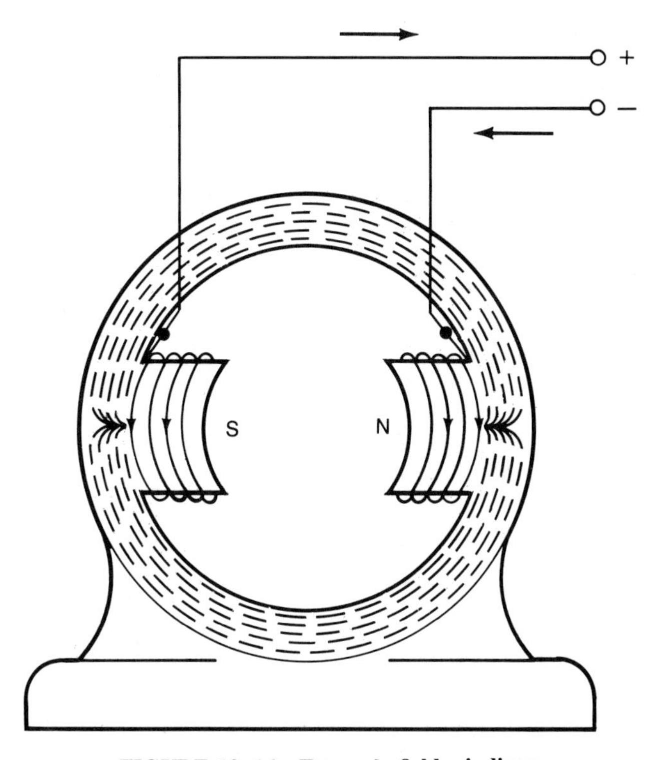

FIGURE 19–16 Two-pole field windings

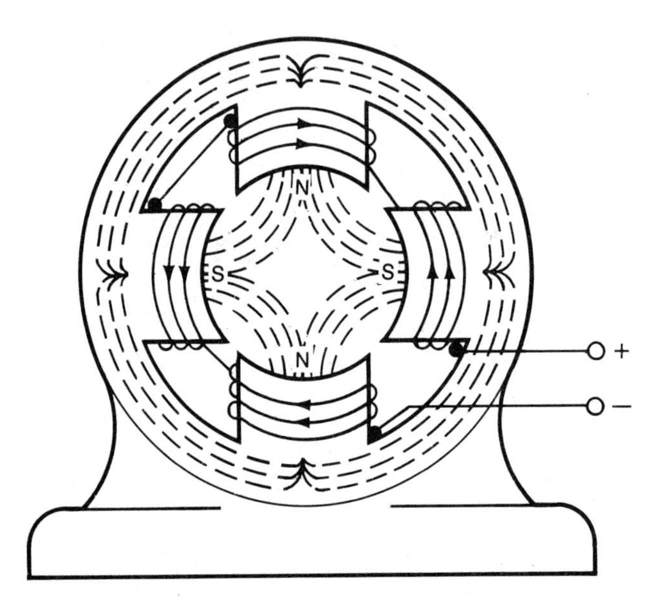

FIGURE 19–17 Four-pole field windings

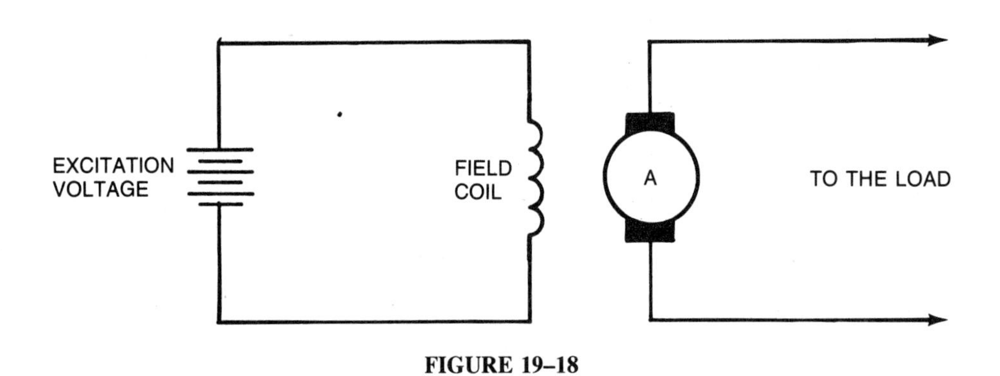

FIGURE 19–18

Field Connections for Self-excited Generators

There are three possible field connections for self-excited generators, energized by current generated in the armature of the same machine.

1. *Series:* The field coils can be in series with the external load circuit. Series coils consist of relatively few turns of large wire, since they must carry the entire output current. Of the three types, this series generator is used least often, Figure 19–19.
2. *Shunt (or parallel):* The field coils are connected across the brushes of the generator, which puts the coils in parallel with the external load. Shunt coils consist of a large number of turns of small wire and carry only a small current, Figure 19–20.

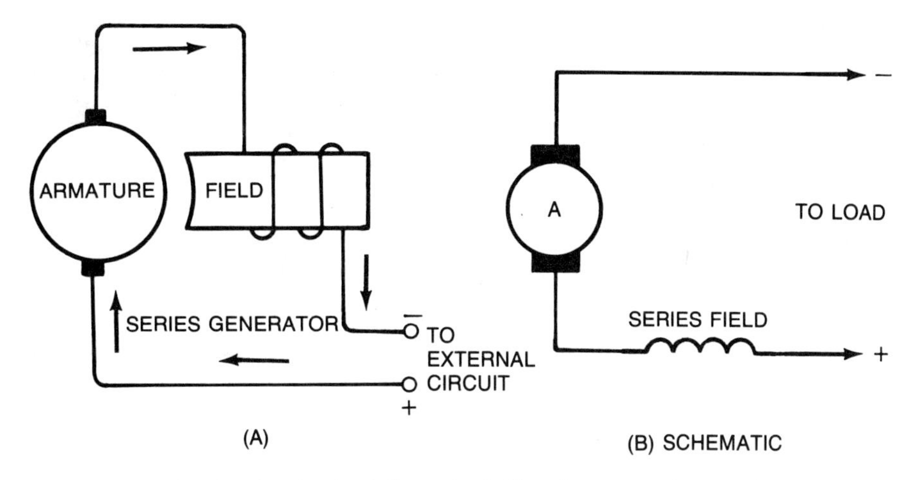

For simplicity in showing the circuit relations, the field is shown as only a single magnet. In actual practice, the armature is surrounded by two, four, or more field poles.

FIGURE 19–19 Series generator

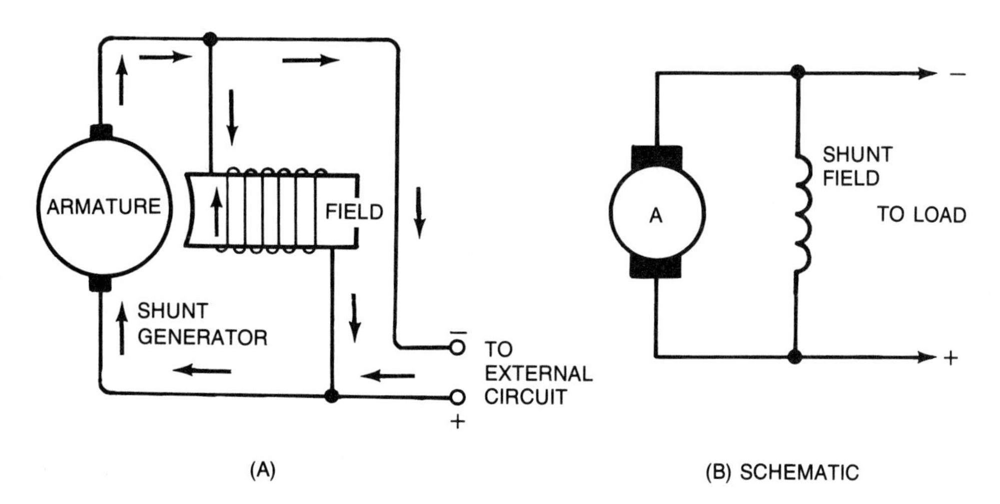

FIGURE 19–20 Shunt generator

3. *Compound:* The field-magnet iron is magnetized by the combined effect of two sets of coils. One set of low-resistance coils is in series with the external load circuit, and one set of high-resistance coils is in parallel with the load circuit.

In both of the compound generators shown in Figures 19–21 and 19–22, the series field aids the shunt field in magnetizing effect. This is the usual arrangement, called *cumulative compounding*. In a less common arrangement called *differential compounding,* the fields are connected so that they oppose each other.

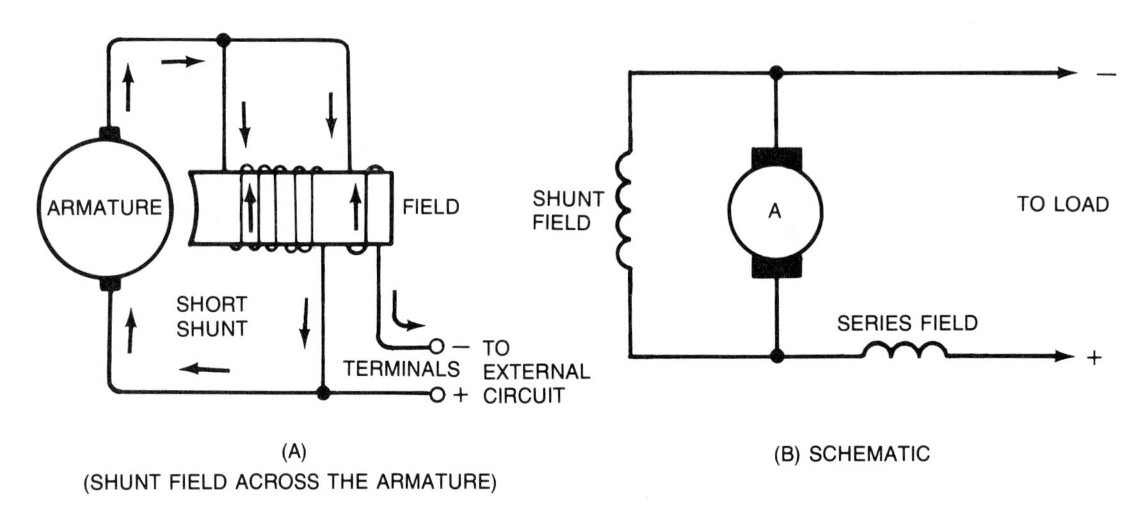

FIGURE 19–21 Short-shunt compound generator

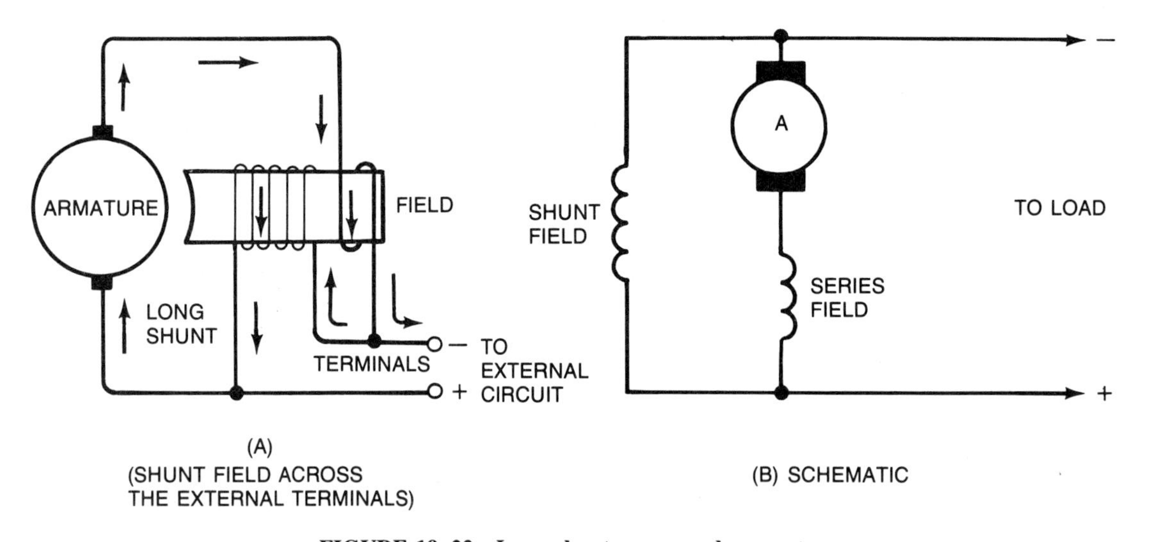

FIGURE 19–22 Long-shunt compound generator

19–4 ARMATURE REACTION

Ideally, the magnetic field in a generator has a straight, uniform pattern, Figure 19–23A. But the current generated in the armature causes another magnetic field, shown in B. Both magnetic fields combine (main field and armature field), making the total magnetic field take the direction shown in C. The distortion, or bending, of the main magnetic field of the generator, caused by the magnetic field of the current in the armature, is called *armature reaction*. Unless the distortion is corrected when the armature is producing current, the actual field in the generator is twisted, Figure 19–23D.

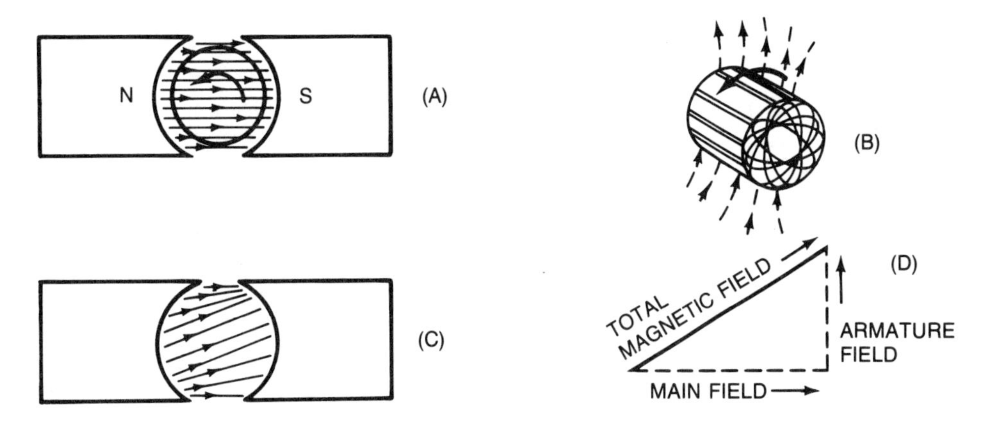

FIGURE 19–23 Armature reaction figure

The Ill Effects of a Twisted Field

The bunching of the lines at the corners of the field poles causes an irregularity in the voltage output. More importantly, the field iron is not used effectively, and the total flux is less, making the average voltage output low.

Furthermore, the twisted field changes the timing of the current reversals in the armature coils. In the explanation of Figures 19–13 and 19–14, it is stated that no harm is done by the brushes at certain instants when emf is not generated in the coil connected to the pair of segments involved. That statement is true only if the magnetic field is not disturbed. When the field is distorted, there is an emf between the commutator segments at the instant when both touch the same brush. This emf generates a brief, high current that causes excessive sparking and arcing as the commutator rotates, Figure 19–24.

Remedies

Rotation. The first remedy used for field distortion was to rotate the brush holder by an equal amount to the twisting of the field. The rotation caused commutator segments to break contact with the brush at the instant of no emf. This remedy was unsatisfactory because the amount of the field distortion changes whenever the armature current (load current) changes. To improve commutation in a generator, the brush holder is turned

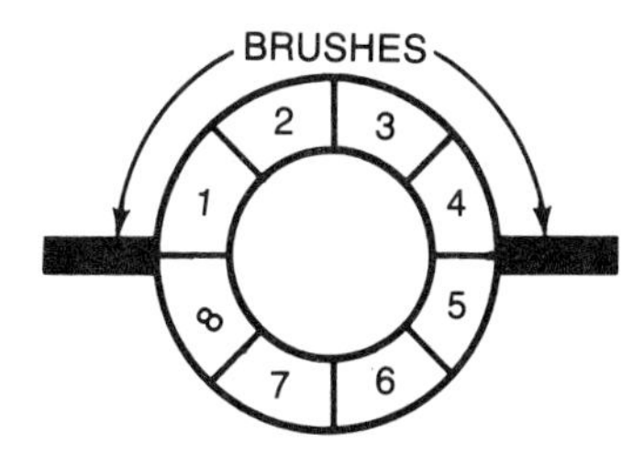

FIGURE 19–24

forward in the direction of rotation of the armature. To improve commutation in a motor, the brush holder is turned backward.

Interpoles. A better remedy is the addition of small field poles, called *interpoles,* or commutating poles, between the main field poles. Previous sketches show the armature current causing a vertical upward flux that tips the main magnetic field. The interpoles create another downward flux that tends to tip the main field back where it belongs. The interpole coils of the generator are connected into the circuit so that the interpoles have the same polarity as the main poles directly ahead of them (ahead in the sense of direction of armature rotation). In Figure 19–25, if rotation is reversed, the polarity of the interpoles must be reversed also.

To make the strength of the interpoles appropriate for their changing duty as the armature and load current changes, the interpoles are energized by coils in *series* with the armature. They therefore carry the same current as the armature. Since interpoles take care of the commutator difficulties, stationary brush holders can be used, with the brushes at the geometrical axis as shown originally.

Compensating Winding. However, these interpoles overcome field distortion only in their immediate neighborhood; much of the overall field-weakening effect is still pres-

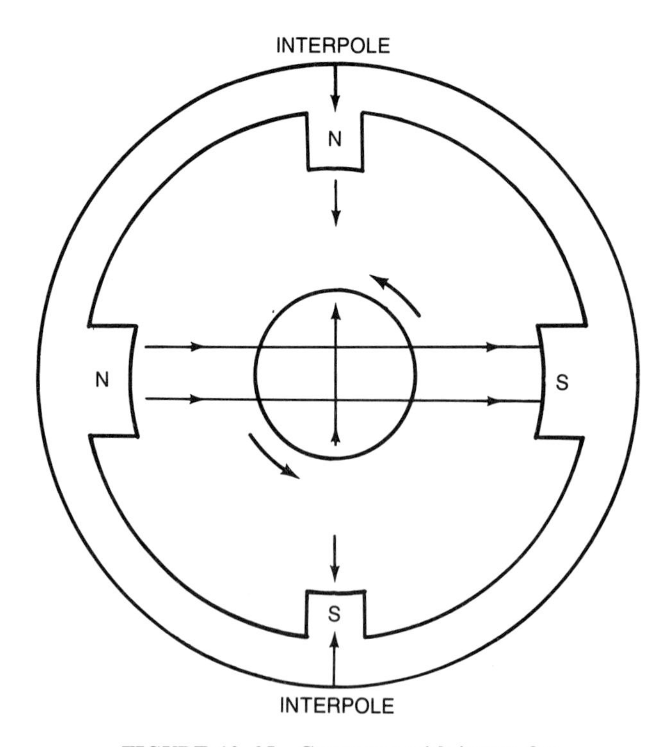

FIGURE 19–25 Generator with interpoles

ent. Large generators carry their output current through a few wires lying in the pole face placed parallel to the armature wires. This pole-face winding, called a *compensating winding,* is the most complete way of overcoming the field-weakening effect of armature reaction.

Advantages. Armature reaction is advantageous in automotive DC generators because these generators must operate over a wide range of speeds. When field strength is constant, emf is proportional to rpm, a condition highly undesirable in the automotive generator. However, at moderate current output, the armature current distorts and weakens the average field sufficiently to help keep the emf at a reasonable value at high speed.

19–5 BUILDUP OF SELF-EXCITED FIELDS

The successful starting up of a generator depends on the existence of residual magnetism in the field iron; that is, a little magnetism remains from the effect of previous current in the field coil. When the armature of a shunt or compound generator starts rotating, a very low voltage is generated in the armature. This voltage is caused by the weak field in which the armature rotates. The low voltage causes a small current in the shunt field coils, increasing the strength of the field slightly. The increased field strength, in turn, causes the generated voltage to increase slightly. This increase causes more current in the field, increasing the field strength and therefore the armature voltage still more. The maximum amount of voltage, current, and field strength that can be built up is shown on the graph in Figure 19–26.

The *magnetization curve,* like that shown in Figure 19–26, shows the increase in field strength as the field current increases. Assume we have a generator, rated 120 volts output, that has a 40-ohm field coil. When the generator is started, the magnetization starts at a point above the zero line. This point represents a residual flux density of, say, 5,000 lines per square inch. According to the scales at the left, 5,000 lines per square inch causes a generated voltage of 7.5 volts when the armature is rotating at rated speed. The small current in the field (I = 7.5/40 = 0.19 amperes) adds to the field strength, and the buildup continues to a field strength of 80,000 lines per square inch. By this time, the generated 120 volts is putting 3 amperes through the field. This 3 amperes is needed to maintain the field at 80,000 lines per square inch in order for the 120 volts to be generated. At 3 amperes, because no more than 120 volts can be generated, the buildup stops. This limit is indicated by the *resistance line* on the graph. (Points on this resistance line give values of volts and amperes for 40-ohms resistance.)

The previous condition assumes constant speed. The generator output can be increased or decreased by increasing or decreasing the rpm. Also, the operating voltage of this shunt generator can be lowered to some other value (possibly 100 volts) by putting a little more resistance in the field circuit by using a rheostat in series with the field coil. Voltage output on a shunt generator is commonly controlled by such a field rheostat.

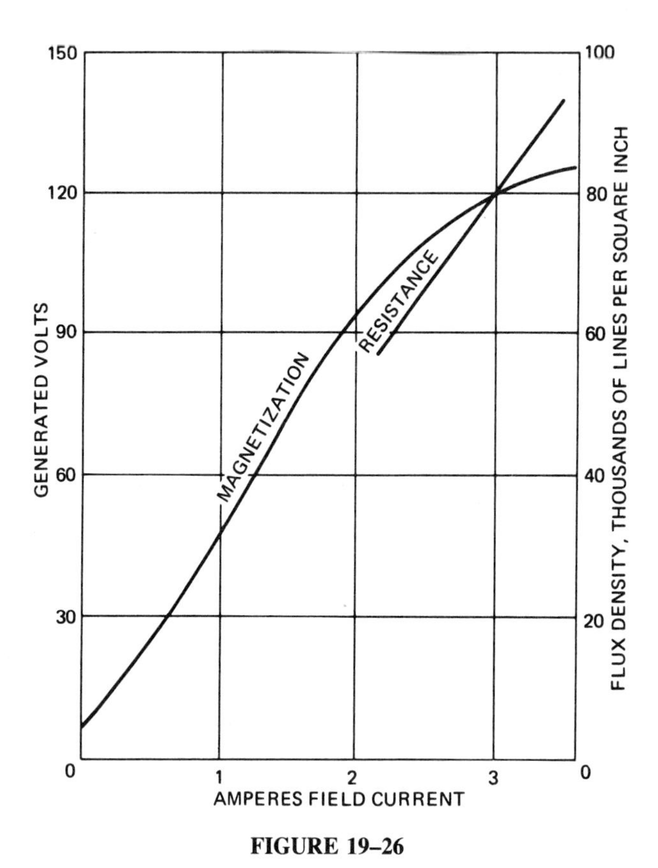

FIGURE 19–26

Failure to Develop Voltage

A generator without residual magnetism will fail to build up the magnetic field that is necessary to develop an output. Any of the following conditions may cause a self-excited generator to fail in producing the desired voltage:

- The direction of rotation may be such that it produces a magnetic field *in the same direction* as the residual magnetism. If the rotation is accidentally reversed, the generated field will *oppose* the residual magnetism and thereby obliterate it.
- A generator should not be started under load. If a load is attached before the generator develops its rated output, the terminal voltage may drop enough to cause a corresponding reduction in the field current, which in turn may cause a further decrease in the terminal voltage. This cycle of events continues until there is virtually no voltage output.
- Almost all generators have a rheostat connected in series with their shunt field winding for changing the output voltage by varying the field strength (excitation). Such a rheostat should be adjusted to its minimum resistance value during the start-up process, while the generator is building up its field.

- Also, accidental application of alternating current (AC) to the field coils will result in the loss of the residual magnetism that is so necessary for the buildup of the generator's field. (Recall from Chapter 15 that AC acts as a demagnetizer.) In the event that the residual magnetism has been lost, it can be restored by a process known as *flashing the field*. This procedure requires a separate DC source to be applied briefly to the field coils. The voltage used for flashing the field should be nearly as high as the rated voltage for the generator. A 10-second to 20-second application is generally sufficient.

19–6 THREE TYPES OF SELF-EXCITED GENERATORS (SERIES, SHUNT, AND COMPOUND)

The Series Generator

Look back at Figure 19–19 and recall that the field of the series generator carries the entire load current to the external circuit; therefore, the greater the load current, the greater the magnetic field strength, Figure 19–27. If the generator is started with the external circuit disconnected from the generator terminals, there is no buildup of field. The small voltage due to residual magnetism cannot produce any current at all in an open circuit. If a small load current is taken from the generator, its output voltage is low. If a reasonably high load current is taken, the output voltage is high. If an ordinary parallel-wired lighting circuit is used as the load and two or three lamps turned on, the

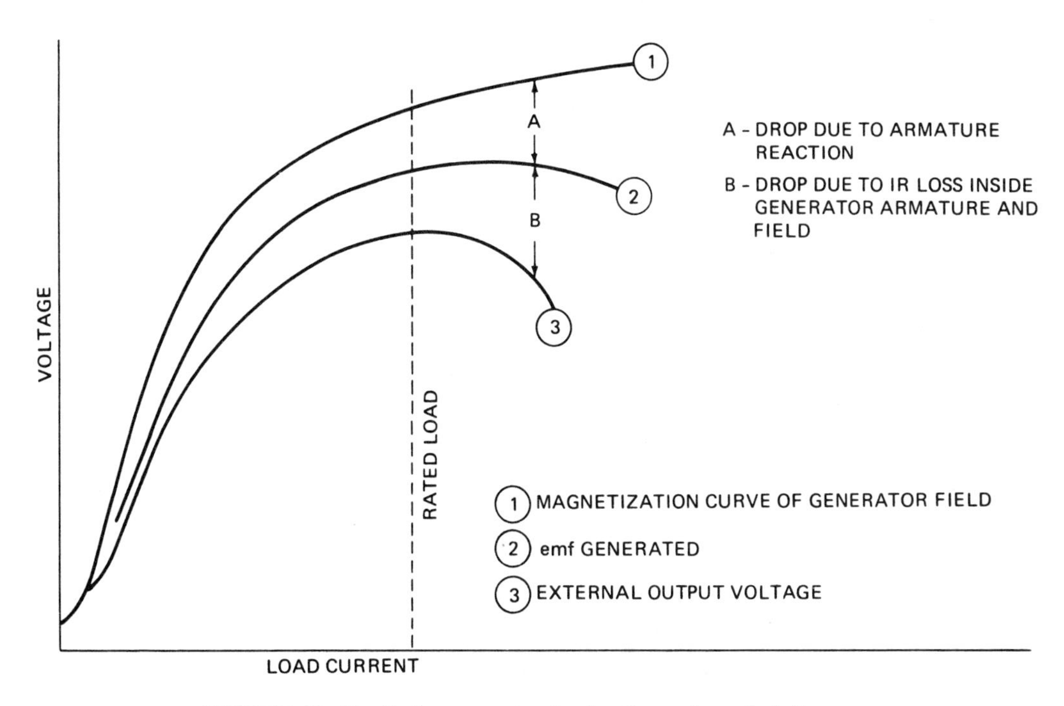

FIGURE 19–27 Series generator load-voltage characteristic curve

lamps are dim. The more lamps that are turned on, the brighter each lamp becomes. However, a little of this voltage-increasing effect can be a good thing; it illustrates the effect of the series field of a cumulative compound generator, Figure 19–27. The series generator, impractical for most jobs, does have one interesting application: It can be used in a simple motor-control system, driving a series motor at nearly constant speed with changing load on the motor (Chapter 21).

The Shunt Generator

We have seen, in Figure 19–20, that a shunt generator has its field winding connected parallel to the armature. Comparing this arrangement with that of the series generator, we find that the shunt field coil is constructed of many turns of relatively thin wire. Recall that magnetic field strength is proportional to the number of ampere-turns; thus, the shunt field coil requires relatively little current, generally amounting to less than 3% of the total current supplied by the armature.

With the field coil parallel to the armature, it is reasonable to expect a stable output voltage as long as the generator is driven at constant speed. This is true, indeed, except that the voltage is gradually reduced as the load on the generator is increased, Figure 19–28.

In our discussion we distinguish between the generated emf (when the generator is unloaded) and the terminal voltage (delivered to the load). The difference between these two values represents the drop in output voltage that accompanies the increase in load current.

This reduction in voltage is due to two factors, as follows:

1. *Armature reaction* weakens the field, thereby decreasing the generated emf, Figure 19–29 (drop A).

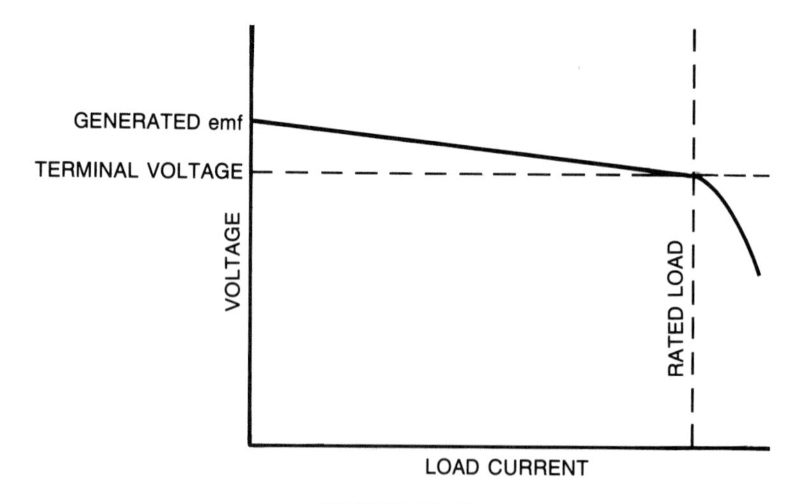

FIGURE 19–28

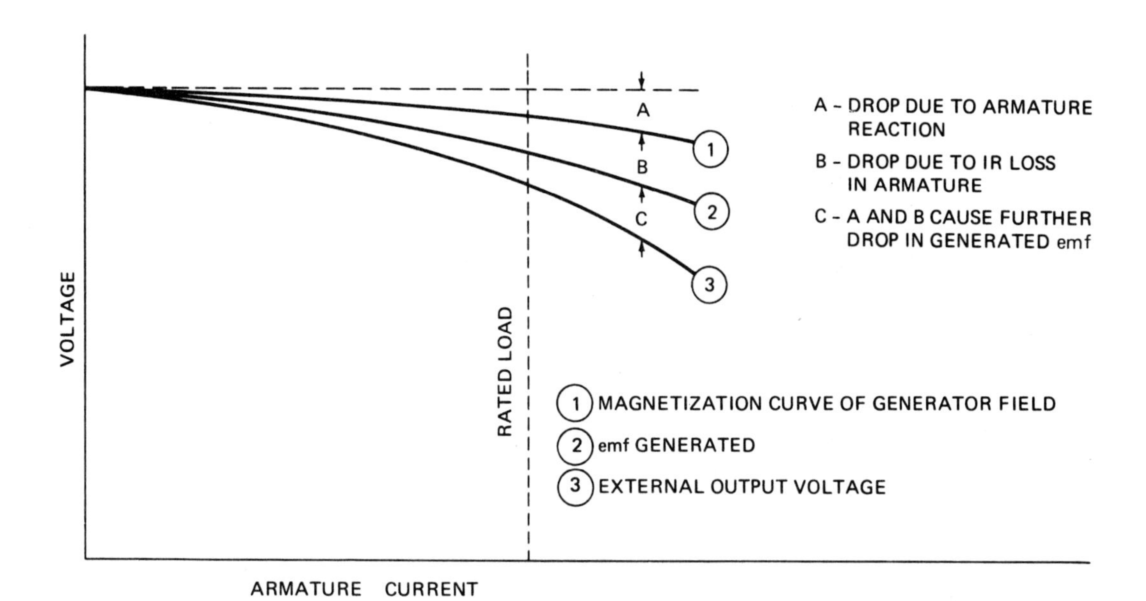

FIGURE 19–29 Shunt generator load-voltage characteristic curve

2. The *IR drop* of the armature. There is only a little resistance in the armature winding but enough to cause a voltage drop (*IR* drop). This *IR* drop, which varies proportionally with the load current, is lost to the load and causes a further reduction in output voltage, Figure 19–29 (drop B).

Both of these effects reduce the output voltage. Reduced output voltage causes reduced field current and therefore field flux. With reduced field flux, there is a further reduction in generated emf, Figure 19–29 (drop C).

Assume that we have a shunt generator with armature resistance equal to 0.4 ohm, rated 10-ampere output current. When operating with no load (external circuit open), we find that the emf is 121 volts. When operated at its rated 10-ampere load, there is a 4-volt drop, or waste, in the armature. ($E = IR = 10 \times 0.4 = 4$ V. Four volts is used in pushing electrons through the generator itself.) This 10 amperes in the armature also distorts and weakens the field so that there is a decrease in the generated emf of 3 volts. Because of all these voltage deficiencies, the field current and flux are reduced; thus, the generated emf is less by another 4 volts. Because of armature reaction and reduced field current, $121 - 7 = 114$ volts is generated. After subtracting the 4 volts used in the armature, there is an output voltage at the terminal of 110 volts.

The characteristic of shunt generators dropping their voltage with increased load is not all bad. In extreme cases of overload, such as a short circuit, the generator will protect itself from damage by reducing the field excitation and, correspondingly, the output voltage. For example, if the generator is overloaded to 15 amperes instead of 10 amperes, increased armature reaction and the resulting reduction in field current bring the generated emf down to 109 volts. Fifteen amperes through the 0.4-ohm armature uses 6 volts, so the terminal voltage is down to 103 volts.

Voltage Regulation

As a measurable quantity, the term *voltage regulation* means the percentage change of voltage from rated-load to no-load conditions.

$$\text{Regulation} = \frac{\text{Open-circuit voltage} - \text{voltage at rated load}}{\text{Voltage at rated load}}$$

Using the shunt generator as an example

$$\text{Regulation} = \frac{121 - 110}{110} = 0.1 = 10\%$$

The Compound Generator

The compound generator is designed to compensate for the voltage losses that characterize shunt generators under the influence of increased load. Recall that in a series generator, just the opposite phenomenon occurs: The voltage *increases* with increasing load. By adding a series winding to the shunt generator, we can combine the features of both windings to provide a more stable output voltage.

Let us examine the effect of adding series field coils to the shunt generator described previously. At a 10-ampere load with a series field coil present, we still have the field-weakening effect of armature reaction and the 4-volt *IR* drop on the armature, plus a possible 1-volt additional *IR* drop in the series field. All of these effects can be overcome by adding enough turns of the wire to the field. These turns are connected in series with the load; thus, the 10 amperes in these turns can add to the magnetizing effect of the shunt field coil, increasing the generated emf. The generated emf can readily be brought up to 126, which, less the 5-volt internal *IR* drops, makes the output voltage 121, the same as at no load. A generator with open-circuit voltage equal to rated load voltage is called *flat compounded,* Figure 19–30.

A generator can be overcompounded. Enough series ampere-turns can be provided to make the output voltage rise above no-load voltage as the output current increases. (Compare series generator.) Usually, compound generators are built with enough series turns to accomplish overcompounding. The user can adjust the current in the series winding to suit specific operating conditions. Adjustment of the diverter rheostat allows some load current to bypass the series coils, Figure 19–31, and thereby to change the degree of compounding. For instance, if the effect of the series coil is severely limited, the shunt coil will dominate the output characteristics of the generator. Such a machine is said to be undercompounded.

The same circuit of Figure 19–31 is shown schematically in Figure 19–32 to illustrate two points.

1. The generator is connected as a short-shunt generator. Remember, compound generators can be connected either in a short-shunt or long-shunt configuration.
2. The drawing shows a rheostat in the shunt field circuit for the purpose of varying the output voltage. As explained earlier in this chapter, with this rheostat, the operator can weaken or strengthen the magnetic field, thereby changing the output voltage of the generator.

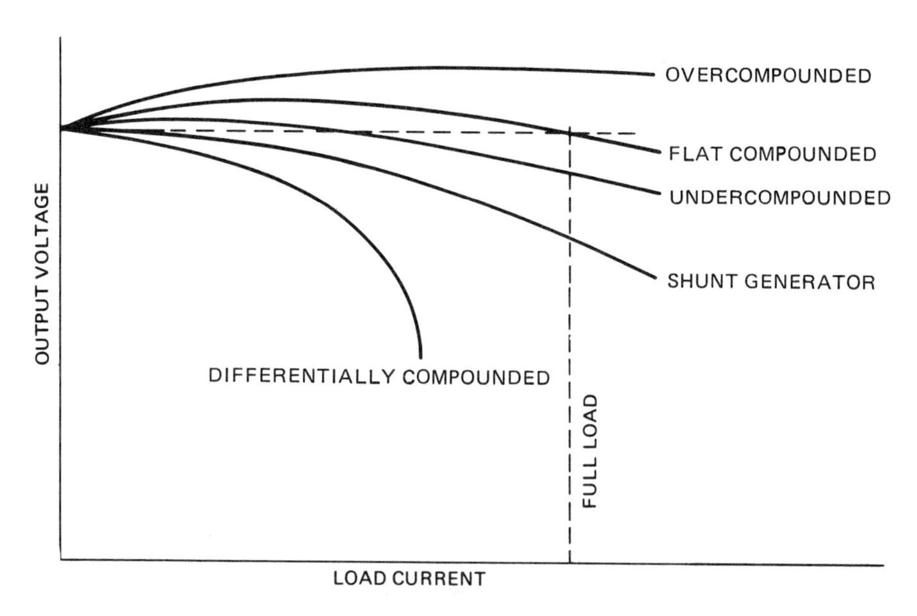

FIGURE 19–30 Compound generator load characteristics

Before leaving the subject of compound generators, let us have another look at Figure 19–30, and note the term *differentially compounded*. Most generators are *cumulatively* compounded; that is, the two windings on the pole pieces (series and shunt) are wound in the same direction, so that their magnetic fields will reinforce each other, Figure 19–31. In a differentially compounded generator, by contrast, the two magnetic fields are opposing each other. Not many generators are differentially compounded.

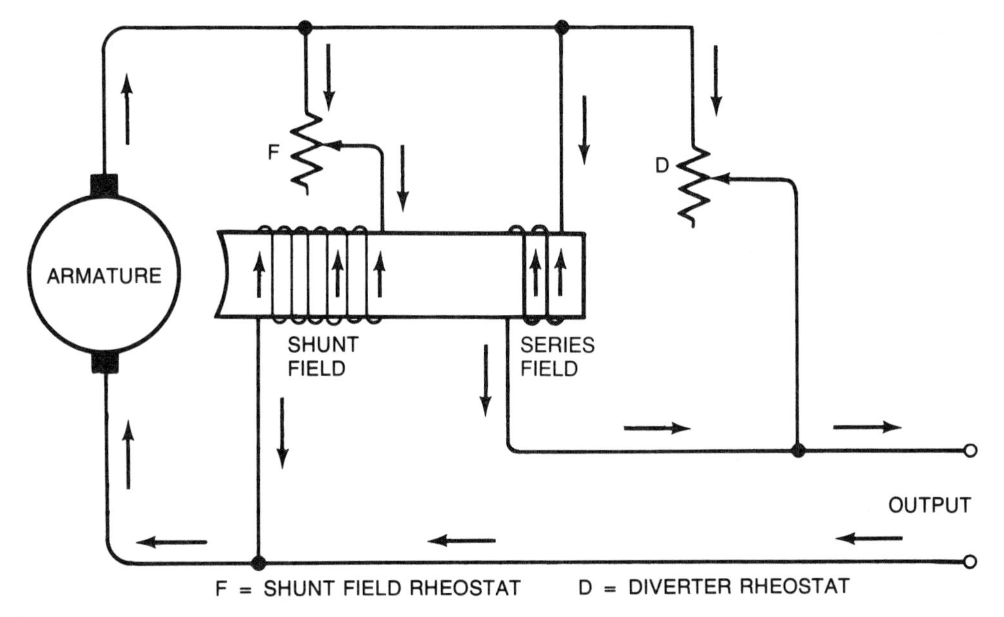

FIGURE 19–31 Compound generator with diverter rheostat

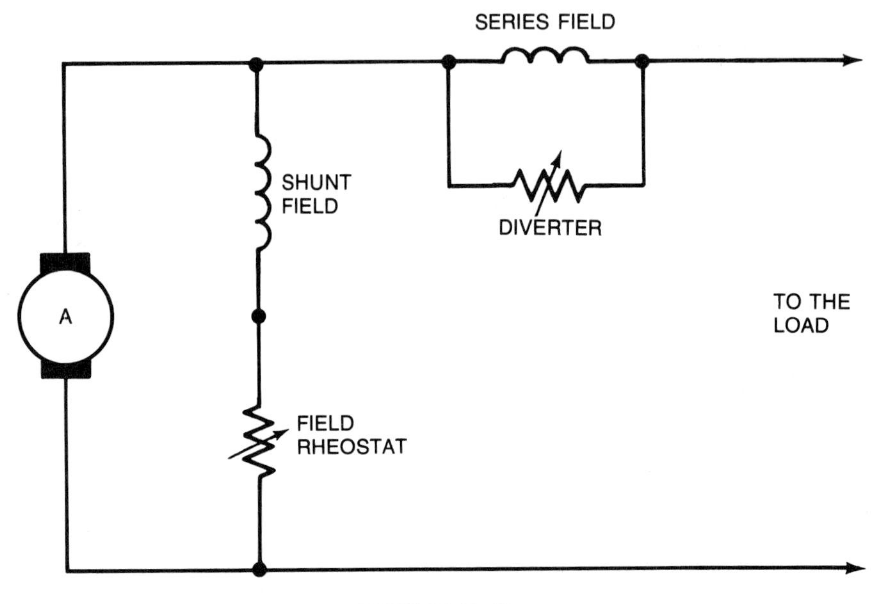

FIGURE 19–32

19–7 SEPARATELY EXCITED GENERATORS

Two general ways of providing a magnetic field for a generator have been mentioned.

1. Permanent magnet (very limited application)
2. Self-excitation—shunt, series, or compound (the most widely used method)

A third possibility is a separate current source energizing the field coils of the generator, Figure 19–33. Normally, this method is used only as a part of a specialized motor-control circuit, such as Ward Leonard system, Rototrol system, or Amplidyne control system, to name a few. The purpose of these systems is to permit the operator to select any specific speed, after which the system holds the motor at that speed regardless of variations in the load on the motor. Separate excitation systems of this type are used in mine hoists, steel-mill rolling mills, paper machines, diesel-electric locomotives, and other similar devices.

19–8 GENERATOR CALCULATIONS

The following paragraphs are not intended to provide enough details to illustrate all of the factors that must be calculated by a generator designer. They are intended only to point out the basic principles of the energy conversion that goes on in a generator.

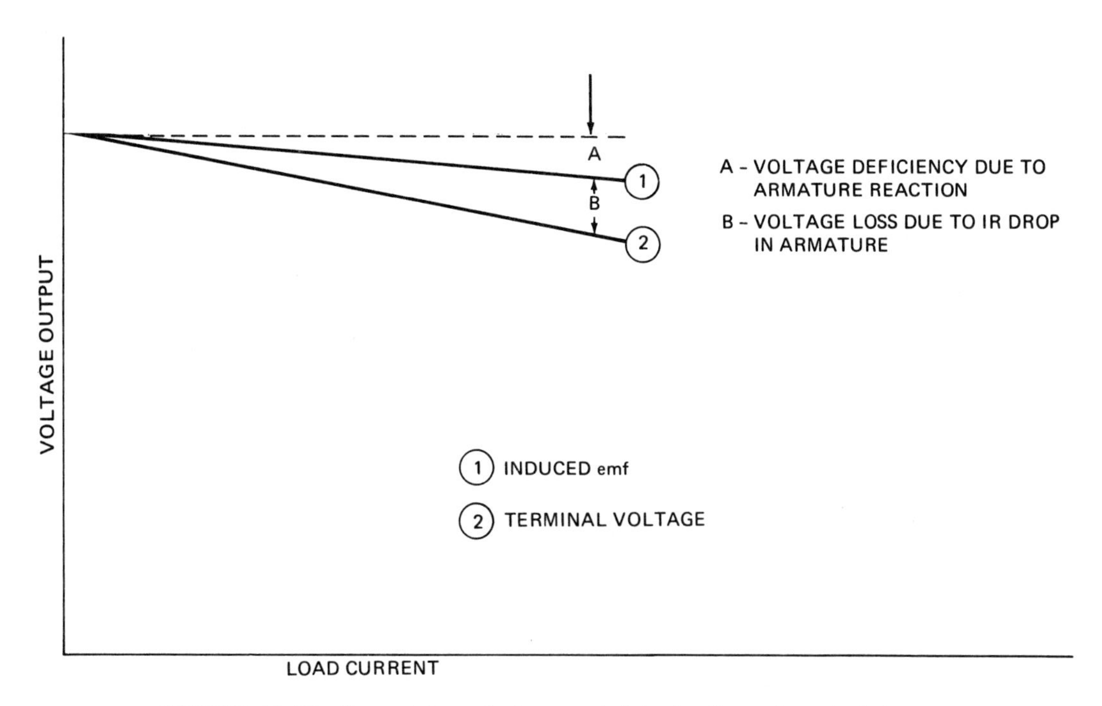

FIGURE 19–33 Separately excited generated load-voltage characteristic curve

Generator emf

As stated before, generator emf is proportional to field strength, number of wires on the armature, and rpm of the armature. When the field strength, armature windings, and rpm of the armature are known, the number of lines of force cut per second can be found. The cutting of 100,000,000 lines by wire each second makes 1 volt.

$$\text{emf} = \frac{\text{Lines cut by wires per second}}{100{,}000{,}000}$$

The following example has been chosen to demonstrate that

$$\begin{array}{c}\text{Total cutting of lines}\\ \text{per second}\end{array} = \frac{\text{Total armature turns}}{\text{Number of parallel paths}} \times 4 \times \text{Flux} \times \frac{\text{rpm}}{60}$$

and

$$\text{Generated emf} = \frac{\text{Turns} \times 4 \times \text{Flux} \times \text{rpm}}{\text{Paths} \times 60 \times 100{,}000{,}000}$$

EXAMPLE 19–1

Given: An armature as shown in Figure 19–34. (This is similar to the armature in Figures 19–13 and 19–14.) The armature has eight coils, each coil consisting of 40

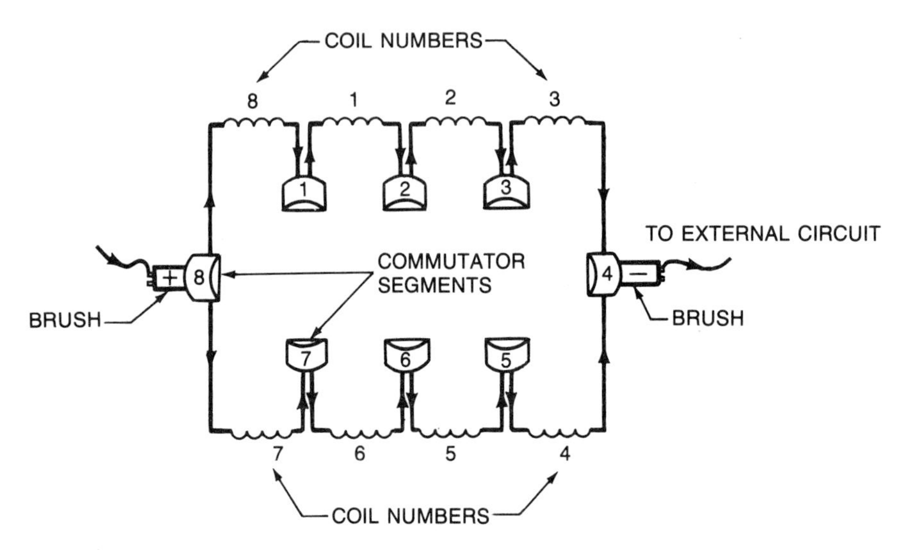

FIGURE 19–34 Eight-coil armature

turns, operating at 1,200 rpm, between two field poles, each pole having 15 square inches of face area and a flux density of 80,000 lines per square inch.

Find: The emf generated.

SOLUTION

Figure 19–34 shows that the coils are in two parallel groups, with four coils in series in each group. The emf is produced by four coils in series, not eight. Therefore, we need to take into account four coils of 40 turns each = 160 turns. Since the purpose of this calculation is to find the average emf, we need not be concerned about differences in instantaneous voltages in the coils.

Each turn of wire in the coil has two sides, both of which cut the entire field twice (once up and once down) during each revolution. The 160 turns, then, have to be multiplied by 4 to give the number of times that the entire field is cut by wire each revolution. The field is cut 640 times.

The total field flux is 80,000 lines per sq. in. × 15 sq. in. = 1,200,000 lines. During one rotation, 1,200,000 lines are cut 640 times. 640 × 1,200,000 = 768,000,000 total lines of force cut.

The coils rotate at 1,200 rpm, which is 20 revolutions per second. The total cutting of lines per second is 768,000,000 times 20. The emf generated is

$$\frac{768{,}000{,}000 \times 20}{100{,}000{,}000} = 153.6 \text{ V}$$

Planning a generator design to produce a given emf involves a sensible choice of turns, flux, rpm, and type of winding, based on both theory and experience, and takes into account any special demands on the generator in use.

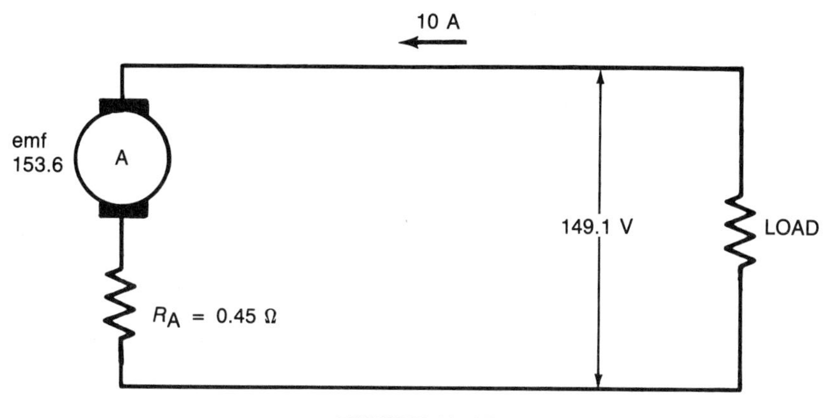

FIGURE 19–35

Emf vs. Terminal Voltage

The emf that we just calculated in the preceding problem is not the same as the voltage delivered to the terminals of the load. To understand why, consider the following: The armature has some resistance, say 0.4 ohm; and the brushes, riding on the commutator, have some resistance of about 0.05 ohm. These resistances are shown as R_A, lumped together outside the armature, in the schematic diagram of Figure 19–35.

Assuming that 10 amps is flowing through the armature, there will be a voltage drop of 4.5 volts in the armature, which is lost to the load. Think of it as a series circuit, and you will see that only 149.1 volts will be available at the terminals of the load.

Kirchhoff's voltage law is proven once again and is stated now like this.

Output voltage = Generated emf − *IR* drop in armature and brushes

19–9 POWER LOSSES

Power losses are of two types, Figure 19–36. *Stray power loss* is mechanical power that is changed to heat power without ever appearing as electrical power generated. *Copper loss,* or I^2R loss, is due to heat production by currents in the armature and field circuits of the machine.

Stray Power Losses. Mechanical losses are caused by friction in bearings and brushes and by friction of the air, called windage. Windage loss varies with speed but is independent of load current. Core losses consist of hysteresis and eddy current losses. Hysteresis loss is due to molecular friction in the iron of the armature because of continual reversal of magnetization. Hysteresis loss increases with increase in flux density and increase in speed. Eddy current losses are due to small circulating, or eddying, currents in the armature iron and pole face iron. Eddy current losses increase as the square of speed and flux density and vary slightly with load current.

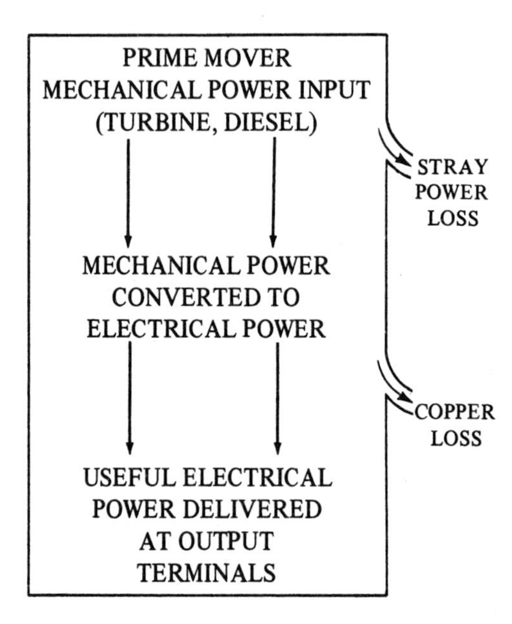

FIGURE 19–36

Copper Losses. These include power lost in heating the shunt field and also in heating the armature circuit, which includes armature wiring, brush contacts, and series fields. The armature circuit loss is proportional to the square of the load current. In other words,

$$\text{Total } I^2R \text{ losses} = I^2R \text{ of armature} + I^2R \text{ of field}$$

EXAMPLE 19–2

Given: A shunt generator with an armature resistance of 1 ohm, delivering 9.5 amperes to a load at a terminal voltage of 120 volts; and the shunt field has a resistance of 240 ohms, Figure 19–37.

Find:

a. Total I^2R losses.
b. The total emf generated.

SOLUTION

a. I^2R in armature $= 10^2(1) = 100$ W
I^2R in field $= (0.5)^2(240) = 60$ W
Total: 160 W

b. emf = IR drop in armature + terminal voltage
emf = 10 V + 120 V
emf = 130 V

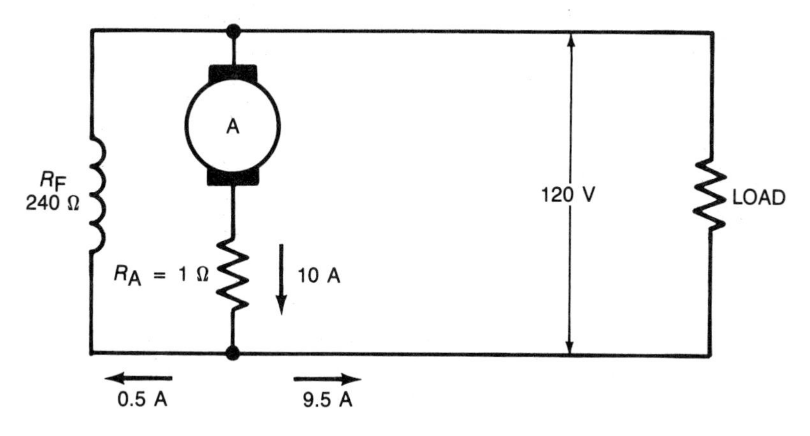

FIGURE 19–37

Efficiency of a Machine

Conversion of energy always incurs losses. In other words, the power output of a machine is always less than its power input; and the difference between the two represents the losses.

Power input = Power output + Losses

The ratio between the power output and the power input is described as the efficiency of the machine. Efficiency is generally expressed as a percentage, as follows:

$$\% \text{ Eff.} = \frac{P_{out}}{P_{in}} \times 100$$

EXAMPLE 19–3

Given: The generator of Example 19–2, Figure 19–37, requires a mechanical input of 2 horsepower.

Find:

a. The efficiency of the machine.
b. The stray power losses.

SOLUTION

a. $P_{out} = E \times I$

$P_{out} = 120 \text{ V} \times 9.5 \text{ A}$

$P_{out} = 1{,}140 \text{ W}$

$$P_{in} = 2 \cancel{\text{hp}} \times \frac{746 \text{ W}}{\cancel{\text{hp}}}$$

$P_{in} = 1{,}500 \text{ W}$

$$\% \text{ Eff.} = \frac{P_{out}}{P_{in}} \times 100$$

$$\% \text{ Eff.} = \frac{1{,}140}{1{,}500}$$
$$\% \text{ Eff.} = 76\%$$

b. $\text{total losses} = P_{out} - P_{in}$
$\text{total losses} = 1{,}500 - 1{,}140$
$\text{total losses} = 260 \text{ W}$

$\text{stray losses} = \text{total losses} - I^2R \text{ losses}$
$\text{stray losses} = 260 \text{ W} - 160 \text{ W}$
$\text{stray losses} = 100 \text{ W}$

19–10 GENERATOR DATA AND RATINGS

The full-load rating of a generator is a statement of conditions that provide efficient operation without exceeding safe limits of speed and temperature. These data, which are supplied on the nameplate of the machine, includes speed, voltage, power input (kW) or current output, and allowable temperature rise.

If operated at very low current output, efficiency is low. At a speed that is too low, air circulation is poor and overheating can result. Higher than normal current over a long time raises temperature and can damage insulation and burn the commutator and brushes. Standard voltages for larger DC generators are 125, 250, 275, and 600 volts.

The *temperature rise* allowed in a machine is a rise above 40°C, which is taken as a standard surrounding temperature. For example, the temperature of Class A insulation (enamel, oil-impregnated paper, and cotton) should not exceed 105°C. A machine with this insulation is rated *50°C temperature rise*. This rating allows the average, or surface, temperature of a coil to be 90°C (40 + 50) while allowing for hot spots in the center of the coil to be 15° higher than 90°, which is the 105°C specified limit.

19–11 MAGNETOHYDRODYNAMIC (MHD) GENERATION

MHD generation, or energy conversion, uses the principles already discussed but in a different manner. Instead of moving a wire carrying electrons through a magnetic field, a stream of electrons and ions is made to move through a magnetic field. The high-speed electrons and positive ions are deflected in opposite directions by the magnetic field. Electrons collect at one plate and positive ions at another. The excess negative and positive charges on the plates cause them to be at a different electrical potential, like the terminals of a battery. Think of the superheater in Figure 19–38 as a place for heating a gas, either by burning it or by applying heat from an external source. The temperature of the superheated gas is so high that gas molecules ionize, forming a cloud of electrons and positive ions.

The ionized gas blows through a nozzle, entering the magnetic field area at a temperature of about 2,000°C and a velocity of over 1,500 miles per hour. As shown in Figure 19–38, the magnetic lines of force are directed at right angles to the page, as from an N-pole above the page to an S-pole behind the page. Just as electrons in a wire in an ordinary generator are forced to move in a direction relative to the magnetic field

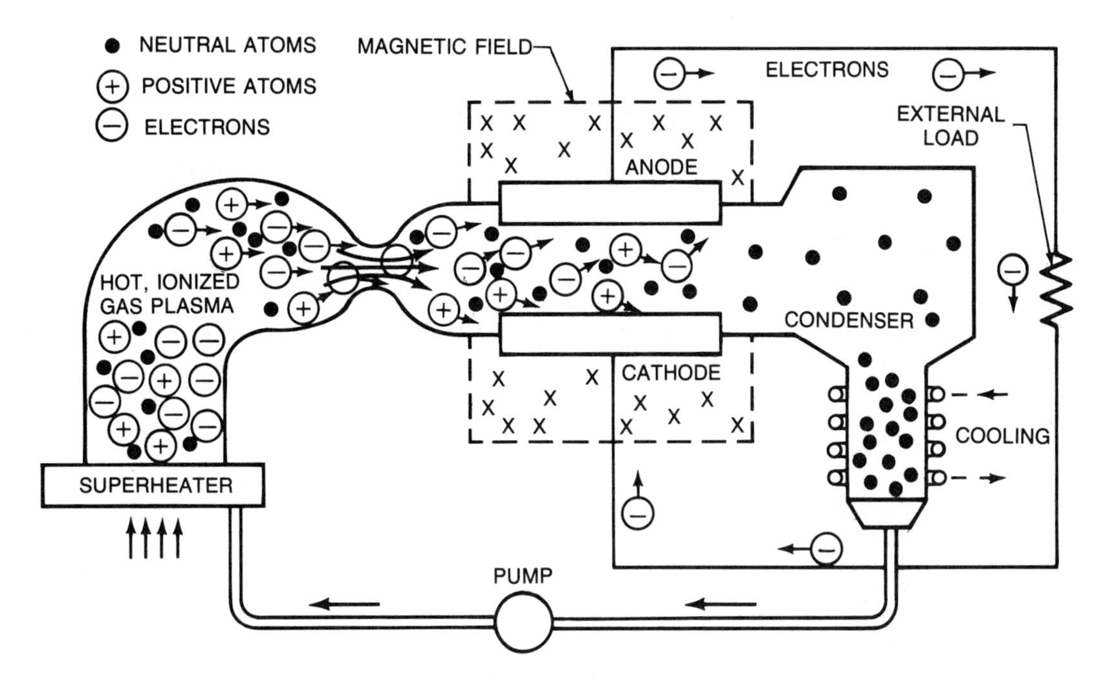

FIGURE 19–38 MHD generator

and to their direction of motion, electrons in this device are forced upward toward the anode.

The anode may be negatively charged already because of the previous accumulation of electrons. As more electrons come roaring by, the magnetic field deflects them. Their high kinetic energy slams them onto the anode, building up the negative charge still more. Positive ions are deflected in the opposite direction. They collide with the cathode and pick up electrons from it; thus the cathode maintains a positive charge (electron deficiency). Anode and cathode connect to the useful external circuit.

When and if MHD converters can be built to produce huge amounts of electrical energy, efficiency much higher than that for conventional steam turbine generator plants is possible. The size of the converter may be relatively small; Westinghouse has built an MHD generator that produces 2.5 kW from a unit about the size of this book.

Serious problems must yet be overcome. Extremely high temperatures are needed for the gases to ionize sufficiently to stay ionized rather than to recombine. Heat sources capable of producing the extremely high temperature must be developed. New materials must be found to resist critically high temperatures, or better use must be made of materials now available.

SUMMARY

- DC armatures carry coils in a series-parallel arrangement. There are two or more parallel paths through the armature, and each path consists of several coils in series.

- The magnetic field of the generator is usually supplied by current from the generator itself. This process is called self-excitation.
- Shunt field coils, in parallel with the load circuit, produce field magnetism that decreases slightly as the current output of the generator increases.
- Series field coils, in series with the load circuit, produce field magnetism that increases in proportion to the output current.
- Compound generators have both series and shunt fields.
- DC generators that are a part of specialized motor-control circuits are usually separately excited.
- Armature reaction, which is field distortion due to armature current, causes commutation difficulties that can be corrected by interpoles. It also causes field weakening, which can be corrected by compensating windings.
- $\text{Generated emf} = \dfrac{\text{Lines cut by wires per second}}{100{,}000{,}000}$
- In a drum armature, $\text{emf} = \dfrac{\text{Turns} \times 4 \times \text{Flux} \times \text{rpm}}{\text{Paths} \times 60 \times 100{,}000{,}000}$
- $\text{Efficiency} = \dfrac{\text{Power output}}{\text{Power input}}$

Achievement Review

1. A four-pole generator field has a flux density of 75,000 lines per square inch. Each pole face is 5 inches square. Calculate the total flux.

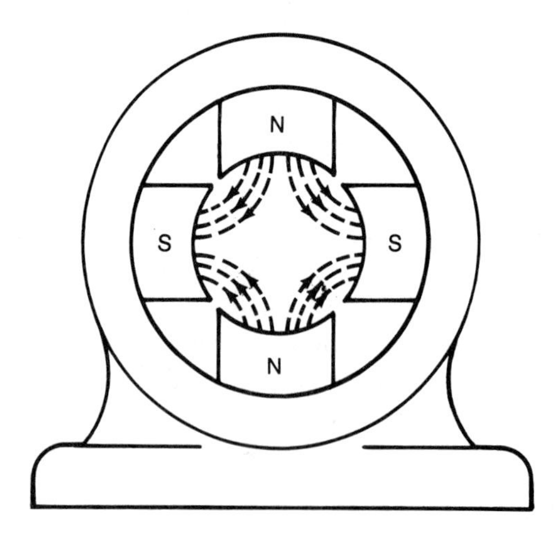

2. Name three types of field coil arrangements for self-excited generators. Which gives the steadiest voltage output?
3. Name the three factors of greatest influence in determining the emf produced in a generator armature.

4. What is armature reaction? Name the two main disagreeable effects of armature reaction. How can both of these effects be corrected?
5. After standing idle for several weeks, an engine-driven DC compound generator is started up. It runs well but produces no voltage. Suggest reasonable causes and remedies.
6. If the efficiency of the generator is the only consideration, should the armature resistance be high or low? Should the shunt-field resistance be high or low? Should the series-field coil be high or low resistance?
7. Draw typical load-voltage characteristic curves for
   a. A shunt generator
   b. An undercompounded generator
   c. An overcompounded generator
   d. A series generator
8. Give three different reasons why the terminal voltage of a shunt generator decreases with an increase in load current.
9. A shunt generator is rated 125 volts, 25 kilowatts; armature resistance is 0.08 ohm; shunt-field circuit resistance is 25 ohms. Determine
   a. The induced emf in the armature at rated load
   b. The watts loss in the armature
   c. The watts loss in the shunt-field circuit
   d. The total power generated in the armature
10. A 10-kilowatt, 120-volt DC generator has an output of 120 volts at rated load. With no load, a voltmeter across the output terminals reads 110 volts. Determine the voltage regulation. Is this machine a shunt, cumulative compound, or differential compound generator? How can you tell?
11. A 12-kilowatt, 240-volt, 1,500-revolutions-per-minute shunt generator has an armature resistance of 0.2 ohm and a shunt-field resistance of 160 ohms. The stray power losses are 900 watts. Assuming shunt-field current is constant, calculate
    a. The efficiency at rated load
    b. The efficiency at half rated load
12. Explain how interpoles accomplish their purpose in a DC generator. State the polarity rule for interpoles.
13. A 10-kilowatt, 230-volt, long-shunt compound DC generator has efficiency = 82%, armature resistance = 0.15 ohm, series field = 0.1 ohm, shunt field = 100 ohms. At rated load, calculate
    a. The armature current
    b. The voltage across the brushes
    c. The generated emf
    d. The total copper losses
    e. The horsepower of the prime mover
14. A separately excited 6-kilowatt generator has a terminal voltage of 135 volts at no load. At full load, the terminal voltage is 120 volts with speed and field excitation unchanged. Armature resistance = 0.25 ohm. Find
    a. The amount of voltage decreases caused by armature reaction
    b. The voltage regulation

15. Complete the internal and external connections for the compound generator illustrated in the sketch below. This generator is to be connected as a cumulative compound long-shunt machine. The interpole field windings are to be a part of the armature circuit terminating at the connection points A-1 and A-2. Be sure the connections for all main field poles and interpoles are correct so that the proper polarities will be obtained.

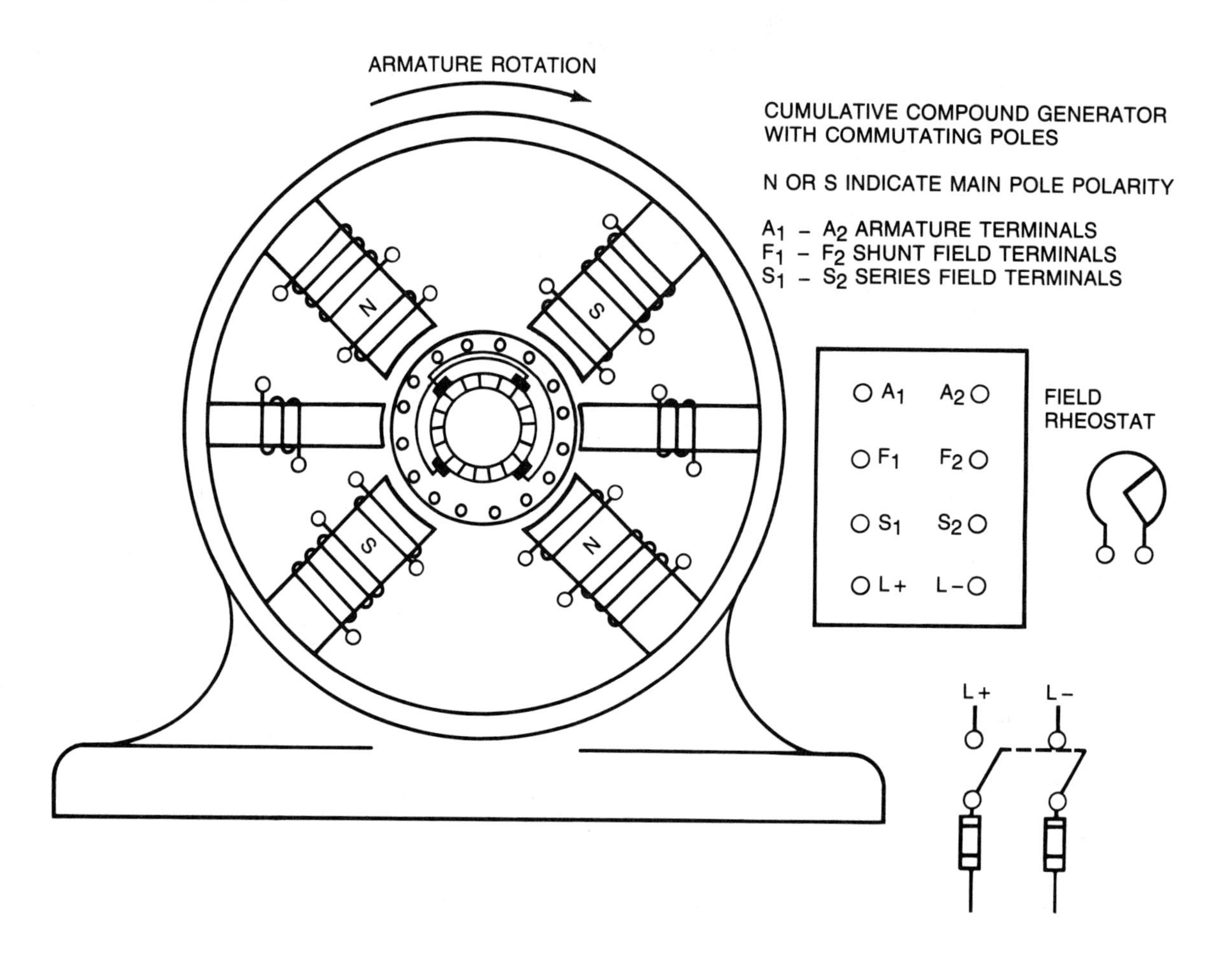

20
Mechanical Motion From Electrical Energy

Objectives

After studying this chapter, the student should be able to

- Define and explain the new technical terms introduced in this chapter

torque	Prony brake
electrodynamometer	tachometer
neutral plane	torque arm

- Describe how torque is developed in an elementary motor
- Explain the purpose of commutators in DC motors
- Differentiate between the left-hand rule for generators and the right-hand rule for motors
- Determine the direction of rotation of an armature, given the direction of the magnetic field and the current through the armature
- Perform correct calculations involving torque and horsepower

20–1 BASIC MOTOR ACTION

The term *motor action* was described briefly in Section 16–4. The sketch accompanying this explanation is reproduced for your convenience, Figure 20–1.

Figure 20–1 illustrates that a wire carrying electrons across a magnetic field will be pushed sideways. The current in the wire has a magnetic field of its own. The wire's magnetic field combines with the externally applied magnetic field. The result of these combined fields causes the closely bunched, distorted lines of force to push the wire into the area that is less densely occupied by flux lines. This same idea is demonstrated, in a slightly different manner, by the sketch in Figure 20–2.

To apply this effect in an electric motor, examine the effect of an external magnetic field on a rectangular loop of wire. Assume that the wire is supplied with DC from a battery and that the external magnetic field is supplied by a permanent magnet, Figure 20–3, page 332. The sections of the loop that lie parallel to the field are not affected

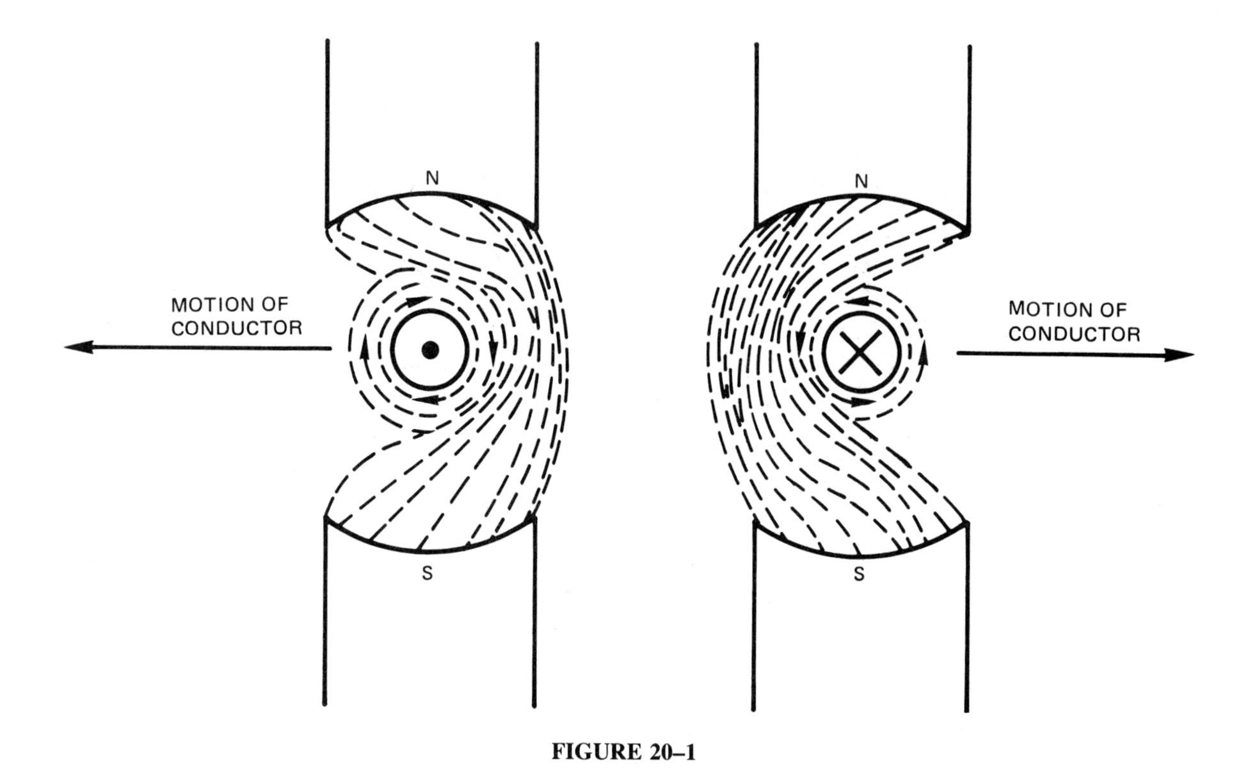

FIGURE 20–1

by the field. The side of the loop marked A is pushed upward; the side marked B is pushed downward. Let us see why this happens.

Figure 20–4 represents a vertical cross section view of the loop in the magnetic field shown in Figure 20–3. The heavy circles represent wires A and B in Figure 20–3. The X in the A circle represents electrons moving away from the observer (like the tail of an arrow flying away). The dot in the B circle represents electrons moving toward the observer (the point of an approaching arrow). The circular patterns around A and B represent the magnetic field of the current in these wires. (Using the left-hand rule, check the correctness of the direction arrows.)

Underneath A, the N ⟶ S magnetic field combines with the field of the wire, making a strong field under the wire. Above A, the field of the magnet and the field of the wire are in opposite directions. They cancel each other, making the weak field represented by the less concentrated lines above A. Wire A is lifted by the strong magnetic field beneath. To account for this lifting effect, think of the N ⟶ S lines of force as both repelling each other and attempting to straighten themselves. A similar effect at B pushes wire B downward, much as a round stick is pushed down by a string, Figure 20–5, page 333.

The direction of motion of the wire can be found quickly by using a three-finger right-hand rule: With first finger, middle finger, and thumb at right angles to each other, *f*orefinger = *f*ield, *c*enter finger = *c*urrent, thu*m*b = *m*otion, Figure 20–6, page 333.

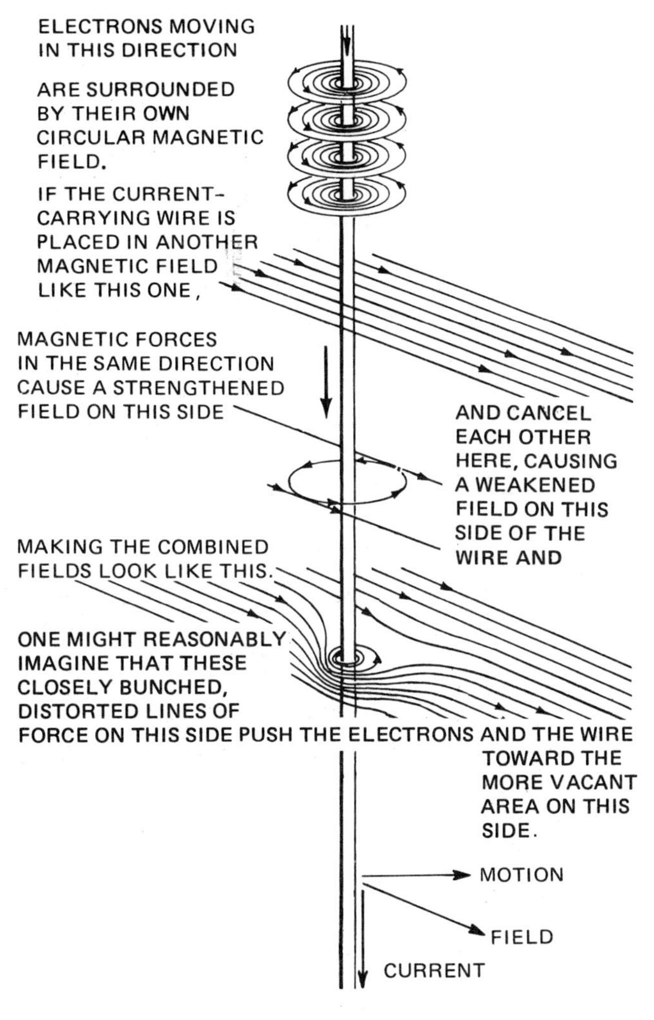

ELECTRONS MOVING
IN THIS DIRECTION
ARE SURROUNDED
BY THEIR OWN
CIRCULAR MAGNETIC
FIELD.
IF THE CURRENT-
CARRYING WIRE IS
PLACED IN ANOTHER
MAGNETIC FIELD
LIKE THIS ONE,
MAGNETIC FORCES
IN THE SAME DIRECTION
CAUSE A STRENGTHENED
FIELD ON THIS SIDE
AND CANCEL
EACH OTHER
HERE, CAUSING
A WEAKENED
FIELD ON THIS
SIDE OF THE
WIRE AND
MAKING THE COMBINED
FIELDS LOOK LIKE THIS.
ONE MIGHT REASONABLY
IMAGINE THAT THESE
CLOSELY BUNCHED,
DISTORTED LINES OF
FORCE ON THIS SIDE PUSH THE ELECTRONS AND THE WIRE
TOWARD THE
MORE VACANT
AREA ON THIS
SIDE.
MOTION
FIELD
CURRENT

FIGURE 20–2

Notice the similarity with Fleming's left-hand rule for generators, as described in Section 18–3. The two rules may appear identical to the casual observer, but you should note the important difference, namely that

The *left-hand rule* is used with *generators*, and the *right-hand rule* applies to *motors*.

20–2 TORQUE AND ROTARY MOTION

Torque, also known as *a moment of force*, is a measure of the twisting effect that produces rotation about an axis. Torque is measured, like work, in foot-pounds (or in similar units), and is calculated by multiplying the applied force by the radius of the

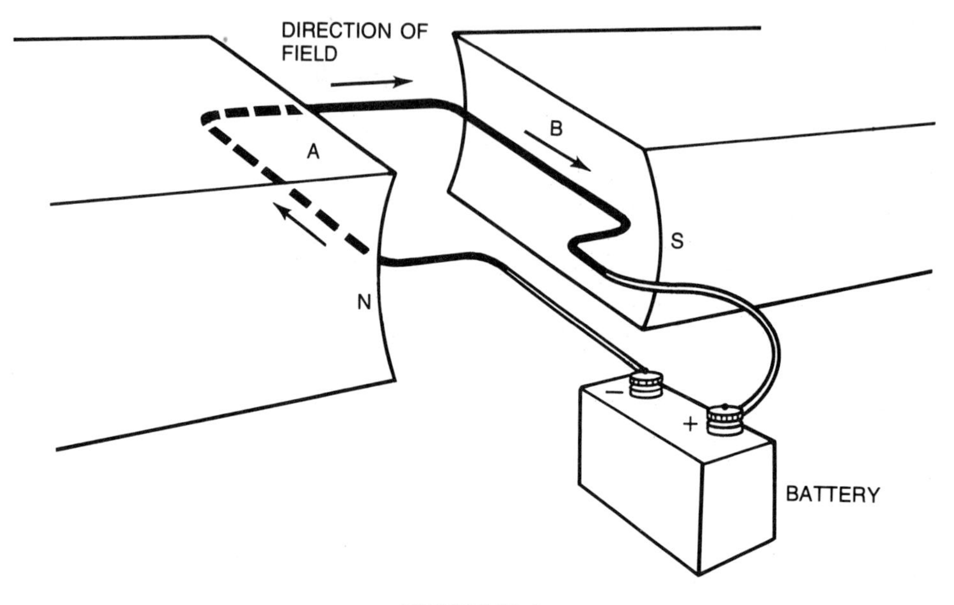

FIGURE 20–3

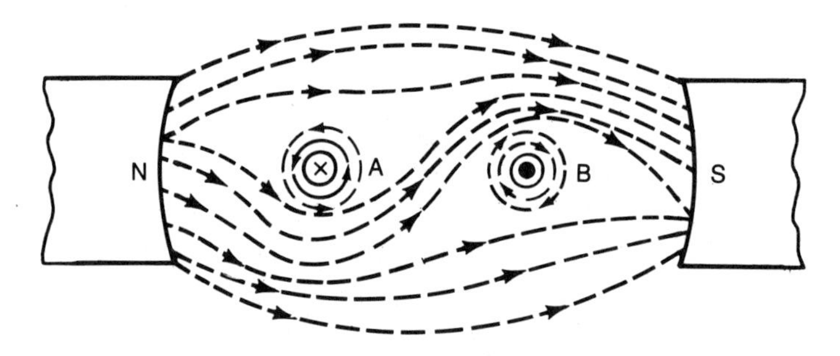

FIGURE 20–4 Diagram of magnetic field

turning circle. Figure 20–7 gives examples of torque. Figure 20–7A represents a wagon wheel with a 4-foot diameter and the weight of a person (150 pounds) pressing downward on one of the spokes. Applying what we know about calculating torque,

2 ft × 150 lb = 300 ft-lb of torque

Note: Remember, radius is one-half the diameter.

Figure 20–7B shows two hands on a steering wheel with an 8-inch radius applying 5 pounds of pressure. Thus,

2/3 ft × 5 lb = 3 1/3 ft-lb of torque

The torque delivered by a motor is a more useful quantity than the single force pushing against the two vertical sides of the coil. The torque of an electric motor is

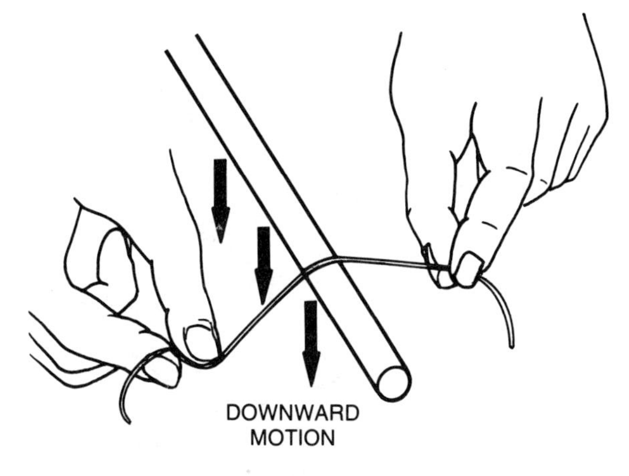

FIGURE 20–5

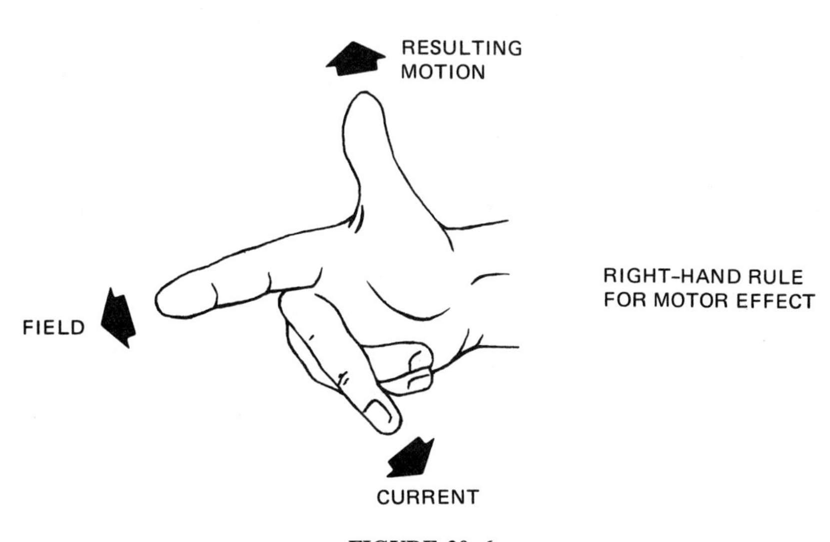

FIGURE 20–6

actually caused by the interaction of two magnetic fields, namely the main magnetic field (in our examples provided by the permanent magnet) and the magnetism created by the current flowing through the armature. Torque is a necessary quantity in the calculation of a motor's horsepower, but we will discuss more about that later.

20–3 THE NEED FOR COMMUTATION

The lifting of one side of the loop and the pushing down of the other side creates a turning effect, or torque, on the loop of the wire. This combination of forces is the torque that turns the armature of electric motors. This same method also explains the turning effect on the moving coil in a voltmeter or ammeter.

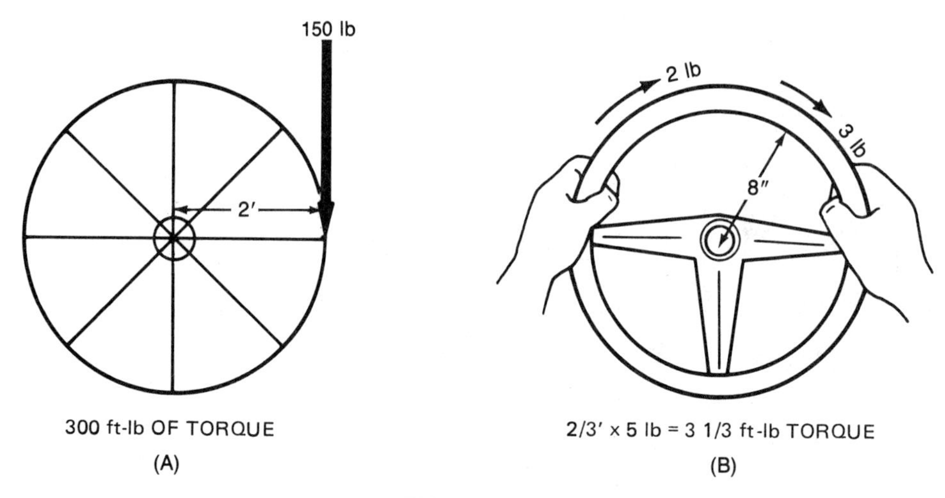

FIGURE 20–7

Imagine that the loop has rotated clockwise 90 degrees from the position shown in Figure 20–3, until wires A and B are vertically aligned, Figure 20–8. Continued lifting of wire A and pushing down on wire B are useless. The loop is now perpendicular to the magnetic field and is said to be in the *neutral plane*.

When the loop is in the neutral plane, no torque is produced. To achieve continued rotation, the current direction in the loop must be reversed just as the loop enters the neutral plane. Since the loop is in motion, its inertia will carry it past the neutral plane position and, thus, continued rotation is assured.

This necessary reversal of the current is accomplished, for a single coil, by a two-segment commutator, Figure 20–9. At the instant shown, electrons flow from the negative brush through A and return to the positive brush through B. As A is lifted to the top of its rotation, the commutator segment that supplied electrons to A slides away

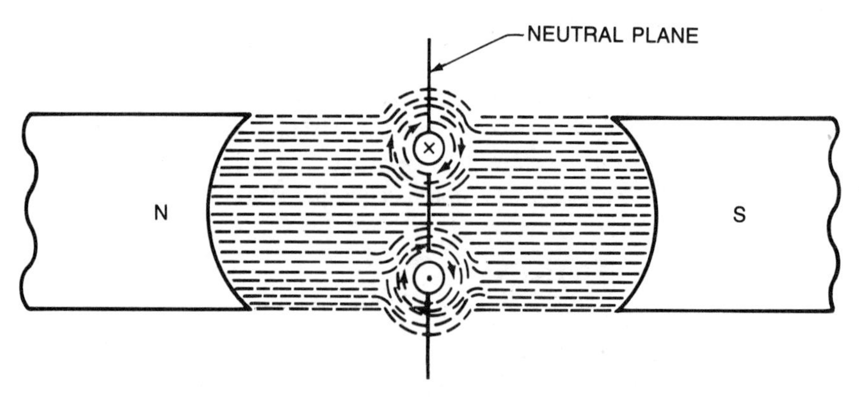

FIGURE 20–8

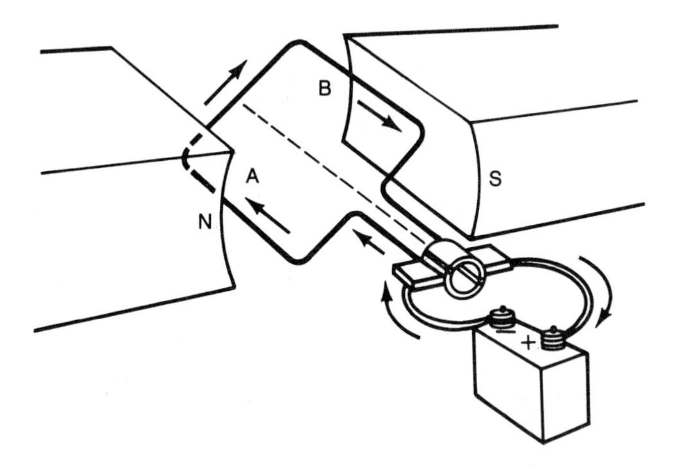

FIGURE 20–9 How a commutator reverses current to produce rotary motion

from the negative brush and touches the positive brush; thus, the current in the loop is reversed. Momentum carries the loop past the vertical position. A is then pushed downward to the right, B is lifted to the left, and another half-turn of rotation continues. This description may help the student understand the need for commutators to reverse the current through the loop each time the loop passes through the *neutral plane*.

An additional function of the commutator is, of course, to carry the current from the supply line, via the brushes, into the rotating armature.

20–4 THE NEED FOR ADDED ARMATURE COILS

This impractical single-coil armature has a variety of faults, one of which is the irregularity of the torque that it produces. When the loop is horizontal, the force on the wire is at its greatest; when the loop is vertical, there is no force on the wire. This is illustrated by the graph in Figure 20–10. Note the similarity of this graph with that of Figure 19–8, which shows the voltage fluctuations of a simple generator. Other drawings in this chapter, too, will look familiar and will help to convey the idea that the design and the internal functions of DC motors are very similar to those of DC generators.

A steadier torque is achieved by using the several coils of a drum-wound armature. The armature shown in Figure 20–11 is just as appropriate in a motor as it is in a generator.

Figure 20–11 shows a similar but simpler drum winding. Assume that the S-pole of the field magnet is at the left. (It has been omitted from the diagram in order to show the windings clearly.) Trace incoming electrons from the negative brush through commutator segment 1 through coil A to segment 2, then through B to segment 3, where the positive brush leads to the completed circuit (not shown) through a source of DC and back to the negative brush. From the negative brush, another circuit can be traced through armature coils C and D to the positive brush. Compare this drum winding with that shown in Figure 19–13 to note the similarity in principle.

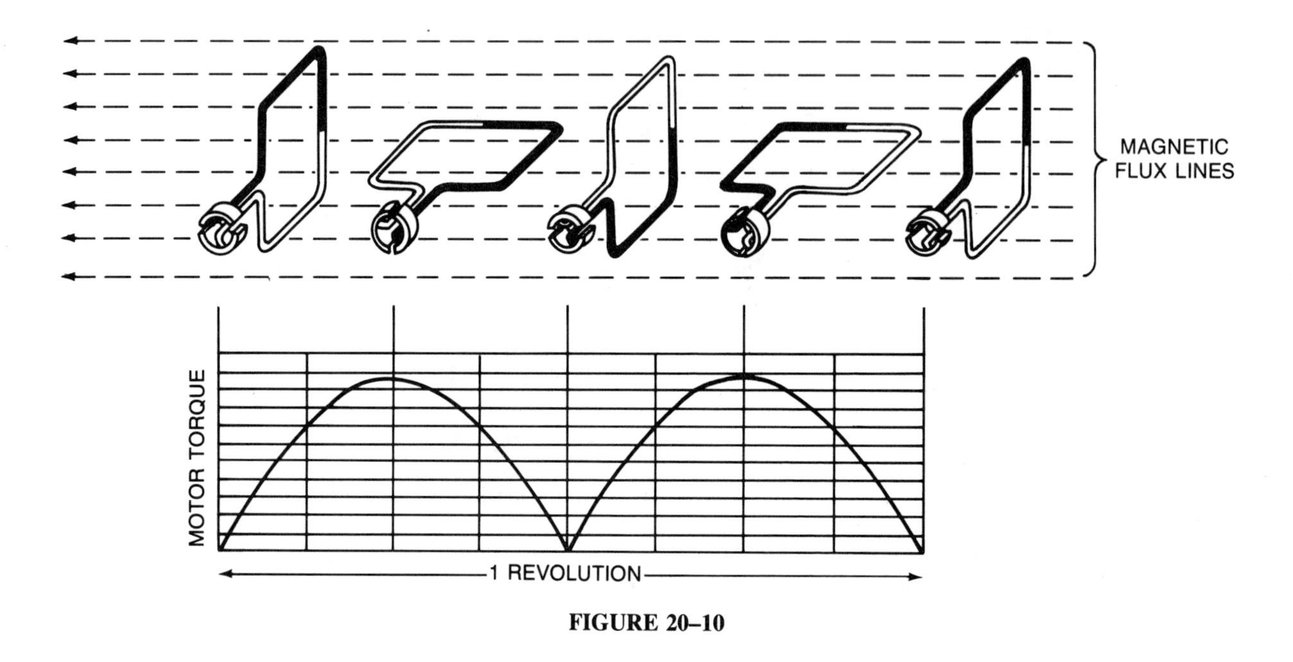

FIGURE 20–10

The circuitry of this drum-wound armature is redrawn two different ways in Figure 20–12 to reveal the series-parallel arrangement of the four-loop armature. Note that with this arrangement, only one additional pair of commutator segments is needed. All four coils, receiving current from the power source simultaneously, contribute their torque to aid rotation of the armature.

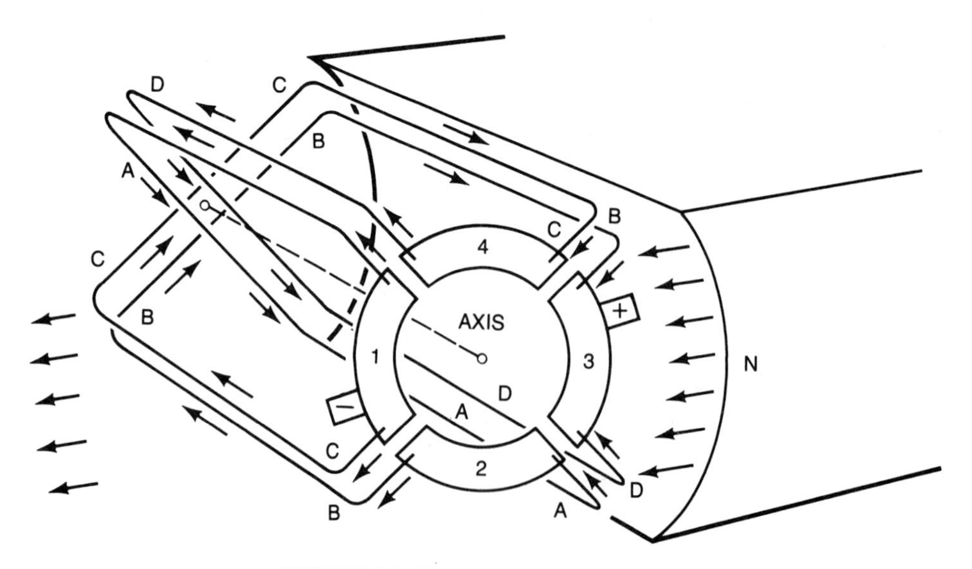

FIGURE 20–11 Drum-wound armature

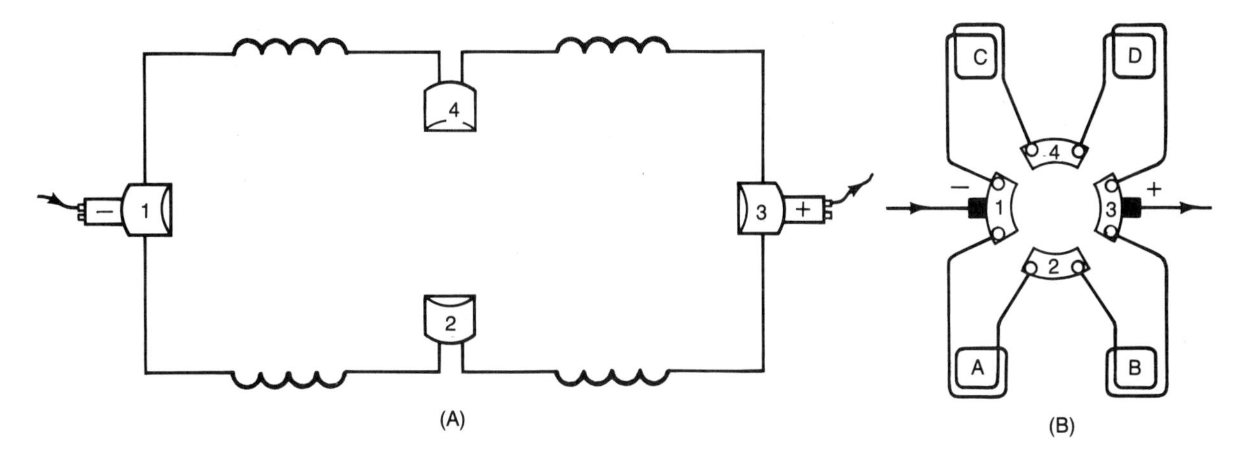

FIGURE 20–12 The four-loop armature redrawn

In the preceding section it was explained that commutation must take place at the moment a loop enters the neutral plane. In a practice DC motor, the brushes are so arranged that they will short circuit the loop as it passes through the neutral plane. This will eliminate undesirable sparking between the brushes and the commutator.

As Figure 20–13 shows, there is no current in coil C because both ends of it touch the same brush. Similarly, there can be no current in coil B. This condition exists at the instant when coils B and C are in a vertical position. Even if there were current in coils B and C, no torque would be produced. Therefore, the absence of current in the coils during this short interval is no disadvantage. At this same instant, coils A and D are in horizontal position, where they are producing maximum torque.

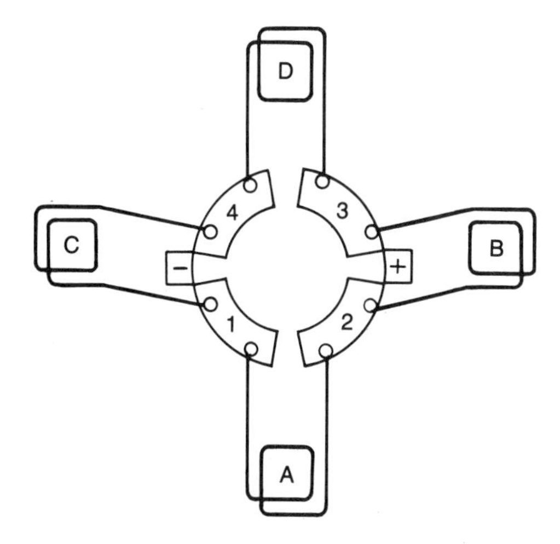

FIGURE 20–13 Armature

20–5 FROM TORQUE TO HORSEPOWER

Some students may wonder about the significance of torque, since this term has been used so frequently in our discussion of mechanical energy.

Torque is an essential component used in the calculation of the horsepower rating of a machine. Torque by itself does not tell the power of a machine; speed must be taken into account. We can say that

Horsepower is proportional to *torque* and *speed.*

Let us see what needs to be done in order to determine the amount of horsepower developed by a four-coil armature such as the one shown in Figure 20–11. First we need to determine the torque exerted on the armature. This can be done two different ways: experimentally and mathematically.

Experimentally, the torque of a motor is measured directly by a device called a *Prony brake,* Figure 20–14. Tightening the bolts makes the brake tend to turn along with the motor pulley. The brake arm is restrained by the stationary spring scale; thus, a torque load is placed on the motor. The torque is the product of the effective length *(l)* of the torque arm in feet times the net force (F) on the scale in pounds.

A more sophisticated device, known as an *electrodynamometer,* can be used to measure torque. Such a device, resembling an electric motor, utilizes the interaction of magnetic forces to yield readings of torque when coupled to the shaft of a motor.

Mathematically, the torque exerted on the armature of a motor can be found by use of the following equation:

$$\text{Torque (ft-lb)} = \frac{\phi \times Z \times I_a}{425{,}000{,}000 \times m}$$

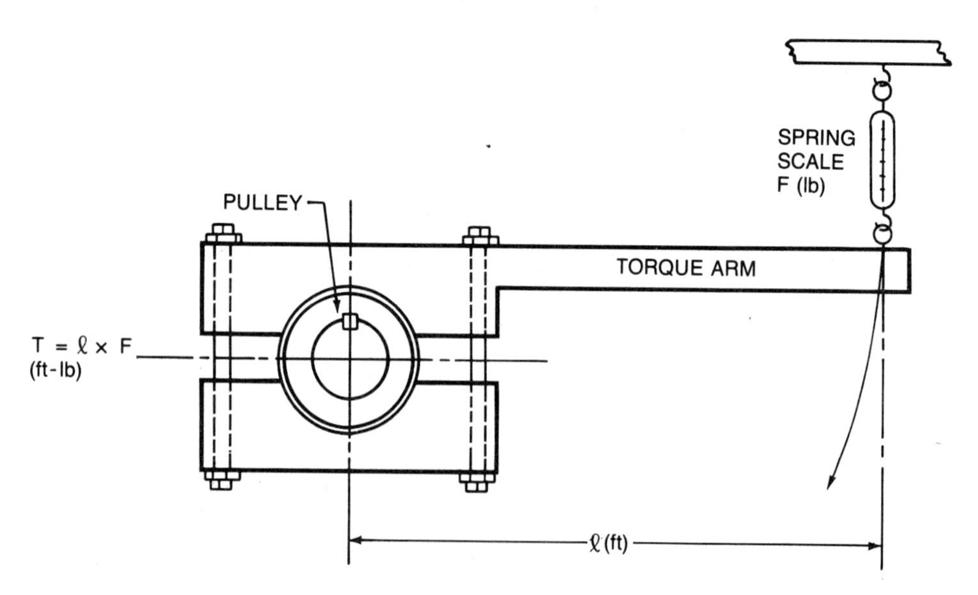

FIGURE 20–14 Prony brake

where:

ϕ is the flux passing through the armature
Z is the number of wires in the armature
I_a is the total armature current
m is the number of parallel paths through the armature

EXAMPLE 20–1

Given: An armature, 6 inches long, placed between magnetic poles as shown in Figure 20–15. The armature carries 200 turns. Flux density equals 50,000 lines per square inch. Current through the armature equals 10 amperes. The armature is designed like the one shown in Figures 20–11 and 20–12.

Find: Torque exerted by the armature.

SOLUTION

200 turns = 400 wires = Z in the formula (every loop has two edges that cut the flux lines). Total number of flux lines to be cut = 1,200,000 (24 × 50,000).

$$\text{Torque} = \frac{1{,}200{,}000 \times 400 \times 10}{425{,}000{,}000 \times 2} = 5.65 \text{ ft-lb}$$

Calculation of Horsepower

Remember, horsepower is a function of torque and speed. Specifically

$$\text{hp} = \frac{2\,\pi \times \text{Torque} \times \text{rpm}}{33{,}000}$$

Speed of the motor can be found with a revolutions counter and stopwatch, or a tachometer, or a calibrated stroboscopic light. Tachometers can read the rotational speed directly in revolutions per minute (rpm). Thus, the true mechanical horsepower output (or brake horsepower) can be determined from measurements of torque and speed.

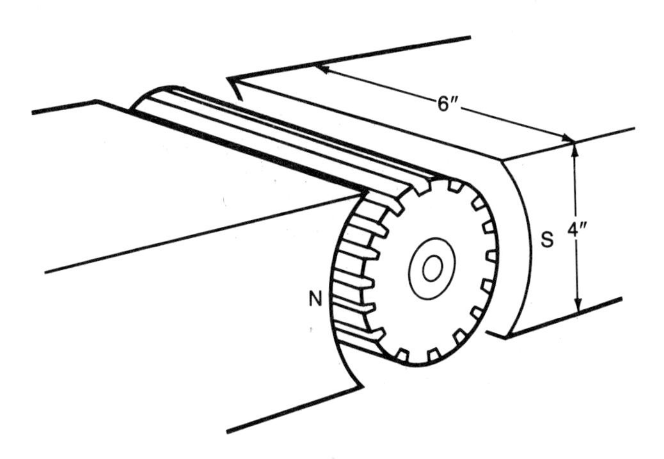

FIGURE 20–15 Armature between poles of magnet

EXAMPLE 20–2

Given: The armature of the preceding example developing 5.65 foot-pounds, turning at 1,250 revolutions per minute.

Find: The horsepower output.

SOLUTION

$$\text{hp} = \frac{2\,\pi \times \text{torque} \times \text{rpm}}{33{,}000}$$

$$\text{hp} = \frac{2 \times 3.14 \times 5.65 \times 1{,}250}{33{,}000}$$

$$\text{hp} = 1.34$$

One horsepower equals 746 watts, and 1.34 horsepower equals 1,000 watts. Assume that this armature has a current of 10 amperes. If it is 100% efficient, a power input of 100 volts × 10 amperes = 1,000 watts or 1.34 horsepower.

If the horsepower output is known, from electrical data, torque can readily be calculated for any known value of rpm.

$$\text{Torque} = \frac{\text{hp}}{0.00019 \times \text{rpm}}$$

(This derives from the fact that $2\,\pi \div 33{,}000 = 0.00019$.)
Therefore,

$$\text{hp} = 0.00019 \times \text{torque} \times \text{rpm}$$

or

$$\text{Torque} = \frac{\text{hp}}{0.00019 \times \text{rpm}}$$

SUMMARY

- Electrons moving through a magnetic field are pushed sideways in a direction at right angles to the field and to their original direction.
- This fact explains the motion of armatures in DC and AC motors and the controlled movement of electron streams in electric arcs or cathode-ray tubes.
- Fleming's right-hand rule is used to describe the motor effect.
- Torque is the twisting effect that produces rotation about an axis.
- The quantity of torque is needed for the computation of horsepower.
- DC motors require commutators for the reversal of current through the armature loop every 180 degrees.
- Reversal of current must occur when the wire loop enters the neutral plane.
- The drum armature is an effective way of arranging coils to produce a continuous torque in a motor.

Achievement Review

1. In the sketch below, how should magnet poles be placed so that the wire is moved upward when the switch is closed?

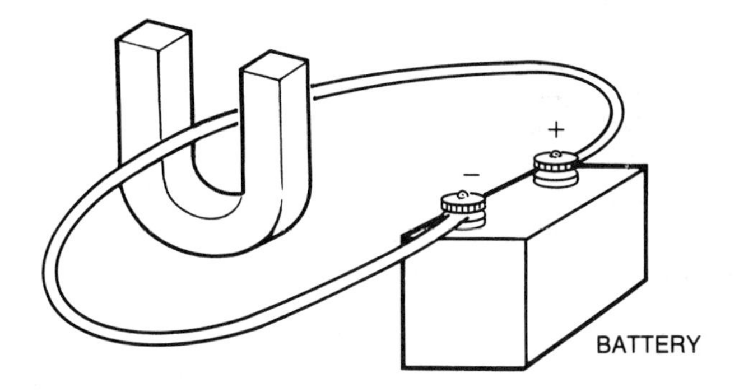

2. What happens to the wire in question 1 if an alternating current (60 cycle) is sent through the wire?
3. A brass strip is fastened to the left terminal in the sketch below and rests in loose contact with the right-hand terminal. What happens when a battery is connected to the terminals
   a. With the left one negative?
   b. Reversed, with the right one negative?

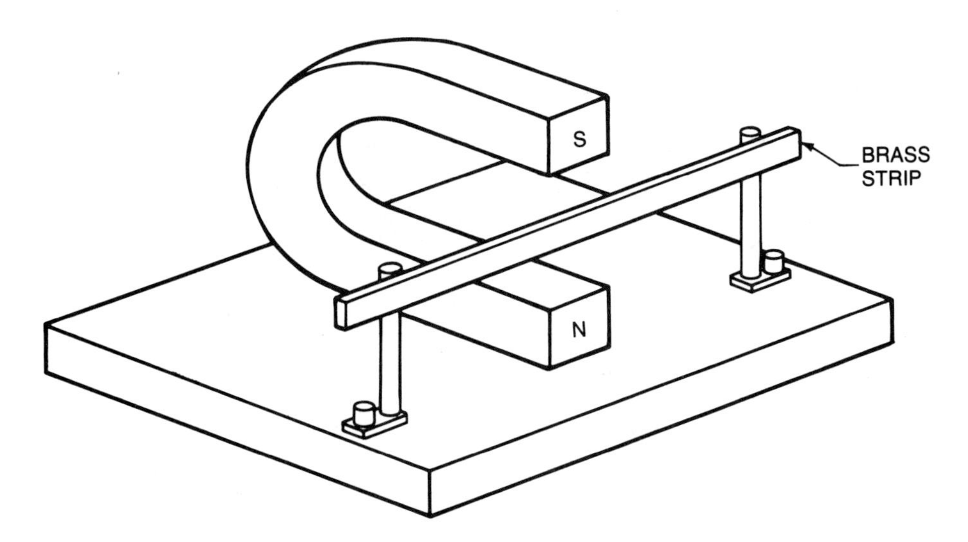

4. Determine the direction of rotation of the armature in Figure 20–11.
5. a. What is torque?

b. What factors determine the amount of torque on a coil placed in a magnetic field?

6. a. Determine the necessary battery polarity to produce rotation as indicated in the sketch below.
   b. Is the polarity the same as that found for question 7 of Chapter 18?
   c. Is the current direction in the loop of wire the same in this sketch as it was in the sketch for question 7, Chapter 18? Why or why not?

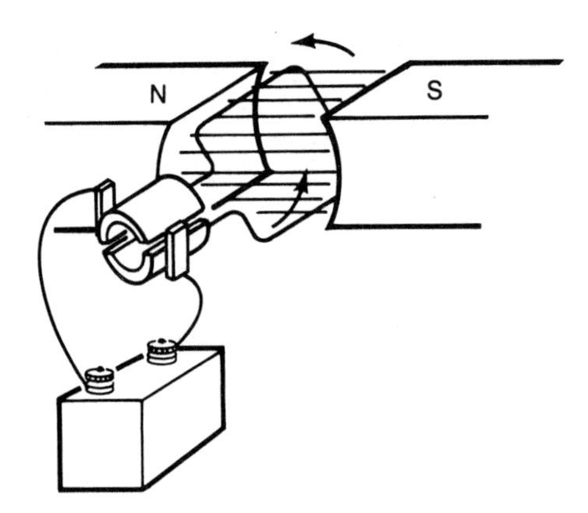

7. The center of a 6-foot-long plank is placed on a fulcrum as shown. How much torque is developed if a force of 75 pounds is applied at the end of the plank?

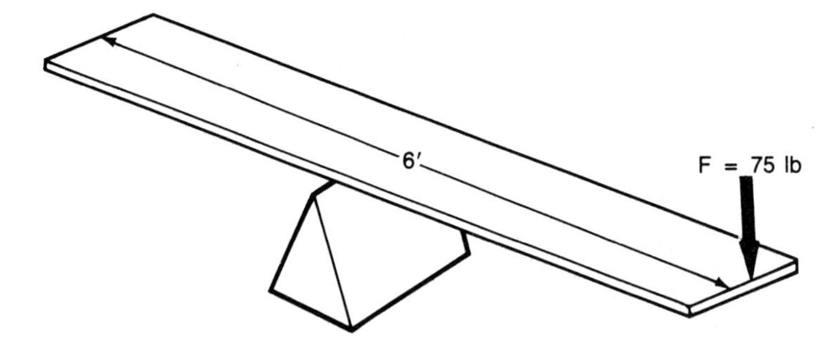

8. Calculate the torque produced on a lap-wound armature of 300 wires. Total armature current is 40 amperes; flux is 2,125,000 lines, m = 2.
9. How much force does the torque in question 8 produce at the rim of a 4-inch-diameter pulley?
10. The armature in question 8 rotates at 990 revolutions per minute. Calculate the horsepower.

21
DC Motors

Objectives

After studying this chapter, the student should be able to

- Define and explain the new technical terms introduced in this chapter:

normal speed	above-normal speed
below-normal speed	SCR
speed regulation	PM motor
DPDT switch	

- Explain the differences and similarities between DC motors and generators
- Calculate the counter-emf generated in the armature of the motor
- Describe the performance characteristics of DC shunt, series, and compound motors
- Assign correct polarity to interpoles installed in DC motors
- Control the speed of various DC motors
- Reverse the rotation of any DC motor
- Determine the speed regulation of DC motors
- Calculate power losses and horsepower output of DC motors

21–1 DC MACHINES: MOTOR OR GENERATOR

The brief introduction to motors provided by Chapter 20 points to great similarities between motors and generators. Certainly their physical design features are so similar that often it is impossible, without close inspection, to tell a motor from a generator. In fact, there are a few machines that can be used either as a motor or as a generator.

All electrical machines are essentially energy converters. It is the direction of the energy flow through the machine that determines its name and function. Figure 21–1 illustrates this idea.

Both machines are look-alikes with their armature, field poles, commutators, and such. Where they differ is in the opposite application of electromagnetism. This fact is emphasized by the contrast between Fleming's *left*-hand rule for *generators* versus the *right*-hand rule for *motors*. If you compare Figure 20–6 with Figure 18–4, you may notice that for any two quantities that are the same, the third one is exactly the opposite. The same idea is conveyed by Figure 21–2. This fact has some significant consequences, as we shall see in the following section.

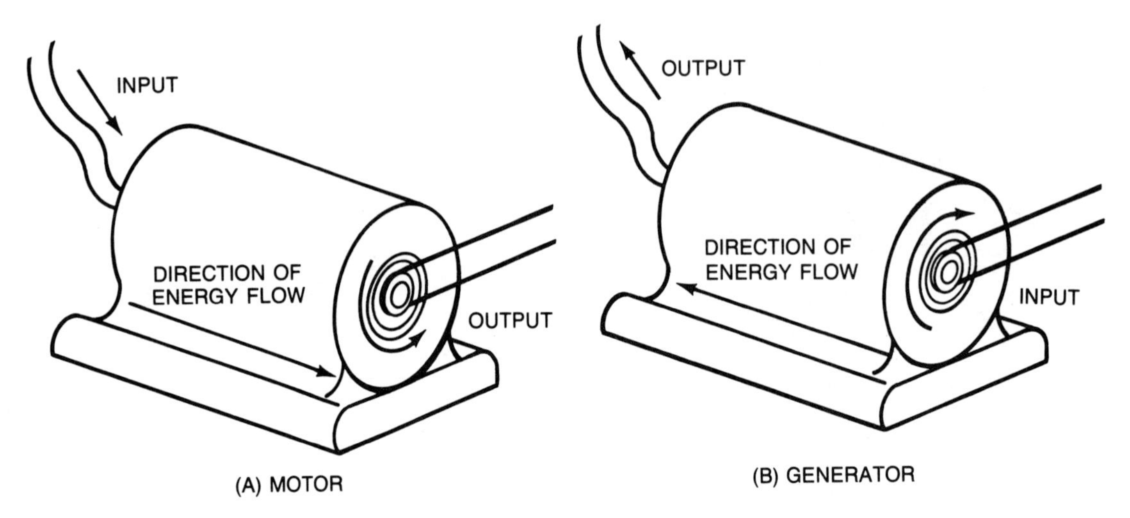

FIGURE 21–1

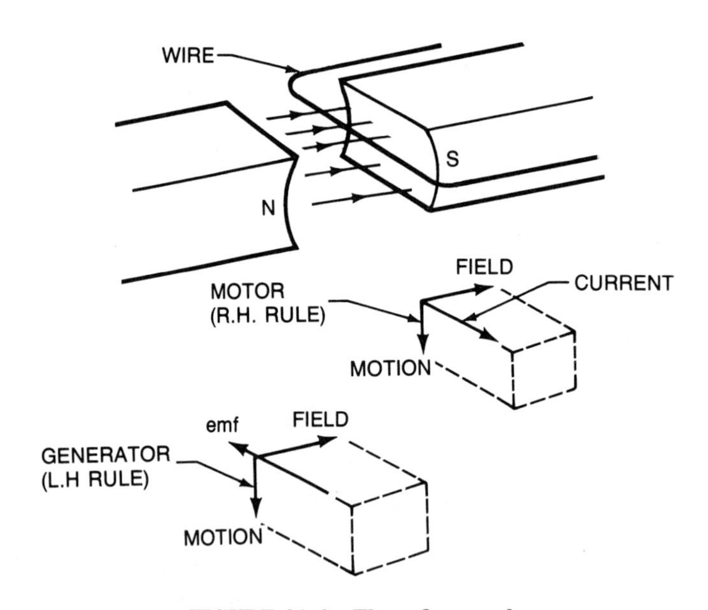

FIGURE 21–2 Three-finger rules

21–2 THE COUNTER-emf IN A MOTOR

When a coil of wire is rotated within a magnetic field, it generates an emf, regardless of what causes its motion. It does not necessarily take an external agent supplying muscle energy to the crank on a shaft, as implied by Figure 21–1B. It can be electro-

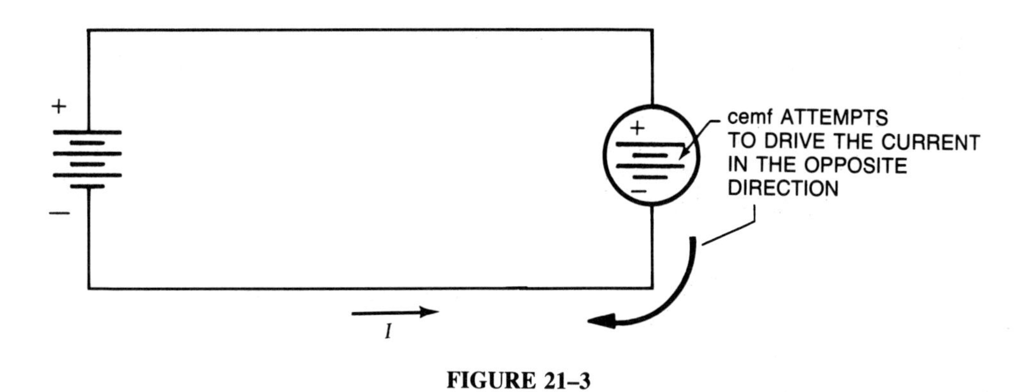

FIGURE 21–3

magnetic motor action that sets the armature spinning and starts to generate an emf. In other words, every motor inadvertently acts like a generator and produces a voltage within itself while it is running. This voltage is known as *counter-emf* (cemf), because its polarity is such that it opposes the applied line voltage, Figure 21–3.

Like the output voltage from a generator, the amount of cemf depends on two factors.

1. The strength of the magnetic field
2. The speed of rotation

We can, in fact, use the generator formula, first introduced in Section 19–8, to calculate the emf generated within the motor.

EXAMPLE 21–1

Given: The identical four-loop armature first discussed in Example 20–1. Its specifications are repeated here: 200 turns; Φ = 1,200,000 lines; I = 10 amperes; speed = 1,250 rpm.
Additional specification: $E_{applied}$ = 120 volts.
Resistance of the armature = 2 ohms.

Find: The cemf generated while the motor is turning at rated speed.

SOLUTION

$$\text{cemf} = \frac{\text{Turns} \times 4 \times \text{Flux} \times \text{rpm}}{\text{Paths} \times 60 \times 100{,}000{,}000}$$

$$\text{cemf} = \frac{200 \times 4 \times 1{,}200{,}000 \times 1{,}250}{2 \times 60 \times 100{,}000{,}000}$$

$$\text{cemf} = 100\ \text{V}$$

It may surprise some students to see that the cemf is nearly as high as the applied voltage. In fact, it is sometimes difficult for a beginner to realize the value of the emf that a motor generates. *The generated emf is a measure of the useful mechanical energy obtained from the electrons passing through the armature.*

As stated before, this emf is called *back-emf* or *counter-emf* because it opposes the voltage applied to the motor. This is a useful sort of opposition. The generated *back-emf* opposes the movement of the incoming electrons in the same way that the weight of a sack of groceries opposes the efforts of a boy carrying the sack up a flight of stairs. If there is no opposition, no useful work is done. If the boy leaves the groceries on the sidewalk, he can run upstairs faster, but the groceries are not delivered. An electric heater element generates no back-emf, the electrons can run through it rapidly, but they produce no mechanical work.

It is easy to make an adjustment on the electric motor so that it generates no counter-emf. Merely bolt down the armature so that it cannot turn. Then electrons run through the motor faster and more easily, and the motor is converted into an electric heater.

The example we used before had a line voltage of 120 volts applied. This power line has two jobs to do: First, it must supply 100 volts, which is converted to mechanical energy. Second, it must supply enough additional voltage to force the 10-ampere current through the 2-ohm wire resistance. It takes 20 volts to put 10 amperes through 2 ohms; thus, the line voltage must be 100 + 20 = 120 volts.

Total voltage applied to the armature = Counter-emf generated in the armature + Voltage *(IR)* used to overcome wire resistance

Carry this calculation one step further. Multiply the total applied voltage (120 volts) by the current (10 amperes) and find the total power input, 1,200 watts.

Generated voltage (100 volts) times 10 amperes equals the useful mechanical power output, 1,000 watts. This same figure was found once before: the 1.34 horsepower output that was found from torque and rpm.

The power input is 1,200 watts; the output is 1,000 watts. Where did the other 200 watts go? By multiplying the 20 volts used on resistance by the 10 amperes, we get 200 watts. This represents the rate at which energy is converted into *heat* in the armature. The rate of heat production is 200 watts.

Total input power (applied volts × amperes) = Useful mechanical power output (counter-emf × amperes) + Heating rate (*IR* × amperes)

The counter-emf does not produce useful mechanical energy, but the useful power produced can be calculated from the counter-emf and the current. This is just like saying that the useful power (working rate) accomplished by the grocery boy can be calculated from the weight of the sack and the speed with which he lifts it. The weight is not *doing* the work; but the greater the weight, the more work done by the boy.

EXAMPLE 21–2

Given: A 5-ohm armature taking 6 amperes from a 115-volt line when operating at its normal rating.

Find:

a. Counter-emf generated
b. Power input
c. Useful power output
d. Heating rate
e. Efficiency
f. Current and heating rate if the motor is stalled

SOLUTION

a. $6\text{ A} \times 5\ \Omega = 30\text{ V}$ used on resistance
$115 - 30 = 85\text{ V} = \text{cemf}$
b. Power input $= 115 \times 6 = 690\text{ W}$
c. Power output $= 85 \times 6 = 510\text{ W}$

$$\frac{510\text{ W}}{746\text{ W/hp}} = 0.68\text{ hp}$$

d. Heating rate $= 30\text{ V} \times 6\text{ A} = 180\text{ W}$
e. $\text{Armature efficiency} = \dfrac{\text{Power out}}{\text{Power in}} = \dfrac{510}{690} = 0.74 = 74\%$
The armature efficiency can also be found from

$$\frac{\text{Counter-emf}}{\text{Line volts}} = \frac{85}{115} = 0.74$$

f. If the motor stalls, current becomes

$$\frac{115\text{ V}}{5\ \Omega} = 23\text{ A}$$

Power input becomes $115 \times 23 = 2{,}645\text{ W}$, the heating rate for the stalled motor.

21–3 ARMATURE WINDINGS

As pointed out in Section 19–2, two main types of drum armature windings are in use: the lap winding and the wave winding. In Figure 21–4 (as in Figure 19–14), the innermost circle of numbered rectangles represents an end view of commutator bars. The next circle of twelve distorted rectangles represents sections of the face of the armature. (A different view of the face sections is shown in Figure 21–5.)

Outside of the rectangles in Figure 21–4, representing the cylindrical armature face, are lines showing the coil connections at the back of the armature. (This infor-

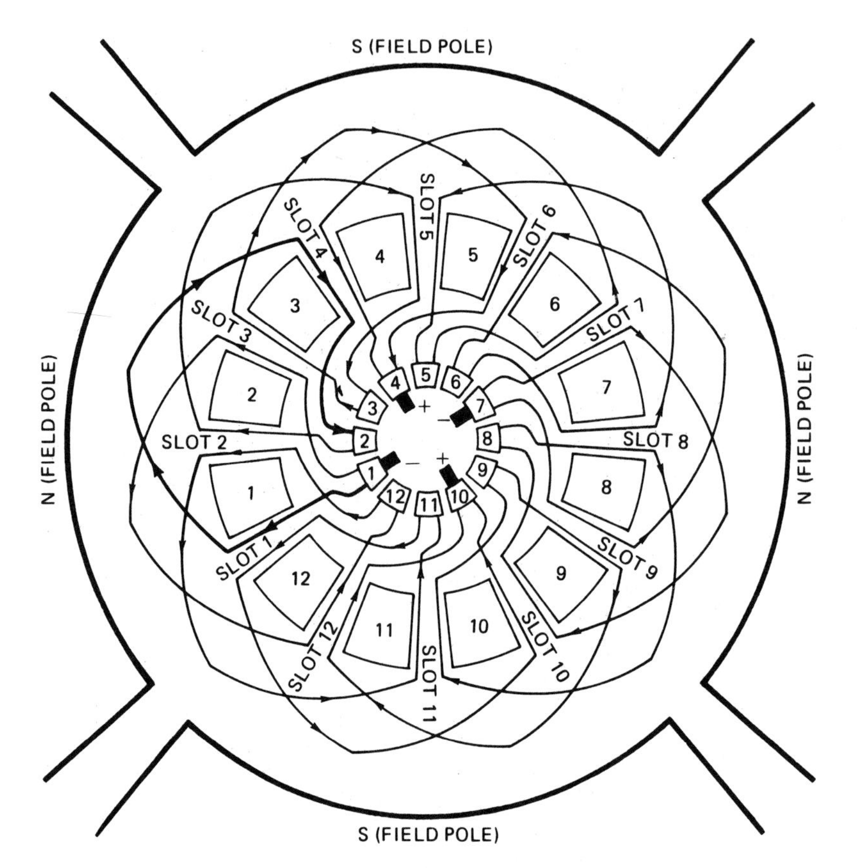

FIGURE 21–4 Lap winding, flattened view

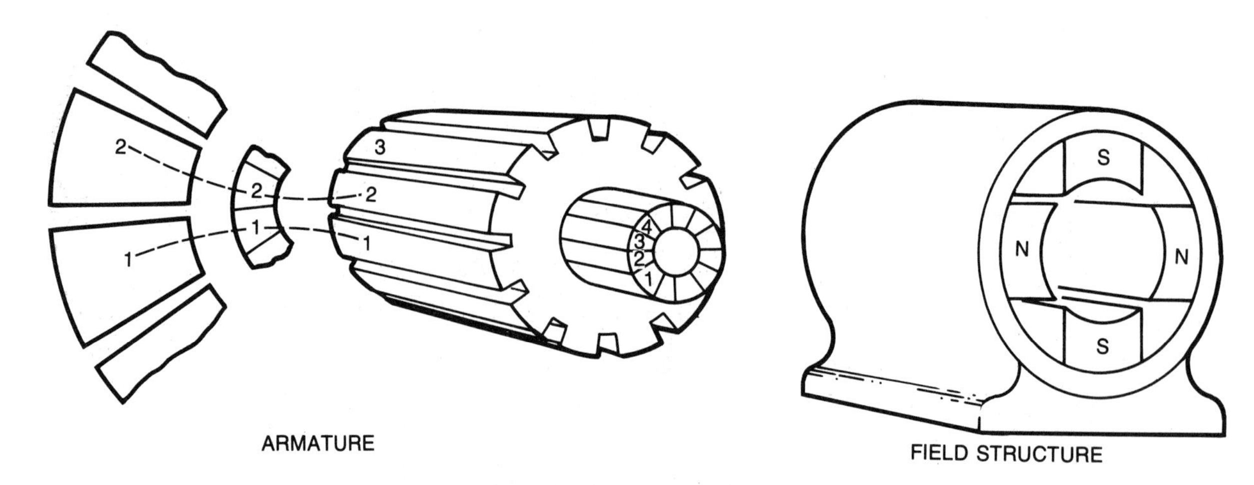

FIGURE 21–5 Simplified view

mation is not visible in a diagram such as Figure 21–5.) The field poles represent a four-pole field in which this armature is to rotate. (Note the simplified view of the field structure in Figure 21–5.)

The Lap Winding

A few minutes spent in tracing the wiring in Figure 21–4 makes the pattern seem less complicated, since the twelve coils on this armature are all alike. Start with the negative brush touching bar 1. Electrons flowing from the DC supply line to bar 1 can escape into the winding by either of two paths. (The brushes are shown as if inside, rather than in their actual position outside.)

Follow the heavily shaded wire through slot 1 and back through slot 4. The main feature of the lap winding is that the finish end of this coil, coming through slot 4, is connected to commutator bar 2, next to the one from which we started (1). From bar 2 the current path is through slot 2, back by slot 5 to bar 3. Then the current path goes out slot 3 and back by slot 6 to bar 4. Bar 4 is in contact with a positive brush that completes the circuit to the DC power source. Returning again to bar 1, another path can be traced through slots 3, 12, 2, 11, 1, and 10, back to the other positive brush.

Note that under both N-poles, the current is away from the commutator. Under both S-poles, the current is toward the commutator. With the help of Figure 21–5 and the three-finger right-hand rule, determine the direction of rotation of the armature.

Figure 21–6 is a schematic view of the circuits through this four-pole armature. Each numbered rectangle corresponds to a numbered commutator bar, Figure 21–4. The purpose of this sketch is to show that there are four parallel paths through this armature.

There are so many variations in armature windings that the subject can merely be introduced in these few pages; for the whole story, one should consult a text dealing only with armatures.

In general, lap windings require

- As many brushes as there are poles
- As many parallel paths through the armature as there are poles

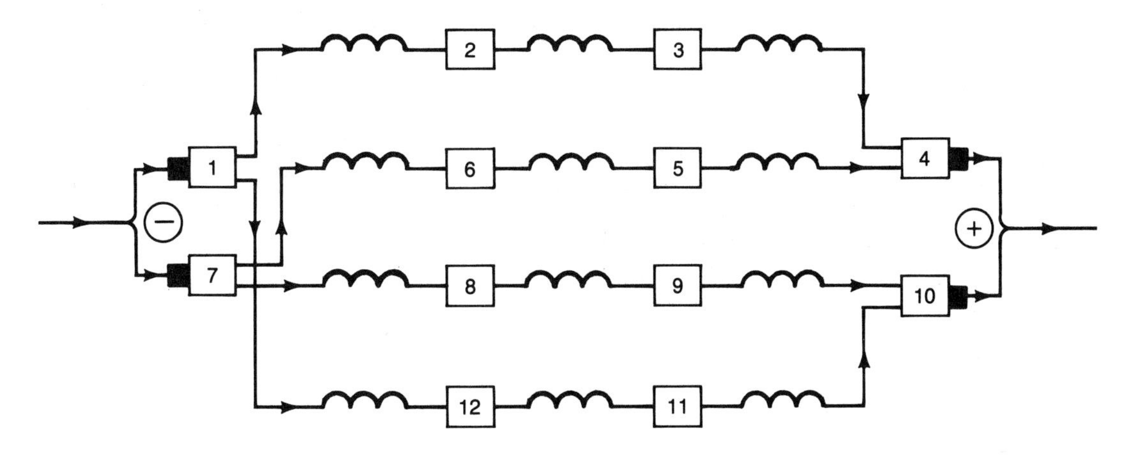

FIGURE 21–6 Lap-wound armature

The Wave Winding

Figure 21–7 shows another pattern called the wave winding. Compare it with the lap winding. Notice that the wave winding accomplishes a similar control of current direction in the wires under each field pole, so that rotation is produced. (One need not be concerned because the wires in slot 7 carry currents in opposite directions, which oppose and therefore cancel each other. Because slot 7 is between field poles at the instant shown, there is no force on the wires anyway.)

To see the difference between the wave winding, Figure 21–7, and the previous lap winding, Figure 21–6, observe the heavily shaded wire that starts from bar 1. The ends of the coil are spread apart, connecting to widely separated bars in the wave winding rather than to adjacent bars as in the lap winding.

Eleven bars and slots are used in the wave winding, rather than twelve, because a four-pole wave winding does not fit on a twelve-bar, twelve-slot armature. The reason for this difference, as well as the reasons for the locations of the commutator connections, involves calculations not worked out here.

Trace through or (better yet) redraw the winding, starting with a sketch that does not show the wiring. Begin at bar 1, through slot 1, over to slot 4, to bar 6, out slot 6, to slot 9, to bar 11, continuing around the armature until bar 4 is finally reached. There

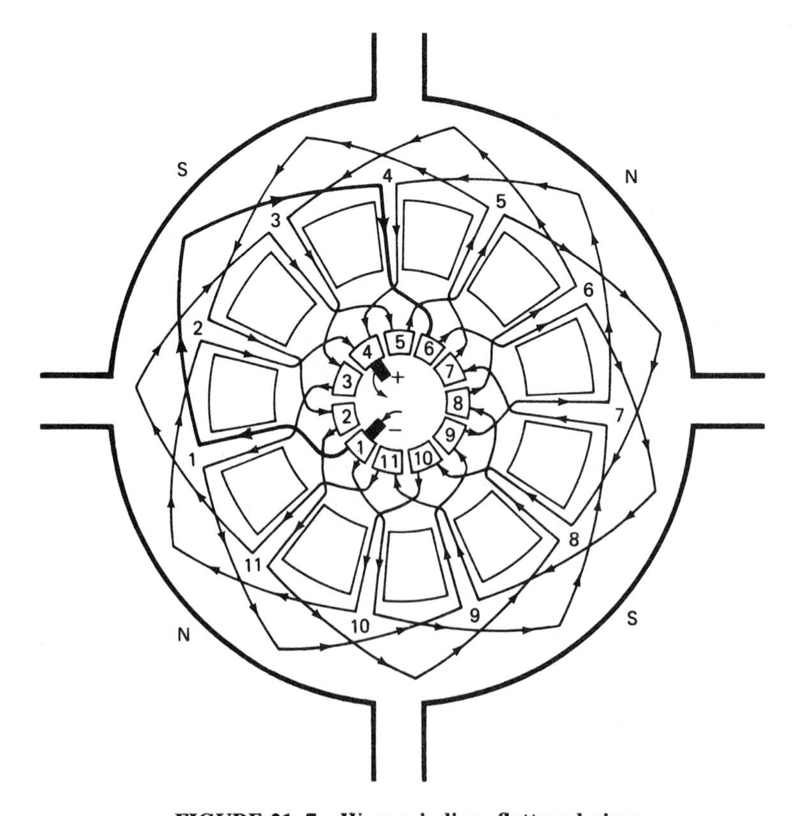

FIGURE 21–7 Wave winding, flattened view

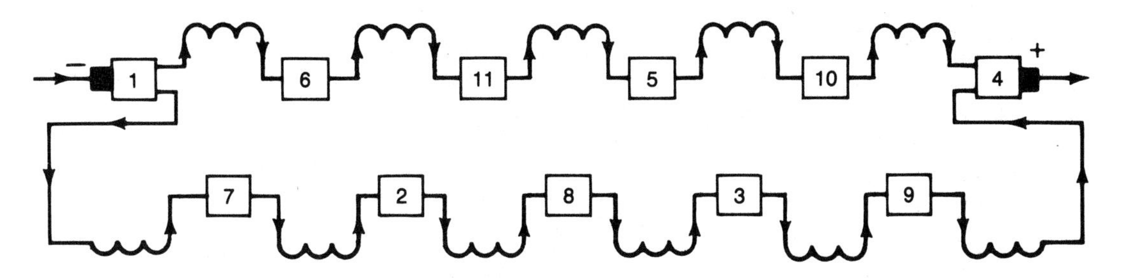

FIGURE 21–8 Wave-wound armature

is another path from bar 1, out slot 10, in slot 7, and so forth, again finally arriving at bar 4.

Figure 21–8 shows, in schematic fashion, that there are only two parallel paths through this armature. (The numbered rectangles correspond to the commutator bars in Figure 21–7.)

Wave windings require

- A minimum of two brushes but a maximum of as many brushes as poles
- Only two parallel paths through the armature in one complete wave winding, regardless of the number of brushes

Comparative Uses

Lap windings are good for high-current, low-voltage motors because they have more parallel paths for the current. Wave windings are good for high-voltage, low-current motors and generators because they have more coils in series.

Up to this point, we have assumed that the armature rotates in the field of a permanent magnet. Actually, permanent magnet fields are used in only a few motors. Electromagnets are used for the field structure of the great majority of useful motors.

Field magnet coils may be in series, or in parallel, or one in series and one in parallel with the armature, as in compound-wound motors, which have two sets of windings. The behavior of these magnetic field windings and their effect on motor performance is the subject of the next chapter.

21–4 FIELD DISTORTION AND THE NEED FOR INTERPOLES

In our study of generators we learned of a concept known as *armature reaction* (Section 19–4), which results in a distortion of the magnetic field. This distortion causes the neutral plane to shift *forward* (with respect to the direction of rotation).

In electric motors, as in generators, the current of the armature produces a magnetic field that interacts with and distorts the magnetic field in which the armature rotates. However, the magnetic action of motors is opposite to that of generators and, consequently, the neutral plane is shifted *backward* with respect to the direction of rotation, Figure 21–9.

To counteract this warping of the magnetic field, large DC motors are often built with interpoles, or *commutating poles*. The windings of these poles are in series with

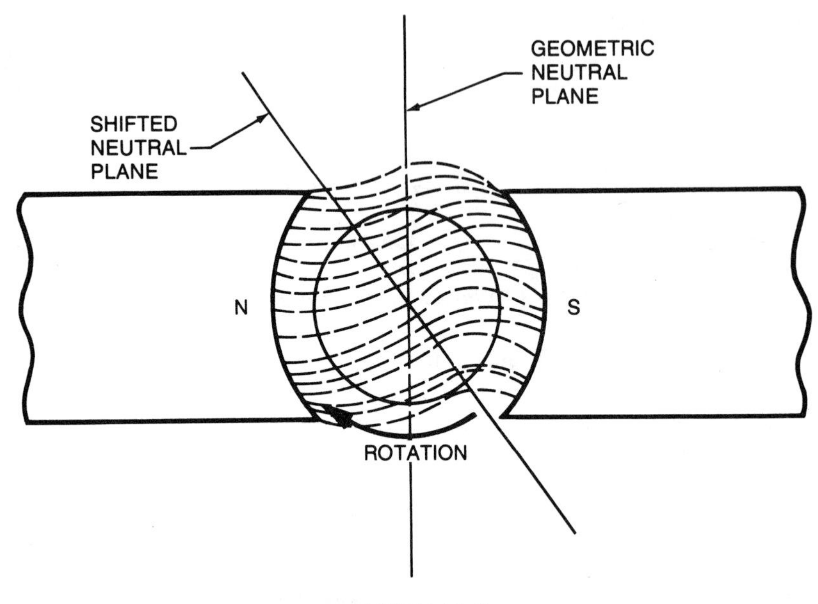

FIGURE 21–9

the armature. The poles produce a field that counteracts the armature field. Interpoles increase efficiency and control excessive sparking at the commutator when the motor is operated under conditions requiring high armature current. In the motor shown in Figure 21–10, the two large poles are the main field poles, and the two small ones are the interpoles.

In a motor, the interpoles must have the same polarity as the main poles directly in back of them (back in the sense of direction of rotation of the armature). In a generator, the interpoles have the same polarity as the main poles directly ahead of them, Figure 21–11.

In Figure 21–10, assuming a *clockwise* rotation of the armature (not shown), the polarities of the four main field poles and their interpoles would be as indicated.

21–5 THE SHUNT MOTOR

DC motors, like DC generators, are classified by the way their field coils are connected; thus, we differentiate between shunt motors, series motors, and compound motors.

Shunt motors, as their name implies, have their field coils connected in parallel to their armature and the power supply, Figure 21–12. The shunt field, therefore, consists of many turns of fine wire and maintains a steady magnetic field as long as the line voltage is constant. The torque of the motor is therefore solely a function of the armature current.

Let us assume that the load on a motor is increased. As a result, the motor slows down. Consequently, the cemf goes down also, the armature current increases, and the motor increases its torque to meet the new demand of the load.

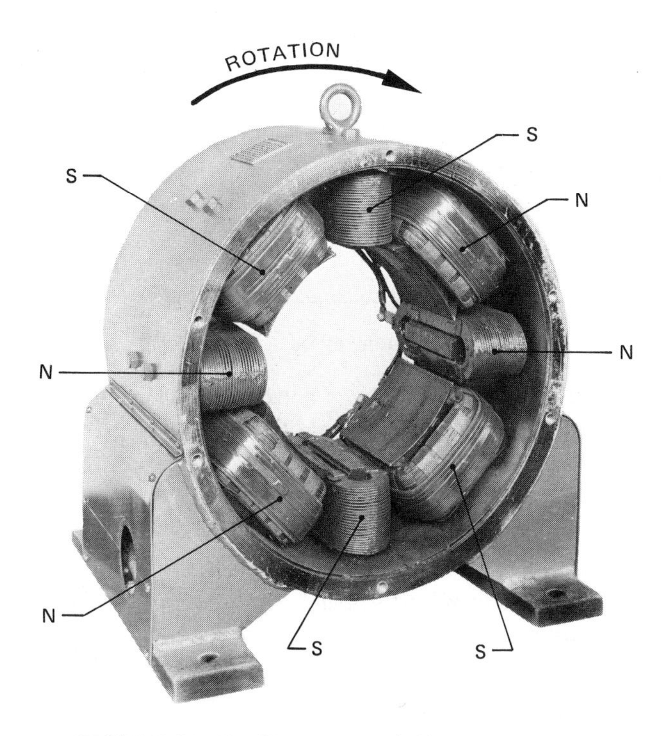

FIGURE 21–10 Shunt motor field with interpoles

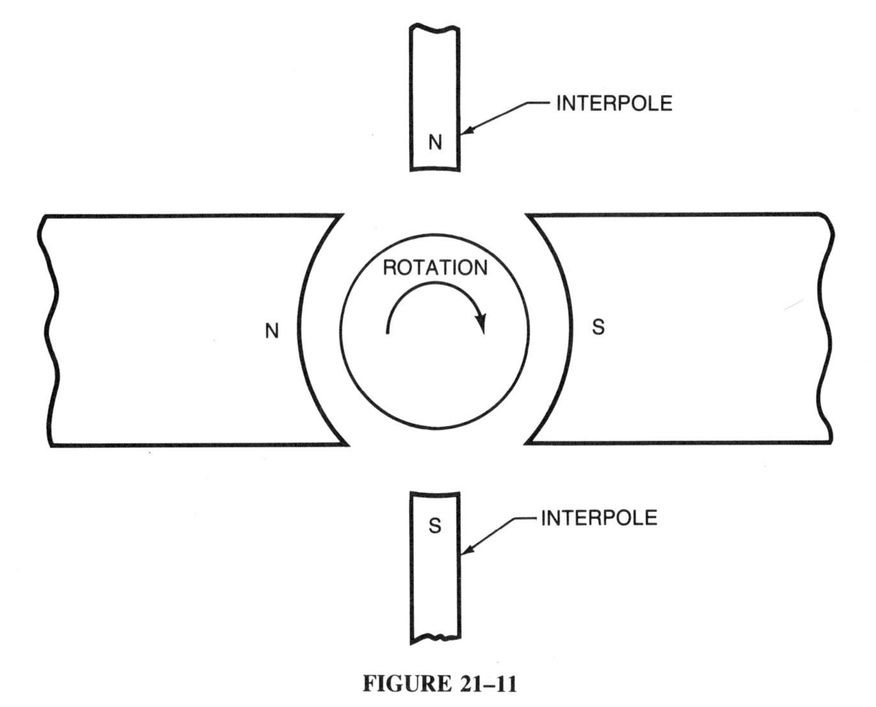

FIGURE 21–11

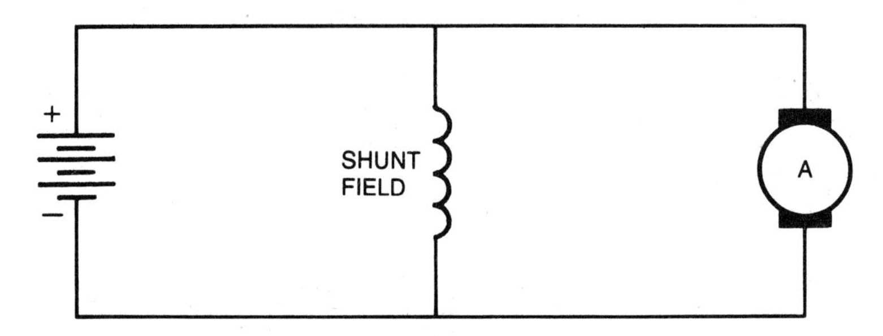

FIGURE 21–12 Schematic of shunt motor

This action can be summarized by a form of shorthand notation, in which arrows pointing *up* indicate an *increase* of the quantity and vice versa, like this.

Load $\uparrow$ rpm $\downarrow$ cemf $\downarrow$ I_{arm} $\uparrow$ Torque $\uparrow$ rpm $\uparrow$

Speed Regulation

When the torque increases to meet the new load demand, the speed will readjust itself; thus, the motor maintians a fairly constant speed. We say that the motor is *self-regulating*. This self-regulating effect, called *speed regulation,* is a characteristic of the motor itself. The speed regulation is numerically expressed as

$$\text{Speed regulation} = \frac{\text{No-load speed minus full-load speed}}{\text{Full-load speed}}$$

Note: The concept of *speed regulation* is not just for shunt motors but applies to all types of electrical motors. The lower the number of regulation, the more constant the speed of the motor.

EXAMPLE 21–3

Given: A motor turns 1,620 rpm at rated load but speeds up to 1,750 rpm when the load is removed.

Find: The speed regulation.

SOLUTION

$$\text{Speed regulation} = \frac{1{,}750 - 1{,}620}{1{,}620} = 0.08 = 80\%$$

The concept of regulation allows us to compare the speed change characteristic of various motors. A low-percentage regulation indicates a fairly constant speed. Example 21–3 is typical of shunt motors, which makes this type of motor desirable for industrial applications with constant speed requirements.

Shunt motors have a peculiar characteristic. When resistance is added to their shunt-field circuits, thereby decreasing the current and the magnetic flux, the motors will speed up. This feature is explained in greater detail later in this chapter.

This fact must be understood to appreciate that shunt motors, especially the large ones, must be protected against an accidental loss of their magnetic field. If, for some reason, the shunt field should open up, the loss of the magnetic field would cause the motor to accelerate to dangerously high levels.

If you wonder how a motor can run without a magnetic field, remember that there is a sufficient amount of *residual magnetism* to cause the motor to run away. Runaway motors can eventually destroy themselves due to the physical stress caused by centrifugal force.

Note: Study groups desiring a more detailed analysis of shunt motor characteristics are referred to the Appendix.

21–6 THE SERIES MOTOR

Unlike the shunt-motor field in which magnetization is attained by a small current in many turns, the series-motor field carries the entire current in a low-resistance coil of few turns, Figure 21–13. Shunt-field magnetization remains constant whether or not armature current changes; series-field magnetization changes as the motor changes under varying load.

Unlike the shunt motor, the series motor does not have a constant speed characteristic. With every change in load, the current through the field coil changes correspondingly, causing tremendous speed variations.

Unlike in the shunt motor, the torque and speed of the series motor are inversely proportional. In our accepted shorthand notation, this can be expressed as

Torque $\downarrow$ Speed $\uparrow$

In other words, whenever the load on a series motor is reduced, the motor will speed up. In fact, if the load should be completely disconnected, the series motor might run away and destroy itself.

One might reason that the brisk acceleration would cause the cemf to increase rapidly enough to shut off the torque-producing current. This, however, is not possible due to the sharp decrease in the magnetic flux. Thus, the torque accelerates the motor further, theoretically without limit. With the load removed, speeds can easily rise to

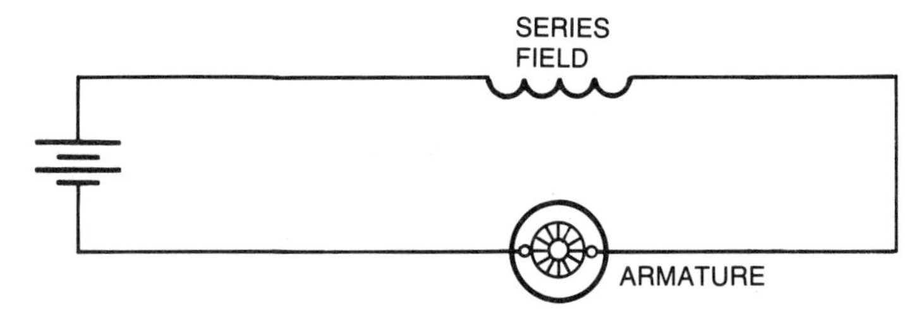

FIGURE 21–13 Series motor

10,000 rpm, and if the friction in the motor is less than equivalent to the torque produced, speeds build up even further.

In actuality the top speed of series motors is limited.

- In small motors, friction from bearings, brushes, and windage limits the speed. At 10,000 rpm, the entire power input can be expended on friction, and there is no further increase in speed.
- In large motors, high speed produces inertial forces that burst the bands holding the armature coils in place. At 5,000 rpm, the surface of an 8-inch-diameter armature is traveling at about 2 miles per minute, and each ounce of copper wire in the slots requires a force of 180 pounds to hold it in place.

For these reasons, it is recommended that series motors be used only in applications where the load is geared directly to the shaft of the motor. Belt drives, which are prone to slip or break, are not suitable for use with series motors.

A motor with such severe limitations must have some other strong advantages to recommend itself and, indeed, it does. The series motor has the ability to provide high levels of torque at startup or whenever a sudden overload condition places a heavy demand on the motor. Let us see why this is so.

Torque is an interaction of the two fields produced by the armature and the series field. Let us assume the current through the motor is doubled. This, in turn, doubles the magnetic flux of the series field as well as the magnetic flux of the armature. As a result, the torque will increase by a factor of $(2)^2$, or 4. We say that the torque of the series motor is proportional to the square of the current. (Of course, this statement is made with the assumption that the magnetic core is not saturated. Nevertheless, it should make the point that the series motor is ideally suited for any industrial application where extremely high torque is required and where very heavy overload is suddenly applied during operation. Examples of such applications include cranes, hoists, electric locomotives for railways, and other electrical vehicles.

But let us remember that this type of motor cannot be used where a relatively constant speed is required from no load to full load. Because the series motor has poor speed regulation, it can reach a dangerously fast speed when the load is removed.

Note: Study groups desiring a more detailed analysis of series motor characteristics are referred to the Appendix.

21–7 THE COMPOUND MOTOR

Compare the advantages of series and shunt motors. The shunt motor has a more constant speed, but the series motor (of the same power rating) can exert a much greater torque without a great increase in current. These two desirable features can be obtained in the same motor by placing both a series-field winding and a shunt-field winding on the field poles of the motor, which is now a *compound motor*. Compound motors combine the desirable features of both the shunt and series motors.

Consider the effect of adding a few series-field turns to an existing shunt motor. At heavy loads, when the motor slows down, the increased current through the series field boosts the field strength, which gives added torque, and speed.

Or consider the effect of adding a shunt field to a series motor. At light loads, when the motor tends to overspeed because of decreased field flux, the added constant-flux shunt field provides enough flux to put a reasonable limit on the top speed.

Combining the two fields within one motor results in a machine that retains the excellent starting torque of the series motor without the excessive speedup when the load is removed. These beneficial characteristics are derived only when the two field coils, the shunt and the series coils, produce magnetic fields aligned in the same direction (Figure 21–14). Wound in this manner, the windings are *aiding* each other in producing flux, and the motor is called a *cumulative compound* motor, Figure 21–14. Most compound motors are wound in this manner and, therefore, the word *compound motor* implies that it is cumulatively compounded.

Infrequently, a motor is connected with its series field opposing the shunt field; in other words, their magnetic fields are opposing each other. This can be done either by winding the two coils in different directions on the field pole, Figure 21–15, or simply by reversing the current through the coils, which are arranged as in Figure 21–14. This type of motor is said to be *differentially compounded*.

When the shunt field is connected directly to the line, the connection is called *long shunt,* Figure 21–16A. When the shunt field is connected directly across the armature, the connection is called *short shunt,* Figure 21–16B. The long shunt connection is generally used, but the type of connection makes no particular difference in motor performance.

If a given long-shunt motor is reconnected to a short shunt, slight changes do occur. At no load, with small armature current, the shunt field current passing through the series field slightly increases the total field flux. As a result, the maximum rpm of the motor is reduced. At heavy overload, with high armature current, the voltage drop on the series field reduces voltage and current available to the shunt field. As a result, to maintain a high torque, the motor can take 1% more current and run 1% faster than when connected long shunt.

Excellent speed regulation can be obtained with this type of motor. The motor runs with practically constant speed under varying load conditions.

In summary, compound motors are generally connected long shunt and cumulative compound. Such motors develop a high torque when the load is suddenly increased.

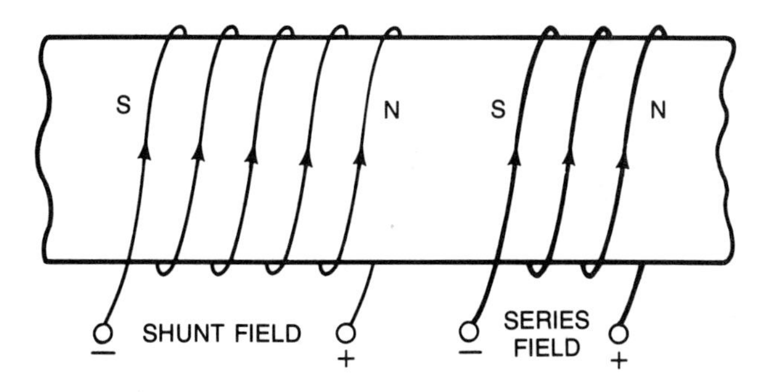

FIGURE 21–14 Fields aiding

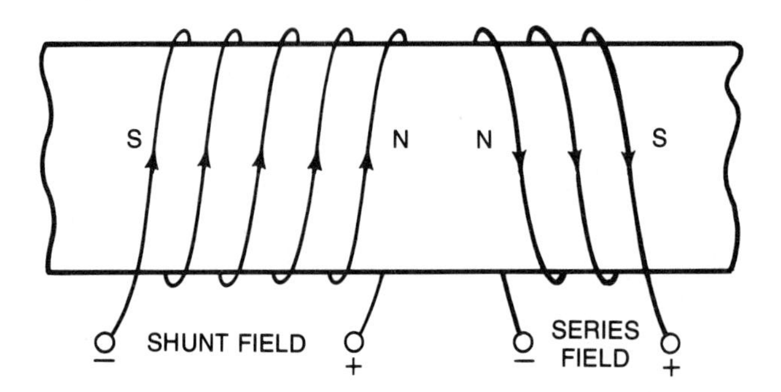

FIGURE 21–15 Fields opposing

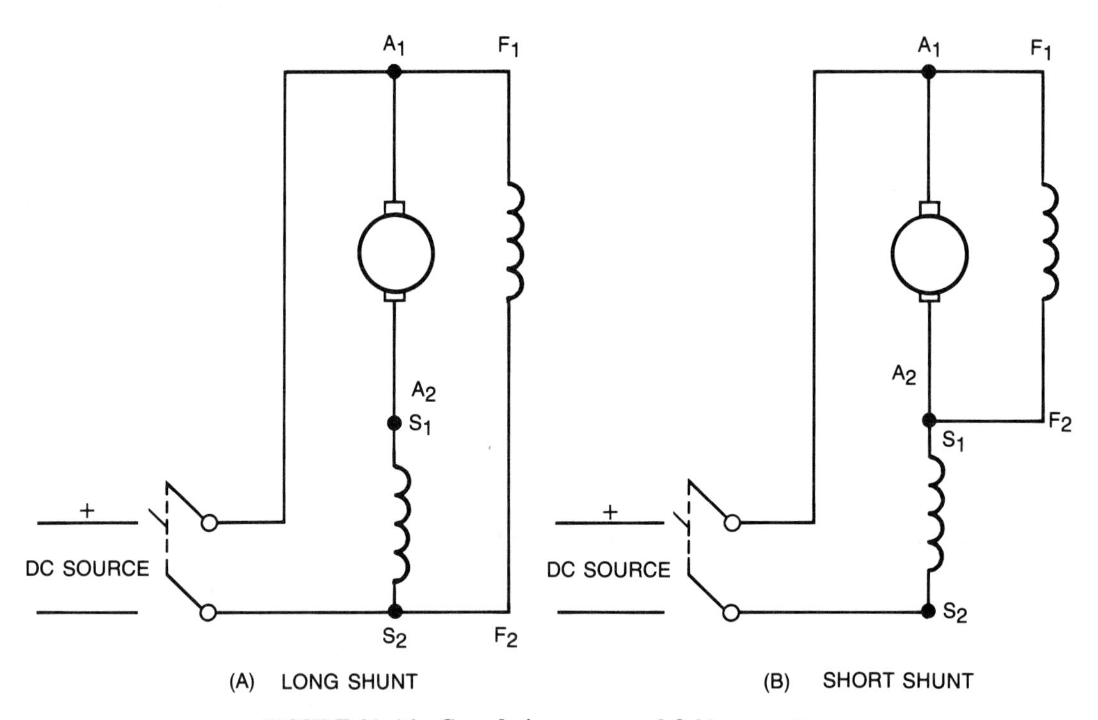

FIGURE 21–16 Cumulative compound field connections

This motor also has another advantage: It does not race to an excessively high speed if the load is removed.

Some of the industrial applications for this motor are drives for passenger and freight elevators, for metal stamping presses, for rolling mills in the steel industry, for metal shears, and in similar applications.

The graphs in Figure 21–17 compare the characteristics of the three types of motors: series, shunt, and cumulative compound. Compound motors can be built with characteristics approaching either the series or the shunt characteristics, depending on the relative division of ampere-turns between series and shunt coils.

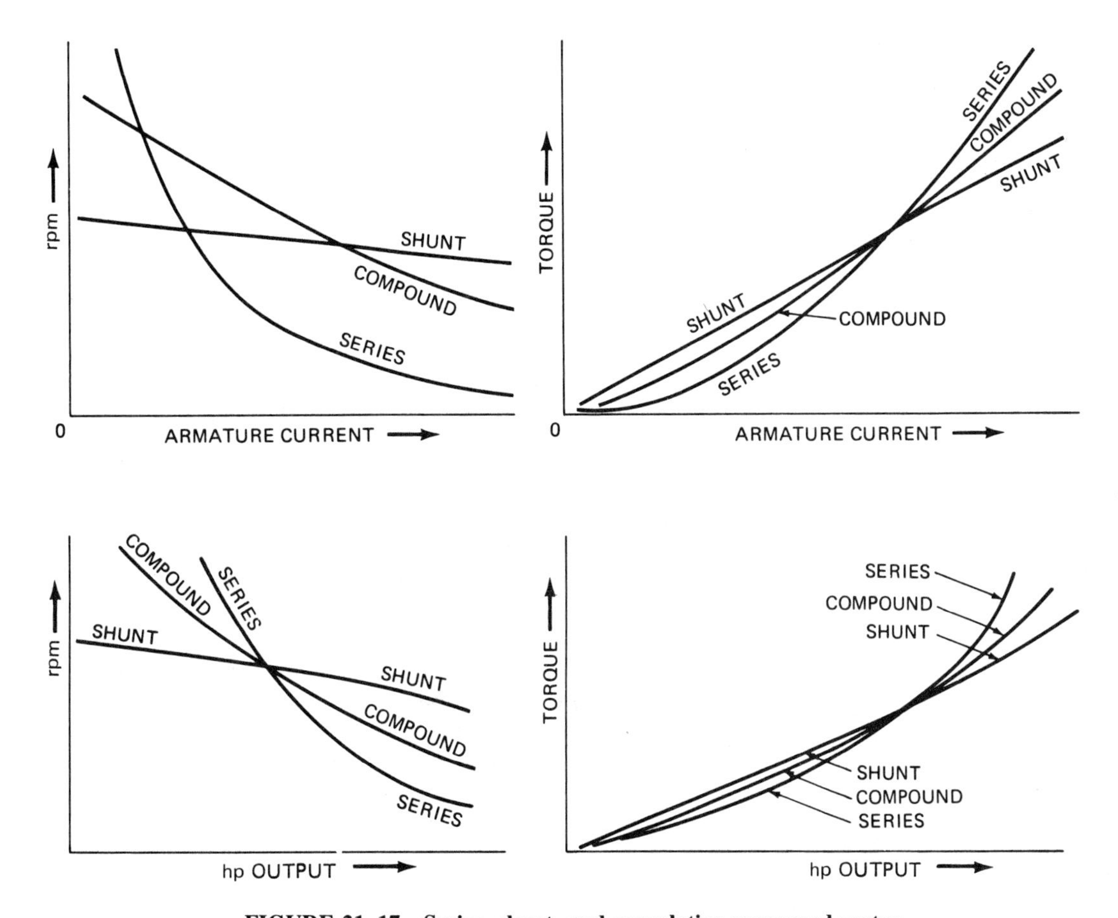

FIGURE 21–17 Series, shunt, and cumulative compound motor

21–8 SPEED CONTROL OF DC MOTORS

Not all electric motors are suited to have their speed controlled by smooth acceleration or deceleration. *DC motors do have this ability.*Shunt motors are easily controlled in their speed by placing a rheostat in series with the shunt field, thereby varying the current and, consequently, the flux, Figure 21–18.

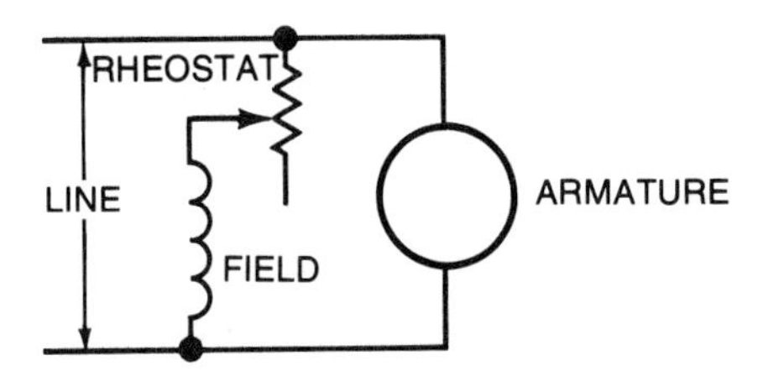

FIGURE 21–18 Shunt-field control

Many students of electricity make the wrong assumption that increased resistance will slow the motor. This is not so; just the opposite will occur. When the resistance is increased, the current and therefore the magnetic flux are decreased; thus, the motor speeds up. This happens because a reduction in flux will reduce the cemf, which in turn increases the armature current. As we have seen earlier, these events lead to increased torque and increased speed.

Our shorthand notation would describe this chain of events as follows:

$$R_{field}\uparrow \quad \Phi_{field}\downarrow \quad \text{cemf}\downarrow \quad I_{arm}\uparrow \quad \text{Torque}\uparrow \quad \text{Speed}\uparrow$$

Needless to say, the opposite actions occur when the field resistance is decreased.

It would also be possible to vary the speed of the motor by adding a rheostat into the armature circuit. This, however, is not a desirable practice, because armature circuits generally draw a lot of current, and the resulting I^2R losses would be prohibitively large. This also results in poorer speed regulation.

There are occasions, however, when a motor must be operated at *below-normal speed,* at which time it becomes necessary to insert resistance into the armature circuit. (The term *normal speed* refers to the speed of the motor without any speed control mechanism in place.)

We have already seen that added resistance in the shunt circuit will result in *above-normal speed*. Should it become necessary to slow the motor to below-normal speed, resistance must be added into the armature circuit. This may not be a happy solution to the problem, but it surely reduces the speed by decreasing the armature current. In this respect, there is no difference between the shunt motor and the compound motor.

Note: The practical and preferred way to vary the speed of a DC motor is by varying the resistance in the shunt-field circuit.

Unfortunately, like everything else in life, this method presents us with a mixed bag of blessings. In other words, controlling speed by varying the field's magnetic strength has some drawbacks, too. With the reduction of magnetic flux, there is a corresponding reduction in the production of torque. This means that a desired speed increase brings with it a reduction in torque.

In this respect, armature control appears to be superior. In order to speed up the motor you would increase the current through the armature (by varying a rheostat). The motor reacts to limit the current increase by speeding up and producing more counter-emf. Since the strength of the magnetic field remains unchanged, the torque will not suffer.

On the other hand, rheostats in the armature circuit would have to be unreasonably large to handle the armature current. As mentioned before, the resulting I^2R losses would be costly, and the efficiency and speed regulation of the motor would suffer. A solution to this dilemma would call for armature control without rheostats. This innovation came about with the introduction of electronic devices known as *thyristors*.

Electronic Speed Control

The rapid development of solid-state devices during the last few decades has brought about some extraordinary changes in the control of electric power. Semiconductor devices, known as *thyristors,* can control vast amounts of power with only very small amounts of input power. Such devices have ushered in a new era of industrial electronics, aimed at the control of electrical power and machines.

The *silicon controlled rectifier* (SCR) is the most popular device among the thyristors. Thyristors can be defined as electronic switches. The SCR can be switched rapidly, thousands of times per second, without any moving parts. The graphic symbol of the SCR is shown in Figure 21–19, where it is shown connected, like a switch, in series with the armature of a shunt motor. An electrical pulse at the gate (G) will trigger the device into its *on* position.

You may have already noticed that the DC motor is shown connected to an AC source. In spite of this, the motor does operate on DC. The field receives DC from a bridge rectifier, and the armature receives rectified DC from the SCR. In addition to providing rectification, the SCR controls the average armature voltage and current by pulsing the current through the armature in response to the trigger pulses received by the gate. This action, in turn, controls the speed of the motor. The trigger pulses at the gate are regulated by special electronic circuits, thereby controlling the amount of energy delivered to the motor.

As you progress in your studies of electronics, you will encounter such devices time and again, and you will then have to study their principles of operation.

For the time being, it will suffice for you to realize that such electronic devices and techniques have invaded the domain of industrial electricity, and that electronic devices, such as the SCR, will be used increasingly to replace rheostats in the use of motor control.

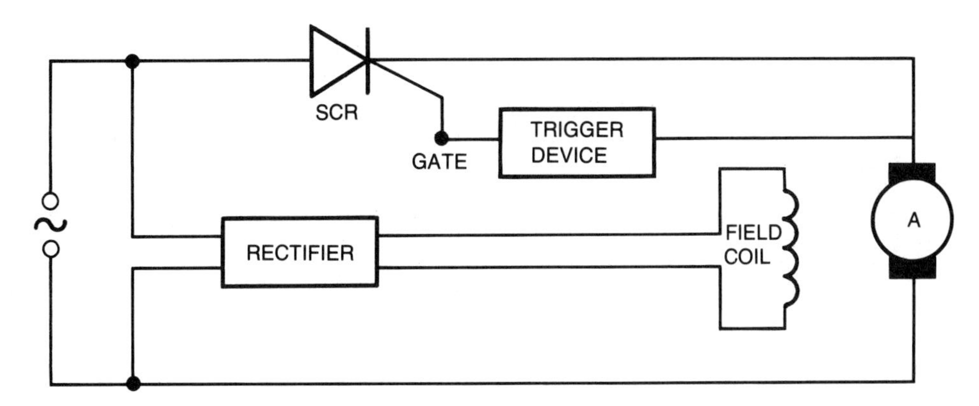

FIGURE 21–19

21–9 REVERSAL OF ROTATION

If you recall that torque in our electric motor depends on the interaction of two magnetic fields, you may easily understand that the *direction* of rotation depends on the magnetic polarity these fields have with respect to each other.

Figure 21–20 illustrates that the forces of repulsion and attraction can be changed by reversing the polarity of the armature with respect to the polarity of the field coils, and vice versa. This will also explain why it is generally not possible to affect reversal simply by changing the leads of the power source. This technique works only with motors in which the field is provided by permanent magnets, or permanent magnet motors (PM motors). Small, electric toy motors are generally of this kind and can be easily reversed by merely switching the power leads.

By contrast, all motors with electromagnetic fields can be reversed by changing the connection of the wires at the terminal box of the motor. The armature wires are generally labelled A_1 and A_2, and the shunt-field winding is marked F_1 and F_2. Let us assume that a given motor is turning clockwise when its A_1 terminal is connected to F_1 and A_2 is connected to F_2, Figure 21–21A. If it should become necessary to change the direction of rotation, it is only necessary to reconnect the wires A_1 to F_2 and A_2 to F_1, Figure 21–21B.

When compound motors are to be reversed, it is best to switch just the connections to the armature. If the design of the machine makes it easier to change the field coils, it must be remembered to change both the shunt and the series field. (The terminals of the series fields are labelled S_1 and S_2.) If only one of the two field windings were switched around, the motor would change its characteristic of a cumulative compound motor to that of a differentially compounded machine, or vice versa.

When frequent reversal of a motor is desired, the installation of a double-pole, double-throw (DPDT) switch will facilitate the task. Figure 21–22 shows how the switch changes direction of the current through the shunt field and, consequently, changes direction of the motor.

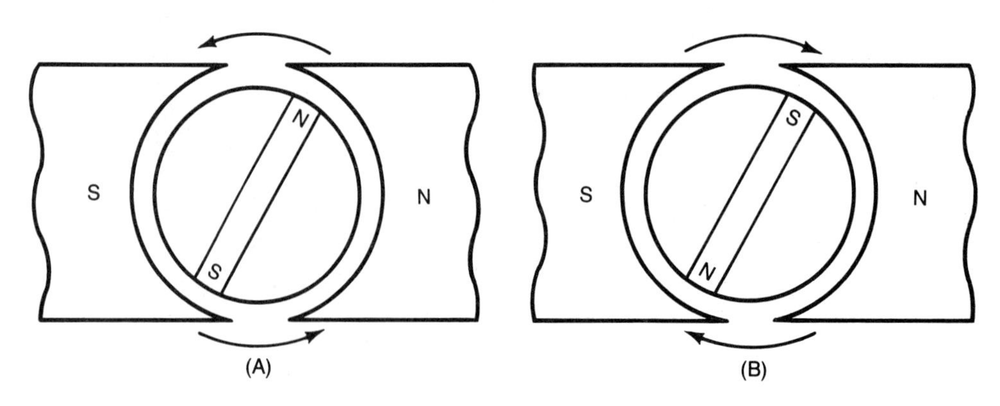

FIGURE 21–20

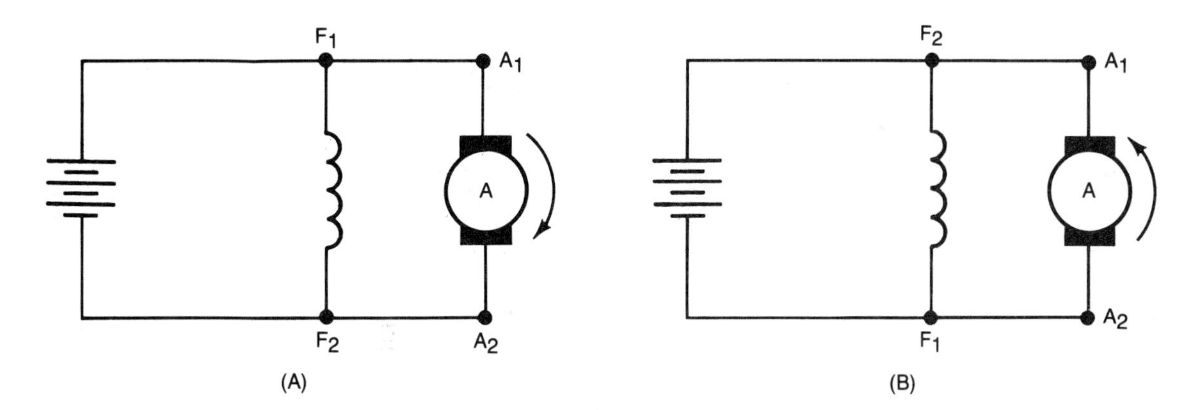

FIGURE 21–21

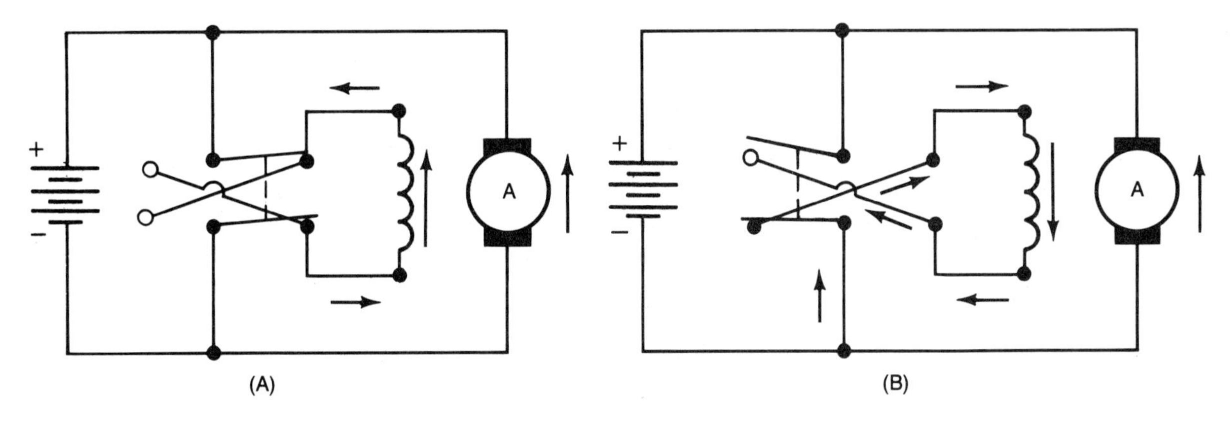

FIGURE 21–22

21–10 POWER LOSSES

Power losses in electric motors are classified, just as they were for generators, as I^2R losses and stray power losses. This was explained in Section 19–9, which you may want to review at this time.

The drawing used in conjunction with that explanation is similar to Figure 21–23, which illustrates the relationship between useful and useless energies in a motor.

The drawing presents an analogy between a machine and a conduit, both of which have an input on one side and an output on the other. The conduit has some leaks through which various losses are incurred, thereby causing the output to be less than the input. This ratio of output to input is, of course, known as *efficiency*. This, too, may be reviewed in Section 19–9.

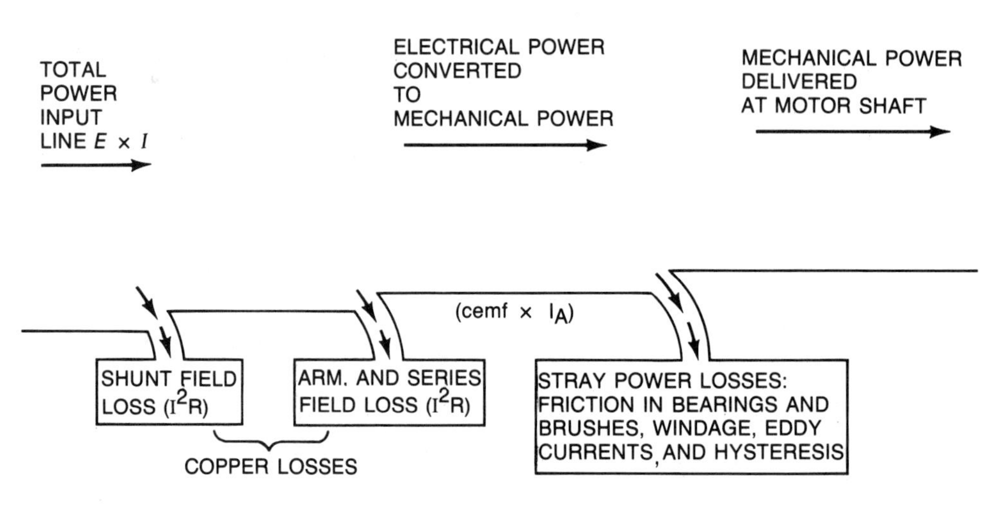

FIGURE 21–23

EXAMPLE 21–4

Given: A shunt motor with a 200-ohm field and a 0.4-ohm armature. At no load, the motor takes 3 amperes on a 220-volt line.

Find: The stray power losses.

SOLUTION

Field current: 220/200 = 1.1 A
Armature current: 3.0 − 1.1 = 1.9 A
Armature circuit: 1.9 A × 200 V = 418 W used
I^2R in armature: 1.9 × 1.9 A × 0.4 Ω = 1.44 W
Stray power losses must be 418 − 1.4 = 416.6 W

SUMMARY

- The left-hand rule applies to generators, while the right-hand rule applies to motors.
- Torque is proportional to flux and armature current. Increase of either increases torque.
- Counter-emf is proportional to flux and rpm. Increase of either increases emf.
- The amount of counter-emf determines the useful mechanical output from the motor.
- Armature windings of motors are similar to those of generators. Lap windings are suitable for high-current applications, while wave windings sacrifice current for high-voltage applications.
- In motors, the interpoles have the same polarity as the main pole directly in back of them (with respect to the direction of rotation). This is opposite to the placement of interpoles in generators.

- The shunt motor has its field coils in parallel with the armature. Its constant field strength limits input current and speed at no load.
- A shunt motor accelerates to excessive speeds when the shunt field is lost.
- The series motor has its field coil in series with the armature. At no load, reduced field permits enough current to produce torque that accelerates the motor greatly. At heavy load, the torque is high, the speed is low.
- A series motor accelerates to excessive speeds when the load is lost.
- When the advantages of series and shunt motors are compared, it is seen that the shunt motor has the more constant speed, but a series motor of the same horsepower rating can exert a much greater torque, when necessary, without a large increase in current.
- Compound motors have both series and shunt fields. The relative ampere-turns of the two fields determine the speed and torque characteristics of the motor.
- Generally, compound motors are cumulatively compounded and long-shunt connected.
- Weakening the field speeds up the shunt motor; reduced armature current slows the shunt motor.
- Speed control can be achieved by either changing the current in the field or in the armature.
- When rheostats are used for speed control, the preferred placement of the rheostat is in the shunt-field circuit.
- Armature control is the preferred method when electronic components are used to vary the speed.

Achievement Review

1. An armature, resistance 0.47 ohm, is supplied with 124 volts DC and 40 amperes. Calculate
   a. The volts used to overcome the resistance of armature
   b. The volts used to make mechanical energy (counter-emf)
   c. The horsepower output
   d. The efficiency
   e. The watts heating rate
2. Calculate the current in the armature of question 1 if the armature stalled because of overload.
3. Determine the direction of the rotation of the armatures shown in Figures 21–4 and 21–7.
4. What is the polarity of interpoles in reference to the polarity of main field poles in a direct current motor?
5. a. Show with characteristic curves the speed and torque performance for a shunt motor.
   b. List four industrial applications for a shunt motor.
6. a. Show with characteristic curves the speed and torque performances for a series motor.

b. List four industrial applications for a series motor.

7. a. Show with characteristic curves the speed and torque performance for a cumulative compound motor.
   b. List four industrial applications for a cumulative compound motor.
8. A shunt motor is required to carry an additional load. List the sequence of steps showing how this motor adjusts itself to carry the additional load.
9. The armature of a DC shunt motor carries 15 amperes. The resistance of the armature circuit is 0.7 ohm. The line voltage is 220.
   a. Find the counter-emf generated in the armature.
   b. Find the power output in watts.
   c. Find the power output in horsepower.
10. A 3-horsepower, 120-volt shunt motor takes 23 amperes at full load and 3 amperes at no load. Shunt resistance is 150 ohms; armature resistance is 0.25 ohm; and no-load speed is 1,600 rpm. Find
    a. Counter-emf at no load
    b. Counter-emf at full load
    c. Full-load speed
    d. Percentage speed regulation
11. Explain the meaning of the following terms:
    a. Speed regulation
    b. Speed control
12. Explain what happens under each of the following conditions:
    a. The load is removed from a series motor
    b. Resistance is added to the armature circuit of a shunt motor
    c. The rheostat in series with the shunt field of a shunt motor operating at no load becomes open circuited.
    d. The field rheostat in series with the shunt field of a shunt motor is adjusted so that all the resistance of the rheostat is cut out and the shunt-field current increases.
13. a. Itemize the losses that reduce the efficiency of a motor, and state in what parts of the motor each loss occurs.
    b. State which of these losses are constant and independent of the load.
    c. When the constant losses are grouped together, what term is used?
14. A motor with an input of 1,000 watts delivers an output of 1 horsepower. If the copper losses are 134 watts, what are the constant or fixed losses, which are independent of the load?
15. A 50-horsepower, 220-volt shunt motor has a full-load efficiency of 83%. Field resistance is 110 ohms; armature resistance is 0.08 ohm. At full load determine
    a. Total power input in watts
    b. Line current
    c. Total copper losses
    d. Stray power losses
16. A DC shunt motor takes 40 amperes at full load when connected to a 115-volt line. At no load, the motor takes 4.4 amperes. Shunt-field resistance = 57.5 ohms; armature circuit resistance is 0.25 ohm; full-load speed is 1,740 rpm.

Find

a. Stray power loss
b. Copper loss at full load
c. Efficiency at full load
d. Horsepower output at full load

17. The generator of a motor-generator set is delivering its full-load output of 10 kilowatts. The generator has an efficiency of 88.5%; the motor operates on a 230-volt line. Determine
    a. The output of the motor in horsepower.
    b. The number of amperes the motor is taking from the line when the generator is delivering at rated load. The overall efficiency of the motor-generator set is 80%.
18. A motor operating on a 120-volt DC supply drives a direct current generator, which is delivering 1 kilowatt at 240 volts. Under these conditions the motor has 78% efficiency and the generator has 85% efficiency. Determine
    a. The output of the motor in horsepower
    b. The line current drawn by the motor
19. A 220-volt, 20-horsepower compound motor (long shunt, Figure 21–16A) has an armature resistance of 0.25 ohm, series-field resistance of 0.19 ohm, and shunt-field resistance of 33 ohms.
    a. Calculate the current taken by the motor at the instant of starting if it is connected directly to the 220-volt line.
    b. Calculate the current when the motor is running if the armature is developing 184 volts counter-emf.
20. Complete the external and internal connections for the shunt motor in the sketch on page 368 for counterclockwise rotation. The interpole field windings are to be a part of the armature circuit, which terminates at the connection points marked A_1 and A_2. Be sure to have the proper connections for the interpole field windings and the main field windings so that their polarities are correct for counterclockwise rotation.

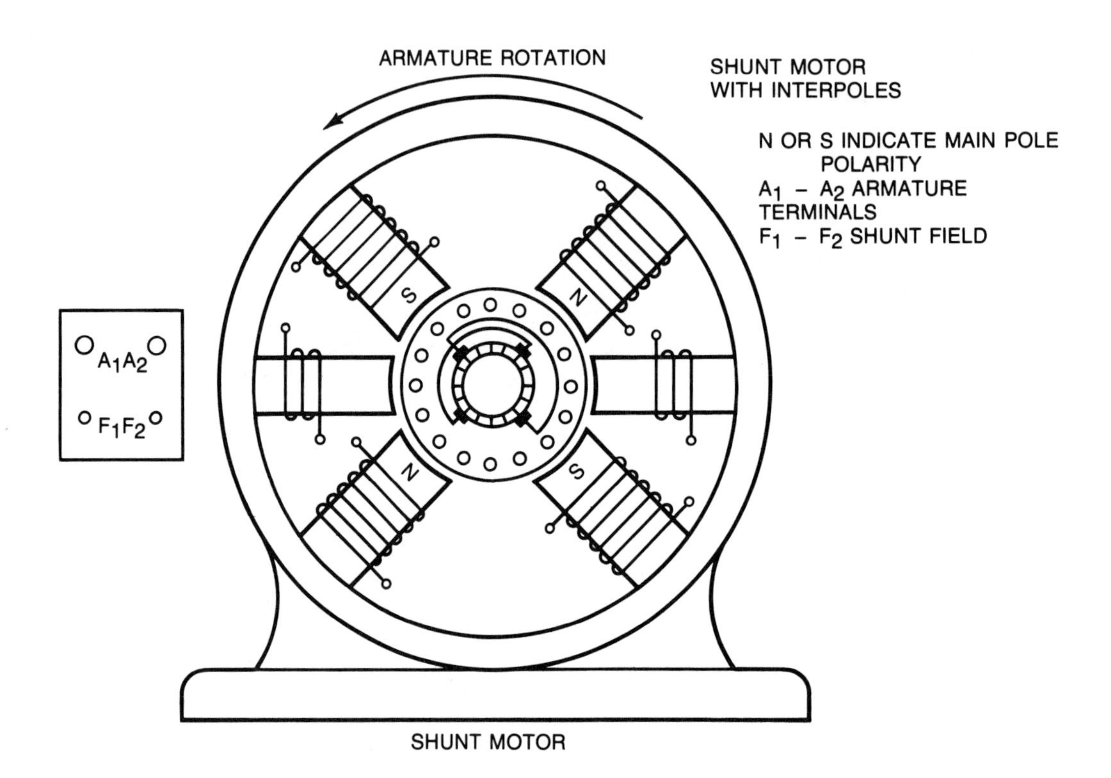
ARMATURE ROTATION
SHUNT MOTOR
WITH INTERPOLES
N OR S INDICATE MAIN POLE POLARITY
A1 – A2 ARMATURE TERMINALS
F1 – F2 SHUNT FIELD
A1A2
F1F2
S
N
N
S
SHUNT MOTOR

22
Starters and Speed Controllers

Objectives

After studying this chapter, the student should be able to

- Define and explain the new technical terms introduced in this chapter

  three-terminal starting rheostat
  no-voltage protection
  holding coil
  drum controller
  push-button station
  magnetic blowout coil
  dynamic braking
  lockout relay
  electrical interlock
  logic gates
  above- and-below-normal speed controller
  four-terminal starting rheostat
  no-load protection
  no-voltage release
  contactor
  sealing circuit
  thermal overload
  cemf controller
  timing relay
  ladder diagram
  programmable controller

- Explain the reasons for using reduced-voltage starting
- Differentiate between starters and controllers
- Trace the current flow through starters and controllers
- Show the connections for a starting rheostat connected to a DC motor
- Discuss the advantages and disadvantages of starting rheostats
- Complete diagrams showing DC motors connected to above-and-below-normal speed controllers
- Explain some specific applications for drum controllers
- Draw, from memory, a simple control circuit with a push-button station and self-sealing contact
- Explain the sequence of operation for the three basic types of controllers covered in this chapter

22–1 THE NEED FOR REDUCED-VOLTAGE STARTING

For starting small DC motors (up to 2 horsepower), the motor is simply connected directly to the DC power line. But the sudden connection of a large motor to a DC line would cause unreasonably high current in the line and armature, since, at the moment of starting, no counter-emf exists to limit the current.

For example, in question 19 in Chapter 21, the armature and series field have a total resistance of 0.44 ohm. With no opposing emf at the instant of starting, the armature current is 220/0.44 = 500 amperes. Without the addition of external resistance, this high current puts a great stress on armature windings, burn brushes, and commutators and causes line voltage drop, which can interfere with other machines on the line.

For the gentle starting of large motors, a *motor starter* is used. It is merely a variable resistance placed in series with the armature. Its primary purpose is to limit the armature current to a safe value during the starting and accelerating period. Along with the starting rheostat, there is usually some arrangement for automatically disconnecting the motor (and leaving it disconnected) if the line voltage fails.

The two common types of manual starting rheostats, or *starting boxes,* used with shunt and compound motors are the *three-terminal* and *four-terminal starting rheostats*.

22–2 MANUAL STARTERS

Three-Terminal Starting Rheostat

The three-terminal starting rheostat, Figure 22–1, has a tapped resistor enclosed in a ventilated box. Contact buttons, located on a slate panel mounted on the front of the

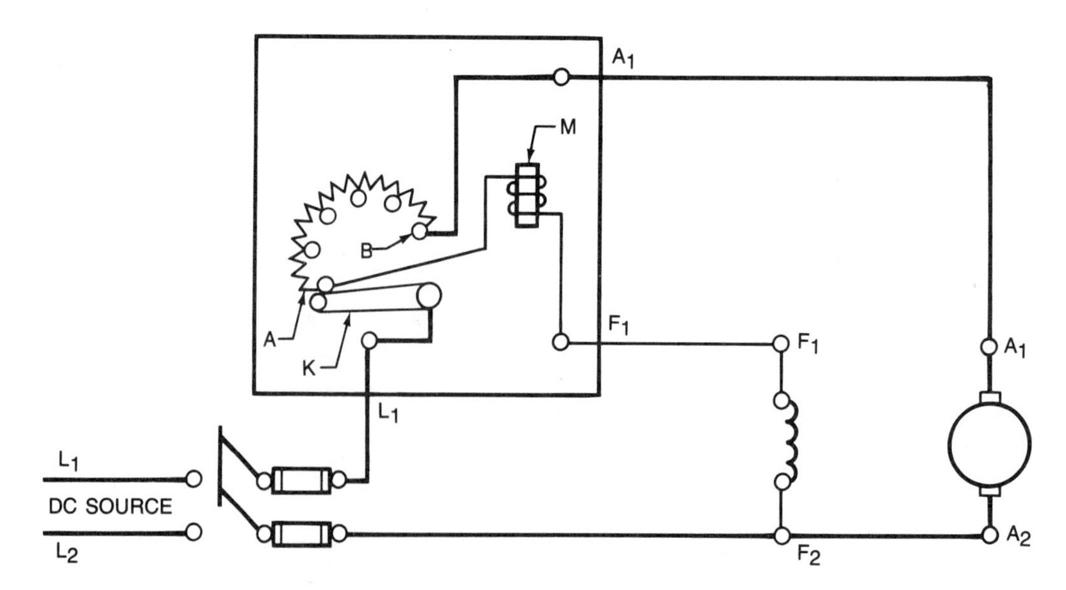

FIGURE 22–1 Three-terminal starting rheostat with shunt motor

box, are connected to the tapped resistor. A movable arm K with a spring reset can be moved over the contact buttons to cut out sections of the tapped resistor.

After the line switch is closed, the arm K is moved to the first contact, A. The shunt field is now connected to the line at full strength. All of the starting resistance is in series with the armature. This resistance, in accepted practice, is calculated to limit the starting current to 150% of the full-load current rating of the motor.

As the motor speeds up, the operator moves the arm gradually toward contact B. The time required depends on the time needed for the machine to build up speed. At B the armature is connected directly across the source voltage. The *magnetic holding coil,* M, holds the arm in the full *on* position. A spring (not shown) tends to return the arm to the *off* position. If the shunt-field current is much reduced while the armature circuit remains connected, the motor races. However, shunt-field current reduction is prevented by having the holding coil in series with the shunt field. Reduced current in the holding coil lets the arm fly back to the off position. The *holding coil* also releases the arm if the line voltage is interrupted. The motor then has to be restarted when line voltage is restored.

The starting resistance is in series with the shunt field when the arm K is in the *on* position, at contact B. This additional resistance is so small, when compared with the field resistance, that it has practically no effect on field strength and speed.

Figure 22–2 illustrates the connections for a three-terminal manual starting rheostat used with a cumulative compound motor. Note that the only difference in this circuit and the connections for a shunt motor is the addition of the series field.

Starting rheostats are designed to carry the starting current for only a short time; they are not intended for speed control. An attempt to obtain below-normal speed by holding the arm K on an intermediate contact is likely to burn out the starting resistor.

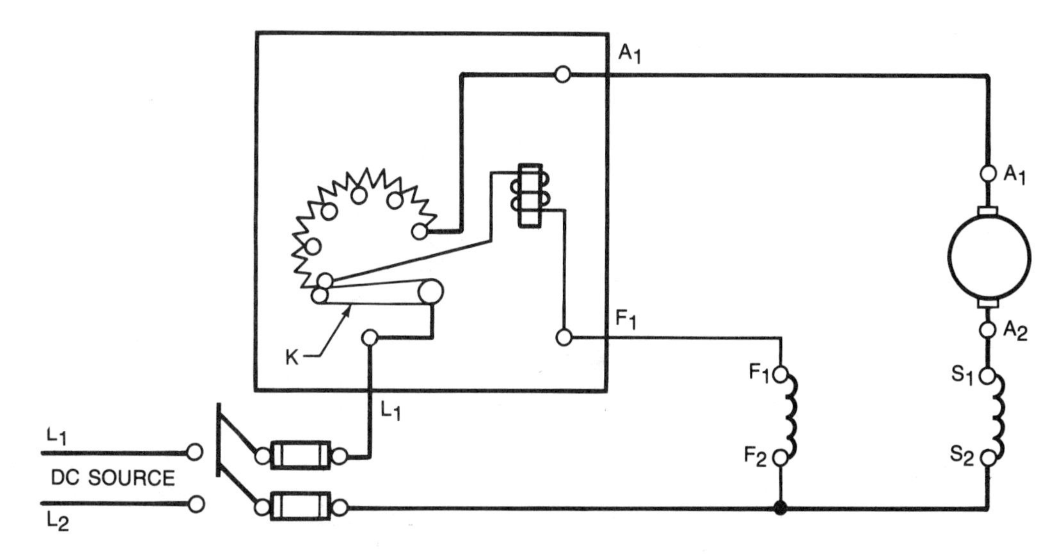

FIGURE 22–2 Three-terminal starting rheostat connected to a compound-wound motor

The three-terminal starting box is not suited for use where a field rheostat is used to obtain above-normal speeds. The reason is that a reduced field current can release the arm and shut down the motor. With field control, a slightly different arrangement, called a four-terminal starting box, is used.

Four-Terminal Starting Rheostat

Four-terminal manual starting has two functions in common with three-terminal starting rheostats: (1) to accelerate a motor to rated speed in one direction of rotation and (2) to limit the starting surge of current in the armature to a safe value. However, this starting rheostat can be used along with a field rheostat. The field control is used to obtain above-normal speeds. Figure 22–3 represents a four-terminal starting box connected to a shunt motor.

Note that the holding coil is not connected in series with the shunt field, as it is in the three-terminal starting box. In this four-terminal starter, the holding coil, in series with a resistor, is connected directly across the source voltage. The holding coil current is independent of field current but still serves as a *no-voltage release*. If line voltage drops, the attraction of the holding coil is decreased, and a reset spring (not shown) returns the movable arm to the *off* position.

A motor is started with a four-terminal starter in the same manner as with a three-terminal starter. Any desired above-normal speed of the motor is obtained by adjustment of the field rheostat in series with the shunt field.

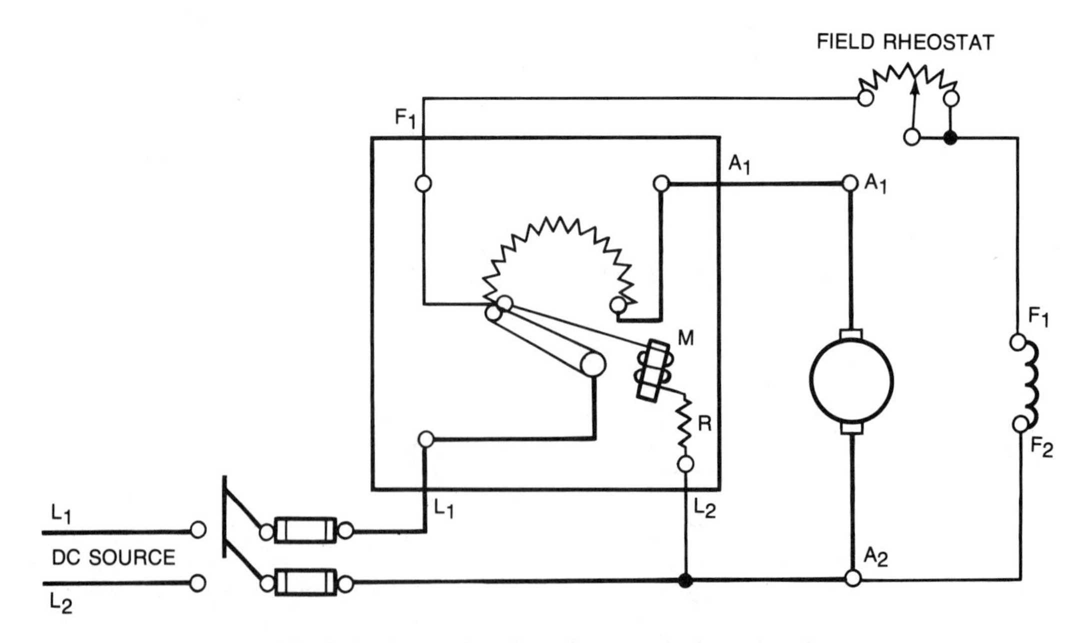

FIGURE 22–3 Connections for a four-terminal starting rheostat

When the motor is to be stopped, all resistance in the field rheostat should be cut out, so that motor speed decreases to its normal value. Then the line switch should be opened. This procedure ensures that the next time the motor is started, it has a strong field and resultant strong starting torque.

22–3 MANUAL SPEED CONTROLLERS

It is often necessary to vary the speed of DC motors. As pointed out before, above-normal rating speeds are obtained by adding resistance to the shunt-field circuit. Below-normal rating speeds are obtained by adding resistance to the armature circuit.

Two types of manual speed controllers are used with shunt and cumulative compound motors, *above-normal speed controllers* and *above-and-below-normal speed controllers*.

The National Electrical Manufacturers' Association (NEMA) defines a manual speed controller as a device for accelerating a motor to normal speed with the additional function of varying speed. (Manual speed controllers must not be confused with manual starting rheostats, which simply accelerate a motor to normal speed.)

Above-Normal Speed Controller

This controller combines the functions of a starter and a field rheostat. The starting resistance is used in the armature circuit only during the starting period. This limits the armature current while the motor accelerates to normal speed. The field control circuit is effective only after the motor is brought up to normal speed. After normal speed, insertion of resistance weakens the field and produces higher speed. The controller illustrated in Figure 22–4, then, has three functions.

1. To accelerate the motor to rated speed by reducing the resistance in the armature circuit
2. To limit the current surge in the armature circuit to a safe value
3. To obtain above-normal speed control by varying the resistance in series with the shunt field

Two rows of contacts are mounted on a slate panel, Figure 22–4. The top row of small contact buttons connects to a tapped resistor, which is the field rheostat. The bottom row of larger contacts connects to a tapped resistor in series with the armature. The control arm K connects to both sets of contacts.

In the start position, arm B bypasses the field rheostat; thus, the full-line voltage is applied to the shunt field. Arm K, when moved clockwise, cuts out starting resistance as the motor accelerates. When arm K approaches the normal-run position, pin C pushes arm B counterclockwise until it is secured against the holding coil. The motor is now accelerated to normal speed.

In Figure 22–5 note that arm B is removed from the field circuit; thus, it no longer short circuits the field rheostat. Instead, arm B now bypasses the starting resistance, providing a direct path from the supply line to the armature.

If it is necessary to increase the speed of the motor to some value above normal, arm K is moved counterclockwise. This has no effect now on armature current, but it

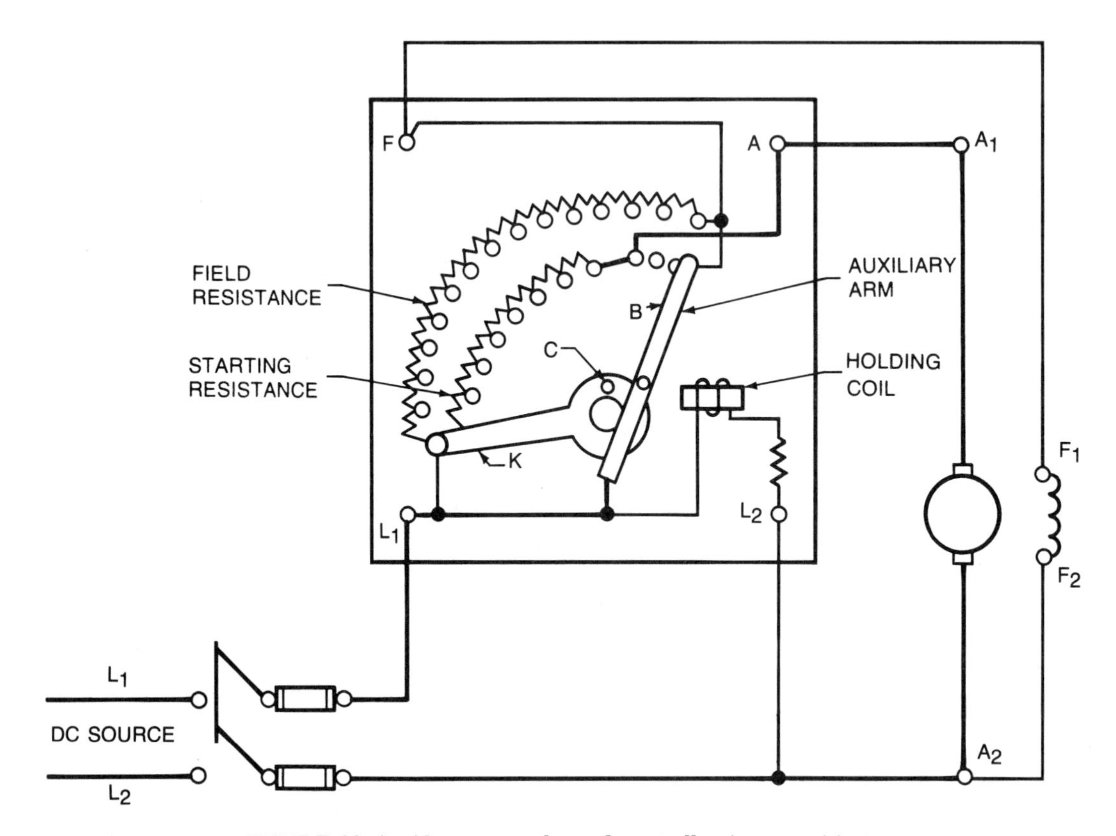

FIGURE 22–4 Above-normal speed controller (start position)

does result in resistance being inserted in the shunt-field circuit. Motor speed now increases. Arm K can be left in any intermediate position to obtain desired above-normal speed.

When the line switch is opened, the holding coil releases arm B, which is returned to its original *on* position by a spring. Pin C is now released and permits arm K to return to the *off* position. K is returned by a reset spring.

This type of controller can be used with either a shunt or a compound motor.

Above-and-Below-Normal Speed Controller

In some motor installations it is necessary to have a wide range of speed control, including both above-normal and below-normal speeds. A typical above-and-below-normal controller is illustrated in Figures 22–6 and 22–7. The movable arm K connects to two rows of contacts. The lower row of contacts connects to taps on the armature circuit resistor, and the upper row connects to taps on the field resistor. The contacts are mounted on the front of a slate panel, while the armature and field resistors are housed in a ventilated box in back of the panel. Continued clockwise movement of the arm results in continued increase of speed. This increase is accomplished first by removing armature circuit resistance and then by inserting resistance in the field circuit.

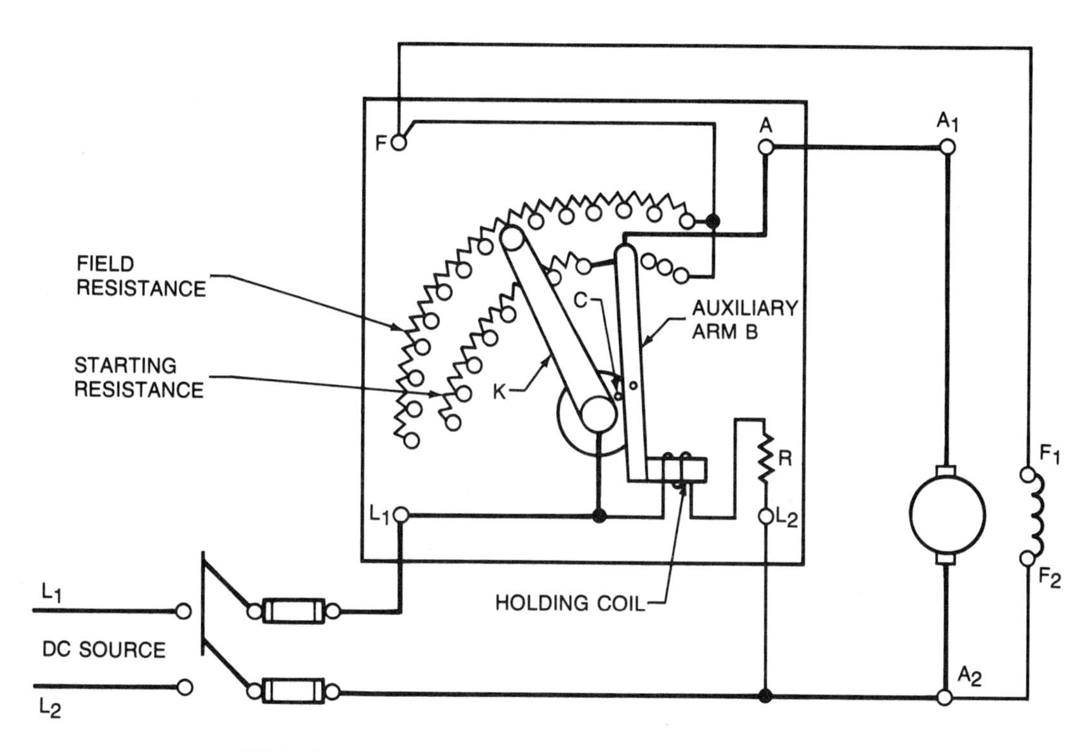

FIGURE 22–5 Above-normal speed controller (run position)

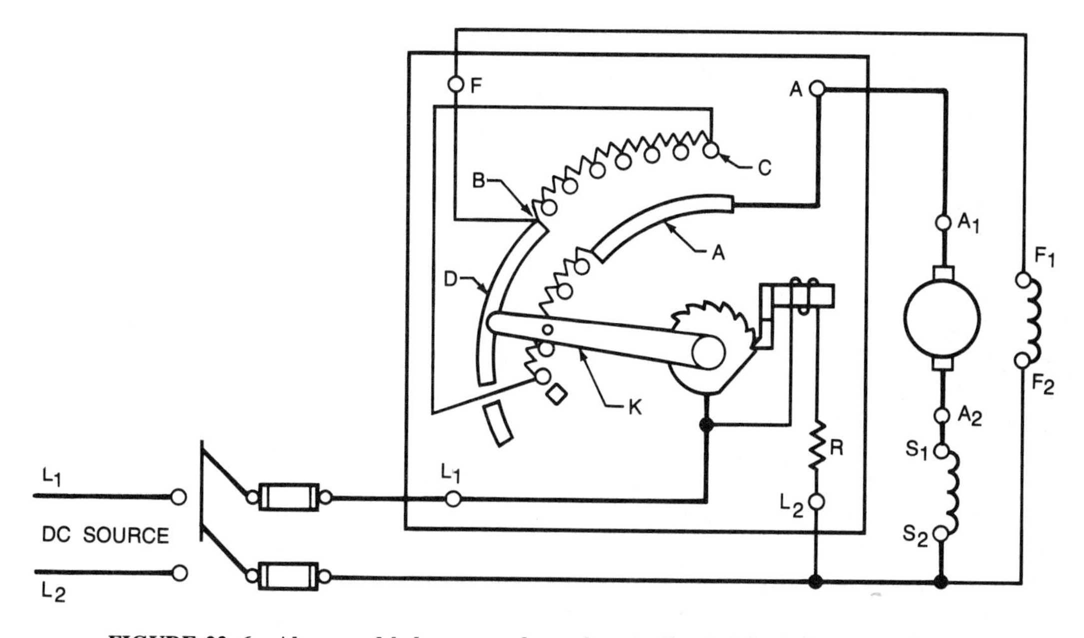

FIGURE 22–6 Above-and-below-normal speed controller (set for below-normal speed)

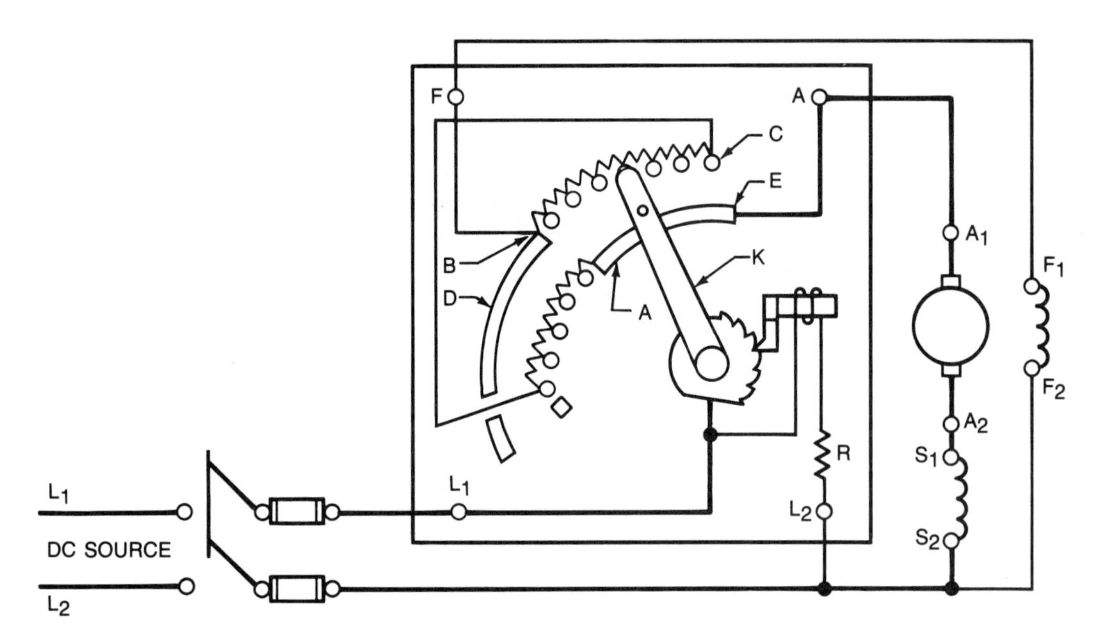

FIGURE 22–7 Above-and-below-normal speed controller (set for above-normal speed)

In the position shown in Figure 22–6, there is considerable resistance in series with the armature. The arm K also contacts the radial conductor D, which connects full-line voltage to the shunt field. With the arm in this position, the speed is below normal. Once the movable arm is set on any contact point, it locks in that position until moved to some other point. This is done by a unique gear and latch system operated with the aid of the holding coil.

When a motor is operating under heavy load at slow speed, there is considerable current in the armature circuit. This large current requires the armature resistors to be of large size in order to radiate the heat produced by the large current. Large resistors make the physical size of this controller larger, for a given horsepower rating, than an ordinary manual starting rheostat.

As the arm is slowly moved clockwise to the upper end of the armature rheostat, it still contacts conductor D (at point B). The arm K also comes in contact with the curved conducting strip marked A. This is the normal speed position. Full-line voltage is applied to both the armature and the shunt field.

In the above-normal speed position, full-line voltage is still applied to the armature through strip A-E. The outer end of the control arm K now contacts a point on the field rheostat; thus, the resistance between the arm and point B is inserted into the field circuit. If the arm is moved to point C, all of the field rheostat is in use, producing maximum speed by field weakening. When the line switch is opened, the holding coil releases the latch, and the reset spring returns the arm to the *off* position.

This type of controller can be used with either a shunt or a compound motor. Connections for a shunt motor differ only by the omission of the series field.

22–4 STARTERS FOR SERIES MOTORS

Series motors require a special type of manual starting rheostat. It is called a *series motor starter*. These starting rheostats serve the same purpose as the three- and four-terminal manual starting rheostats used with shunt and compound motors. However, series motor starters have different internal and external connections. There are two types of series motor starters. One type of starter has *no-voltage protection,* whereas the other has *no-load protection*.

A series motor starter with no-voltage protection is illustrated in Figure 22–8. The holding coil is connected across the source voltage. This starter is used to accelerate the motor to rated speed. In case of voltage failure, the holding coil no longer acts as an electromagnet. The spring reset then quickly returns the arm to the off position to protect the motor from damage.

Another type of series motor starter, Figure 22–9, has no-load protection. The holding coil is in series with the armature. Because of the large current in the armature circuit, the holding coil consists of only a few turns of heavy wire.

The same care is used in starting a motor with this type of starting rheostat as is used with three- and four-terminal starting rheostats. The arm is slowly moved from the *off* position to the *on* position, pausing on each contact button for a period of one to two seconds. The arm is held against the tension of the reset spring by means of the holding coil connected in series with the armature. If the load current to the motor drops to a low value, the holding coil weakens and the reset spring returns the arm to the *off* position. This is an important protective feature. A series motor can reach a dangerously high speed at light loads; therefore, this type of starting rheostat protects the motor from damage caused by excessive speeds.

22–5 DRUM CONTROLLERS

Series and cumulative compound motors are often used on cranes, elevators, machine tools, and other devices where the motor is under the direct control of an operator and where frequent starting, varying speed, stopping, and reversing are necessary. A

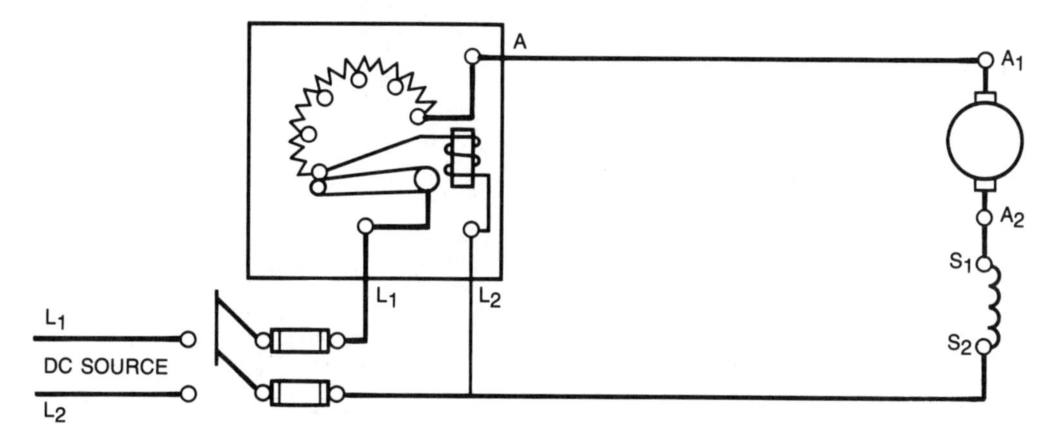

FIGURE 22–8 Series motor starter with no-voltage protection

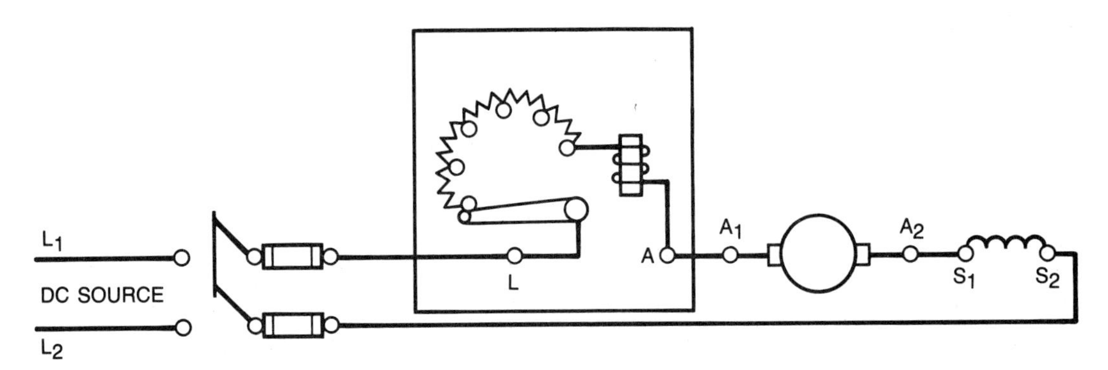

FIGURE 22–9 Series motor starter with no-load protection

manually operated controller that is more rugged than a starting rheostat is used in these applications. This starting rheostat is called a *drum controller*.

A typical drum controller is illustrated in Figure 22–10. Inside the switch is a series of contacts mounted on a movable cylinder. These contacts, insulated from the cylinder and from each other, are the movable contacts. There is another series of contacts,

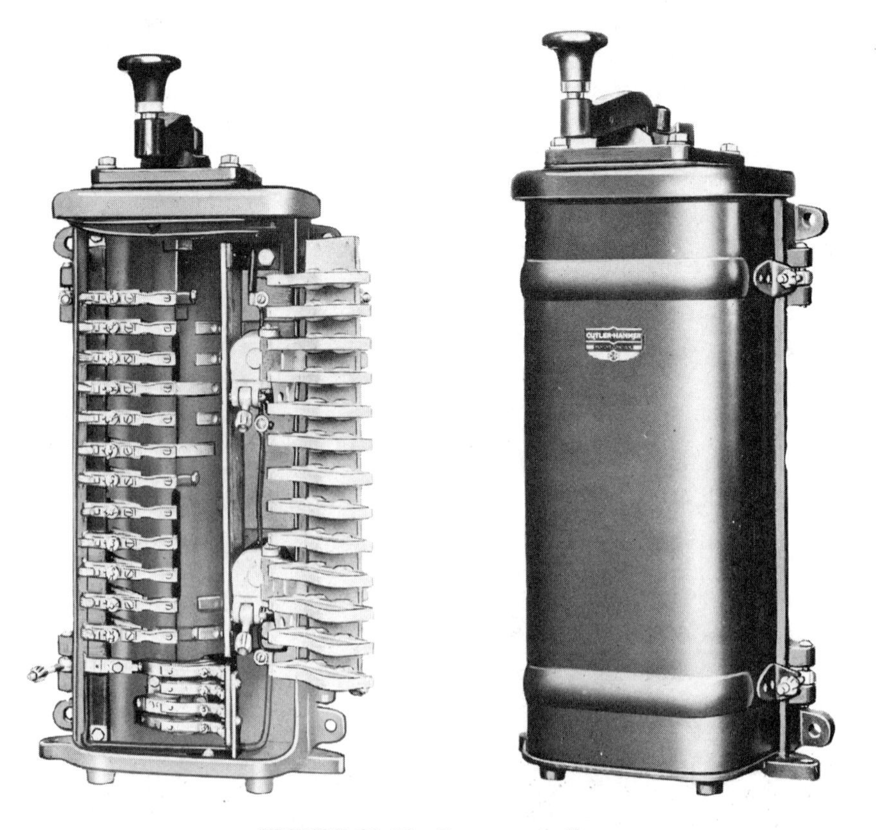

FIGURE 22–10 Drum controller

located inside the controller, called stationary contacts. These stationary contacts are arranged to make contact with the movable contacts as the cylinder is rotated. On top of the drum controller is a handle that is keyed to the shaft for the movable cylinder and contacts. This handle can be moved in either a clockwise or a counterclockwise direction, providing a range of speed control in either direction or rotation. Once set, a roller and notched-wheel arrangement keeps the cylinder and movable contacts stationary until the handle is turned by the operator.

A schematic diagram of a drum controller having two steps of resistance is shown in Figure 22–11. In this wiring diagram, the contacts are shown in a flat position to make it easier to trace connections. For operating in the forward direction, the movable contacts on the right connect with the center stationary contacts. For operation in the reverse direction, the movable contacts on the left touch the stationary contacts in the center.

There are three forward positions and three reverse positions in which the controller handle can be set. In the first forward position, all resistance is in series with the armature. The circuit for the first forward position is traced as follows:

1. Movable fingers A, B, C, and D contact the stationary contacts 7, 5, 4, and 3.
2. The current path is from 7 to A, from A to B, from B to 5, and then to armature terminal A_1.

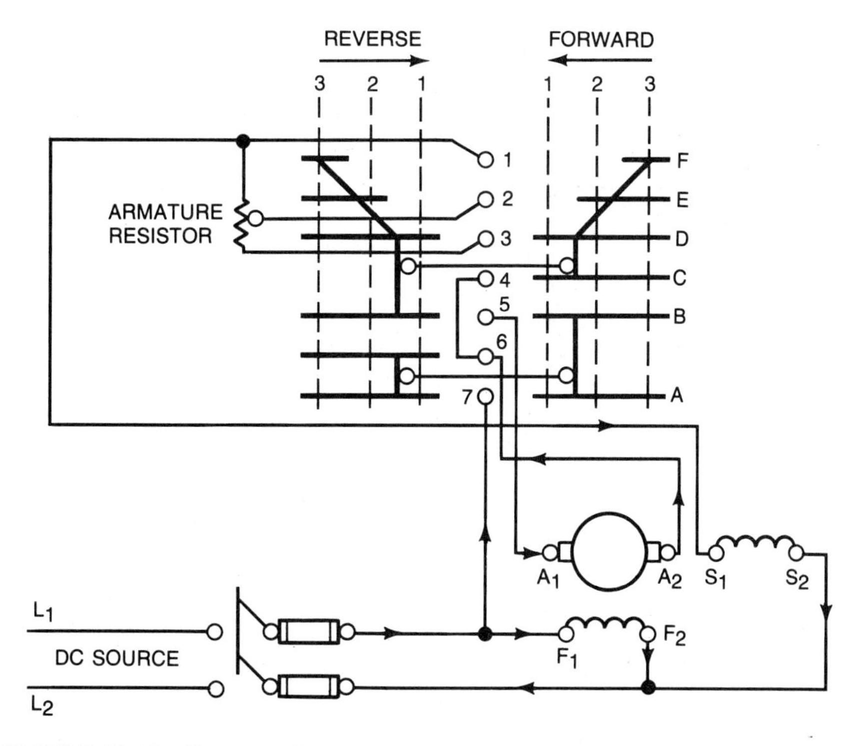

FIGURE 22–11 Cam-type drum controller connected to a compound-wound motor

3. From A_1, the current path is through the armature winding to terminal A_2, then to stationary contact 6, and then to stationary contact 4.
4. From contact 4, the current path is to contact C, to D, and then to 3.
5. From 3, the current path is through the entire armature resistor, through the series field, and then back to the line.

In the second forward position, part of the resistance is cut out by the connection from D to E. The third forward position bypasses all resistance and puts the armature circuit directly across the source voltage.

In the first reverse position, all resistance is again inserted in series with the armature. Figure 22–12 illustrates the first position of the controller for the reverse direction.

The current in the armature circuit is reversed. However, the current direction in the shunt and series fields is the same as for the forward direction. As shown earlier, changing the direction of the current in only the armature changes the direction of rotation. In the second position, part of the resistance circuit is cut out. The third reverse position cuts out all resistance and puts the armature circuit directly across line voltage.

There are more elaborate drum controllers with more positions and a greater control of speed; however, they all use practically the same circuit arrangement.

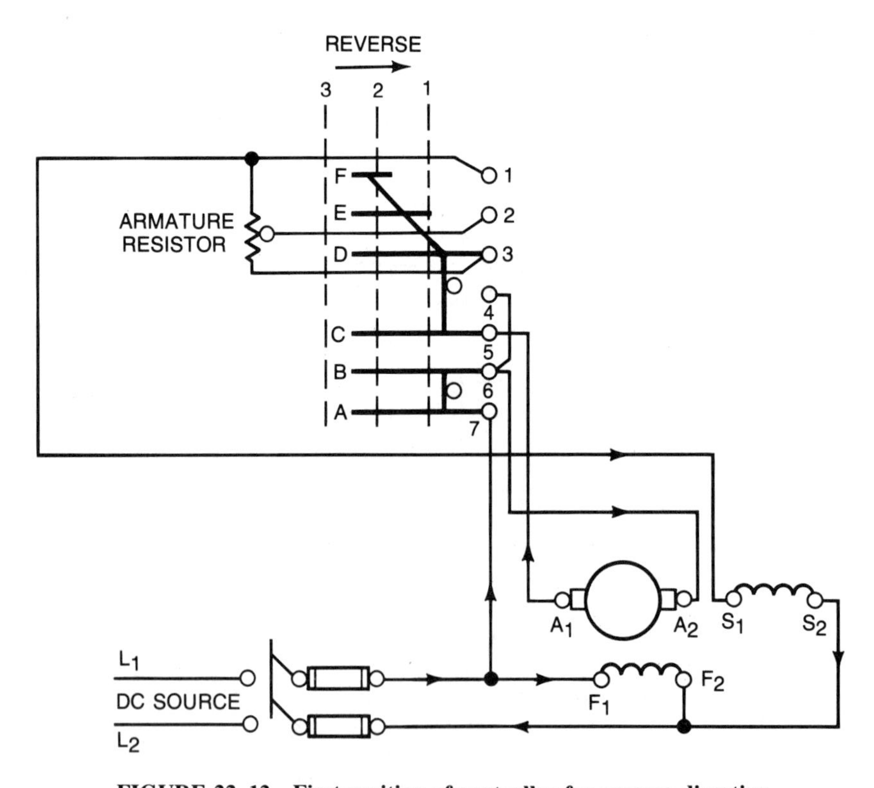

FIGURE 22–12 First position of controller for reverse direction

22–6 MAGNETIC CONTROLLERS

The manual starters and controllers described in the foregoing sections have been increasingly replaced by magnetic starters with push-button or automatic control. This type of equipment is convenient and has the added advantage of reducing damage caused by human misjudgment. Many of these automatic systems control DC motors that are used because of their wide speed range and excellent torque characteristics.

The type of schematic diagram used to describe motor control circuitry is often referred to as a *ladder diagram* or *relay ladder logic*. The word *ladder* is derived from the fact that all control components are arranged like the rungs of a ladder between two vertical lines that represent the control voltage.

Standard symbols have been established for control circuit components, such as: relay coils, contactors, push-button stations, overload devices, and limit switches. Such symbols, often known as JIC (or Joint Industrial Council) standards, conform to standards established by the National Electrical Manufacturers' Association, Figure A–12.

The following sections will serve as introduction to the concept of automatic control of DC motors.

A contactor operated by a relay is an important part of any automatic motor controller. A *contactor* is a switch that is closed or opened by the magnetic pull of an energized relay coil. Figure 22–13 shows a relay coil with contactors. A relay coil connected in series in the circuit is normally represented by a heavy line, Figure 22–14. A relay coil connected in parallel is represented by a light line. A contactor that is open when the coil is de-energized is known as a *normal open* (N.O.) contactor and is indicated by two short parallel lines, Figure 22–14. A contactor that is closed when the coil is de-energized is known as *normally closed* (N.C.) and is indicated by a diagonal line drawn across two parallel lines, Figure 22–14. Letters are added to show which contacts are operated by a given coil.

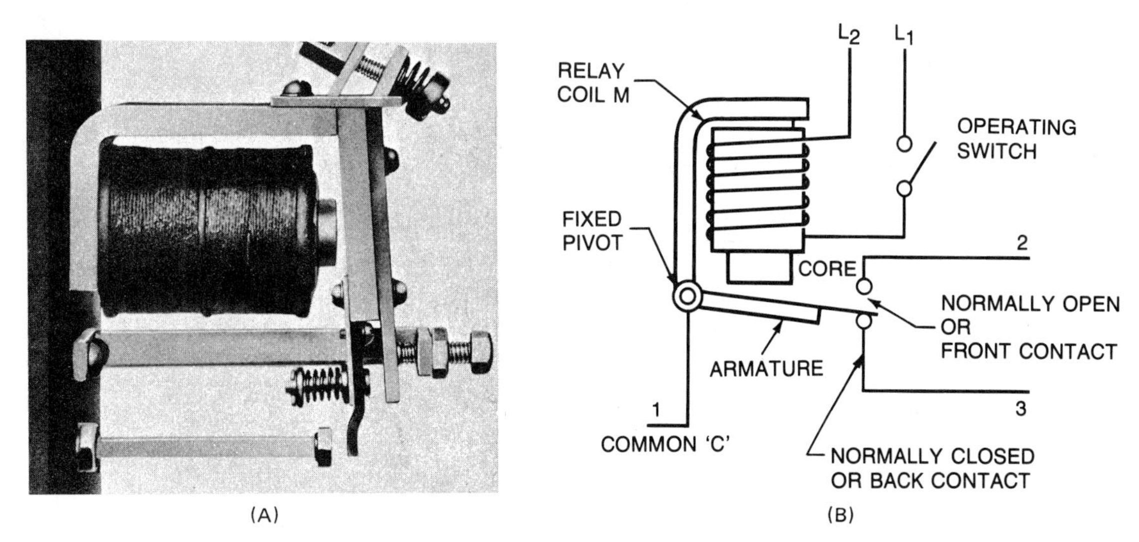

(A) (B)

FIGURE 22–13

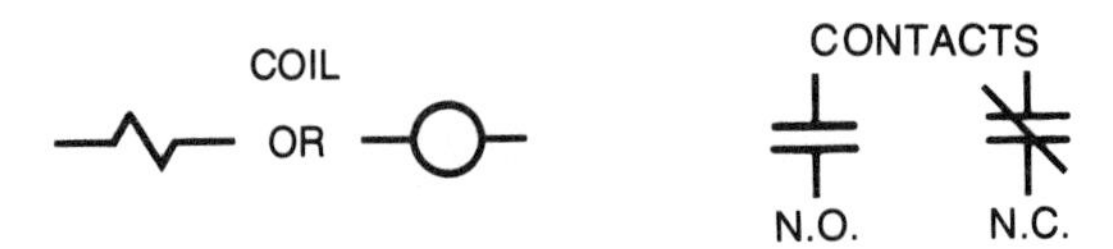

FIGURE 22–14 Symbols of contactor elements

When contactors interrupt a large current, a severe arc forms. This arc can burn the surface of the contactor. To reduce this burning effect, a *magnetic blowout coil* is added in series with the contactors to extinguish the arc by electromagnetic action, Figure 22–15. An arc is, after all, nothing but a stream of electrically charged particles (similar to current in a wire) and, therefore, can be deflected by a magnetic field. Figure 22–15 shows how the blowout coil sets up a magnetic field that serves to force the arc off the contacts by deflecting it. An arc chute is provided for the protection of surrounding equipment. Figure 22–16 shows a typical relay so equipped, and Figure 22–17 represents its corresponding schematic diagram.

Push-button stations, like the ones shown in Figure 22–18, are used to provide control of the motor. The push buttons are really spring-controlled switches, which are classified as being either normally open or normally closed (N.O. or N.C.), Figure 22–19.

Pressure on a normally open start button closes the switch contacts momentarily. When the button is released, the spring reopens the switch, Figure 22–19A.

By contrast, a stop button is a normally closed switch. Finger pressure opens the contacts, which close again when pressure is released, Figure 22–19B.

Many push buttons have two sets of momentary contacts, Figure 22–19C, one of which opens when the other set is closed.

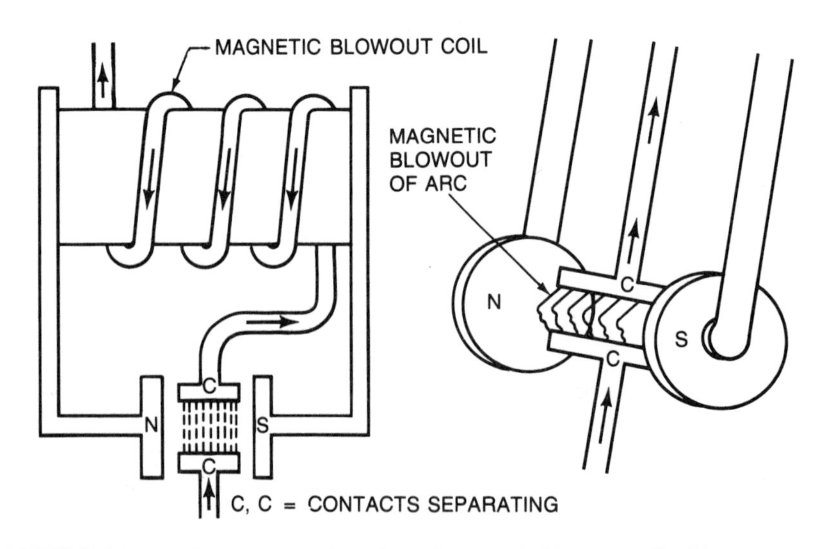

FIGURE 22–15 Electromagnetic action of magnetic blowout coil with contactors

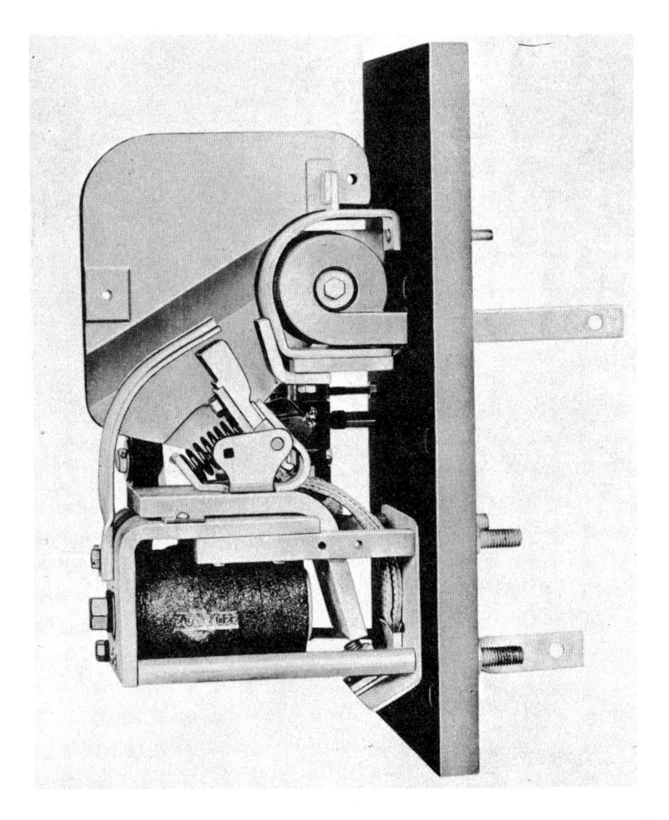

FIGURE 22–16 Magnetic contactors with blowout coil and arc chute

These push buttons generally are used in connection with a relay coil, as shown in Figure 22–20. This diagram represents part of an elementary control circuit.

When the start button is pressed, closing contacts 2–3, there is a circuit from line L_1 through normally closed contacts 1–2, through 2–3, and through relay coil M to supply line L_2. The current in relay coil M causes contact M to be held closed. Hence, when the start button is released (opening contacts 2–3), there is still a circuit through the stop button 1–2, through contact M, and through coil M to L_2. This arrangement is called a *sealing circuit* and is a part of the control circuits soon to be described. Mo-

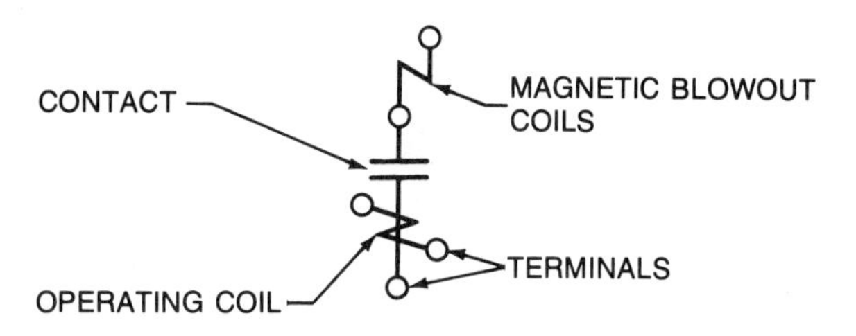

FIGURE 22–17 DC magnetic contactor symbols

FIGURE 22–18 A typical push-button station

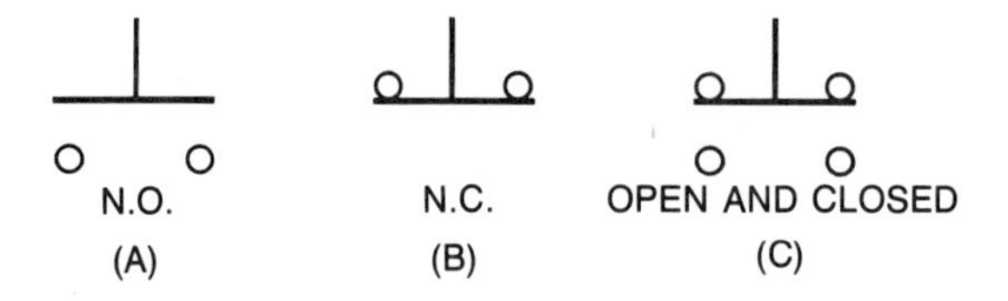

FIGURE 22–19 Push-button symbols

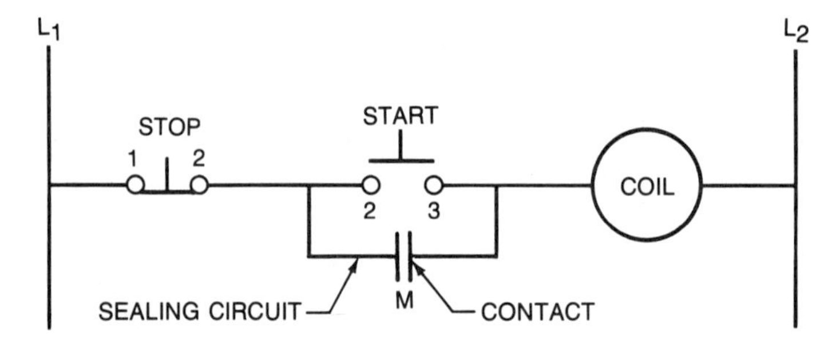

FIGURE 22–20 A control circuit with start and stop buttons and sealing circuit

mentary pressure on the start button energizes the relay coil. The sealing contact M keeps the coil energized. In control circuits, coil M also closes other contactors as well as the sealing contact.

When the stop button 1–2 is momentarily pressed, the circuit is broken. Coil M loses its magnetic pull and contact M opens. The release of the stop button (closing 1–2) does not reestablish the circuit. Both contact M and start button 2–3 are open. Consequently, coil M cannot be energized until the start button again closes the circuit.

There are so many types of automatic controllers for special applications that it is impossible to cover all of them in this chapter; therefore, three standard types are described in some detail: the counter-electromotive force controller, the voltage drop acceleration controller, and the definite time limit controller.

22–7 THE COUNTER-ELECTROMOTIVE FORCE MOTOR CONTROLLER

This type of controller, shown in Figure 22–21, is a commonly used method for the automatic acceleration of a DC motor. First, the line switch is closed. When the start button is pressed, relay coil M is energized. This control circuit remains energized because the sealing contacts are closed, as previously described.

When coil M in the control circuit is energized, it closes a heavy pair of contactors, M, 6–7, in the power circuit. The closing of these contactors establishes a circuit from line 1 through the overload device 1–6. Such an overload device generally operates when excessive heat develops due to overload. This device is commonly known as *thermal overload protection*. The circuit continues through the M contactors, through the current-limiting resistor in series with the armature, and through the armature wind-

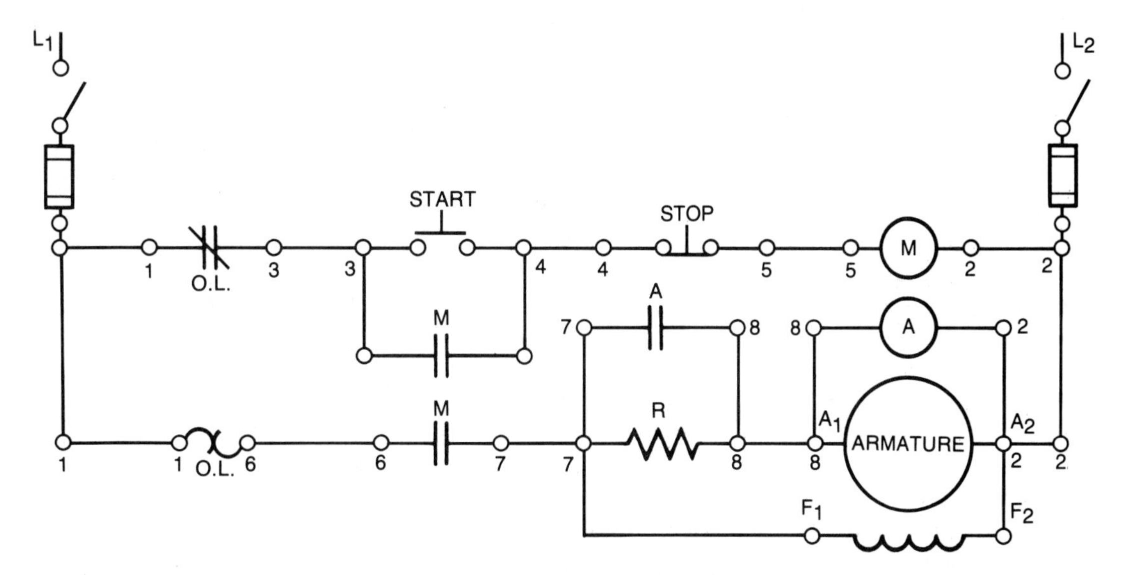

FIGURE 22–21 Connection for counter-electromotive force motor controller to a shunt motor

ings to the other side of the line. The shunt field is directly across the full-line voltage and insures maximum starting torque.

At the instant the motor circuit is energized, the counter-emf is 0 and nearly all of the line voltage is expended on the current-limiting resistor in series with the armature. As the armature accelerates, the counter-emf increases in proportion to the speed. With increased counter-emf, more of the line voltage appears across the armature terminals and the accelerating relay coil A. This coil is in parallel with the armature terminals. Relay coil A is calibrated to close its contactors when about 80% of the rated line voltage is applied to the coil. When the voltage across the armature terminals reaches this predetermined value (80% of the line voltage), coil A closes contactors A. These contactors then shunt out the resistor in series with the armature. The armature is now connected directly across line voltage, and the motor accelerates to its rated speed.

Pressing the stop button breaks the control circuit, and both sets of M contactors open to disconnect the motor from the line. As the armature slows down, coil A cannot hold its contactors closed. With the A contactors open, the current-limiting resistor is again connected in series with the armature. Now the motor is again ready to be started.

Starting protection for this controller, or any DC automatic controller, consists of fuses or circuit breakers rated at 150% of the full-load current of the motor. Running overload protection is provided by the overload heater unit (O.L. 1-6). Overheating of this unit causes a bimetallic strip to trip open the normally closed contactor (O.L. 1-3) in the control circuit. The heater unit is rated at 125% of the full-load current rating of the motor. A continued overload on the motor for 45 to 60 seconds brings it into operation. The 150% starting current surge of 3 to 4 seconds does not produce enough heat to cause the thermal element to open its contactors.

Dynamic Braking

In some installations there is a need for quick stopping and immediate reversal of rotation. Dynamic braking is a method for quickly using up the mechanical energy of motion of the armature and its mechanical load after the armature circuit is opened. *Dynamic braking* is achieved by connecting a resistor across the armature at the instant the armature circuit is disconnected. This disconnected armature, rotating on its own momentum, is cutting flux and acts as a generator. The generated current quickly dissipates this mechanical energy by heating the resistor; thus, the motor stops quickly.

Figure 22–22 shows how dynamic braking facilities can be added to the previously described counter-emf controller. At starting, the same sequence of events takes place as described before.

Relay coil M and the dynamic braking coil, DBM 6-7, operate on one pivoted armature, as shown in Figure 22–23. When the start button is pressed, coil M closes the normally open contactors and pulls open the one set of DBM contactors. As the motor accelerates, relay coil A shunts out the current-limiting resistor as before. Although the DBM coil is now energized by the full-line voltage, coil M has already tipped the pivoted armature clockwise to open the DBM contact. Coil DBM is not strong enough to bring the armature back to the position shown in Figure 22–23; thus, the DBM contactors remain open while the motor operates normally.

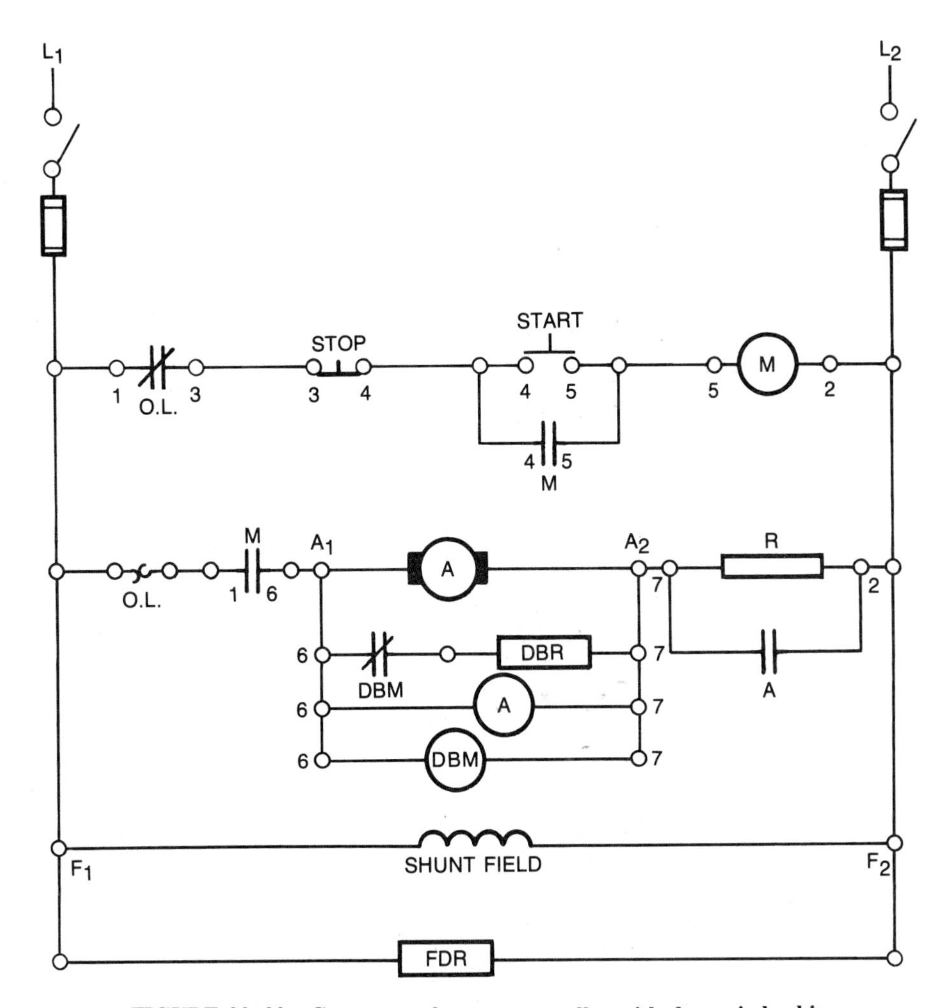

FIGURE 22–22 Counter-emf motor controller with dynamic braking

When the stop button is pressed, coil M releases the pivoted relay armature. Since the shunt field of the motor is still connected to the line, the rotating armature of the motor generates a current that keeps coil DBM energized. Coil DBM is now able to pull contactors DBM to their normally closed position, coil M being de-energized. (Small coiled springs, not shown in Figure 22–23, help make this action more positive.) With DBM contactors closed, the dynamic braking resistor is connected directly across the motor armature. Now the armature, acting as a generator, converts its mechanical energy into electrical energy that is quickly dissipated in the DBR resistor, and the motor armature comes to a quick stop. Coil A releases contactors A, reinserting the current-limiting resistor in series with the armature.

In the circuit of Figure 22–21, the pressing of the stop button disconnected the entire motor circuit from the line. In that case, the collapsing magnetic field of the

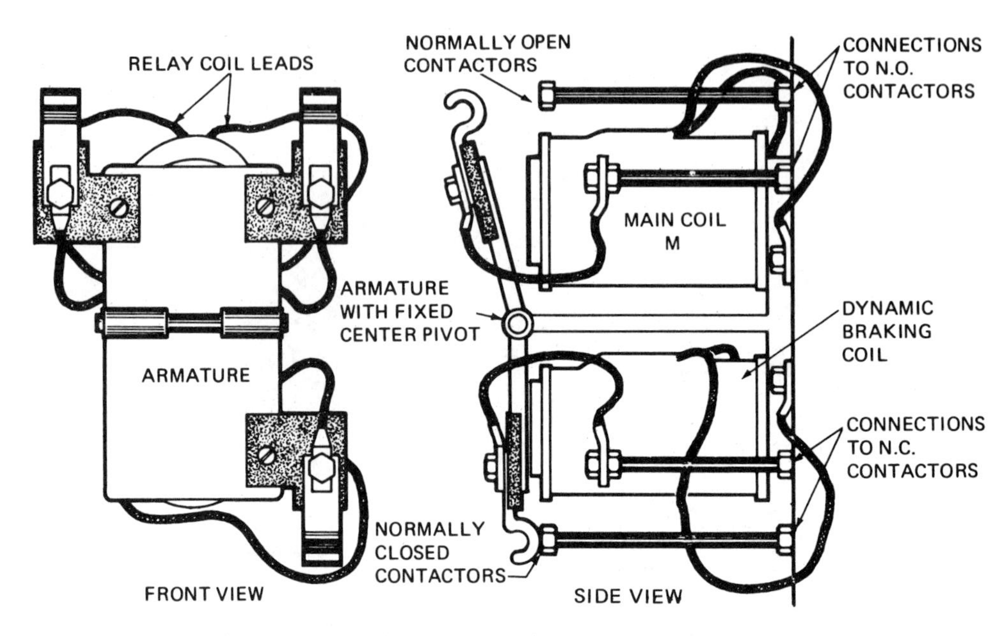

FIGURE 22–23 Direct current contactor assembly

shunt-field winding delivered its energy to the armature circuit. In the circuit of Figure 22–22, the armature is disconnected from the field when the stop button is pressed; therefore, a field discharge resistor (FDR) is connected across the field to dissipate the field energy when the line switch is opened.

Counter-emf Controller with Reversing and Dynamic Braking

In many applications, it is necessary not only to bring a motor to a quick stop but also to reverse the direction of rotation immediately. This is usually done by reversal of armature connections, as shown in the circuit in Figure 22–24.

When the forward button is pressed, the normally closed contact 4-7 opens and the normally open contact 4-5 closes. Relay coils 1F and 2F become energized, closing the 1F sealing contact 4-5 and the armature current contactors 1F and 2F. These normally closed contactors (1F) DB_2 are held open by a relay like that of Figure 22–12. The (1F) DB_2 contactors are held open so that the dynamic braking resistor is disconnected when the motor is energized. The normally open contactors DB_2 (1F) at points 9-11 close and connect the accelerating relay coil A across the armature. As the motor accelerates, relay coil A closes contactors A. Closing contactors A puts full-line voltage on the armature so the motor operates at rated speed.

Pressing the stop button opens the control circuit. The contactors 1F and 2F then open and disconnect the armature from the line. Contactors DB_2 (1F) also open, de-energizing relay coil A, which opens contactors A. At the same time, the normally closed (1F) DB_2 recloses and connects the dynamic braking resistor across the armature so that the motor stops quickly.

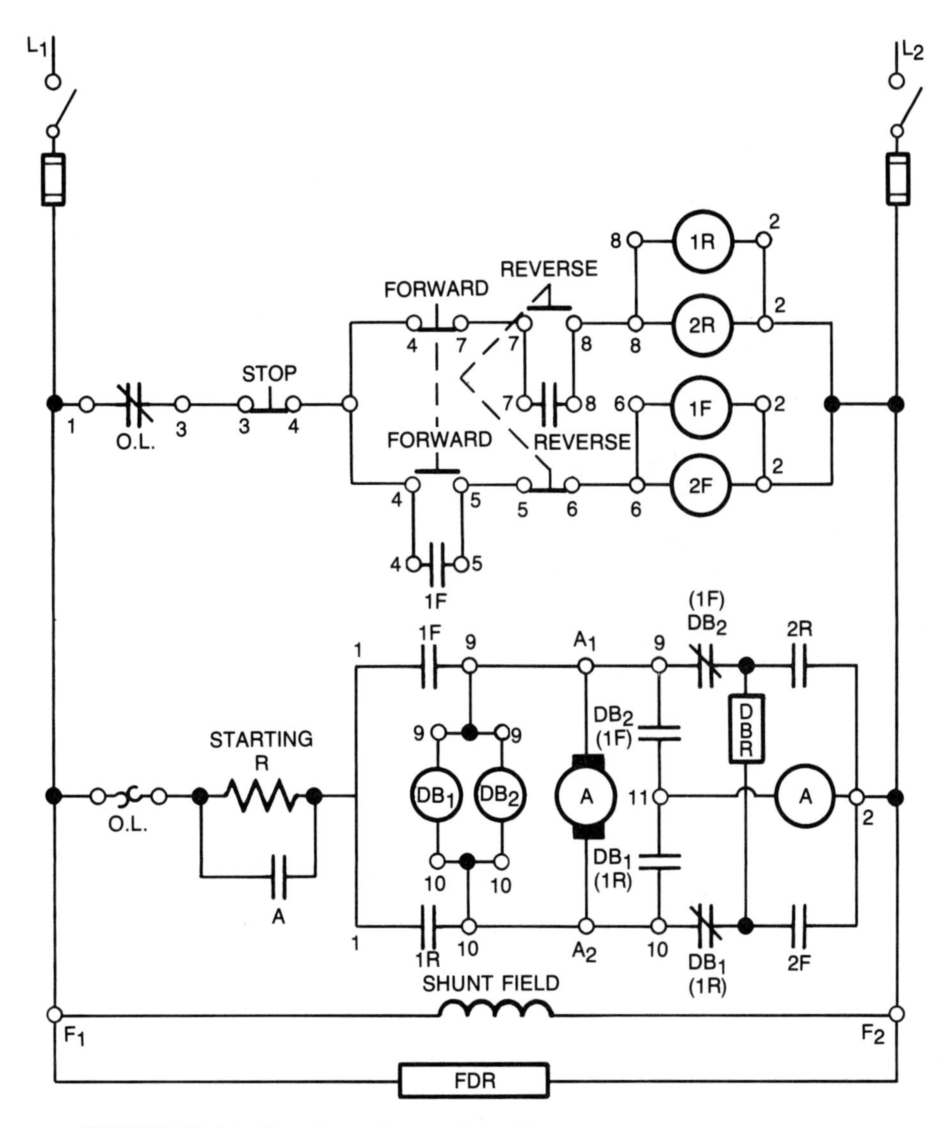

FIGURE 22–24 Counter-emf controller with reversing and dynamic braking

When the reverse button is pressed, the direction of current in the armature is reversed. The motor accelerates in the reverse direction, and when the armature emf is high enough, coil A closes contactors A and the motor operates at rated speed in the reverse direction. Use of the stop button opens contactors A and inserts the dynamic braking resistor into the circuit as before.

In the circuit in Figure 22–24, the forward and reverse push buttons each has a normally closed contact and also a normally open contact. This circuit arrangement makes it impossible to energize the reverse relays 1R and 2R until the forward relays

1F and 2F are de-energized. For example, if the reverse button is pressed, it first breaks contact at points 5-6 and de-energizes coils 1F and 2F before closing across points 7-8 and energizing relay coils 1R and 2R. The same protection exists if relay coils 1R and 2R are energized and the forward button is pressed. This type of connection arrangement, called *electrical interlocking,* is often used in control circuitry so that when one set of devices is operating, the circuit to a second set of devices cannot be energized at the same time.

22–8 THE VOLTAGE DROP ACCELERATION CONTROLLER (LOCKOUT ACCELERATION)

Large DC motors require controlled steps of acceleration. A series of resistors connected to *lockout relays* provide the means for smooth and uniform motor acceleration.

Like the counter-emf controller, the voltage drop acceleration controller makes use of these facts.

1. At the instant of starting, armature current is high and voltage across the armature is low. Voltage losses across each of the current-limiting resistors, connected in series, is high.
2. As the motor accelerates, the counter-emf increases and the armature current decreases; therefore, the voltage drop across the current-limiting resistors, in series

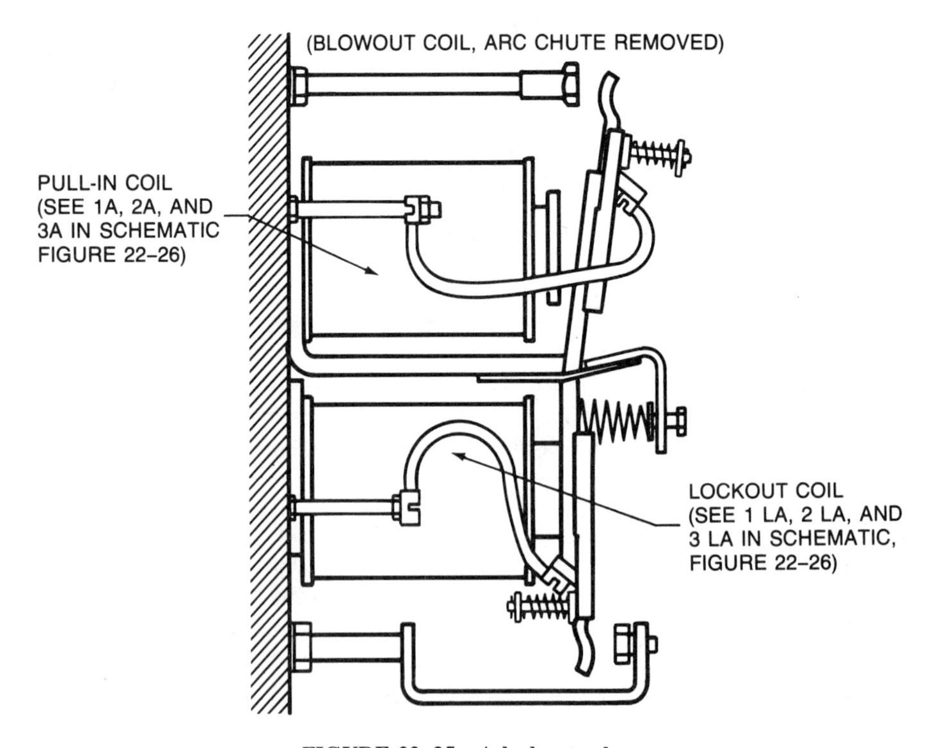

FIGURE 22–25 A lockout relay

with the armature, decreases. Relays, connected across these resistors, are calibrated to operate and shunt out the starting resistors in a definite sequence as the armature speed increases.

The sketch in Figure 22–25 shows a typical lockout relay. Two coils affect the one pivoted armature. The normally open contacts can be held open by the lockout coil, even if the pull-in coil is energized. With reduced current in the lockout coil, the energized pull-in coil can tip the armature and close the contacts. Each of the three relays in the schematic diagram, Figure 22–26, is the same as in Figure 22–25. Relay coils marked 1A, 2A, and 3A are pull-in coils. Relay coils marked 1LA, 2LA, and 3LA are lockout coils.

Figure 22–26 shows a voltage drop acceleration controller connected to a cumulative compound motor. Since there are three resistors, there are three steps of acceleration. After the line switch is closed, pressure on the start button energizes relay coil M

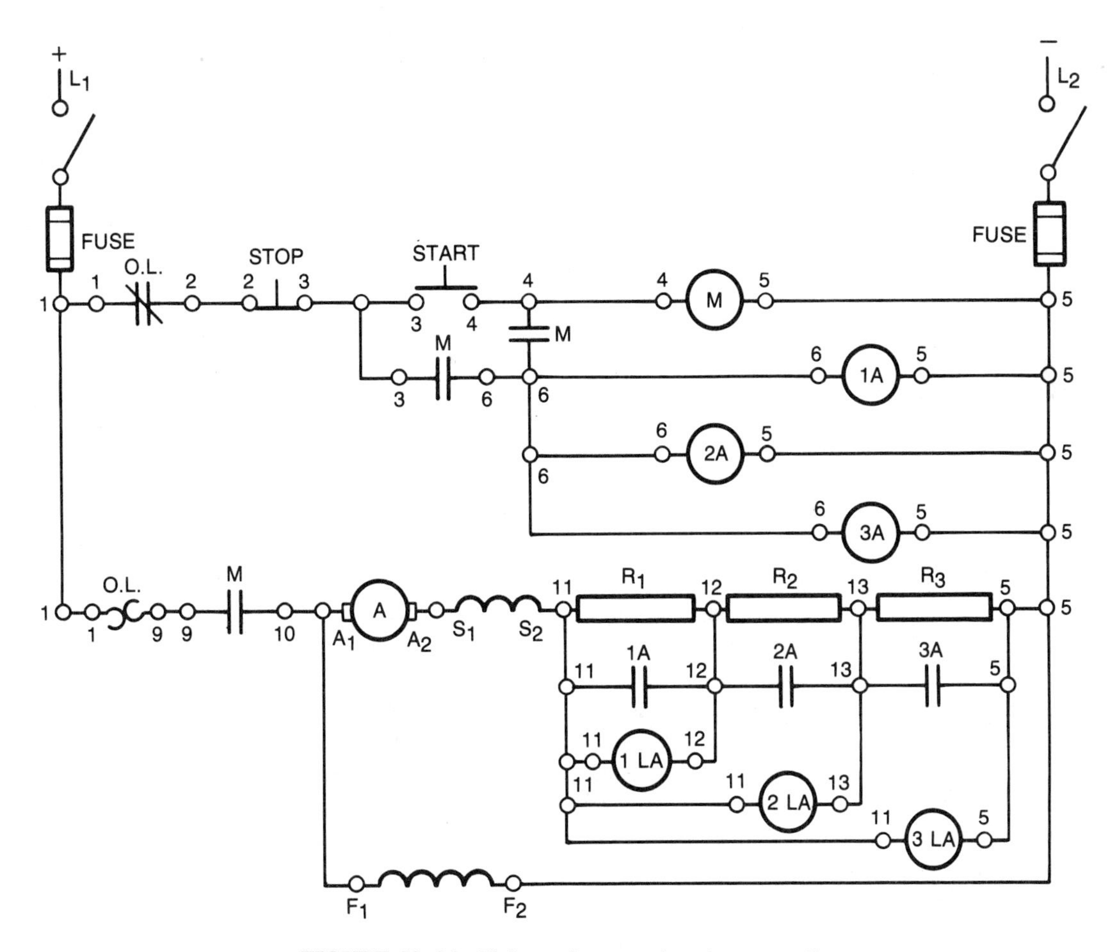

FIGURE 22–26 Voltage drop acceleration controller

in the control circuit. Coil M closes main contactors M 9-10, which both closes the armature circuit and connects the shunt field across the line.

The initial current through the starting resistors R_1, R_2, and R_3 produces a relatively large voltage drop across each section of starting resistance; therefore, the lockout coil (LA) of each relay has a relatively high voltage across it and can hold the accelerating contactors 1A, 2A, and 3A open. At the instant of starting, coil M also closes the sealing contactors M 3-6 and M 6-4. The short time interval required for closing these contacts insures that the pull-in coils 1A, 2A, and 3A become energized no sooner than the lockout coils.

The lockout relays are calibrated to operate and shunt out sections of starting resistance in a definite sequence as the armature accelerates. As current through the series resistors decreases during acceleration, less voltage is impressed on lockout coil 1LA. Its pull on the movable contactor becomes less than that of pull-in coil 1A; therefore, pull-in coil 1A can close contactors 1A and shunt out resistor R_1. As R_1 is cut out of the circuit, current increases but decreases again as the motor continues to accelerate. Soon the voltage across 2LA is low enough to allow pull-in coil 2A to close contactors 2A and shunt out R_2. Then R_3 is cut out in the same manner as R_1 and R_2. Thus, the motor is accelerated to rated speed in three steps.

22–9 DEFINITE TIME CONTROLLER

The function of this controller is to short out starting resistance, thus connecting the armature across line voltage at a predetermined time. The closing of contactors at a definite time after the pressing of the start button is controlled by either a very small constant-speed motor or by one of several types of magnetic timing devices. The schematic diagram, Figure 22–27, shows a definite time controller using a magnetic timer escapement. The magnetic timer escapement controls the closing of contactors by a solenoid type of relay.

Closing the start button completes the control circuit from line 1 to point 5, through N.C. contactors (which shunt out one part of the time relay coil), and through the rest of the time relay coil to line 2. As the energized solenoid relay coil TR starts to pull its iron plunger up into the coil, Figure 22–28, contactors TR_1 close, energizing an entirely different relay marked M. Relay coil M closes the heavy M contactors, establishing a circuit path through the series current-limiting resistor and the armature and also connecting the shunt field across line voltage. The motor now accelerates toward rated speed.

Meanwhile, the solenoid relay coil TR is pulling the plunger up into the coil at a rate determined by the time escapement mechanism. Finally, the plunger is pulled up as far as possible, closing contactors TR_2. Contactors TR_2 energize relay coil A, which closes contactors A. Contactors A shunt out the current-limiting resistor and place the armature directly across line voltage. In the definite time interval between the closing of contactors M and contactors A, it is assumed that counter-emf becomes sufficient to permit applying full-line voltage to the armature.

Full movement of the plunger into the solenoid coil opens the N.C. contactors TR_3, which bypassed part of the solenoid relay coil TR. After the plunger moves fully

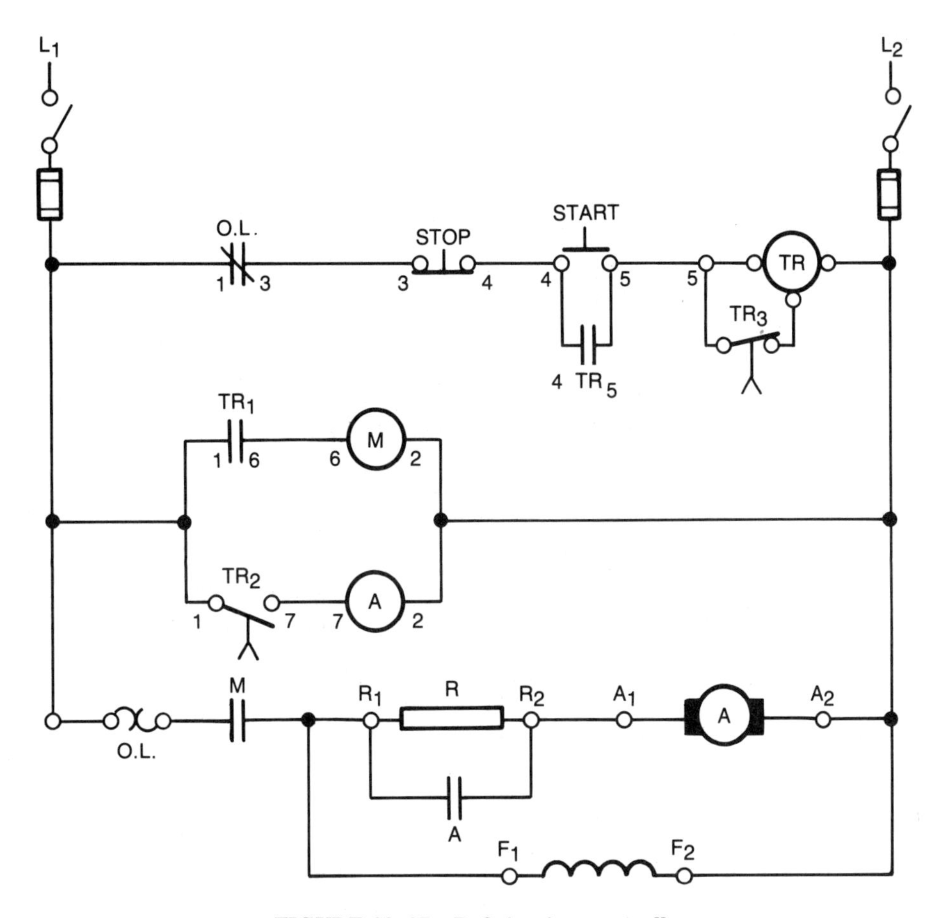

FIGURE 22–27 Definite time controller

into the coil, considerably less current is required to hold the plunger in operating position than when the TR coil was initially energized; therefore, the current in coil TR is decreased considerably by cutting in a high-resistance section of the coil, once the plunger is in its final operating position. This decrease eliminates unnecessary energy loss.

The escapement mechanism in Figure 22–28, left, controls the time required for the solenoid coil to pull up the plunger. TR_1 closes first, energizing relay coil M. Then after the definite time interval, contactors TR_2 close, energizing coil A. The front view and right side views in Figure 22–28 show the normally open TR contactors, which act as sealing contactors around the start button. Nearby are the normally closed contacts, which are connected across part of the TR solenoid coil.

The starting overload and running overload protection used with this controller functions practically the same as for the other types of automatic controllers previously described.

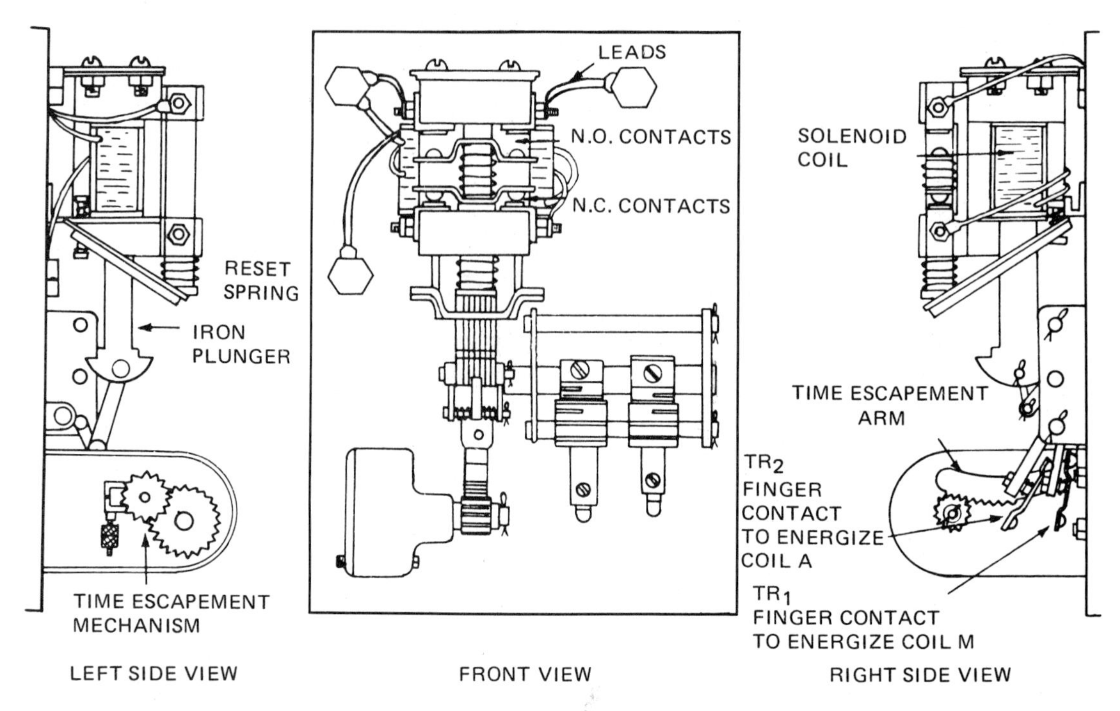

FIGURE 22–28 Definite time delay mechanism

22–10 ELECTRONIC CONTROLLERS

In automated machine operations, DC motors are driven by electronically controlled rectifiers that get their power directly from an AC line. We have seen, in Section 21–8, how silicon-controlled rectifiers can be employed for speed control by sending triggering pulses to the gate. The implication was that some manual control is provided for the operator to monitor and regulate the performance of the machine. In automated equipment, the driven machine continuously feeds back information to a master control to maintain any reasonable combination of speed and torque desired, Figure 22–29.

The program control is any system based on punch cards, magnetic tape, or microprocessors that causes the machine to follow a prescribed sequence of operations. These kinds of devices are faster acting and more reliable than electromechanical relays and/or timing devices. Such devices were developed in response to the demands of highly specialized, high-speed manufacturing processes. The electronics industry responded with modular designed solid-state devices.

One such system was known as *static control* and embodied circuits known as *logic gates*, such as AND gates, NOR gates, NOT gates, OFF RETURN MEMORY, and J-K flip flops.

With the development of microprocessors and computers, a new product was born: the *programmable controller* (PC). The PC is a solid-state device designed to perform the logic functions previously accomplished by electromechanical relays, drum switches, mechanical timers, and counters. Internally, there are still the logic gates, but

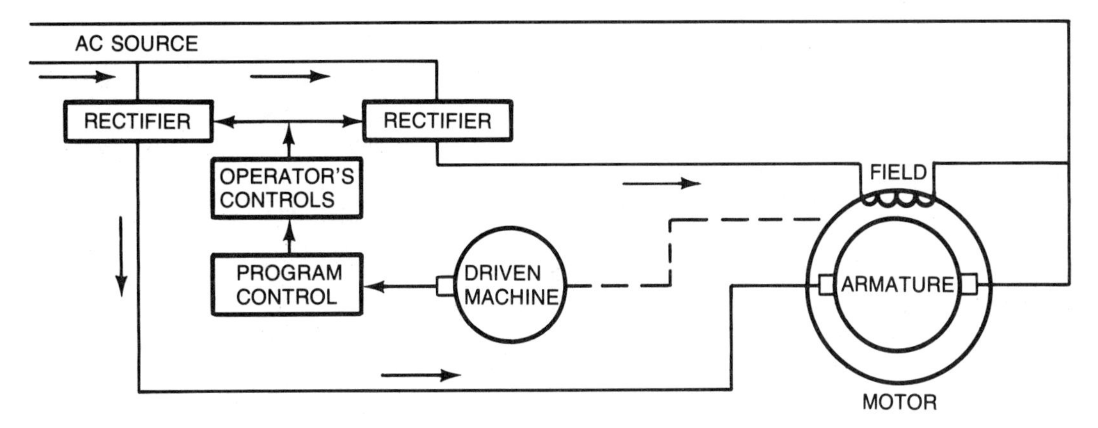

FIGURE 22–29 Electronic devices for motor controls

they have now been wedded with the graphic display of *relay ladder logic,* which is understood by all competent electricians.

To achieve such competencies, you will have to continue your studies and expand your knowledge in the field of industrial electronics.

SUMMARY

- DC motors have an extremely high starting current due to low armature resistance and low counter-emf.
- Starting rheostats are not intended for speed control.
- No-voltage release is a safety feature used to prevent the automatic restarting of a motor at the end of a power failure.
- Speed controllers are designed to accelerate a motor to normal speed and, in addition, vary the speed.
- For above-normal speeds, resistance is added to the shunt field.
- For below-normal speeds, the starting resistance is reinserted into the armature circuit.
- Series motors require a special type of starting rheostat.
- Series motor starters come with either no-voltage protection or no-load protection.
- Drum controllers are used where the motor is under direct control of an operator and when frequent starting, stopping, reversing, and varying of speeds are necessary.
- Review circuitry and sequence of operations for
  a. The counter-emf controller
  b. The voltage drop acceleration controller
  c. The definite time controller
- Dynamic braking stops a motor by making it act as a generator. It converts its rotational energy to electrical energy and then to heat in a resistor.
- Electrical interlocking is a system for insuring that one device is disconnected before an interfering or contradictory device is energized.

Achievement Review

1. Show the connections for a three-terminal manual starting rheostat connected to a shunt motor.
2. Give one advantage and one disadvantage of a three-terminal manual starting rheostat.
3. Show the connections for a four-terminal manual starting rheostat connected to a cumulative compound motor. Include a separate field rheostat in the shunt-field circuit for speed control.
4. Give one advantage and one disadvantage of a four-terminal manual starting rheostat.
5. A three-terminal manual starting rheostat has a resistance of 5.2 ohms in its starting resistor. The holding coil resistance is 10 ohms. This starting rheostat is connected to a shunt motor. The resistance of the armature is 0.22 ohm; the resistance of the shunt field is 100 ohms. The line voltage for this motor circuit is 220 volts.
   a. Determine the starting surge of current taken by the motor.
   b. The motor has a full-load current rating of 30 amperes. National Fire Underwriters requires the starting surge of current to be not greater than 150% of a motor's full-load current rating. Show with computations whether this manual starting rheostat complies with this requirement.
6. Using the data in question 5, determine
   a. The current in the holding coil with the movable arm in the run position
   b. The counter-electromotive force with the movable arm in the on position if the armature current is 20 amperes
7. Explain the difference between a manual starting rheostat and a manual speed controller.
8. Why does one type of manual starting rheostat used with series motors have no-load protection?
9. State the applications of the drum controller. Why is it desirable in these applications?
10. Explain why a shunt motor's direction of rotation does not change if the connections of the two wires are reversed.
11. What is the function of a holding coil in a manual starting rheostat?
12. Explain how to reverse the direction of rotation of
    a. A shunt motor
    b. A series motor
    c. A cumulative compound motor
13. Complete the internal and external connections for the above-normal speed controller and cumulative compound motor shown in the drawing on page 397.

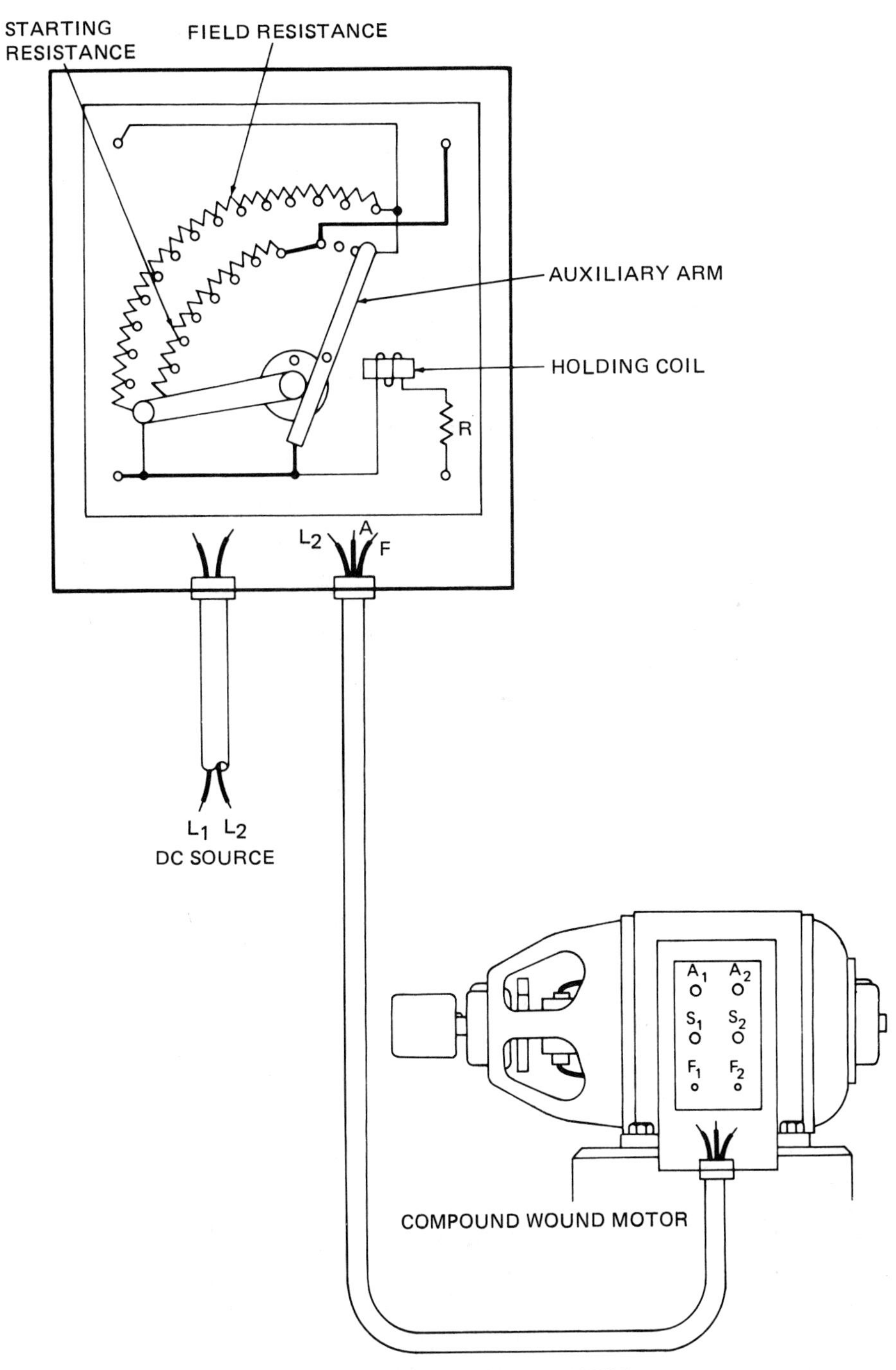

SPEED CONTROLLER AND MOTOR

14. Complete the internal and external connections for the above-and-below-normal speed controller and cumulative compound motor illustrated below.

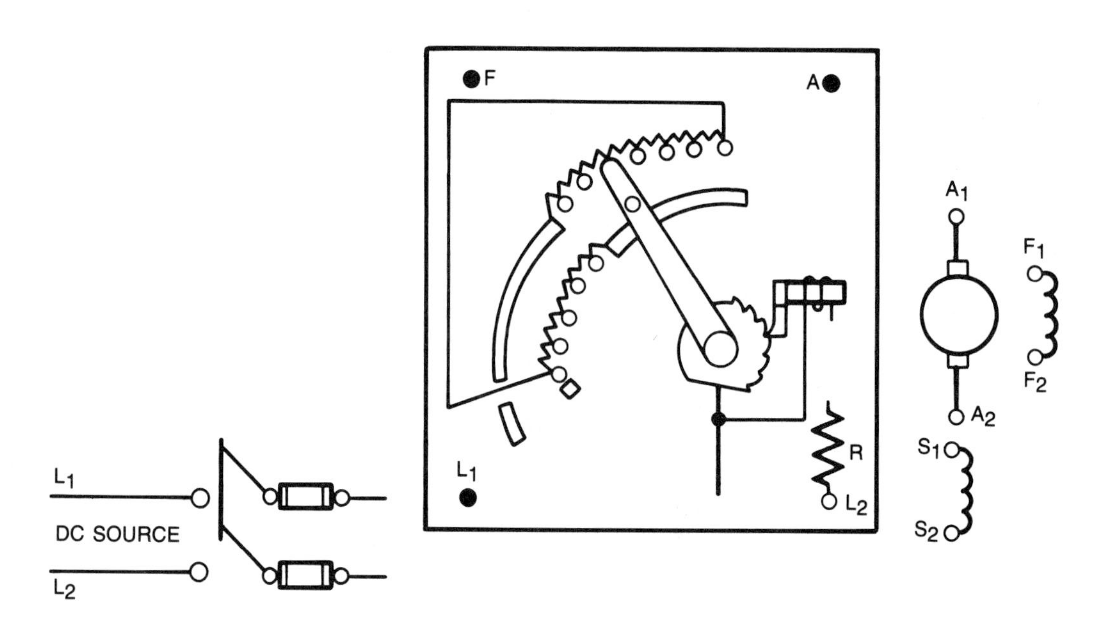

15. Draw the graphic symbols (JIC standards, Figure A–12 in the Appendix) representing the following components:
    a. N.O. limit switch
    b. N.C. limit switch held open
    c. N.O. timer contact (action retarded upon energizing)
    d. N.C. pressure switch
    e. N.O. float switch (for liquid level)
    f. N.C. flow switch (for air or water)
    g. N.O. contact with blowout
    h. thermal overload
16. A DC shunt motor is rated at 40 amps, 115 volts, 5 horsepower. Calculate the proper current rating for a thermal overload unit to install in a cemf controller to serve as proper *running* overload protection. Also, determine the size of the fuses that should be used as *starting* protection.
17. Complete the diagram shown by making all connections necessary to put the motor across the line without a starting resistance. Provide for dynamic breaking when the motor is stopped.

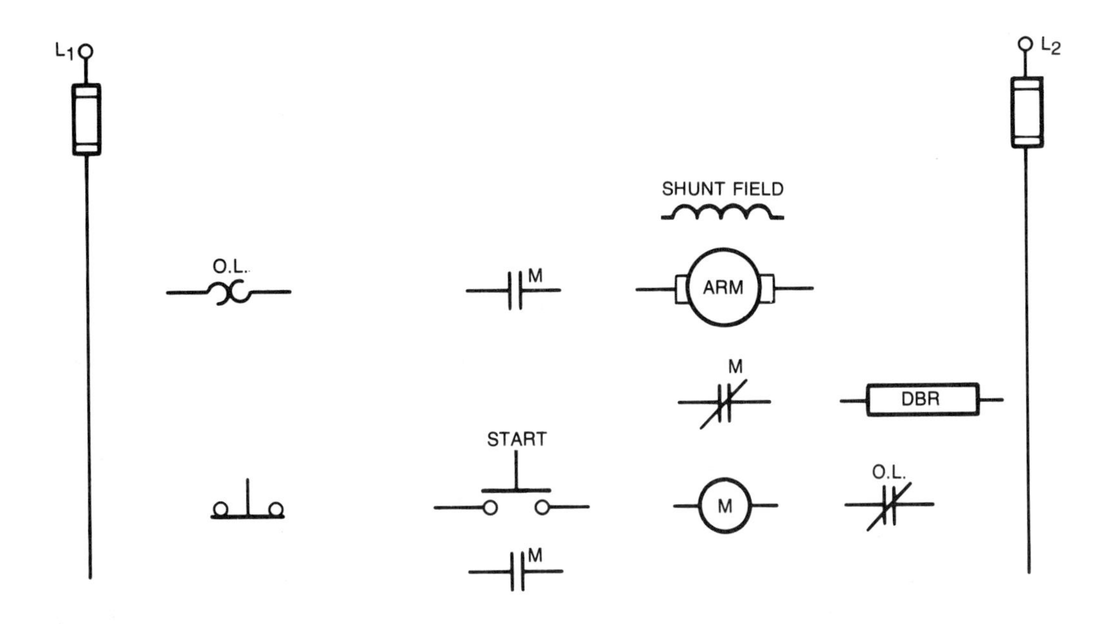

18. a. Which of the three types of controllers discussed in this chapter is represented by this drawing?
    b. Explain the sequence of operation.

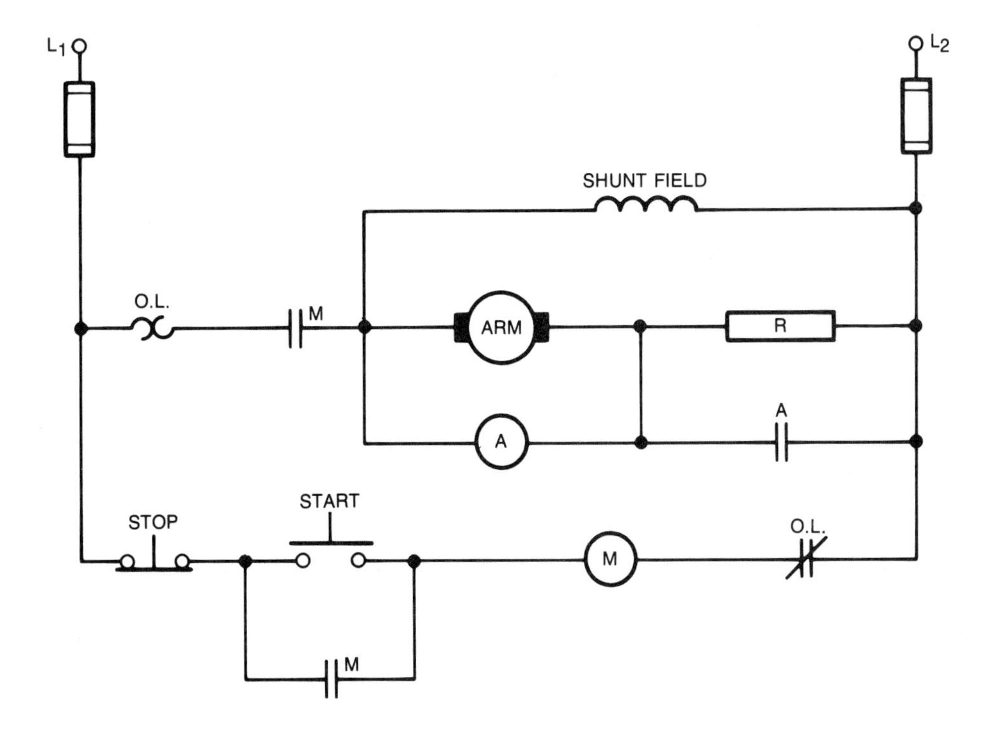

19. a. Which of the three types of controllers discussed in this chapter is represented by this drawing?
    b. What adjustment could be made to change the acceleration speed of the motor?

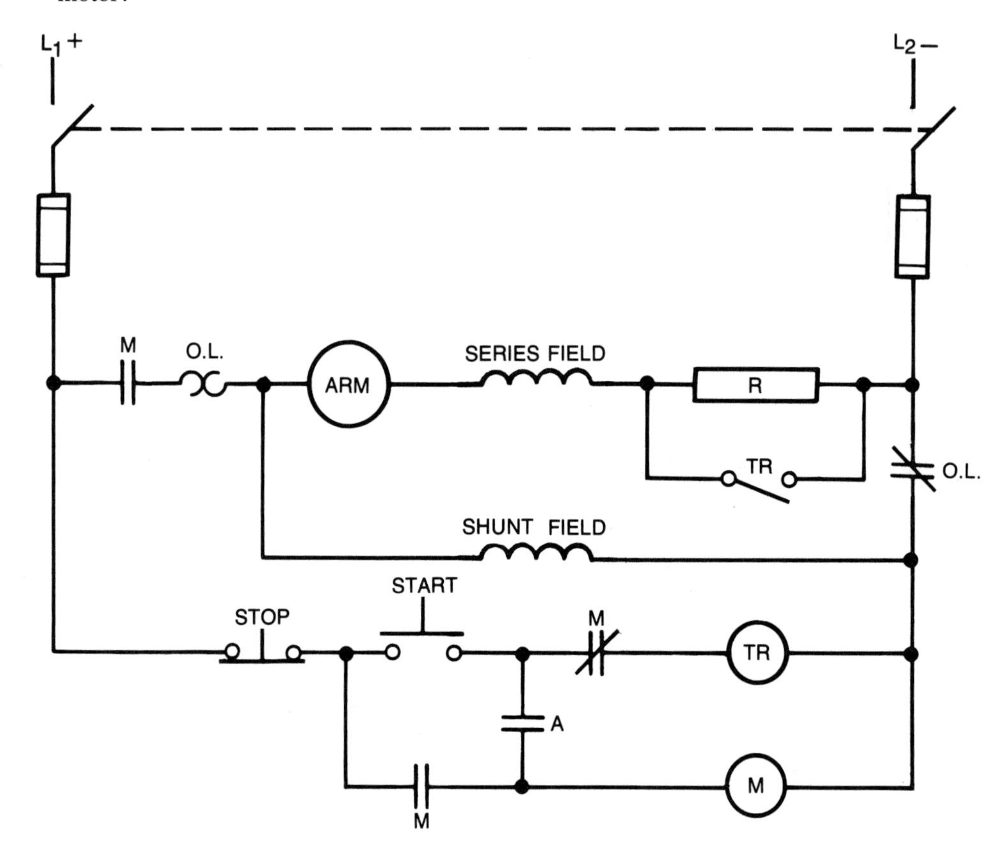

20. a. Which of the three types of controllers discussed in this chapter is represented by this drawing?
    b. Does the voltage drop across the resistors increase or decrease when the motor accelerates? Explain.

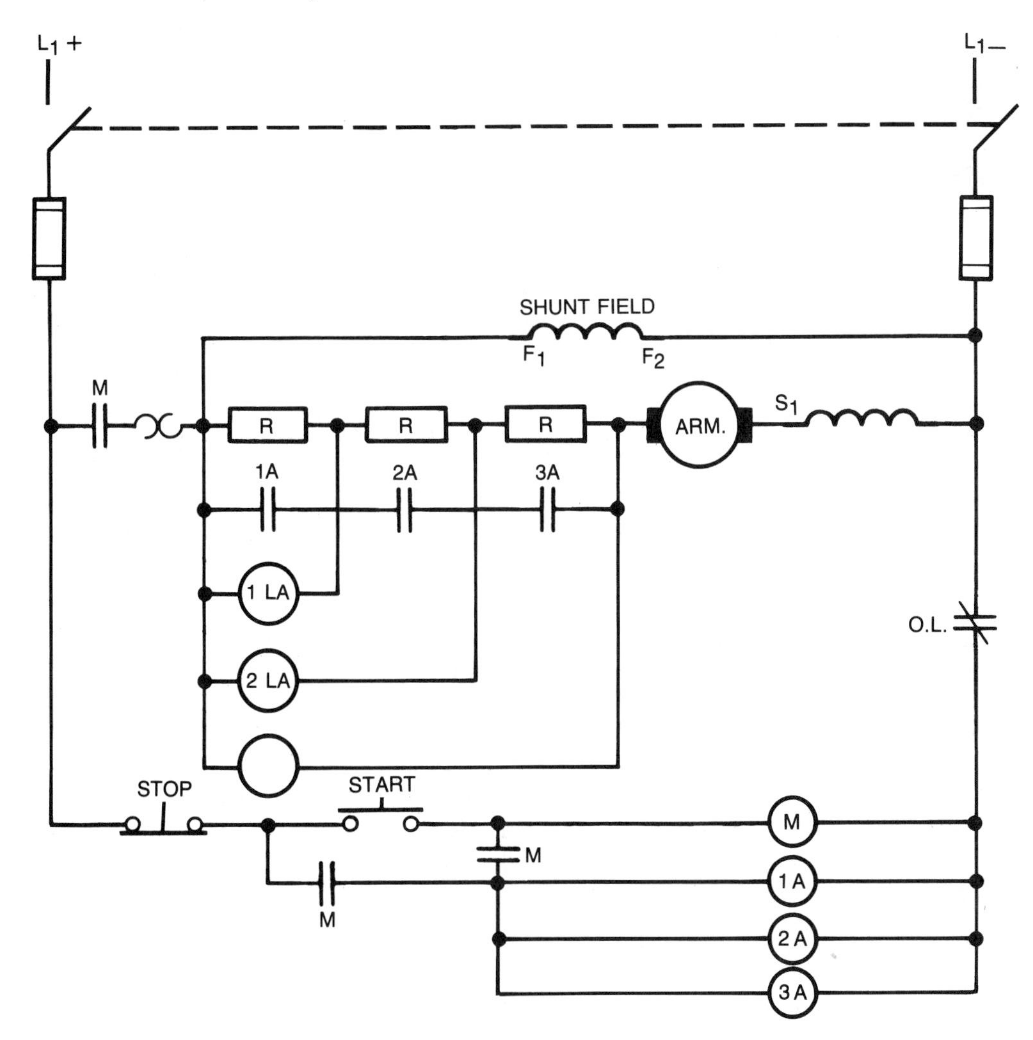

23 Electrical Heating and Lighting

Objectives

After studying this chapter, the student should be able to

- Define and explain the new technical terms introduced in this chapter

  nichrome
  resistance welding
  arc welding
  arc furnace
  induction heating
  dielectric heating
  three-wire, single-phase
  neutral wire
  bimetal
  heat pump
  candela
  foot-candle
  lumen
  light meter
  electromagnetic spectrum
  frequency
  ultraviolet
  infrared
  carbon arcs
  incandescent lamps
  mercury-vapor lamps
  sodium-vapor lamps
  neon lamps
  fluorescent lamps
  electroluminescence
  three-way switch
  four-way switch

- Explain the three-wire, single-phase system
- Describe the cardinal features of different kinds of lighting systems
- Draw diagrams of three-way and four-way switching circuits
- Correctly compute simple mathematical problems related to heating and lighting

23–1 RESISTANCE HEATING

Most common heating devices, such as toasters, irons, and electric ranges, are heated by current in a coil of nickel-chromium alloy wire or ribbon. Electrical devices are widely used for localized production of small amounts of heat because of their ease of control. The appeal of nickel-chromium (nichrome, chromel, etc.) alloys lies in their small size and reasonable relationship of cost to durability.

To design a heavy-duty resistance-type soldering iron that develops heat at a rate of about 180 watts, consult Figure 23–1 for wire recommendations. Number 26-gauge wire might be used.

Use formulas already learned for calculations.

The current in the wire is found from

$$\text{Watts} = \text{Volts} \times \text{Amperes}$$
$$180 = 120X$$
$$X = 1.5 \text{ A}$$

The resistance of the wire is found from Ohm's law:

$$E = IR$$
$$120 = 1.5X$$
$$X = 80\ \Omega$$

The length of the #26 nichrome wire is found from $R = lK/\text{CM}$. For nichrome wire, K is about 600; the circular mils for #26 wire is 254; R is 80 ohms. Calculating l, we find that about 33 feet of wire is needed.

If winding and insulating a coil containing 33 feet of wire results in a bulkier coil than necessary, try #28 wire. No. 28 wire has 160 circular mils instead of 254; $R = Kl/\text{CM}$ equals 21 feet of wire necessary. Therefore, #28 wire is better because the coil is smaller, lighter in weight, and cheaper to build. However, since the 180 watts is produced in a smaller volume of wire, the surface temperature of the wire is higher. Therefore, the #28 wire will oxidize faster and have a shorter life.

Electric stoves often use a heater element consisting of a coiled resistance wire enclosed in a steel tube. The space around the wire inside the tube is packed with magnesium oxide, or similar filler, to insulate the wire from the protecting tube.

Heater Watts (115 Volts)	Wire Gauge #
100–200	26–30
200–350	24–28
350–400	22–26
450–500	20–24
550–650	19–23
700–800	18–22
850–950	17–21
1,000–1,500	16–20
1,200–1,350	14–18
1,400–1,500	12–16
2,000	10–14

FIGURE 23–1 Nichrome wire recommendations

Three-heat hot plates have two heater elements controlled by a four-position switch. On the high-heat setting, both elements are in use, connected in parallel in the 115-volt line. On the medium and low positions, either the larger or smaller of the elements is used individually.

For producing larger amounts of heat in industrial kilns and furnaces, solid rods of silicon carbide are used as heating elements. These so-called *Glo-bar* heaters are made in lengths from 4 inches to 6 feet and are used for temperatures up to about 2,800°F. Their resistances range from 0.4 to 5 ohms.

The trade name *Kanthal* is applied to a group of alloys containing iron, chromium, aluminum, and cobalt. These alloys are useful for heater elements with a high temperature requirement. Such applications include resistor-type soldering irons and cigarette lighters (which require a concentrated heat source) and high-temperature kilns and furnaces. Kanthal builds up a very adherent aluminum oxide coating that resists further oxidation. Various alloys can operate at temperatures from 2,100° to 2,460°F. One powder-metallurgy product, Kanthal Super, containing molybdenum, silicon, and other metals and ceramics, operates at 2,900°F.

23–2 HOME HEATING AND THE NEED FOR DUAL VOLTAGE

Many households employ a variety of electrical heating devices. Portable space heaters are very popular. Some homes even boast complete, permanently installed heating systems.

Flexible cable installed in the ceiling before plastering heats the room by radiation. Baseboard units heat the room by convection. Radiant wall and radiant baseboard units are also available. These heating methods are noiseless and safe and should be long lived if properly built and installed. Electrical heat eliminates the expense of a chimney in a new construction, too (unless neighborhood status requires a fireplace).

Heating circuits like this do require large amounts of electrical energy. So do other large heating devices, such as electric kitchen ranges, water heaters, clothes dryers, and high-Btu air conditioners.

The Three-Wire Supply

It is economically advantageous to operate such high-powered appliances on a higher voltage. Since $P = E \times I$, it stands to reason that by doubling the voltage we cut the current in half. Consequently, by using a higher voltage, we can minimize undesirable voltage drops and I^2R losses in the line without using excessively large wires.

The common distribution system brings two voltages on three wires into your home. Figure 23–2 illustrates such a system as if it came from two series-connected generators. Each generator develops 120 volts; the two in series give 240 volts. The two outside wires at top and bottom provide the 240-volt supply for ranges and large heaters. The middle wire, in practice, is grounded securely to a cold-water pipe. The *neutral wire* carries only the difference in current carried by the two outside wires and therefore need not be as large as the outside wires.

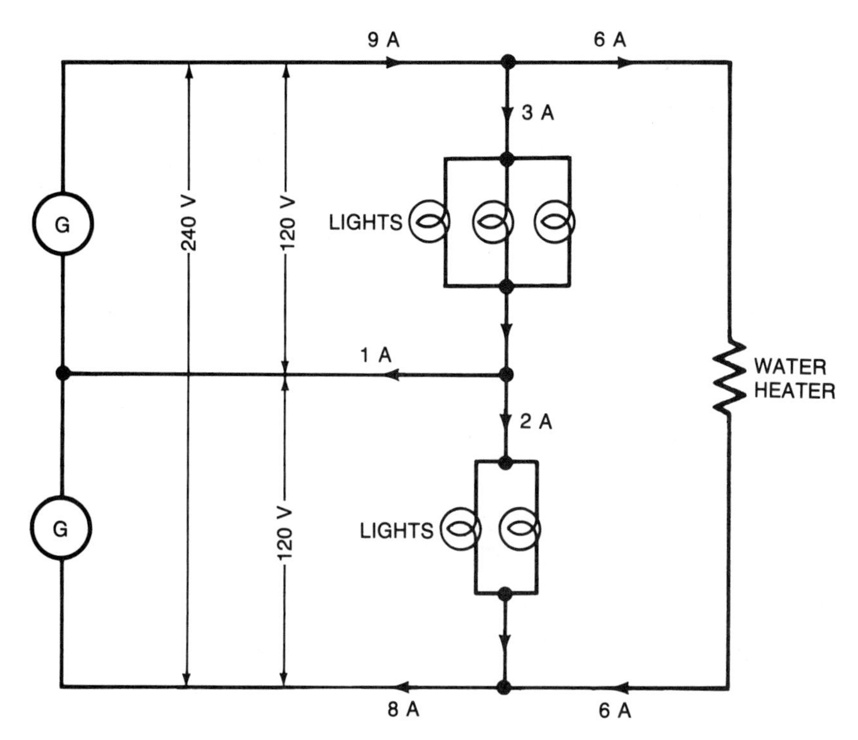

FIGURE 23–2 Three-wire circuit

These relationships of voltages and currents hold true whether the source voltage is DC or AC. On alternating current, the two circles marked G in Figure 23–2 represent the two halves of a 240-volt winding on a distribution transformer that supplies AC energy to a house. Such a wiring arrangement for AC is called *three-wire, single-phase.*

23–3 THE KITCHEN RANGE

Electric ranges connect to the neutral as well as to the 240-volt wires. With both 120 and 240 volts available, a multiple-contact switch allows as many as eight different heating rates to be obtained from various combinations of voltages on two different heating elements in each cooking unit.

For example, one manufacturer uses a six-position switch to obtain five heat settings from two heater elements (46-ohm and 59-ohm) in one range unit, Figure 23–3A.

1. The 46-ohm and the 59-ohm in parallel on 230 volts: 1,150 + 900 = 2,050 watts, Figure 23–3B
2. The 46-ohm on 230 volts: 1,150 watts, Figure 23–3C
3. Both in parallel on 115 volts: 290 + 220 = 510 watts, Figure 23–3D
4. The 46-ohm on 115 volts: 290 watts, Figure 23–3E
5. Both in series on 115 volts: 125 watts, Figure 23–3F

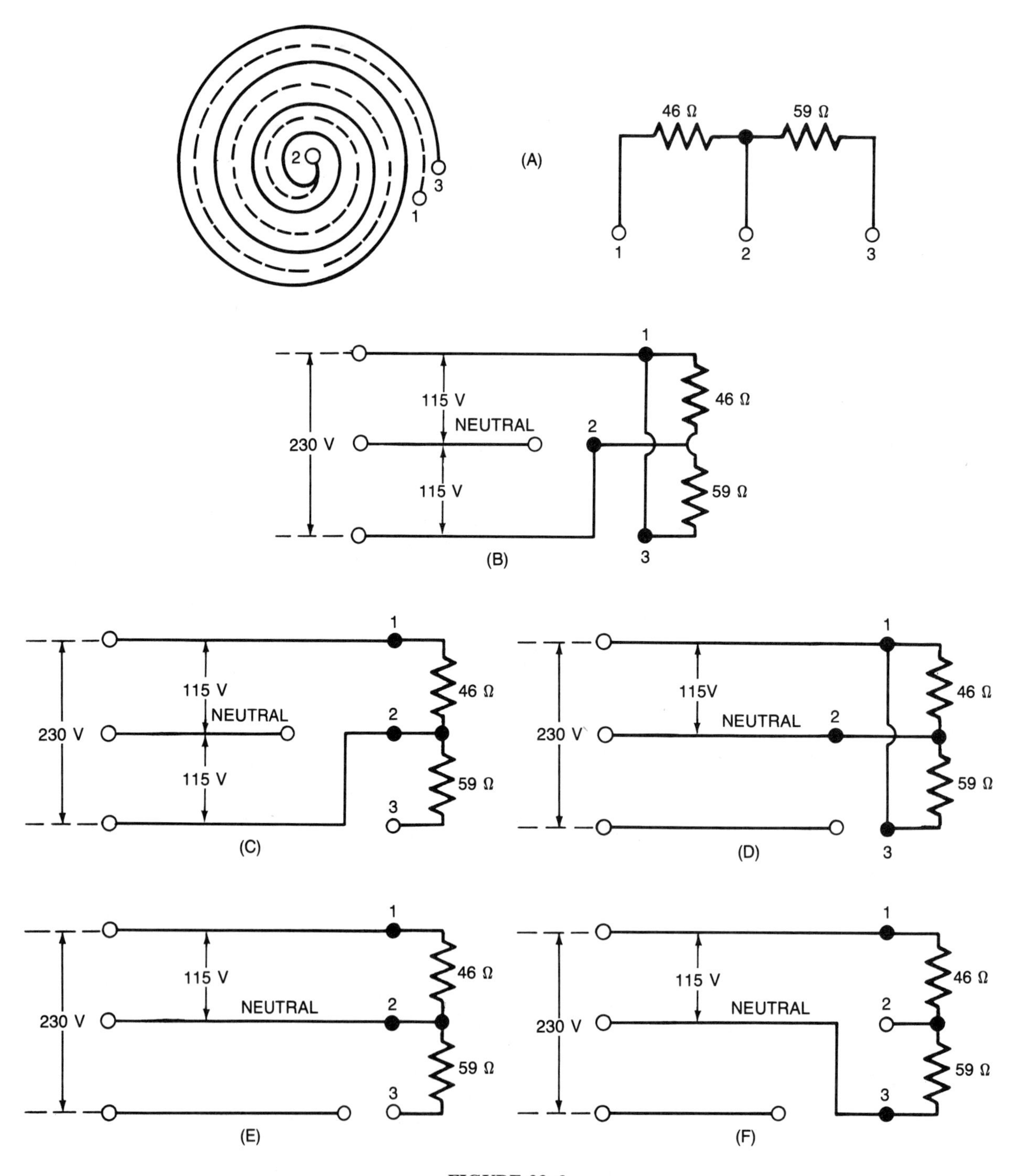

2
3
1
(A)
46 Ω
59 Ω
1
2
3
1
115 V
NEUTRAL
2
230 V
115 V
46 Ω
59 Ω
3
(B)
1
115 V
NEUTRAL
2
230 V
115 V
46 Ω
59 Ω
3
(C)
1
115V
NEUTRAL
2
230 V
46 Ω
59 Ω
3
(D)
1
115 V
NEUTRAL
2
230 V
46 Ω
59 Ω
3
(E)
1
115 V
NEUTRAL
2
230 V
46 Ω
59 Ω
3
(F)

FIGURE 23–3

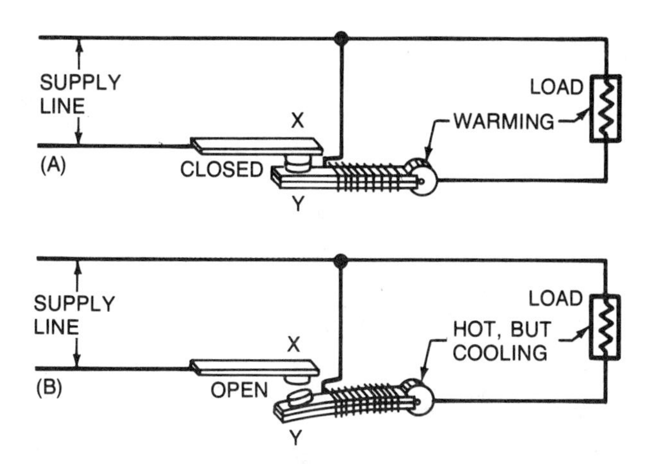

FIGURE 23–4 A bimetal thermostat

For stepless regulation of the average temperature of a heating element, a compound bar can be used to close and open a pair of contacts. The principle is shown in Figure 23–4. A compound bar is a strip of brass welded to a strip of invar. When warmed, the brass side of the bar expands, but the invar alloy side does not. This difference in expansion of the two metals causes the bar to curve. As shown in Figure 23–4A, contacts X and Y are closed. The load resistor (heating element) is warming up, and a coil of thin resistance wire around the compound bar is also warming the bar. When the bar is heated, it bends and disconnects from the supply. The bar soon cools enough to return contact Y to contact X again. This contact restores power to the heating element. Temperature is controlled by a movable cam (or similar mechanical arrangement not shown), which adjusts the position of the contacts. For example, if contact X is pushed down farther toward Y, then the compound bar must be heated to a higher temperature for Y to be pulled away from X. The load resistor is incidentally also heated to a higher temperature.

23–4 THE HEAT PUMP

Figure 23–5 shows the essentials of a system used in refrigeration, in air conditioning, and in house heating. The freezer section of a household electric refrigerator can properly be called the *evaporator*. Inside its double-walled chambers is a liquid that evaporates very readily. Hands wet with gasoline, cleaning solvents, or water cool as the liquid evaporates. The faster the liquid evaporates, the cooler the hands become; that is, the more heat is taken from the hands.

In the refrigerator, when the interior warms up a little, a thermostat starts the electric motor that runs a pump, pumping vapor away from the liquid in the evaporator to give it more chance to evaporate. As the liquid evaporates, it takes heat from the metal evaporator itself and from everything else nearby. As the vapor goes through the pump, it is heated by compression. The hot vapor, under pressure, is cooled enough in

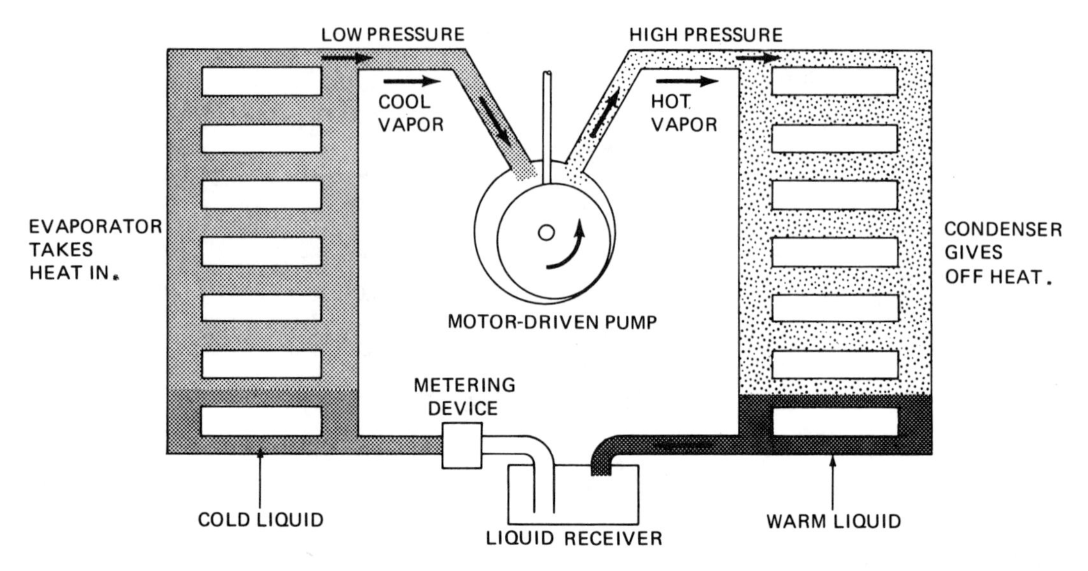

FIGURE 23–5 A heat pump

the condenser to change back to liquid; that is, it gives off the heat that it absorbed back in the evaporator when it became vapor. The air of the room cools the condenser, that is, takes heat from the condenser.

To make this device serve as a house heater, we can put the evaporator outdoors where it cools the outside air, and the condenser indoors where it warms the inside air. In summer, the valves can be reversed so that condensation takes place outdoors and evaporation takes place in pipes inside, cooling the house. This system is especially appropriate in mild climates for winter heating and summer air conditioning.

23–5 RESISTANCE AND ARC WELDING

Resistance Welding

Sheet steel is often welded with heat developed by current in the steel itself. Copper electrodes are pressed against the steel; then the electrodes are connected to a low-voltage, high-current supply for a short time, Figure 23–6. The resistance of the steel is small, but I^2R is large enough to weld the metal. After the current is turned off, electrode pressure is maintained briefly; then the work is released. In production machines, the squeeze-weld-hold-off cycle and the control of current are accomplished by using specialized electronic equipment.

Arc Welding

Motor-driven DC generators supply the power for most industrial arc welding. Welding imposes a peculiar set of operating conditions on the generator. No-load voltage is about 60 volts. Voltage has to be reduced greatly when the arc is struck and must

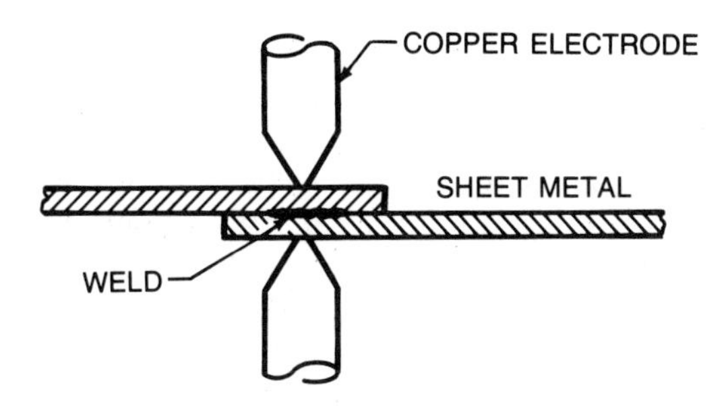

FIGURE 23–6 Resistance welding

be restored at once if the arc is extinguished. During welding, the current should remain constant although resistance of the arc varies.

Figure 23–7 shows one type of welding generator. The exciter generator supplies a fairly constant but adjustable field for the welding generator. The reactor is a low-resistance iron-core coil. The purpose of this coil is to help hold the output current constant. If current in the coil suddenly increases, an emf generated in the coil itself opposes the increase. If current starts to decrease, the generated emf in the coil helps maintain the current (Lenz's law). The series field is connected differentially (in opposition) to the field supplied by the exciter. The purpose of this connection is to lower the output voltage if load resistance is suddenly reduced. The series field can be controlled by a diverter rheostat in parallel with the field instead of by the taps shown previously. For high-current output, the series field is shorted out.

Arc Furnaces

Electric furnaces utilize heat produced by carbon arcs in the production of alloy steels. In the laboratory, a small, high-temperature furnace can be constructed by surrounding a carbon arc with refractory material.

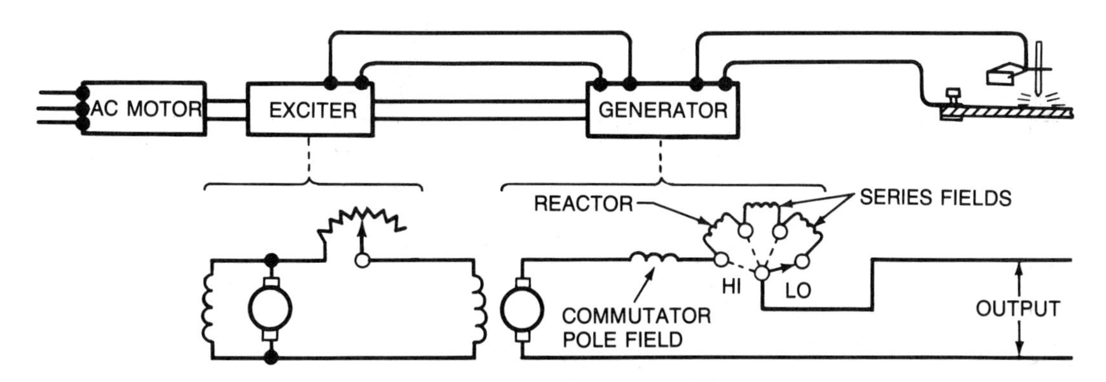

FIGURE 23–7 Welding generator

23–6 INDUCTION AND DIELECTRIC HEATING

Induction Heating

Although induction heating is strictly an alternating current process, it deserves mention in this chapter. If a piece of metal is placed inside a coil that is supplied with high-frequency alternating current, the rapidly changing magnetic field induces eddy currents in the surface of the piece of metal, thereby heating it. This method is often used in surface-hardening heat treatment for gears and similar machine parts.

Dielectric Heating

Heat can be produced throughout the body of an insulating material *(dielectric)* by putting the material between two metal plates and applying a high-frequency alternating voltage to the plates. For example, in the making of plywood, the wood and glue are heated to dry the glue and bond it to the wood. Wood is a poor conductor of heat, and gluing is time consuming if the wood has to be heated by direct contact with hot metal. However, when high-frequency voltage (such as 20 million cycles/second) is applied to metal plates on each side of the wood, electrons in the wood molecules vibrate back and forth in accordance with the rapid positive to negative changes of the metal plates. This electron vibration quickly heats the molecules of wood internally. Dielectric heating can also be seen in use with rubber, plastic, or ceramic materials. One type of electronic stove cooks meat by this process. When applied to people, dielectric heating is called diathermy.

23–7 WHAT IS LIGHT?

Light is energy that is radiated by electronic disturbances in atoms. Electrons in an atom can accumulate energy by absorption from a heated object or an object radiating energy or by bombardment by other electrons, as in a gas conduction tube. Sooner or later, the energy absorbed by the electron is released. The amount of energy an electron can get rid of in one burst depends on where the electron is; that is, what kind of atom it is in and what its location in the atom is.

The energy is radiated as a wave-like pulse of electric lines of force and magnetic lines of force called an *electromagnetic wave*. The vibration frequency of these traveling lines of force is proportional to the amount of energy that the electron gives off. Frequencies in the range from 4.3×10^{14} to 7.5×10^{14} vibrations per second affect electrons in our eyes. We call electromagnetic waves in this frequency range *light*. Waves of a frequency slightly higher than 7.5×10^{14} are called *ultraviolet*. Waves in a lower frequency range, from 10^{12} to 10^{14} vibrations per second are called *infrared* and *heat radiation*. (The incorrectly used term *black light* means either ultraviolet or infrared.) Ultraviolet and infrared differ greatly in their effects and uses.

Low-frequency waves are emitted by the electrons in the outermost rings of the atom. X rays and gamma rays are large bursts of radiation emitted by electrons in the innermost rings of the atoms. A listing of the applications of electromagnetic waves of various frequencies is given in Figure 23–8. The speed of travel of all of these electromagnetic waves, provided there is nothing in the way, is the *speed of light,* 186,000 miles per second.

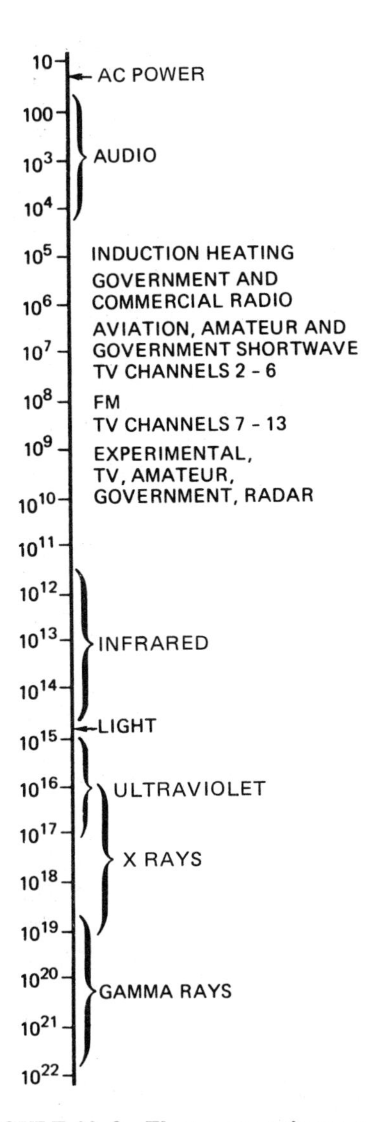

FIGURE 23–8 Electromagnetic spectrum

23–8 LIGHT MEASUREMENT

The term *measurement of light* can refer to either the intensity of a source or the illumination on a surface.

The intensity of a light source can be measured in *candlepower*. For many years this measuring unit was the light of a *standard candle* of specified size and shape, made of sperm wax and burning at the rate of 120 grains per hour. This standard was unsatisfactory on several counts. The present standard, adopted in 1948, is based on the light produced by white-hot powdered thorium oxide in a furnace at 3,216°F. The light coming from a 1/60-square-centimeter hole in this furnace is defined as having a source intensity of *1 candle*. This standard is far more convenient and more precise than candle flame. The term *candela* also means candle or candlepower.

One unit of measure for surface illumination is the *foot-candle,* defined as the illumination from a standard candle on a surface 1 foot away. Another unit is called the *lumen*. One lumen is the rate of flow of visible light energy through a 1-square-foot hole at a distance of 1 foot from a 1-candlepower light source. If we place a 1-square-foot surface at a distance of 1 foot from a standard candle, its illumination is 1 lumen per square foot. One lumen per square foot is the same as 1 foot-candle. Since the surface area of a sphere is equal to $4\pi r^2$, a 1-candlepower source produces 4π lumens.

The surface brightnesses listed in Figure 23–9 make interesting comparisons.

A 100% efficient light source that converts all of its electrical energy into visible light produces 621 lumens per watt. An ordinary 100-watt tungsten lamp has an efficiency of 18 lumens per watt; fluorescent tube lamps provide about 50 lumens per watt; and sodium-vapor and mercury-vapor lamps, 50 to 100 lumens per watt.

Fortunately, the eye adjusts readily to changes in illumination. Outdoor illumination can vary from 8,000 to 10,000 lumens per square foot in clear, bright sunlight to 100 lumens per square foot on a dull, dark day to 0.03 lumen per square foot on a night lighted by a full moon. Recommended illuminations for artificial lighting vary from 0.5 lumen per square foot for sidewalks to 10 or 20 lumens per square foot for classrooms and offices to 100 lumens per square foot for drafting rooms. For ordinary work, illumination is measured by *light meters*. These consist of a photoelectric cell and a microammeter that is scaled to read lumens per square foot, Figure 23–10.

Source, Distance, and Illumination

Figure 23–11 shows the geometry of distributions of light. Light radiating from a single point source spreads so that the surface illumination is inversely proportional to the square of the distance of the surface from the source, Figure 23–11A. The light that falls on a 1-foot-square surface held 2 feet below the lamp covers 4 square feet if allowed to travel twice as far from the source.

Light from a line source, such as a long string of fluorescent tubes, spreads out so that the surface illumination is proportional to the distance from the source, Figure

Surface Brightness in Candles per Square Centimeter	
The sun	160,000
Ordinary carbon arc	13,000
High-intensity carbon arc	80,000+
Melted tungsten at 6,120°F	5,740
Filament of tungsten projection lamp	2,100
Interior of furnace at 3,216°	60
Frosted glass surface of 40-watt lamp	2.5

FIGURE 23–9

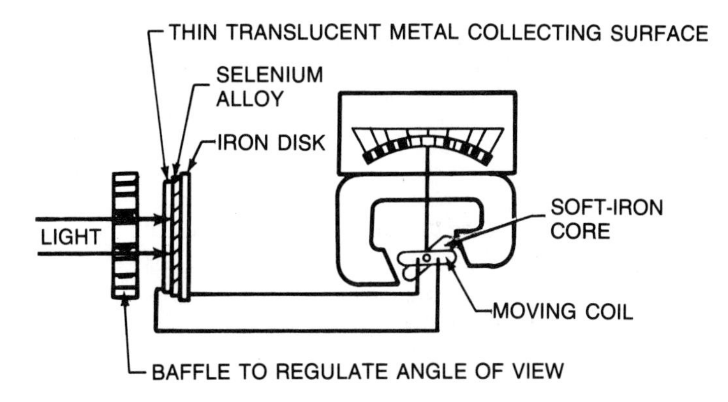

FIGURE 23–10 Light meter

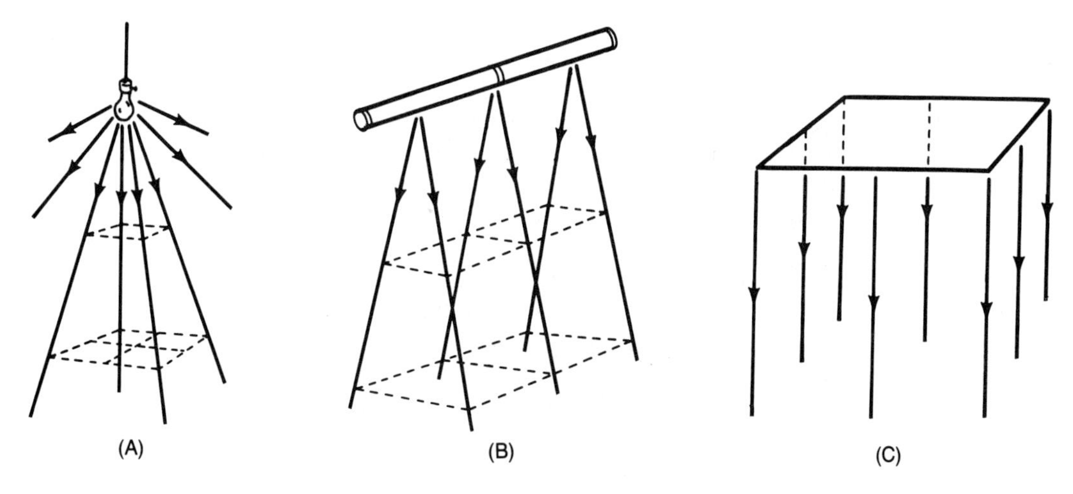

FIGURE 23–11 Geometry of distribution of light

23–11B. Illumination of surfaces below these two light sources is improved by placing reflecting surfaces above the source.

Light from a source that is a flat surface, indefinitely wide in extent, gives an illumination that is independent of distance from the source, Figure 23–11C.

23–9 CARBON ARCS

Carbon arc lamps have one particular feature that, in some applications, is a great advantage. That feature is the extremely concentrated brilliance of the source of light. In some types of light projection equipment, such as 35 millimeter movies and searchlights, carbon arcs are preferred. This concentrated source of light simplifies the optical system.

Despite the high surface brilliance of the carbon arc, so much of the energy supplied to the arc is converted to heat that its efficiency is low (10 to 20 lumens per watt), making it unsuitable for general illumination. At the high temperature of the carbon arc (about 10,000°F), the radiation includes considerable amounts of ultraviolet. Eye irritation noticed after exposure to arcs is caused by this ultraviolet radiation.

23–10 INCANDESCENT LAMPS

The first practical electric light developed was the carbon arc. The carbon filament incandescent lamp followed. (*Incandescent* means glowing due to heat, in this case, whitehot.) Then came the tungsten filament incandescent lamp.

Considering all of the specialized types of lamps that are manufactured, we find hundreds of sizes and types of tungsten incandescent lamps. The reader can learn how they are constructed by smashing a few old lamp bulbs and then making a cross section diagram.

The filament temperature in ordinary 25-watt to 300-watt lamps ranges from 4,200°F to 4,800°F. The average lamp life is 750 to 1,000 hours. Efficiencies range from 10.5 lumens per watt for the 25-watt lamp to 20 lumens per watt for the 300-watt lamp.

What gives more light—ten 40-watt lamps or four 100-watt lamps? Both total 400 watts. Because the efficiency of the 100-watt lamp at 16.8 lumens per watt is higher than that of the 40-watt at 11.7 lumens per watt, four 100-watt lamps give about 42% more light than ten 40-watt lamps. Both sets produce the same total energy, most of it heat.

How long should a lamp last before it burns out? Both the life and the brightness of an incandescent lamp depend on the temperature of the filament. This temperature is controlled by filament dimensions and current. The higher the temperature, the more light the bulb produces and the shorter its life. Extra-long-life lamps do not produce as much light as a standard bulb. One-hundred-watt, 120-volt incandescent lamps can be built for any predetermined life. Ordinary bulbs average 1,000 hours and produce 1,675 lumens. A 100-watt lamp can be built to last 300 hours and put out 2,285 lumens. This value is almost as much as a standard 150-watt lamp. The 100-watt lamp can be made to last 3,000 hours; but if it does, it will give only 1,260 lumens because its filament temperature must be lower to give the longer life. The main purpose of a lamp is to produce light. The energy the lamp uses during its life costs more than the lamp. Of the three lamps just described, the 100-watt, 300-hour lamp produces the most light per dollar.

Originally, incandescent lamp filaments operated in a vacuum to prevent oxidation of the filament. Forty-watt and larger lamps are now filled with nitrogen at near-atmospheric pressure. The gas pressure retards evaporation of the filament and thus permits the filament to be operated at higher temperature and higher efficiency.

23–11 INFRARED LAMPS

About 70% of the energy given to the heated filament of a lamp is radiated away in vibration frequencies too low to be visible as light. This energy is called infrared. When infrared radiation strikes and is absorbed in a material, the material is warmed.

Infrared lamps, then, are often called *heat lamps*. Sizes range from 250 to 5,000 watts. They are used in industry for drying and baking and at home for treating lame shoulders. They are also used commercially for heating people in open areas such as grandstands, warehouse platforms, or outside of store windows. The filaments in infrared lamps operate at lower temperatures than those in lamps intended for lighting; hence, their efficiency as visible light sources is appropriately low and their life is long (about 5,000 hours).

23–12 VAPOR LAMPS—GASEOUS CONDUCTION

A variety of inert gases and metallic vapors is usable in arc lamps. Most such lamps require auxiliary current-limiting equipment in the lamp fixture.

Mercury Lamps

The high efficiency of mercury lamps (about 50 lumens per watt) is the reason for their wide use in street and industrial lighting. The light produced by the glowing gas is not uniformly white because its energy is concentrated at a few frequencies. Colors viewed under mercury light may not appear normal. Yet, the blue light of the mercury arc is an advantage in some photochemical processes. High-pressure mercury lamps for general illumination are available in sizes from 100 to 3,000 watts and require a special fixture. The life of these lamps averages over 16,000 hours.

Ultraviolet Lamps

Hundreds of specialized uses for ultraviolet are found in various industries: foods, minerals, textiles, metal parts inspections, dyes, coatings, oils, and decorations. Most of these uses depend on the ability of many materials to glow when illuminated by invisible ultraviolet. Other uses depend on ultraviolet's ability to initiate chemical changes. One use, for example, is in the photographic copying process.

Low-pressure mercury arcs are sources of ultraviolet radiation. An incandescent lamp, operated at a higher-than-normal filament temperature, produces a little ultraviolet. The most used sources, however, are fluorescent lamps. These lamps operate in principle like the ordinary fluorescents but use a coating that radiates almost all of the energy as ultraviolet. Ultraviolet lamps are sometimes covered or enclosed by a special glass filter that stops visible light but lets the invisible ultraviolet pass through.

Sodium-Vapor Lamps

Sodium-vapor lamps were introduced in a few highway lighting systems some 30 years ago. They had long life and high efficiency (55 lumens per watt) but produced only yellow light. Recent improvements have put a high-pressure sodium arc in a translucent ceramic tube, giving a whiter light at about 100 lumens per watt.

23–13 NEON AND FLUORESCENT LAMPS

Neon Lamps

High voltage applied to neon gas in a long tube is extensively used in neon signs. Also, small neon-glow lamps (0.04 to 3 watts), giving an orange-red light, are often used as indicator lamps. Their life averages over 3,000 hours, and they are usually reliable in that they fail gradually, having no filament to burn out suddenly. The 2-watt NE-34 lamp is appropriate for a night light for home use.

Small argon-glow lamps of similar form produce a pale blue-violet light. They are useful as a small-scale source of ultraviolet.

Fluorescent Lamps

The conductor in the fluorescent lamp, Figure 23–12, is mercury vapor at very low pressure. Current through this low-pressure vapor produces a little blue and violet light, but most of the radiation is invisible ultraviolet. This ultraviolet radiation is absorbed by the coating on the inside of the tube. The coating reradiates some of the energy at lower frequencies that are visible. The color of light radiated by the coating is characteristic of the coating material itself. Substances that have this ability to absorb radiation and immediately reradiate energy are called *fluorescent* materials or *phosphors*. The efficiency of the fluorescent lamp can be as high as 50 lumens per watt, even after 5,000 hours of use.

The sketch in Figure 23–13 illustrates the simplest circuit arrangement for starting and operating a fluorescent lamp. No current is produced by 115 volts applied to the circuit at first, because the resistance of the tube is too high to permit an arc to start. Closing the starting switch permits a current through the filaments at the ends of the tube and through the ballast. The ballast, or reactor, is a coil of copper wire on an iron core. On AC, the bouncing magnetic field in the ballast coil induces voltages that hinder the continually changing current; thus, the amount of current in the circuit is limited. Heating of the filaments in the ends of the tube warms the mercury vapor, and electrons are emitted from the hot filament. The tube is then ready to conduct. When the starting switch is opened, the collapsing magnetic field in the ballast coil induces a momentary high voltage in the circuit. This voltage starts electrons moving through the mercury vapor, the vapor ionizes, and conduction is under way.

During operation, the voltage across the tube is between 50 and 100 volts, depend-

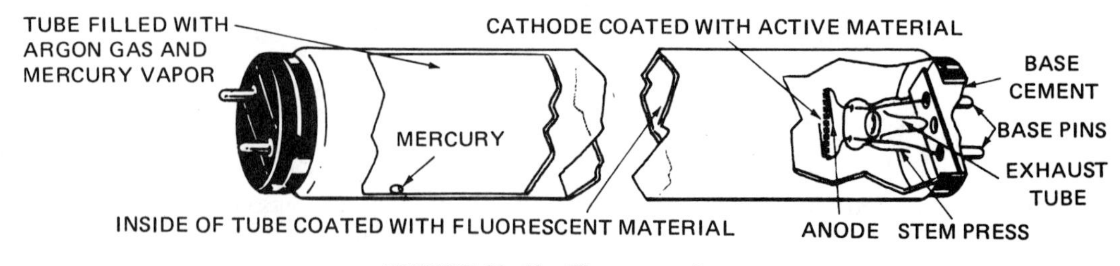

FIGURE 23–12 Fluorescent lamp

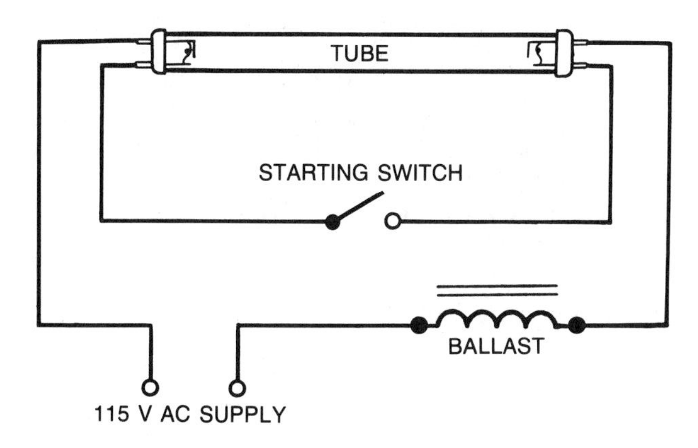

FIGURE 23–13 Fluorescent lamp circuit

ing on the size of the tube. Opposing voltage generated in the inductive ballast coil accounts for the rest of the circuit voltage. The ballast limits current through the tube without causing as much heat as a series resistor would cause. Fluorescent fixtures generally include an automatic starter, Figure 23–14.

The Glow Switch

The glow switch, shown in Figure 23–15, replaces the starting switch shown in Figure 23–13. The glow tube is a neon-filled glass bulb, containing a U-shaped bimetallic strip and a fixed contact, normally open. When the circuit is connected to the 115-volt line, current is small because of the high resistance of the glow tube. There is little

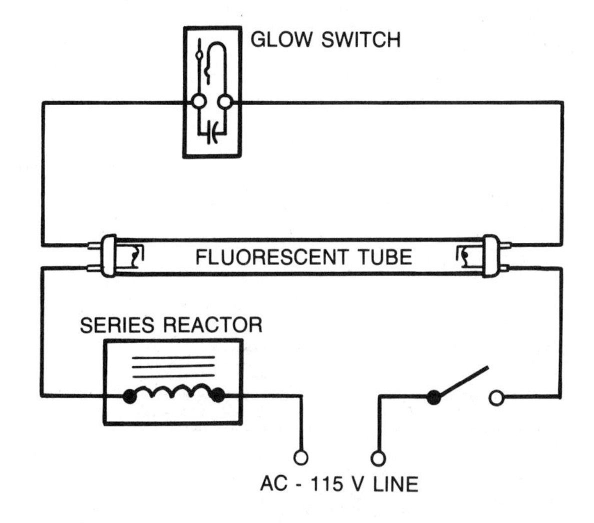

FIGURE 23–14 Circuit for fluorescent tube

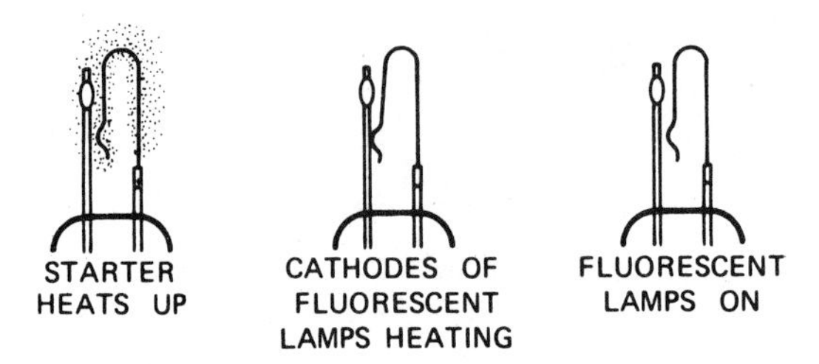

FIGURE 23–15 A glow switch starter

voltage drop across the series reactor. Voltage across the glow tube contacts is enough to start a little arc discharge between the bimetal strip and the fixed contact. This arc heats the bimetal, which bends and touches the fixed contact.

The closing of the starting switch causes the current to heat the tube filaments. The closing of the contact in the glow tube stops the glow tube arc; the bimetal cools, and the contacts open. At the instant of opening, the inductance voltage kick generated in the series reactor coil starts conduction in the fluorescent tube.

Instant-start fluorescent lamps have a cathode that requires no preheating. A special instant-start ballast is used; an arrangment of coils and capacitors provides a high starting voltage, yet limits the current after conduction starts.

23–14 ELECTROLUMINESCENCE

The section on dielectric heating described one effect on materials subjected to alternating voltage between capacitor plates. A few materials have the ability to produce light when sandwiched between metal plates that are supplied with alternating currents at frequencies from 60 to 1,000 cycles, Figure 23–16. This light is not produced by heat. It is produced by electrons in the phosphor in a direct conversion of electrical energy to light. The color of light is determined by the phosphor mix; green, yellow, blue, and white are available.

The most common example of an electroluminescent lamp is the 3 by 3 inch plastic panel, 0.02-watt, used as a night light in homes. It is a long-life device, losing about

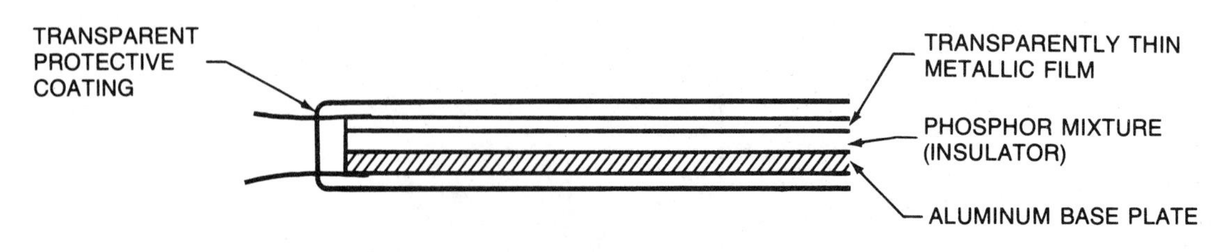

FIGURE 23–16 Electroluminescent lamp

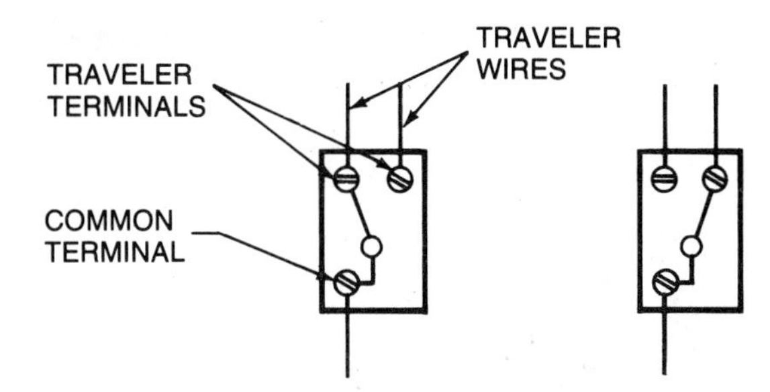

FIGURE 23–17 Two positions of three-way switch

half of its initial brightness after 10,000 hours of use. Luminescent panels can be built in a great variety of sizes and shapes, either rigid or flexible. Instrument lighting applications range from clock radios to aircraft and spaceships.

23–15 SWITCH CIRCUITS

The control of one lamp or group of lamps from two locations is accomplished by the use of two three-way switches. The switch in Figure 23–17 is a three-way single-pole, double-throw switch. The movable blade is always in connection with the common terminal. There is no on or off position marked on the switch because the common terminal is always connected to one or the other traveler terminal.

Figure 23–18 shows two three-way switches in a lamp circuit. As shown, the lamp is on. It can be turned off by either switch. When the lamp is off, either switch can again close a circuit to the lamp, with one or the other of the two traveller wires between the switches.

For control of a lamp from three or more locations, a four-way switch has to be put into the circuit between the two three-way switches. A four-way switch has the same effect in a circuit as a reversing switch. The schematic diagram in Figure 23–19 shows lamps controlled from four locations. A and D are three-way switches. B and C are four-way switches, as in Figure 23–20. Operation of switch B converts its internal circuit to that shown in C. Moving the handle of C makes its connections like those drawn for B. Tracing the circuit through the switches in the diagram, one finds that the circuit is open. Operating any one of the four switches will close the circuit.

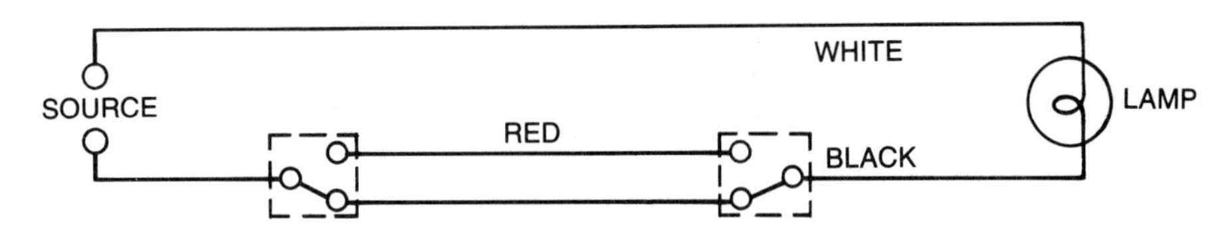

FIGURE 23–18 Circuit with two three-way switch controls

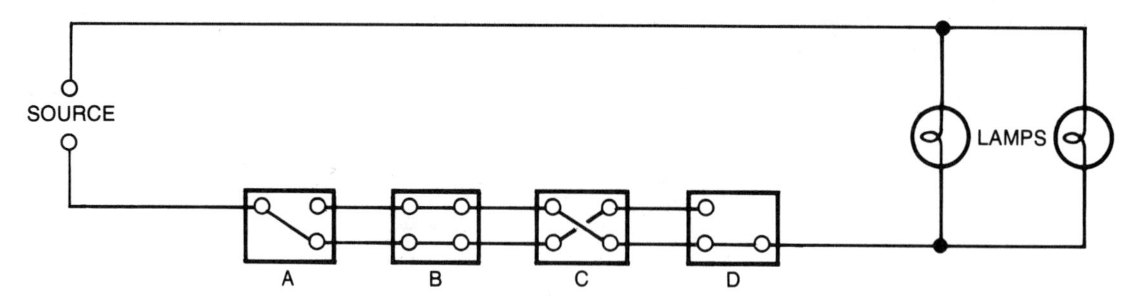

FIGURE 23–19 Four-way switch

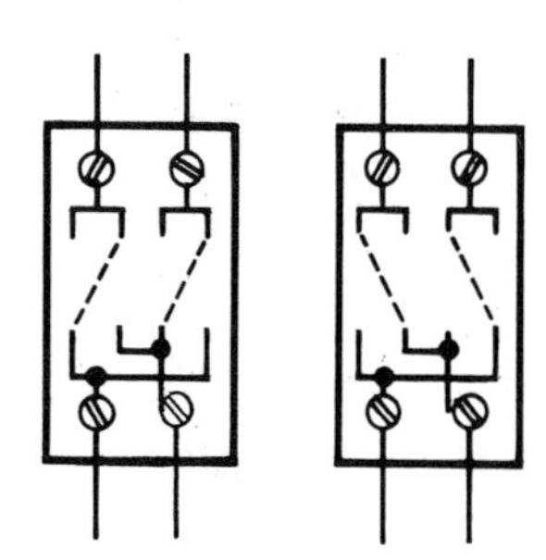

FIGURE 23–20 Two positions of four-way switch

23–16 REMOTE CONTROL SYSTEMS

To control several outlets in a building from each of several different locations, low-voltage operated relays can be installed to turn the 115-volt outlets on or off. The wiring installation for this low-voltage system is less expensive because there is no need to run 115-volt cables to the switches. Low-cost, easily installed #18 or #20 wire is used.

At each controlled outlet, one 24-volt relay is mounted on the outlet box. The relay contains two coils. A momentary current in the *on* coil closes the 115-volt contacts.

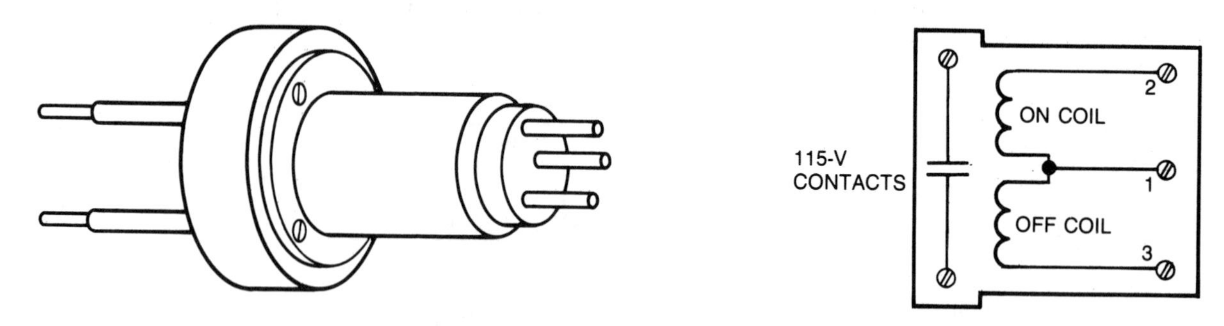

FIGURE 23–21 Relay connections

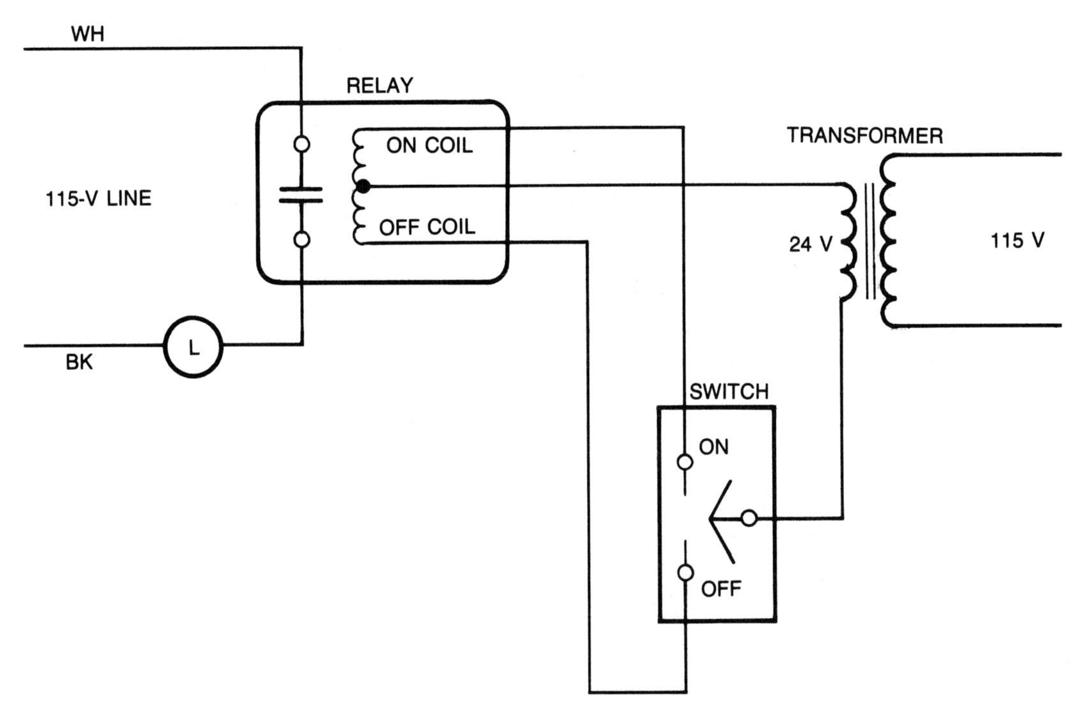

FIGURE 23–22 One light controlled from one switch point

These contacts remain closed until a momentary current in the *off* coil opens them, Figure 23–21.

The relay coils are operated from a 24-volt transformer by normally open momentary contact switches. Several of these switches may be connected in parallel to control one relay. The connection of such a switch in the low-voltage circuit is shown in Figure 23–22.

Low-voltage systems can be expanded to include a variety of automatic lighting controls, timers, and security systems. In some cases the addition of a rectifier in series with the 24-volt transformer output is recommended. The relay coils then operate on half-wave rectified AC.

SUMMARY

- Resistors are inherently 100% efficient as heat producers.
- Heating rate in watts = I^2R.
- The three-wire distribution system conserves energy and copper. Current in the neutral wire equals the difference in the currents of the outside wires.
- Resistance welding uses large currents at low voltages. Arc welding uses moderate current at moderate voltage.

- An induction furnace generates heat-producing alternating current in the metal to be heated.
- Intensity of a source of light is measured in lumens. One candlepower equals 12.57 lumens.
- Intensity of illumination of a surface is measured in lumens per square foot.
- Surface illumination is increased not only by increasing the power of the source but also by reducing the distance between source and surface and by placing reflecting surfaces behind the light source.
- Light is produced by electron disturbances in atoms. Light is one narrow band of vibration frequencies in the entire electromagnetic spectrum.
- Fluorescent lamps and mercury arcs are 2½ to 3 times as efficient as incandescent lamps.
- Large incandescent lamps are generally more efficient than small ones.

Achievement Review

1. Assuming that 10,000 Btu per hour are required to keep a room at 70° when the outdoor temperature is 25°, calculate the necessary equivalent rate of electrical resistance heating in watts.
2. Resistors are connected to a three-wire system, as shown in the sketch below. Calculate the current at each of the lettered points A, B, C, and D.

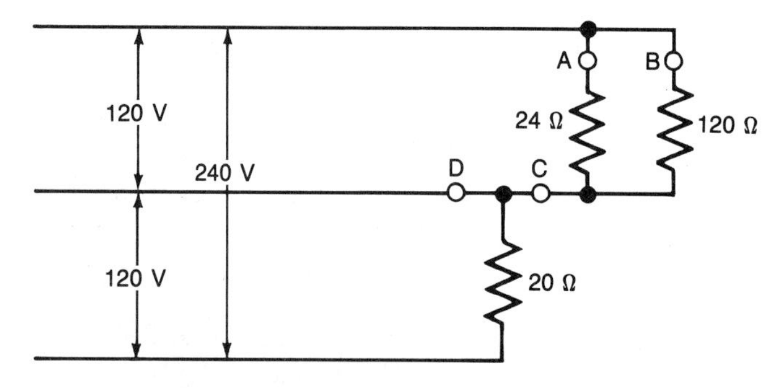

3. Calculate the appropriate amount of nichrome wire for a 440-watt heating element.
4. A certain heater element is rated at 1,600 watts and 220 volts. Calculate its resistance. What would be its power rating in watts if operated at 110 volts?
5. Name two units for measuring the intensity of a light source. State the numerical relationship between them.
6. Name two units of measure for surface illumination.
7. A 25-watt lamp has an efficiency equal to 10.4 lumens per watt. How much light does it produce?
8. A work surface is illuminated by a single incandescent lamp with no reflector placed 5 feet above the surface. If the lamp is lowered to 4 feet above the

surface, the illumination on the work directly below the lamp is increased how many times?

9. State two uses for carbon arc lights.
10. State uses for mercury arcs, sodium arcs, and neon-glow lamps.
11. What is a phosphor?
12. Diagram a simple fluorescent lamp circuit.
13. Diagram a circuit for operating a lamp by switches at three different locations.
14. Under what circumstances are remote control relays appropriate?
15. Shown below are the suggested outlines for five schematic diagrams. Complete these diagrams by drawing the necessary conductors in accordance with the corresponding instructions. Draw neat, straight lines as short as possible.
    a. Three lamps connected *in series* supplied by a battery. A rheostat controls the current (and thereby the light output) of all lamps.

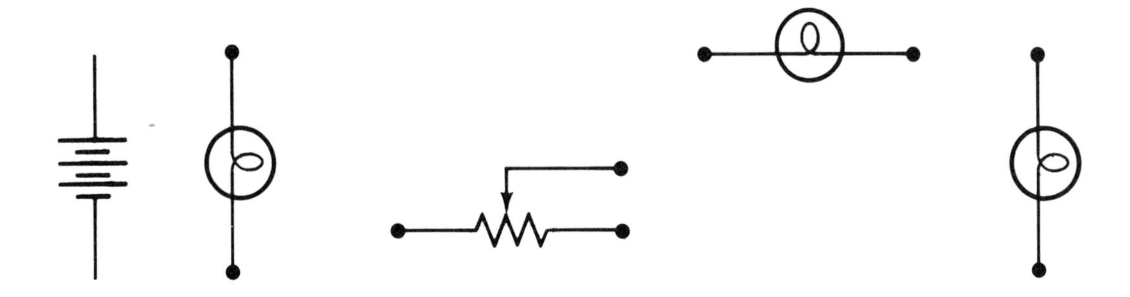

    b. Three lamps connected *in parallel* supplied by a battery. A rheostat is used to dim the second lamp only.

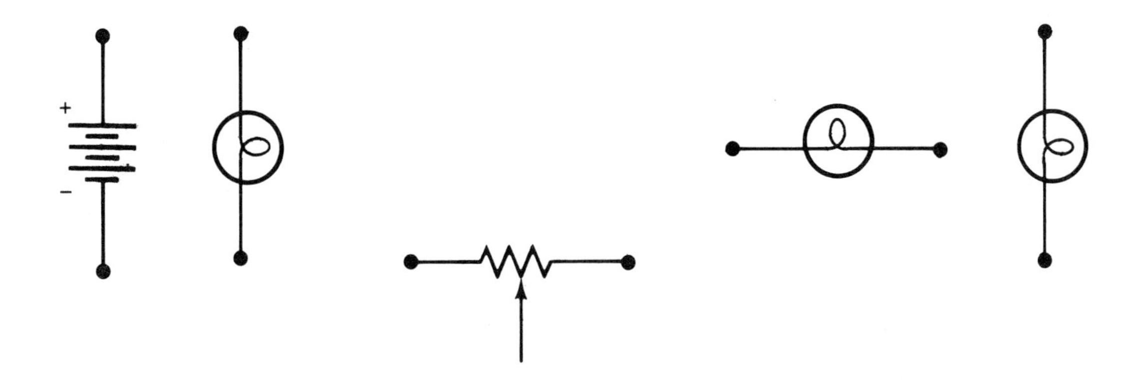

    c. Two motors, supplied from a generator, to be controlled by an SPDT switch, so that only one motor will run at a time. An DPST switch is provided as a master switch to disconnect all power.

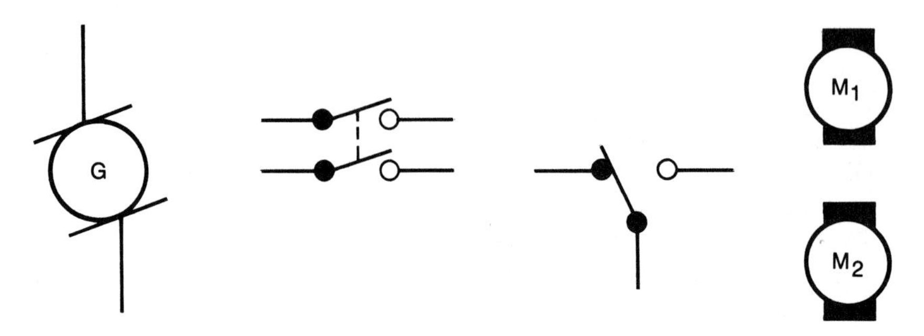

d. An unspecified load resistance to receive variable voltages ranging from 0 to 120 volts from a potentiometer.

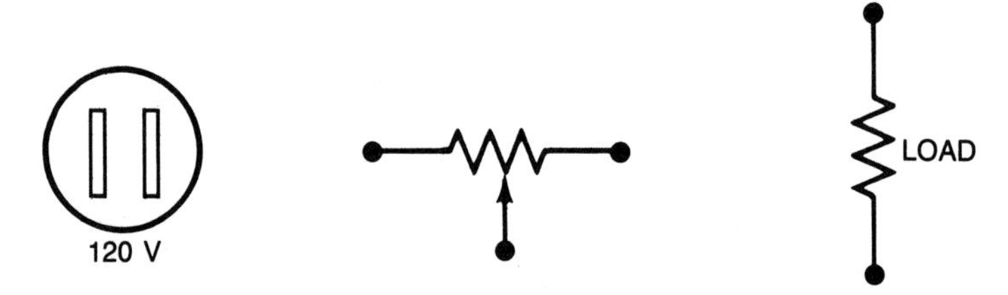

e. Three resistors connected in a series-parallel configuration; R_1 is to be in series with the parallel combination of R_2 and R_3. Two ammeters are shown to read I_T and I_2. The voltmeter is used to read E_1. Provide all meters with polarity markings (+ and −).

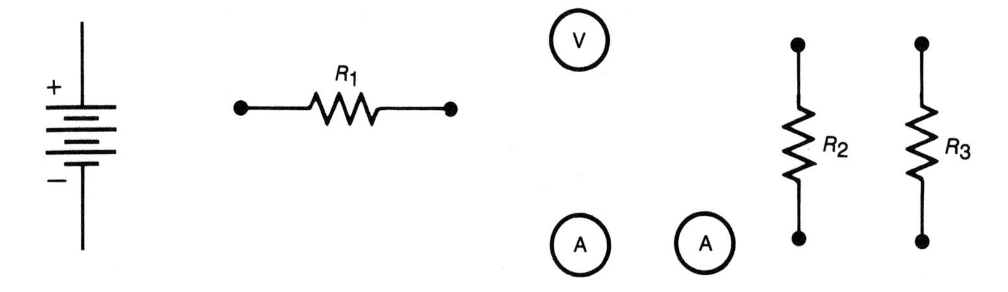

24 Solving DC Networks

Objectives

After studying this chapter, the student should be able to analyze complex circuits by the use of either

- The loop current method
- The superposition theorem
- Thevenin's theorem

A network is any complex arrangement of circuit elements and often involves multiple voltage sources. The use of Ohm's and Kirchhoff's laws alone may not suffice to solve such problems. Analysis of such circuits requires additional techniques that we will explore in this chapter.

Note: It is essential that the student have a firm grip on the concepts covered in Chapter 12. A quick review of Kirchhoff's laws may be in order (see Sections 12–2 and 12–3).

Furthermore, it is assumed that the student can completely solve a system of simultaneous equations with multiple unknowns.

24–1 THE LOOP CURRENT METHOD

This method applies to circuits with a series of interconnected branches forming loops. Such circuits are also known as *mesh circuits*. The circuit in Figure 24–1 is an example of this.

The solution of such circuits requires the following procedure:

1. Identify all the loops in the circuit and identify corresponding loop currents, such as I_1, I_2, and I_3.
2. Assign each current an arbitrary direction.
3. For each loop, write an equation using Kirchhoff's voltage law.

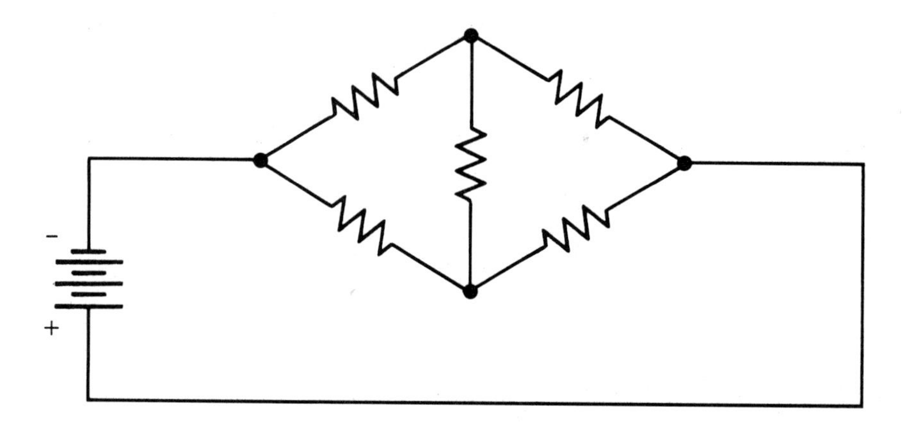

FIGURE 24–1

Note: If a resistor is part of two loops, it will have two different currents flowing in it that may be adding or subtracting each other.

4. Simplify the loop equations by substituting values and by collecting terms.
5. Solve the simultaneous equations for their unknown loop currents.

EXAMPLE 24–1

Given: A Wheatstone bridge circuit as shown in Figure 24–2. (*Note:* Wheatstone bridge circuits are covered in Chapter 17.)

Find: The amount and direction of all branch currents.

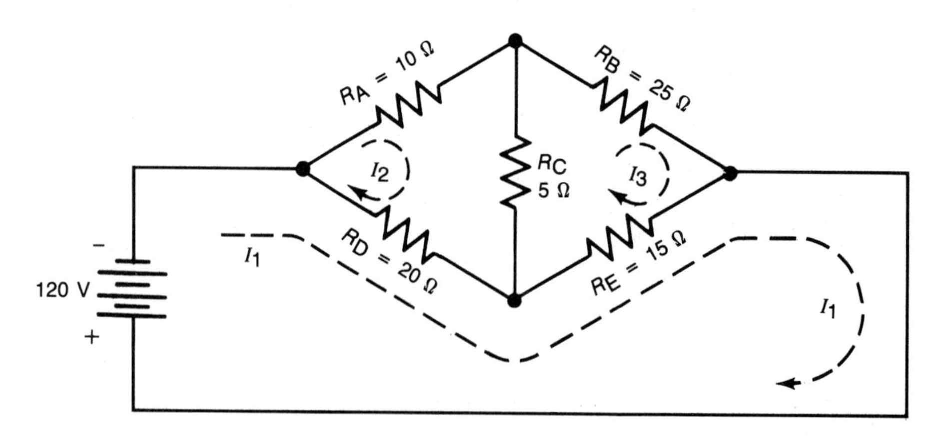

FIGURE 24–2

SOLUTION

Note that we have identified three loop currents as follows:

I_1 flowing clockwise through R_D and R_E
I_2 flowing clockwise through R_D, R_A, and R_C
I_3 flowing clockwise through R_E, R_C, and R_B

Loop 1:

$$R_D(I_1 - I_2) + R_E(I_1 - I_3) = 120$$
$$20(I_1 - I_2) + 15(I_1 - I_3) = 120$$
$$20\ I_1 - 20\ I_2 + 15\ I_1 - 15\ I_3 = 120$$
$$\text{①}\qquad 35\ I_1 - 20\ I_2 - 15\ I_3 = 120$$

Loop 2:

$$R_D(I_2 - I_1) + R_A I_2 + R_C(I_2 - I_3) = 0$$
$$20(I_2 - I_1) + 10\ I_2 + 5(I_2 - I_3) = 0$$
$$20\ I_2 - 20\ I_1 + 10\ I_2 + 5\ I_2 - 5\ I_3 = 0$$
$$\text{②}\qquad -20\ I_1 + 35\ I_2 - 5\ I_3 = 0$$

Loop 3:

$$R_E(I_3 - I_1) + R_C(I_3 - I_2) + R_B I_3 = 0$$
$$15(I_3 - I_1) + 5(I_3 - I_2) + 25\ I_3 = 0$$
$$15\ I_3 - 15\ I_1 + 5\ I_3 - 5\ I_2 + 25\ I_3 = 0$$
$$\text{③}\qquad -15\ I_1 - 5\ I_2 + 45\ I_3 = 0$$

Combining ② and ③ and solving for I_1 yields:

$$I_1 = 2.48\ I_3$$

Substituting the above into ②, we now solve for I_2.

$$-20(2.48\ I_3) + 35\ I_2 - 5\ I_3 = 0$$
$$I_2 = 1.56\ I_3$$

Next, we substitute the above values into ① and solve for I_3.

$$35(2.48\ I_3) - 20(1.56\ I_3) - 15\ I_3 = 120$$
$$I_3 = 2.96\ \text{A}$$

I_3 also $= I_B$

It follows that

$$I_1 = 2.48\ I_3$$
$$I_1 = 2.48(2.96)$$
$$I_1 = 7.34\ \text{A}$$

I_1 also $= I_{LINE}$

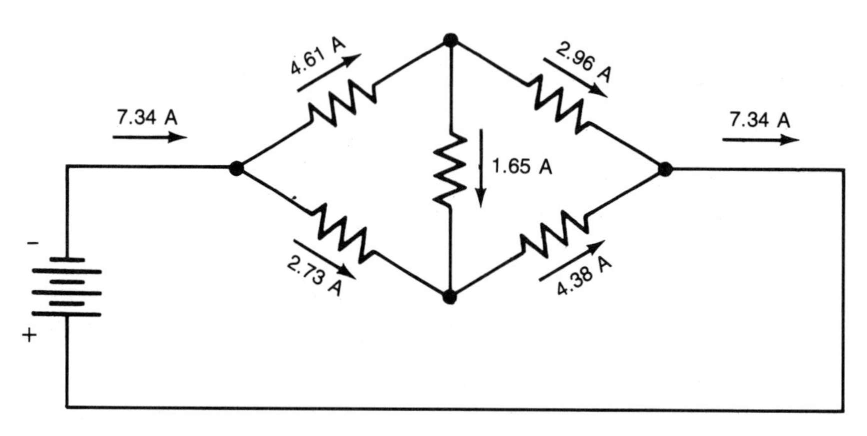

FIGURE 24–3

Likewise

$$I_2 = 1.56(I_3)$$
$$I_2 = 1.56\,(2.96)$$
$$I_2 = 4.61\text{A}$$
$$I_2 \text{ also } = I_A$$

$$I_C = I_2 - I_3$$
$$I_C = 4.61 - 2.96$$
$$I_C = 1.65\text{A}$$

$$I_D = I_1 - I_2$$
$$I_D = 7.34 - 4.61$$
$$I_D = 2.73\text{A}$$

$$I_E = I_C + I_D$$
$$I_E = 1.65 + 2.73$$
$$I_E = 4.38\text{A}$$

Figure 24–3 shows us what the circuit looks like once all currents are assigned.

EXAMPLE 24–2

Given: The circuit shown in Figure 24–4.

Find: All the branch currents, using the loop method.

SOLUTION

All components in this circuit are contained within two loops; therefore, this problem can be solved by using two loop currents only.

The loop currents I_1 and I_2 have been arbitrarily defined as shown in the drawing.

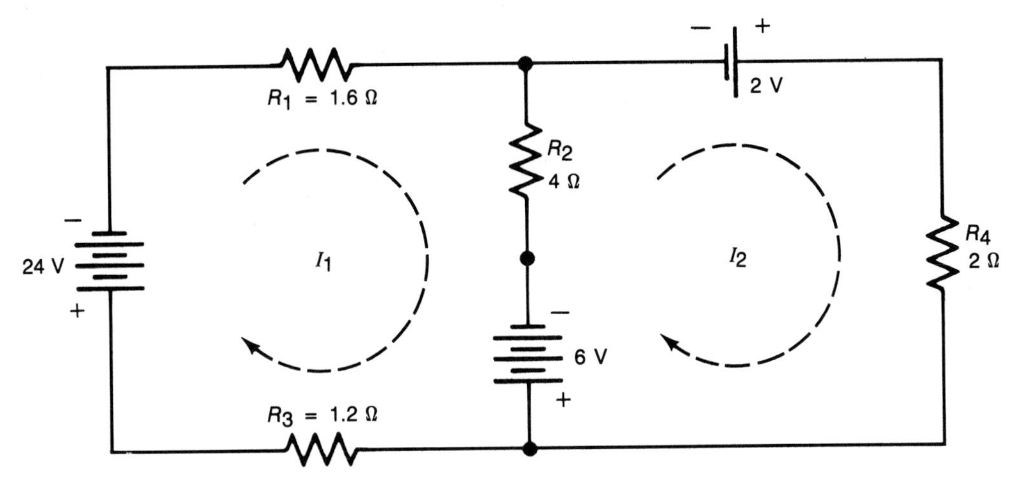

FIGURE 24–4

Loop 1:

$$24\text{ V} = I_1R_1 + (I_1 - I_2)R_2 + 6 + I_1R_3$$
$$24\text{ V} = 1.6\,I_1 + 4\,I_1 - 4\,I_2 + 6 + 1.2\,I_1$$
① $$18\text{ V} = 6.8\,I_1 - 4\,I_2$$

Loop 2:

$$6 = (I_2 - I_1)4 + 2 + 2\,I_2$$
$$6 = 4\,I_2 - 4\,I_1 + 2 + 2\,I_2$$
② $$4 = -4\,I_1 + 6\,I_2$$

Combining ① and ②, we obtain

$$I_1 = 5\text{ A}$$

Substituting the above result into ② and solving for I_2

$$4 = -4(5) + 6I_2$$
$$6\,I_2 = 24$$
$$I_2 = 4\text{ A}$$

It is suggested that you, the student, prove the results to yourself by computing the voltage drops and inserting the answers, with their respective polarities, into the drawing of Figure 24–4. Then, use Kirchhoff's voltage law to confirm your results.

24–2 THE SUPERPOSITION THEOREM

In a circuit with multiple power sources, each power source contributes toward the current flow through the resistors. The *superposition theorem* is developed on the principle that we can determine just how much each power source contributes toward the branch currents. This is done by considering only one power source at a time. For this purpose, the remaining power sources are replaced with a short-circuiting conductor. (To simplify our explanation, we assume that the internal resistance of the power source = 0.)

This analysis is done for each power source, one at a time, and the resulting currents are then algebraically combined to determine the magnitude and direction of each current flow. This can be visualized by superpositioning the results in the original schematic diagram.

The following example will illustrate the procedure.

EXAMPLE 24–3

Given: The circuit shown in Figure 24–5.

Find: The magnitude and direction of all the currents flowing through the resistors.

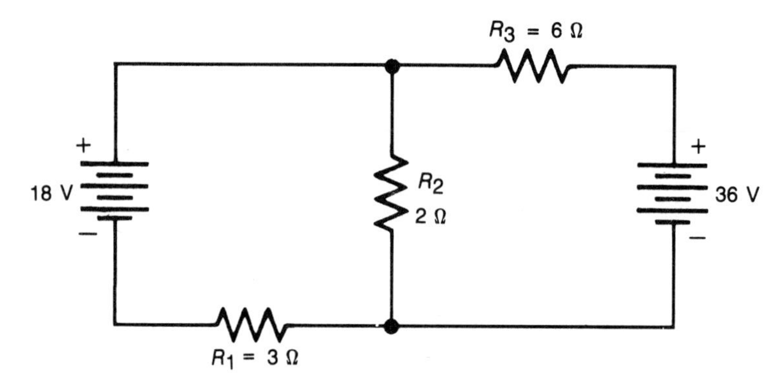

FIGURE 24–5

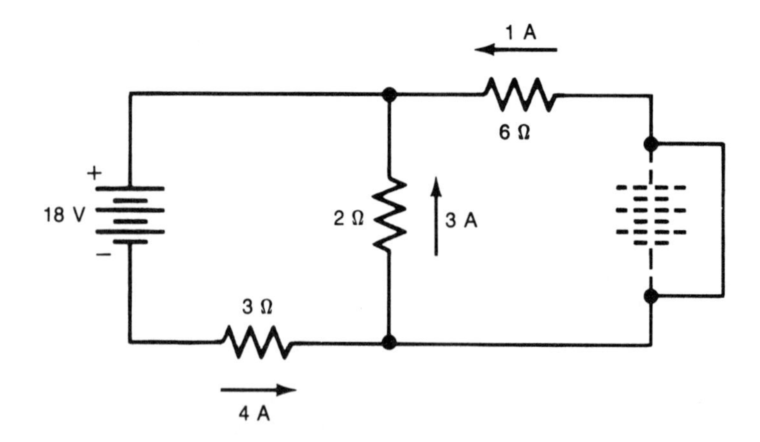

FIGURE 24–6

SOLUTION

1. Omit the 36-volt power source by replacing it with a short circuit.
   R_T = 4.5 ohms. Thus, the battery delivers 4 amperes, distributed as shown in Figure 24–6.
2. Omit the 18-volt battery by replacing it with a short circuit.
   R_T = 7.2 ohms. Thus, the battery delivers 5 amperes, distributed as shown in Figure 24–7.
3. Superposition your results in the original drawing, Figure 24–8.

Inspection of these results will suggest the remainder of the solution. The currents through each resistor must be added algebraically. In other words, currents flowing in the same direction are being added, and currents flowing in opposite directions are being subtracted. The final results of our example are shown in Figure 24–9.

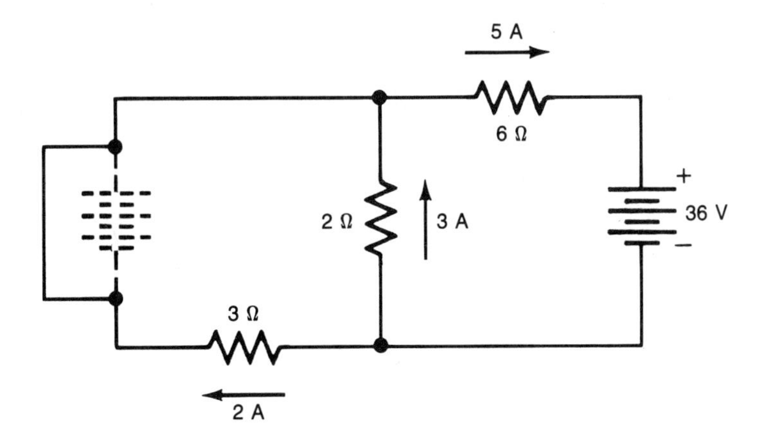

FIGURE 24–7

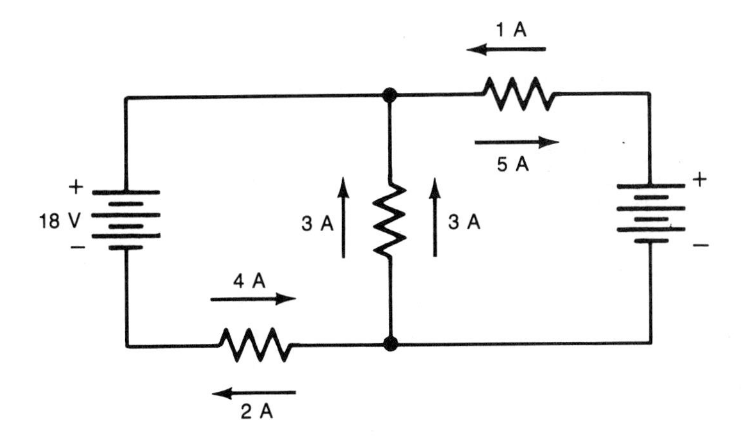

FIGURE 24–8

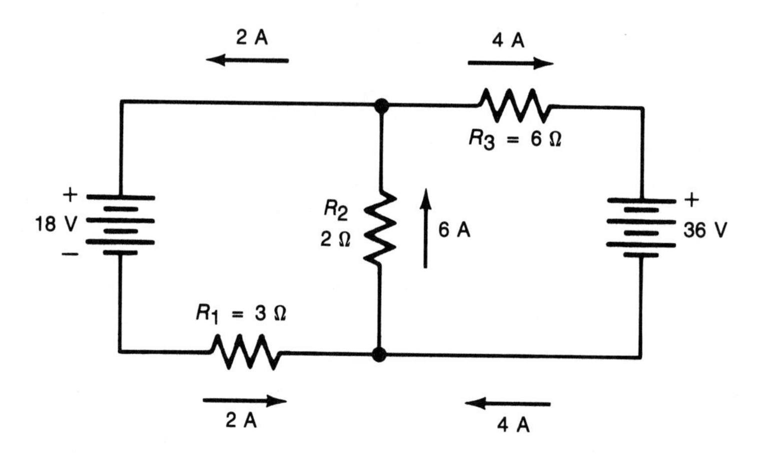

FIGURE 24–9

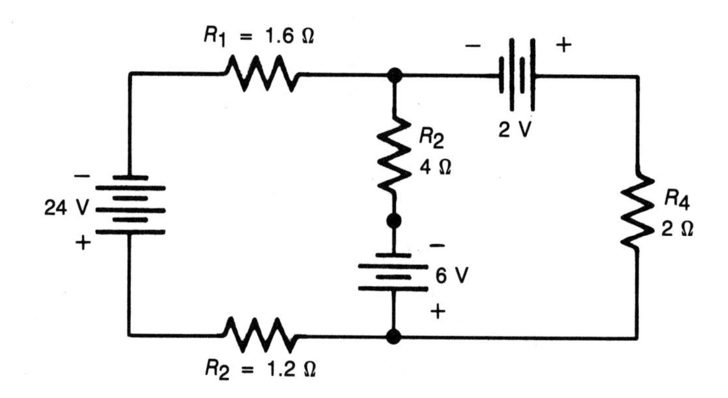

FIGURE 24–10

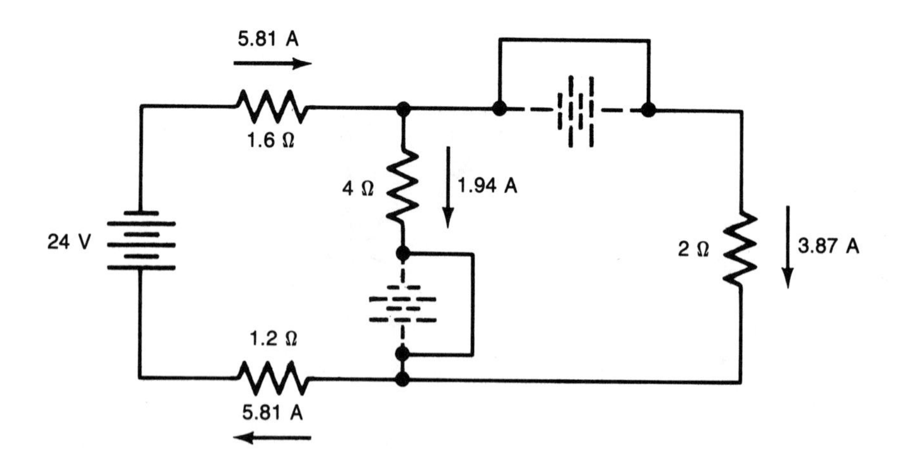

FIGURE 24–11

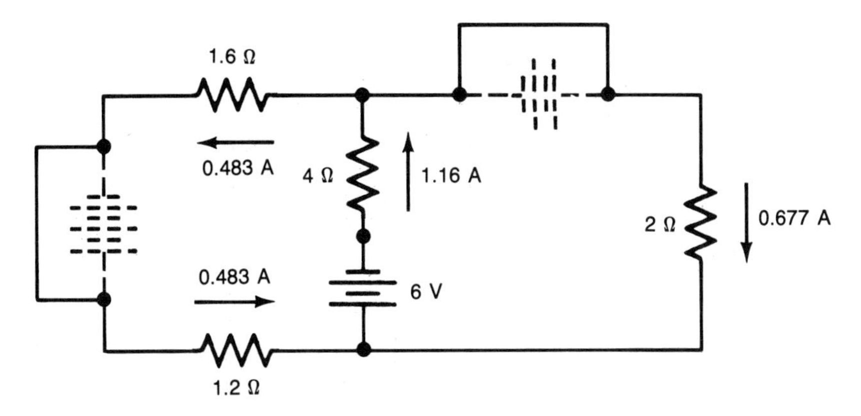

FIGURE 24–12

EXAMPLE 24–4

Given: The circuit shown in Figure 24–10.

Find: The magnitude and direction of all branch currents, using the superposition theorem.

SOLUTION

1. Eliminate the 6-volt and 2-volt batteries by replacing them with a short circuit. R_T = 4.133 ohms. Thus, the battery delivers 5.81 amperes, distributed as shown in Figure 24–11.
2. Omit the 24-volt and the 2-volt batteries and find the current due to the 6-volt power source only, Figure 24–12. Redraw the circuit to help you solve for R_T and

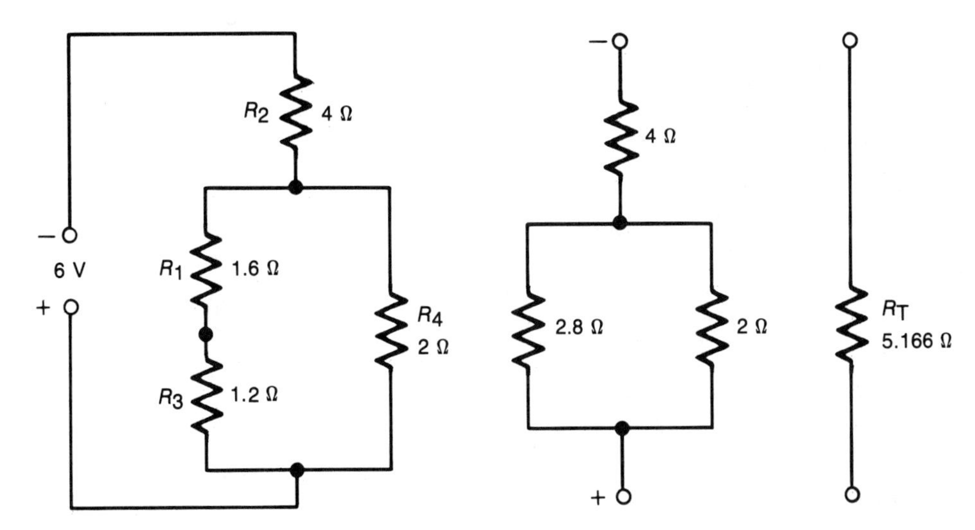

FIGURE 24–13

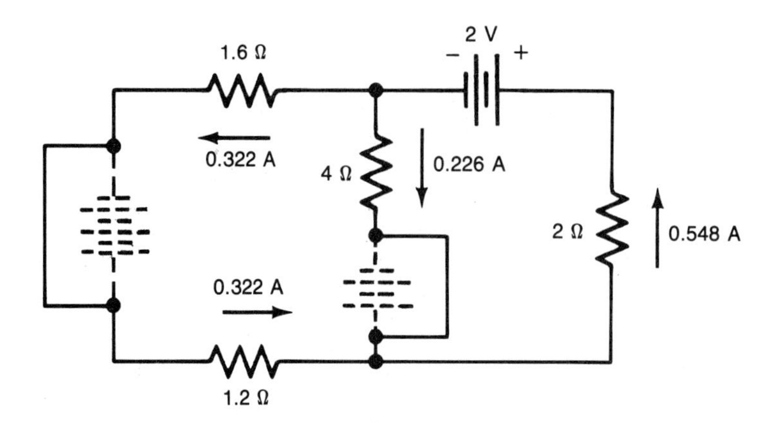

FIGURE 24–14

the branch currents. R_T = 5.166 ohms, and the current distribution is shown in Figure 24–13.

3. Next, determine the branch currents when only the 2-volt source is in the circuit, Figure 24–14. R_T = 3.647 ohms. Thus, the current distribution is as shown in Figure 24–14.
4. The results of the three preceding steps are now incorporated into the original circuit drawing, Figure 24–15. We can now algebraically combine these results to find the net current through each resistor. The final results are graphically presented in Figure 24–16.

Note: In case you have not noticed, this is the identical circuit we solved before by the loop current method in Example 24–2.

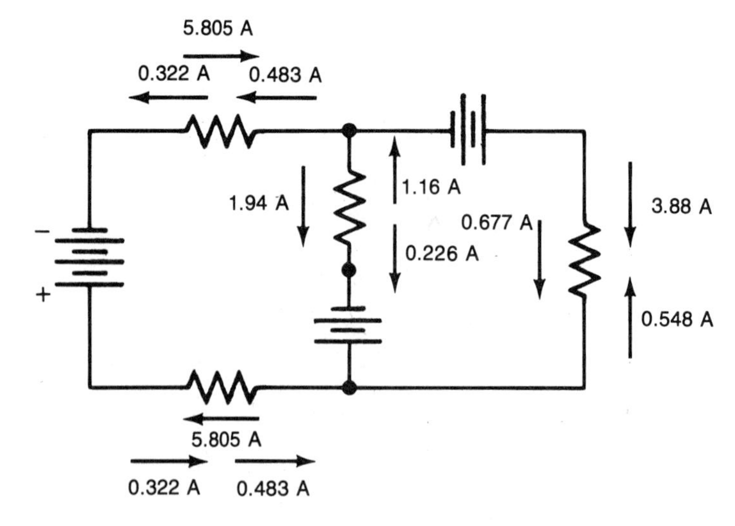

FIGURE 24–15

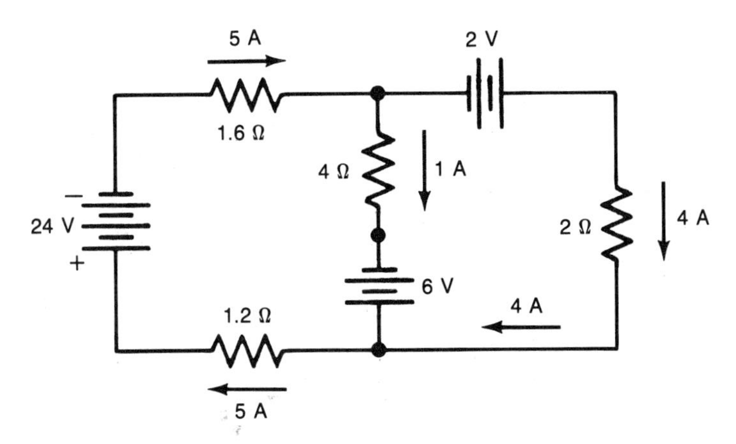

FIGURE 24–16

24–3 THEVENIN'S THEOREM

Occasionally, the electronics technician is confronted by a complex network of emfs and resistors for the purpose of determining the electrical data for only one of the resistors.

For example, let us consider the network shown in Figure 24–17 and assume that we need to know more specific information about the resistor called R_L.

We can simplify our task by reducing the rest of the network to a single equivalent series circuit of one emf and one resistance, known as Thevenin's voltage E_{TH} and Thevenin's resistance R_{TH}.

In other words, we can pretend to place all components, except R_L, into a black box and bring out only the two terminals labelled A and B. The contents within the black box may then be "thevenized," that is, reduced to the simple equivalent circuit as shown in Figure 24–18.

The following example is designed to show how this can be done.

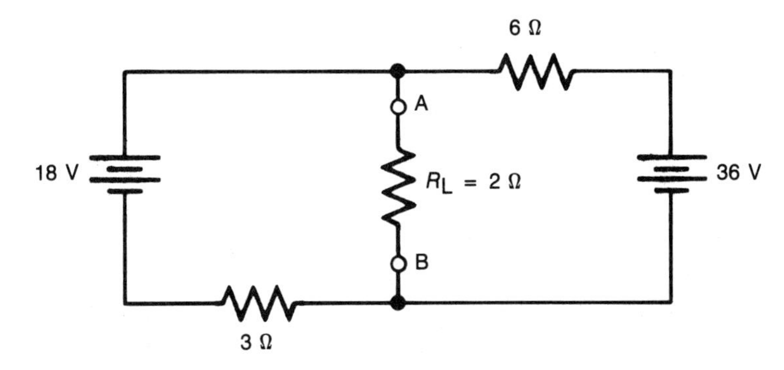

FIGURE 24–17

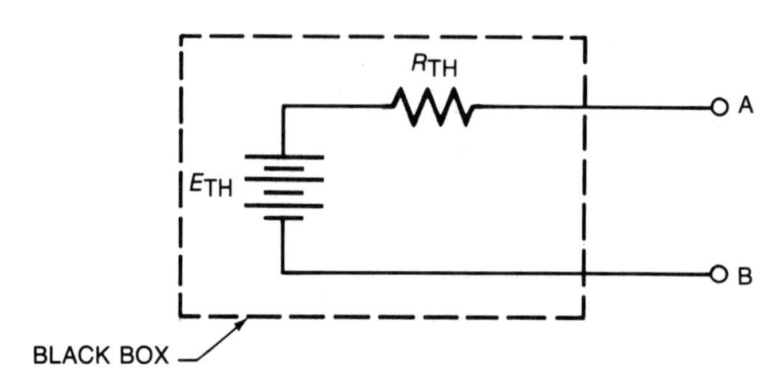

FIGURE 24–18

EXAMPLE 24–5

Given: The circuit shown in Figure 24–17.

Find: The magnitude and direction of the current through R_L.

SOLUTION

1. We thevenize the circuit by the black box principle, and determine the voltage at points A and B, which is Thevenin's voltage E_{TH}. This can be done in the following manner:
   a. Rearrange the circuit without R_L, as shown in Figure 24–19.
   b. Then, if the circuit actually exists, measure the voltage with a high-impedance voltmeter, or
   c. calculate the voltage drop by Kirchhoff's voltage law, as follows: The two voltage sources have subtractive polarity and furnish a combined voltage of 18 volts to the loop containing 9 ohms. This results in a 2-ampere loop current with the resultant voltage drops as shown in Figure 24–19. Consequently, $E_{TH} = 36 - 12$ (or $18 + 6$) $= 24$ volts.
2. Next, determine the resistance within the black box. This is known as Thevenin's resistance, R_{TH}. To do this, first replace the voltage sources with their internal

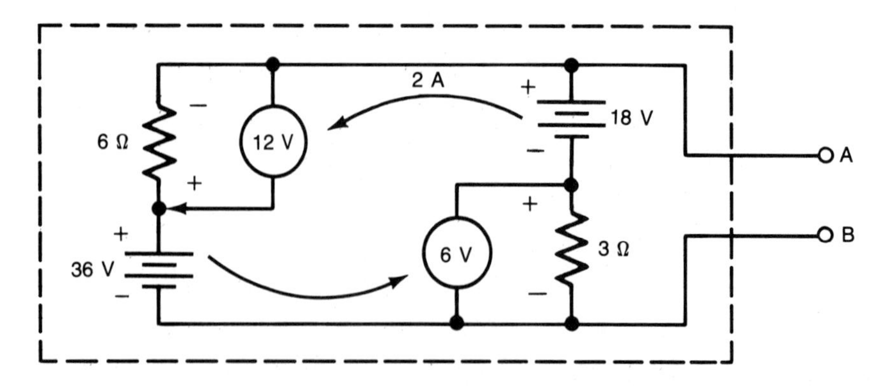

FIGURE 24–19

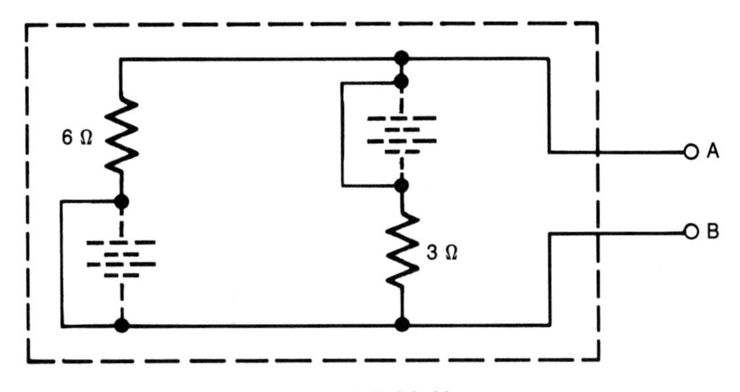

FIGURE 24–20

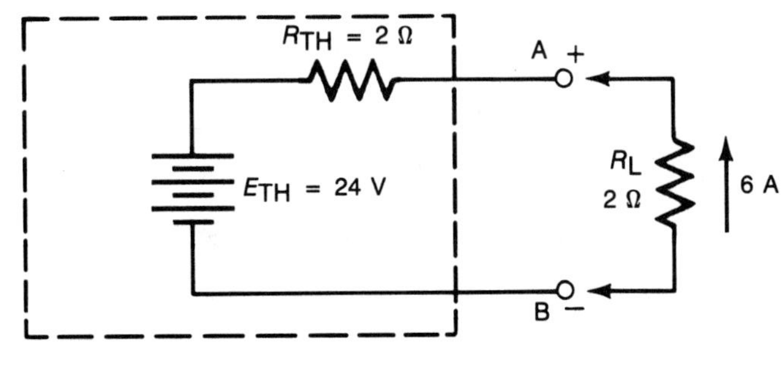

FIGURE 24–21

resistance. If the voltage sources are considered to be ideal (their internal resistance equals 0), simply replace them with a short circuit, Figure 24–20. Now we have two choices.

a. If the circuit actually exists, apply an ohmmeter at points A and B.

b. Otherwise, calculate the resistance within the black box. In this case

$$R_{TH} = \frac{3 \times 6}{3 + 6}$$

$$R_{TH} = 2\ \Omega$$

3. The results from steps 1 and 2 can now be used to show us the equivalent circuit within the black box, Figure 24–21.
When we apply the load resistor ($R_{TH} = 2\ \Omega$) to terminals A and B, it should become obvious that a current of 6 amperes is flowing from point B to point A. Compare this answer with that of Figure 24–9.

Let us reinforce what we have learned by thevenizing one more circuit.

EXAMPLE 24–6

Given: The circuit shown in Figure 24–22.

Find: The magnitude and direction of the current through R_L by thevenizing the network.

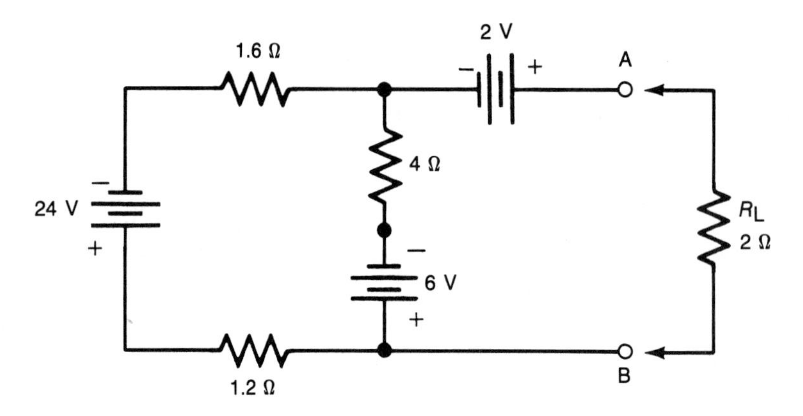

FIGURE 24–22

SOLUTION

1. Rearranging the network within the black box yields the circuit conditions as shown in Figure 24–23.
   The loop current is

   $$\frac{24\text{ V} - 6\text{ V}}{6.8} = \frac{18}{6.8} = 2.647\text{ A}$$

   Therefore, the voltage drop across the 4-ohm resistor is equal to 10.588 volts. Hence, the voltage from A to B, Thevenin's voltage, equals

   $$6\text{ V} + 10.588\text{ V} - 2\text{ V} = 14.588\text{ V} = E_{TH}$$

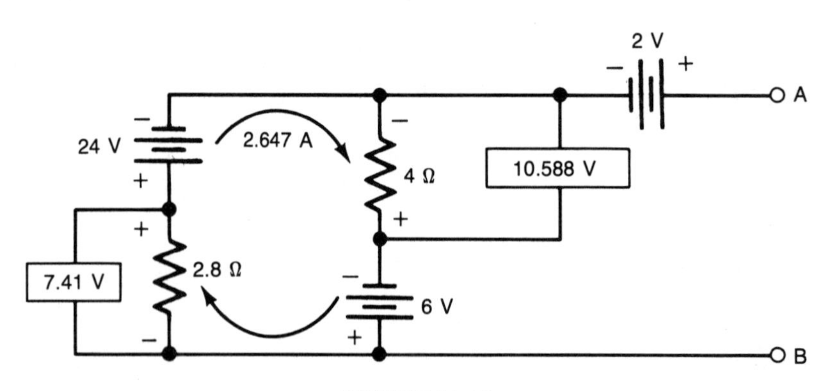

FIGURE 24–23

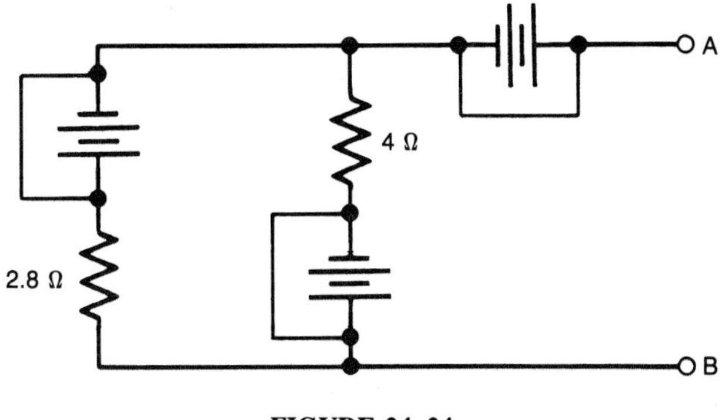

FIGURE 24–24

2. Short-circuiting the voltage sources to find R_{TH}, we obtain the following arrangement, Figure 24–24.

$$R_{TH} = \frac{2.8 \times 4}{2.8 + 4} = 1.647\ \Omega$$

3. Thevenin's equivalent circuit (in the black box) is shown in Figure 24–25. The current through R_L can now be computed as follows:

$$I_L = \frac{E_{TH}}{R_{TH} + R_L}$$

$$I_L = \frac{14.588}{3.647}$$

$$I_L = 4\ \text{A}$$

The direction of the current is *from A to B*.

Note: In case you did not notice, this is the identical circuit we have solved twice before, first by the *loop current* method in Example 24–2 and, secondly, by use of the *superposition theorem* in Example 24–4.

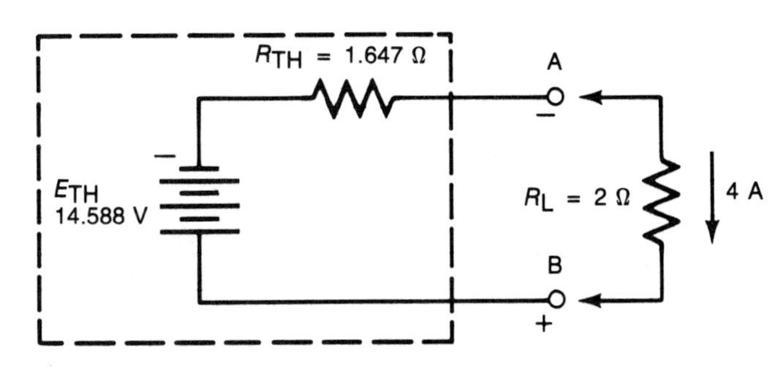

FIGURE 24–25

SUMMARY

Three different methods of analyzing complex mesh circuits have been under discussion, namely

1. The loop current method, involving Kirchhoff's law and the solution of simultaneous equations
2. The superposition theorem
3. Thevenin's theorem

The Loop Current Method

1. Label all terminal points with letters to identify the various paths.
2. Indicate the fixed polarity of all emf sources.
3. Assign a current to each branch of the network. An arbitrary direction may be given to each current.
4. Show the polarity of the voltage developed across each resistor as determined by the direction of the assumed currents.
5. Apply Kirchhoff's voltage law around each closed loop. This step yields simultaneous equations, one for each loop.
6. Apply Kirchhoff's current law at a node that includes all of the branch currents of the network. This step gives one more simultaneous equation.
7. Solve the resulting simultaneous equations for the assumed branch currents.
8. If the solution of the equations yields positive current values, then the directions assumed in step 3 are the actual directions. If any current value is negative, then the actual direction of the current is opposite to the assumed direction. It is not necessary to resolve the network using the true direction of current.

The Superposition Theorem

The superposition theorem simplifies the analysis of networks having more than one emf source. The theorem is stated as follows:

> In any network containing more than one source of emf, the current through any branch is the algebraic sum of the currents produced by each source acting independently.

This theorem is applied in the following steps:

1. Select one source of emf. Replace all of the other sources with their internal resistances. If the internal resistance is 0, replace the source with a short circuit.

2. Calculate the magnitude and direction of the current in each branch due to the source of emf acting alone.
3. Repeat steps 1 and 2 for each source of emf until the branch current components are calculated for all sources.
4. Find the algebraic sum of the component currents to obtain the true magnitude and direction of each branch current.

Thevenin's Theorem

Thevenin's theorem states that any two terminal network containing resistances and sources of emf may be replaced by a single source of emf in series with a single resistance. The emf of the single source of emf, called E_{TH}, is the open circuit emf at the network terminal. The single series resistance, called R_{TH}, is the resistance between the network terminals when all of the sources are replaced by their internal resistances.

When the Thevenin equivalent circuit is determined for a network, the process is known as *thevenizing* the circuit. Thevenin's theorem is applied according to the following procedure:

1. Remove the load resistor and calculate the open circuit terminal voltage of the network. This value is E_{TH}.
2. Redraw the network with each source of emf replaced by a short circuit in series with its internal resistance.
3. Calculate the resistance of the redrawn network as seen by looking back into the network from the output terminals. This value is R_{TH}.
4. Draw the Thevenin equivalent circuit. This circuit consists of the series combination of E_{TH} and R_{TH}.
5. Connect the load resistor across the output terminals of the series circuit. This Thevenin circuit is equivalent to the original network. The advantage of thevenizing a circuit is that the load resistor can be varied without changing the Thevenin equivalent circuit. Changes in R_L affect only the simple series Thevenin circuit. Therefore, the load current can be calculated easily for any number of values of R_L.

Achievement Review

For questions 1 through 4, calculate the magnitude and direction of the current through the resistor labeled R_L. For added practice try any of the methods described in this chapter.

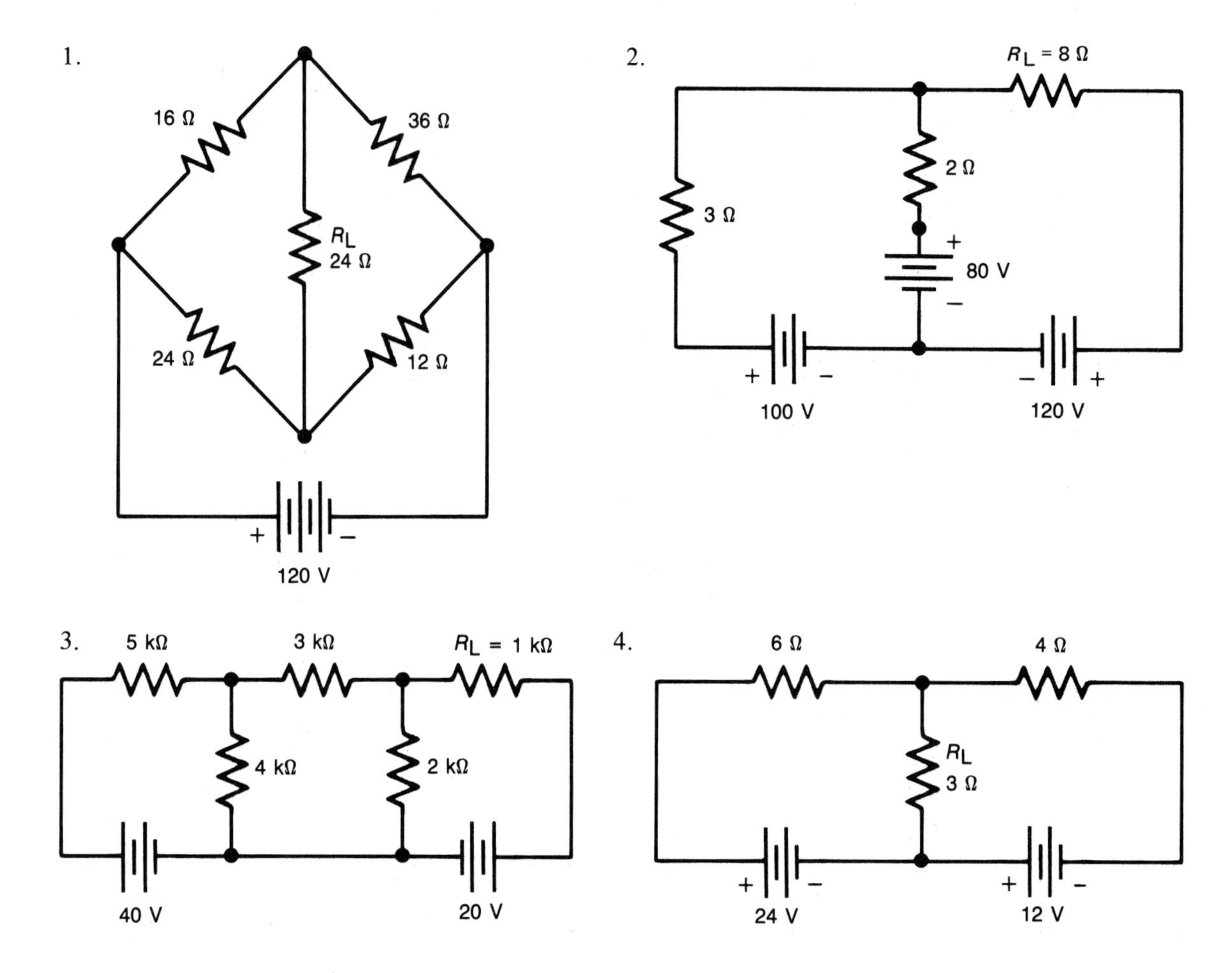

5. Determine the total resistance of the circuit as seen by the voltage source.

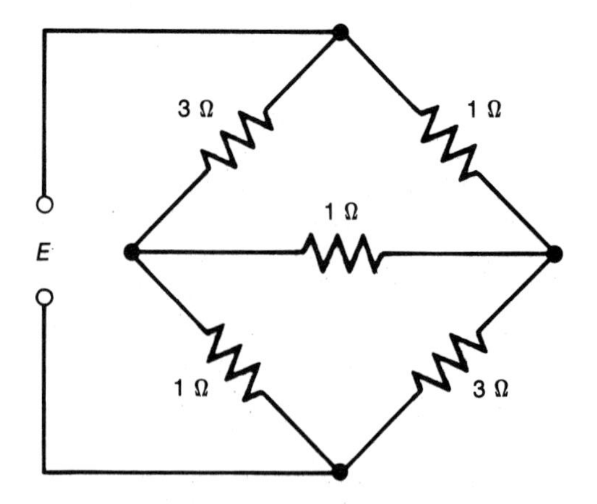

Appendix

RESISTANCE OF METALS AND ALLOYS				
Properties of Materials	**Resistivity in Microhm-cm at 20°C**	**Temp. Coeff. of Resistance**	**Melting Point (0°C)**	**Specific Gravity**
Aluminum	2.83	0.004	659	2.67
Alumel	33.3	0.0012		
Aluminum bronze (90% Cu, 10% Al)	12.6–12.7	0.003	1 050	7.5
Brass (various comp.)	6.2–8.3	0.002	880–1 050	8.4–8.8
Bronze (88% Cu, 12% tin)	18	0.0005	1 000	8.8
Bronze, commercial (CU, Zn)	4.2	0.002	1 050	8.8
Carbon, amorphous	3 500–4 100	−0.0005	3 500	1.85
Carbon, graphite	720–1 000 (for furnace electrodes at 2 500°C)			
Chromel (Ni, Cr)	70–110	0.0001	1 350	8.3–8.5
Constantan (60% Cu, 40% Ni)	44–49	0.0000	1 190	8.9
Copper, annealed	1.724–1.73	0.004	1 083	8.89
Copper, hard-drawn	1.77	0.0039		
German silver (Cu, Zn, 18% Ni)	33	0.004	1 100	8.4
Gold	2.44	0.0034	1 063	19.3

FIGURE A–1

continued

RESISTANCE OF METALS AND ALLOYS—CONT'D				
Properties of Materials	**Resistivity in Microhm-cm at 20°C**	**Temp. Coeff. of Resistance**	**Melting Point (0°C)**	**Specific Gravity**
Invar (65% Fe, 35% Ni)	81		1 495	8
Iron & Steel Iron, pure	10	0.005	1 530	7.8
Iron, cast	60			
Soft steel	15.9	0.0016	1 510	7.8
Steel, glass-hard	45.7			
Steel, 4% silicon	51			
Steel, transformer	11.09			
Lead	22	0.00387	327	11.4
Magnesium	4.6	0.004	651	1.74
Manganin (84% Cu, 12% Mn, 4% Ni)	44–48	0.0000	910	8.4
Mercury	95.7	0.0008	−39	13.6
Monel (Ni, Cu)	42.5	0.00019	1 300	8.9
Nickel	7	0.006	1 452	8.9
Nichrome (Ni, Fe, Cr)	90–112	0.00017	1 350–1 500	8.25
Nichrome V (Ni, Cr)	108	0.00017		8.41
Platinum	11.5	0.003	1 755	21.4
Silver	1.628	0.0038	960	10.5
Sodium	4.4	0.004		

FIGURE A–1 —CONT'D

RESISTANCE OF METALS AND ALLOYS—CONT'D				
Properties of Materials	**Resistivity in Microhm-cm at 20°C**	**Temp. Coeff. of Resistance**	**Melting Point (0°C)**	**Specific Gravity**
Tin	11.5	0.004	232	7.3
Tungsten	5.51	0.0045	3 410	18.8
at 1730°C	60.			
at 2760°C	100.			
Zinc	6	0.004	419	7.1

Resistivity in microhm-cm is the resistance of a cube with sides of 1 cm each, in millionths of an ohm.
To obtain ohms per mil-foot, multiply the resistivity figure as given above by 6.015.

FIGURE A–1 —CONT'D

AMERICAN WIRE GAUGE TABLE							
			Ohms per 1 000 Ft. (ohms per 100 m)			Pounds per 1 000 Ft. (kg per 100 m)	
B & S Gauge No.	Diam. in Mils	Area in Circular Mils	Copper* 68°F (20°C)	Copper* 167°F (75°C)	Aluminum 68°F (20°C)	Copper	Aluminum
0000	460	211 600	0.049 (0.016)	0.0596 (0.0195)	0.0804 (0.0263)	640 (95.2)	195 (29.0)
000	410	167 800	0.0618 (0.020)	0.0752 (0.0246)	0.101 (0.033)	508 (75.5)	154 (22.9)
00	365	133 100	0.078 (0.026)	0.0948 (0.031)	0.128 (0.042)	403 (59.9)	122 (18.1)
0	325	105 500	0.0983 (0.032)	0.1195 (0.0392)	0.161 (0.053)	320 (47.6)	97 (14.4)
1	289	83 690	0.1239 (0.0406)	0.151 (0.049)	0.203 (0.066)	253 (37.6)	76.9 (11.4)
2	258	66 370	0.1563 (0.0512)	0.190 (0.062)	0.526 (0.084)	201 (29.9)	61.0 (9.07)
3	229	52 640	0.1970 (0.0646)	0.240 (0.079)	0.323 (0.106)	159 (23.6)	48.4 (7.20)
4	204	41 740	0.2485 (0.0815)	0.302 (0.099)	0.408 (0.134)	126 (18.7)	38.4 (5.71)
5	182	33 100	0.3133 (0.1027)	0.381 (0.125)	0.514 (0.168)	100 (14.9)	30.4 (4.52)
6	162	26 250	0.395 (1.29)	0.481 (0.158)	0.648 (0.212)	79.5 (11.8)	24.1 (3.58)
7	144	20 820	0.498 (0.163)	0.606 (0.199)	0.817 (0.268)	63.0 (9.37)	19.1 (2.84)
8	128	16 510	0.628 (0.206)	0.764 (0.250)	1.03 (0.338)	50.0 (7.43)	15.2 (2.26)
9	114	13 090	0.792 (0.260)	0.963 (0.316)	1.30 (0.426)	39.6 (5.89)	12.0 (1.78)
10	102	10 380	0.999 (0.327)	1.215 (0.398)	1.64 (0.538)	31.4 (4.67)	9.55 (1.42)
11	91	8 234	1.260 (0.413)	1.532 (0.502)	2.07 (0.678)	24.9 (3.70)	7.57 (1.13)
12	81	6 530	1.588 (0.520)	1.931 (0.633)	2.61 (0.856)	19.8 (2.94)	6.00 (0.89)

FIGURE A–2

continued

AMERICAN WIRE GAUGE TABLE—CONT'D

B & S Gauge No.	Diam. in Mils	Area in Circular Mils	Ohms per 1 000 Ft. (ohms per 100 m)			Pounds per 1 000 Ft. (kg per 100 m)	
			Copper* 68°F (20°C)	Copper* 167°F (75°C)	Aluminum 68°F (20°C)	Copper	Aluminum
13	72	5 178	2.003 (0.657)	2.44 (0.80)	3.29 (1.08)	15.7 (2.33)	4.8 (0.71)
14	64	4 107	2.525 (0.828)	3.07 (1.01)	4.14 (1.36)	12.4 (1.84)	3.8 (0.56)
15	57	3 257	3.184 (1.043)	3.98 (1.27)	5.22 (1.71)	9.86 (1.47)	3.0 (0.45)
16	51	2 583	4.016 (1.316)	4.88 (1.60)	6.59 (2.16)	7.82 (1.16)	2.4 (0.36)
17	45.3	2 048	5.06 (1.66)	6.16 (2.02)	8.31 (2.72)	6.20 (0.922)	1.9 (0.28)
18	40.3	1 624	6.39 (2.09)	7.77 (2.55)	10.5 (3.44)	4.92 (0.731)	1.5 (0.22)
19	35.9	1 288	8.05 (2.64)	9.79 (3.21)	13.2 (4.33)	3.90 (0.580)	1.2 (0.18)
20	32.0	1 022	10.15 (3.33)	12.35 (4.05)	16.7 (5.47)	3.09 (0.459)	0.94 (0.14)
21	28.5	810	12.8 (4.2)	15.6 (5.11)	21.0 (6.88)	2.45 (0.364)	0.745 (0.110)
22	25.4	642	16.1 (5.3)	19.6 (6.42)	26.5 (8.69)	1.95 (0.290)	0.591 (0.09)
23	22.6	510	20.4 (6.7)	24.8 (8.13)	33.4 (10.9)	1.54 (0.229)	0.468 (0.07)
24	20.1	404	25.7 (8.4)	31.2 (10.2)	42.1 (13.8)	1.22 (0.181)	0.371 (0.05)
25	17.9	320	32.4 (10.6)	39.4 (12.9)	53.1 (17.4)	0.97 (0.14)	0.295 (0.04)
26	15.9	254	40.8 (13.4)	49.6 (16.3)	67.0 (22.0)	0.77 (0.11)	0.234 (0.03)
27	14.2	202	51.5 (16.9)	62.6 (20.5)	84.4 (27.7)	0.61 (0.09)	0.185 (0.03)
28	12.6	160	64.9 (21.3)	78.9 (25.9)	106 (34.7)	0.48 (0.07)	0.147 (0.02)
29	11.3	126.7	81.8 (26.8)	99.5 (32.6)	134 (43.9)	0.384 (0.06)	0.117 (0.02)
30	10.0	100.5	103.2 (33.8)	125.5 (41.1)	169 (55.4)	0.304 (0.04)	0.092 (0.01)
31	8.93	79.7	130.1 (42.6)	158.2 (51.9)	213 (69.8)	0.241 (0.04)	0.073 (0.01)
32	7.95	63.2	164.1 (53.8)	199.5 (65.4)	269 (88.2)	0.191 (0.03)	0.058 (0.01)
33	7.08	50.1	207 (68)	252 (82.6)	339 (111)	0.152 (0.02)	0.046 (0.01)
34	6.31	39.8	261 (86)	317 (104)	428 (140)	0.120 (0.02)	0.037 (0.01)
35	5.62	31.5	329 (108)	400 (131)	540 (177)	0.095 (0.01)	0.029
36	5.00	25.0	415 (136)	505 (165)	681 (223)	0.076 (0.01)	0.023
37	4.45	19.8	523 (171)	636 (208)	858 (281)	0.0600 (0.01)	0.0182
38	3.96	15.7	660 (216)	802 (263)	1080 (354)	0.0476 (0.01)	0.0145
39	3.53	12.5	832 (273)	1012 (332)	1360 (446)	0.0377 (0.01)	0.0115
40	3.15	9.9	1049 (344)	1276 (418)	1720 (564)	0.0299 (0.01)	0.0091
41							
42	2.50	6.3					
43							
44	1.97	3.9					

*Resistance figures are given for standard annealed copper. For hard-drawn copper add 2%.

FIGURE A–2 —CONT'D

WIRES PER INCH									
Gauge No.	Wires per Inch (Wires per cm)		Approx. Wires per Square Inch (Approx. wires per cm²)				Feet per Pound (Meters per kg)		
	S.C.C.	D.C.C.	S.C.C.	D.C.C.	P.E.	Formvar	Bare	P.E.	D.C.C.
8	7.4 (2.9)	7.1 (2.8)	55 (8)	50 (8)	58 (9)		20 (13.4)	19.8 (13.2)	19.5 (13)
9	8.2 (3.2)	7.9 (3.1)	69 (11)	63 (10)					
10	9.3 (3.7)	8.9 (3.5)	86 (13)	78 (12)	92 (14)		31.8 (21.4)	31.5 (21.2)	31 (20.7)
11	10.3 (4)	9.9 (3.9)	108 (17)	98 (15)					
12	11.5 (4.5)	10.9 (4.3)	132 (20)	120 (19)	145 (22)		50.6 (33.9)	50 (33.6)	49 (32.8)
13	12.8 (5)	12.1 (4.8)	166 (26)	148 (23)					
14	14.2 (5.6)	13.5 (5.3)	206 (32)	183 (28)	225 (35)		80.4 (54)	79.4 (53.4)	77 (51.8)
15	15.8 (6.2)	14.8 (5.8)	255 (39)	223 (35)					
16	17.9 (7)	16.5 (6.5)	320 (50)	280 (43)	358 (55)	340 (53)	128 (86)	126 (85)	119 (80)
17	20 (7.9)	18.3 (7.2)	400 (62)	340 (53)					
18	22 (8.7)	21 (8.3)	492 (76)	415 (64)	572 (89)	530 (82)	203 (136)	201 (135)	188 (126)
19	24.5 (9.6)	23.5 (9.2)	625 (97)	510 (79)					
20	27 (10.6)	24.5 (9.6)	770 (119)	625 (95)	875 (136)	800 (125)	323 (217)	319 (214)	298 (200)
21	30 (11.8)	26.7 (10.5)	940 (146)	750 (115)					
22	34 (13)	30.2 (11.9)	1 165 (181)	915 (140)	1 332 (206)	1 200 (185)	514 (346)	507 (341)	461 (311)
23	37.5 (14.8)	32.2 (12.7)	1 400 (217)	1 070 (165)					
24	41.5 (16.3)	35.5 (14)	1 700 (264)	1 260 (195)	2 045 (317)	1 820 (280)	818 (549)	805 (541)	745 (501)
25	45.5 (17.9)	38.5 (15.2)	2 065 (320)	1 495 (230)					
26	50 (19.7)	42 (16.5)	2 510 (390)	1 745 (270)	3 090 (480)	2 700 (420)	1 300 (873)	1 200 (861)	1 118 (752)
27	55 (21.6)	45 (17.7)	3 030 (470)	2 020 (315)					
28	60 (23.6)	48.5 (19.1)	3 654 (565)	2 330 (360)	4 670 (725)	4 000 (620)	2 067 (1 389)	2 030 (1 365)	1 759 (1 182)
29	65 (25.6)	52 (20.5)	4 280 (865)	2 690 (420)					
30	71.5 (28.1)	55.5 (21.8)	5 060 (785)	3 050 (470)	6 860 (1 065)	5 500 (850)	3 287 (2 209)	3 20 (2 165)	2 534 (1 704)
31	77.5 (30.5)	59 (23.2)	6 000 (930)	3 480 (540)					
32	84 (33.1)	62.5 (24.6)	7 050 (1 090)	3 900 (600)	10 050 (1 550)	7 700 (1 200)	5 225 (3 515)	5 120 (3 450)	3 317 (2 110)
33	90 (35.4)	66 (26)	8 100 (1 250)						
34	97 (38.2)	70 (27.6)	9 400 (1 450)		14 250 (2 200)	10 500 (1 600)	8 310 (5 590)	8 160 (5 550)	6 168 (1 750)
35	104 (40.9)	74 (29.1)	10 800 (1 650)						
36	112 (44.1)	78 (30.7)	12 800 (2 000)		20 000 (3 100)	14 900 (2 300)	13 210 (8 880)	12 850 (8 650)	7 875 (5 295)
37									
38	127 (50)	84 (33.1)					21 010 (14 130)		
39									
40	143 (56.3)	90 (35.4)					33 410 (22 470)		

S.C.C. = Single Cotton Covered
D.C.C. = Double Cotton Covered
P.E. = Plain Enamel

FIGURE A–3

FULL-LOAD CURRENTS IN AMPERES, DIRECT CURRENT MOTORS		
hp	**120 V**	**240 V**
1/4	2.9	1.5
1/3	3.6	1.8
1/2	5.2	2.6
3/4	7.4	3.7
1	9.4	4.7
1 1/2	13.2	6.6
2	17	8.5
3	25	12.2
5	40	20
7 1/2	58	29
10	76	38
15		55
20		72
25		89
30		106
40		140
50		173
60		206
75		255
100		341
125		425
150		506
200		675

These values of full-load currents are for motors running at base speed.

FIGURE A–4

PERFORMANCE ANALYSIS OF A SHUNT MOTOR

Note the following two formulas:

$$\text{Torque} = a \times \phi \times I_a$$

where

ϕ is the flux through the armature,
I_a is the current in the armature, and

a combines the quantities of the formula $\frac{Z}{425{,}000{,}000\ m}$ (a is a fixed number for a given armature once it is assembled in the motor);

and

$$\text{emf} = b \times \phi \times \text{rpm}$$

where

b is, for a given armature, a constant term that combines the quantities previously expressed as $\frac{\text{Turns} \times 4}{\text{Paths} \times 60 \times 10^8}$.

As long as the line voltage is constant, the unchanging field-coil current maintains a steady magnetic field, regardless of variations in armature current.

To see just how a motor behaves under changing load conditions, examine the information obtained in the following example:

$\phi = 2{,}000{,}000$
$a = 6 \times 10^{-7}$
$b = 8.5 \times 10^{-8}$
line volts = 220
armature resistance = 4 ohms

Using the formula

$$\text{Torque} = a \times \phi \times I_a$$
$$\text{Torque} = 6 \times 10^{-7} \times 2{,}000{,}000 \times I_a$$
$$\text{Torque} = 1.2 \times I_a$$

Using the formula

$$\text{cemf} = b \times \phi \times \text{rpm}$$
$$\text{cemf} = 8.5 \times 10^{-8} \times 2{,}000{,}000 \times \text{rpm}$$
$$\text{cemf} = 0.17 \times \text{rpm}$$
$$\text{Line volts} = \text{cemf} + I_a R_a$$

The current and rpm of the motor can be calculated for several torque loads. For example, we give a motor work to do that requires it to exert a torque of 4.8 foot-pounds. Since torque = $1.2 \times I_a$ for this particular motor, an armature current of 4 amperes is required to produce the 4.8 foot-pounds of torque.

IR for the armature is 4 amperes × 4 ohms = 16 volts
220 line volts − 16 = 204 volts for emf
emf = 0.17 × rpm
204 volts = 0.17 × rpm
rpm = 1,200

The results of a series of such calculations are given in Figure A–5.

VALUES FOR A GIVEN SHUNT MOTOR									
					Armature			Overall	
Torque	I_a	I_aR_a	emf	rpm	Input (watts)	Output (watts)	Efficiency	Input (watts)	Efficiency
0	0	0	220	1294	0	0	—	110	-
1.2	1	4	216	1270	220	216	0.98	330	0.65
2.4	2	8	212	1247	440	424	0.96	550	0.77
3.6	3	12	208	1223	660	624	0.945	770	0.81
4.8	4	16	204	1200	880	816	0.93	990	0.824
6.0	5	20	200	1176	1100	1000	0.91	1210	0.825
7.2	6	24	196	1153	1320	1176	0.89	1430	0.822
8.4	7	28	192	1129	1540	1344	0.87	1650	0.815
9.6	8	32	188	1106	1760	1504	0.855	1870	0.805
10.8	9	36	184	1082	1980	1656	0.84	2090	0.79
12.0	10	40	180	1059	2200	1800	0.82	2310	0.78
13.2	11	44	176	1035	2420	1936	0.80	2530	0.765
14.4	12	48	172	1012	2640	2064	0.78	2750	0.75
15.6	13	52	168	988	2860	2184	0.765	2970	0.735
16.8	14	56	164	965	3080	2296	0.745	3190	0.72
18.0	15	60	160	941	3300	2400	0.73	3410	0.70
33.0	27.5	110	110	647	6050	3025	0.50	6160	0.49

FIGURE A–5

In order to obtain illustrative figures for the total power input to the whole shunt motor, the field coil is assumed to require 0.5 ampere from the 220-volt line. This 110 watt is added to the armature input to obtain total power input.

The values for torque, armature current, *IR* in armature, emf, and rpm were previously calculated. Power input for the armature is 220 volts × the armature current. Power output for the armature is emf × the armature current.

The armature efficiency is

$$\frac{\text{cemf}}{\text{Line volts}} \text{ or } \frac{\text{Power output of armature}}{\text{Power input of armature}}$$

The overall efficiency of the entire motor is

$$\frac{\text{Power output of armature}}{\text{Total power input to motor}}$$

Note from Figure A–5 that, in the problem, at the 4.8 foot-pounds torque used, the motor runs at 1,200 rpm, produces 816 watts of mechanical power, and uses 64 watts (880 minus 816) in heating the armature. Suppose that the load on the motor is suddenly increased, putting a torque drag of 12 foot-pounds on the motor pulley tending to slow the motor. The motor must either develop 12 foot-pounds of torque or stall.

As the shunt motor slows a little, the reduced rpm results in a reduced emf, which permits more current through the armature. When the armature slows down to 1,059 rpm, Figure A–5, only 180 volts emf is generated. The armature current becomes 10 amperes, which produces 12 foot-pounds of torque. Although the power output of the motor has increased, efficiency has dropped to 78%, and the 400 watts (2,220–1,800) heating rate in the armature can raise the armature temperature beyond a safe limit.

One more change: Assume that while operating at 12-foot-pounds torque, the belt breaks, reducing the load on the motor to 0. Armature current of 10 amperes is still producing 12 foot-pounds of torque, which now accelerates the motor armature. If the speed gets to 1,294 rpm, 220 volts emf is generated. Thus the armature current becomes 0 and no torque is produced; therefore, the motor accelerates no further.

Of course, this assumes a motor free of friction. Actually the armature must produce a small torque to overcome friction. Consequently, a few tenths of an ampere is needed in the armature, and the actual top speed of the motor may be 1,290 instead of 1,294 rpm.

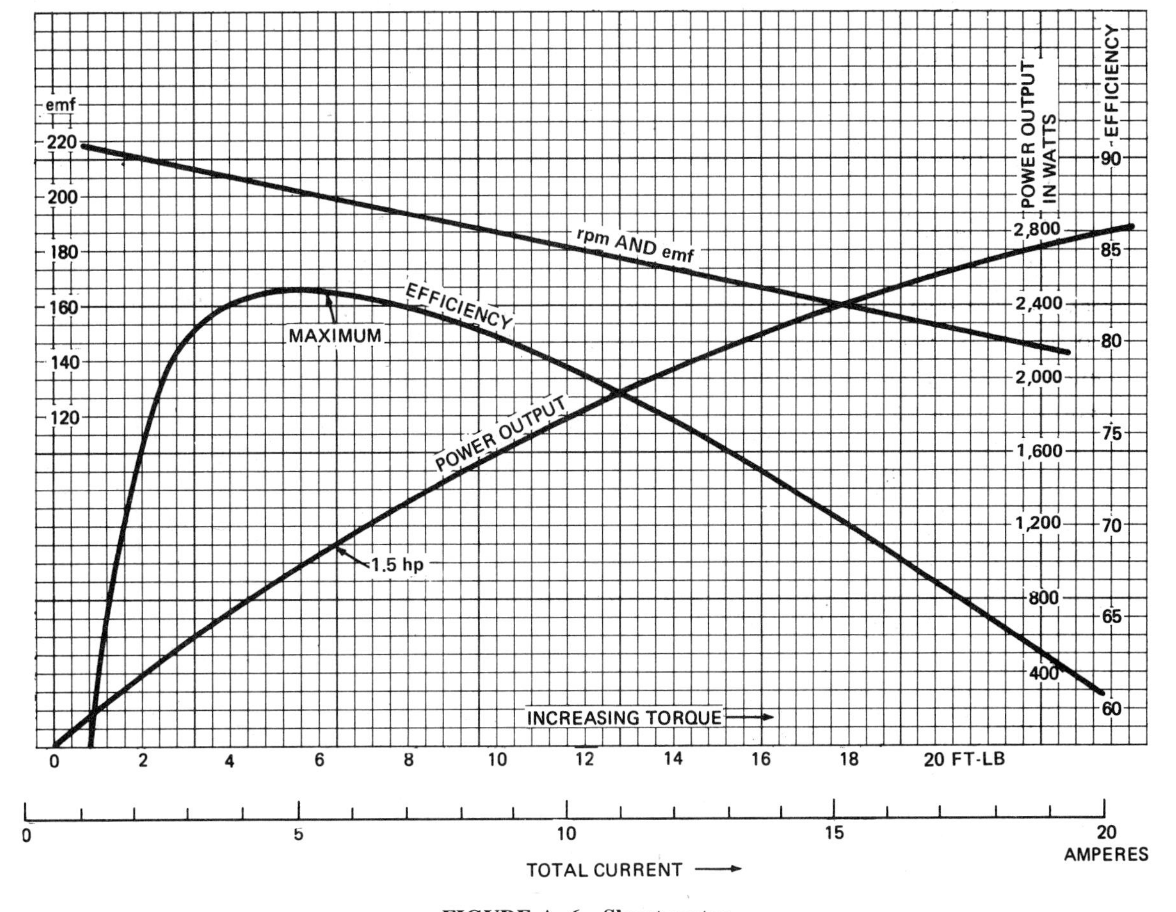

FIGURE A–6 Shunt motor

The graph in Figure A–6 summarizes information from Figure A–5. It shows how rpm, efficiency, and power output change as the torque load (and current) is increased. The scale chosen emphasizes the change in efficiency.

Normally the rated load for the motor is the horsepower output at peak efficiency. For example, a 10-horsepower motor is designed to have its maximum efficiency at 10-horsepower output. The overall electrical efficiency is greatest when I^2R in the armature equals I^2R for the field.

The graph in Figure A–7 summarizes the tabulated information in a different form, relating the changes in rpm, efficiency, and current to the horsepower output. Some portions of the graph are of more theoretical than practical interest, because the graph covers a wider range than the range in which a 1.5-horsepower motor is normally operated. For example, if this motor is operated below 3/4 horsepower, the efficiency is low, but no harm is done. On the other hand, if it is operated above 2 horsepower, current increases rapidly, increasing the heat production in the armature. The maximum possible power output for this particular motor is about four horsepower at 50% efficiency. At still lower efficiency, current increases still further, and horsepower output is reduced. (Anyone planning to operate this motor at such overloads should also plan to replace it with a new one every few minutes.)

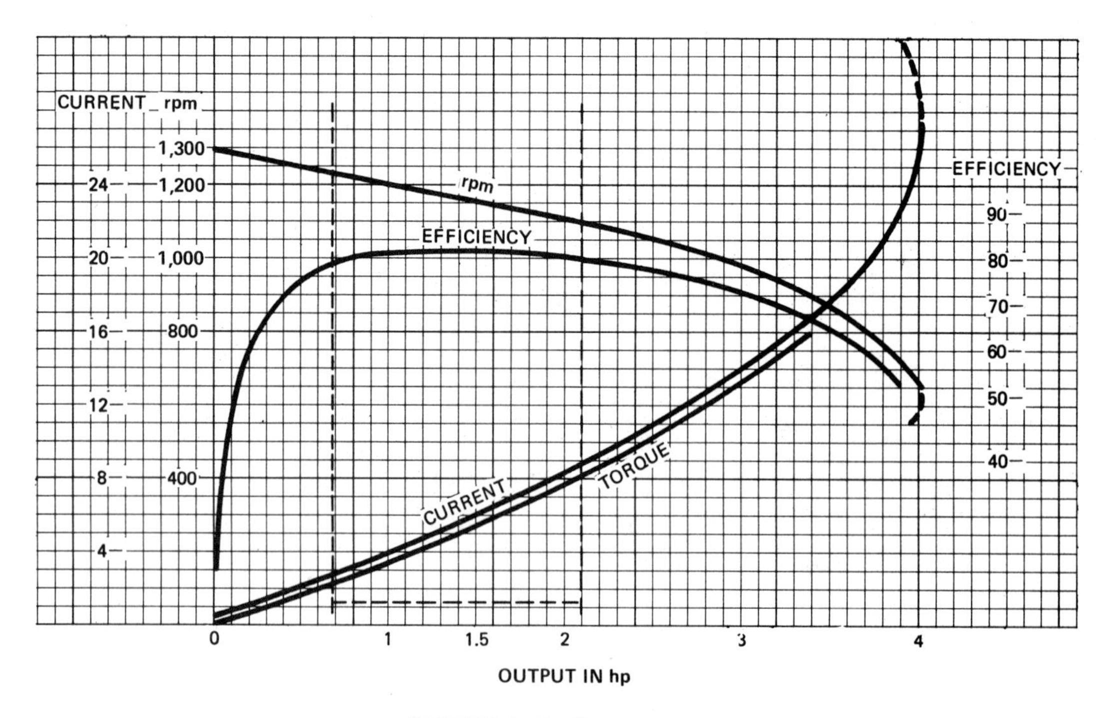

FIGURE A–7 Shunt motor

Field Control

What happens to the speed of a motor when the magnetic field is weakened by inserting resistance into the field circuit? The answer may be surprising.

Assume the motor is operating steadily at 1,176 rpm, taking 5 amperes and producing 6 pound-feet of torque, Figure A–5. Suddenly the field current is reduced from 0.5 amperes to 0.4 ampere, which reduces the flux from 2,000,000 lines to 1,600,000 lines.

When rotating at 1,176 rpm, the armature was originally cutting 2,000,000 lines and generating 200 volts. If speed remains at 1,176 for a moment after flux is reduced, the cutting of 1,600,000 lines generates only 160 volts because emf is proportional to flux cut.

If only 160 volts is generated, 60 volts (220 lines volts − 160 volts) is left to pour 15 amperes through the 4-ohm armature. (Volts = IR, $60 = 4 \times 15$.) This change in current, along with the changed flux, affects the torque.

$$\text{Torque} = a \times \phi \times I_a$$
$$\text{Torque} = 6 \times 10^{-7} \times 1{,}600{,}000 \times 15$$
$$\text{Torque} = 14.4 \text{ lb-ft}$$

This new torque, 14.4, is more than needed to manage the 6 pound-feet load. Thus the armature accelerates. Increased speed increases the generated emf, which in turn reduces the current input. With reduced field, the motor picks up speed until the suddenly larger current is reduced to a value just enough to maintain the 6 pound-feet torque load.

A current of 6.25 amperes produces the 6 pound-feet of torque. Use the torque formula again, with $\phi = 1{,}600{,}000$.

IR in the armature becomes $6.25 \times 4 = 25$ volts.

The generated emf is $220 - 25 = 195$ when the armature arrives at its new steady speed.

This new rpm can be found from

$$\text{emf} = b \times \phi \times \text{rpm}$$
$$195 = 8.5 \times 10^{-8} \times 1.6 \times 10^{6} \times \text{rpm}$$
$$\text{rpm} = 1{,}434$$

Compare conditions before and after the change of field strength.

Flux	Torque	I_a	rpm	emf	Power Input	Power Output
2,000,000	6 lb-ft	5	1,176	200	1,100 watts	1,000 watts
1,600,000	6 lb-ft	6.25	1,434	195	1,375 watts	1,219 watts

Armature Control

Determine the effect of adding 1-ohm resistance in series with a 4-ohm armature.

Torque = 6 lb-ft
I = 5 amperes
ϕ = 2,000,000

At the instant of adding resistance, rpm is still 1,176, generating 200-volt emf that leaves 20 volts available to overcome the resistance of the armature circuit. Since the armature circuit now has 5 ohms instead of 4, the current in the armature is momentarily reduced to 4 amperes, from the previous 5 amperes. Because this reduced current produces less torque (4.8 instead of 6), the motor slows down, unable to maintain the 6 pound-feet load. This slowing reduces the emf. This reduction makes more voltage available for producing current through the armature circuit. As soon as 5 amperes is again available, the 5 amperes produce the 6 pound-feet torque that carries the load at a reduced speed.

Calculate the new speed: 5 amperes through a 5-ohm armature circuit requires 25 volts. Generated emf can be only 220 − 25 = 195 volts. This 195 volts runs the motor at 1,157 rpm. Compare:

Flux	Armature-Circuit Resistance	I_a	rpm	emf	Power Input	Power Output
2,000,000	4 ohms	5 amperes	1176	200 V	1,100 watts	1,000 watts
2,000,000	5 ohms	5 amperes	1147	195 V	1,100 watts	975 watts

Field control is used much more often than armature control because additional resistance in the armature circuit increases both power loss and poorer speed regulation. However, armature control for below-normal speeds and field control for above-normal speeds can both be used on the same machine. For example, manual speed controllers with above-and-below-normal speed control facilities are used with shunt and cumulative compound motors. Also, electronic control units are used with dc motors to obtain above-and-below-normal speeds where the currents are controlled by means other than rheostats.

Shunt motors are used in those industrial applications where a relatively constant speed is required from no-load to full-load and where there are no severe mechanical overload conditions. Shunt motors are often used with fans and blowers, as a prime mover or motor unit for motor-generated sets, and as the motor unit for such metal-working machines as lathes, shapers, milling machines, and grinders. They are also used in the textile industry for spinning frames and looms; or in any industrial application where the speed must be kept relatively constant at various load points.

PERFORMANCE ANALYSIS OF A SERIES MOTOR

Assume we have a series motor in which field resistance is 2.4 ohms, armature resistance is 4.0 ohms (total 6.4 ohms), $a = 48 \times 10^{-8}$, $b = 6.8 \times 10^{-8}$, and $\phi = 2{,}000{,}000$ when $I = 6.25$ amperes (rated current). Line volts = 220.

When $I = 6.25$, the voltage drop due to resistance is $6.25 \times 6.4 = 40$ volts. The emf generated is $220 - 40 = 180$.

$$\text{Torque} = a \times \phi \times I = 48 \times 10^{-8} \times 2{,}000{,}000 \times 6.25 = 6 \text{ ft-lb.}$$

Since emf = $b \times \phi \times$ rpm, we can find rpm from $180 = 6.8 \times 10^{-8} \times 2{,}000{,}000 \times$ rpm; rpm = 1,320.

In order to calculate changes that occur when the torque load on this motor is changed, we must take into account the fact that the field flux changes when the armature current changes. In Figure A–8, the flux curve indicates that the field increases as the current increases, up to about 12 amperes. At this point saturation is reached, and the flux is assumed to be constant for higher values of current (not *exactly* true, but satisfactory for this example).

The values in Figure A–9 are found as follows: For various values of current from 0 to 20, flux is determined from Figure A–8. *IR* is the voltage used on resistance;

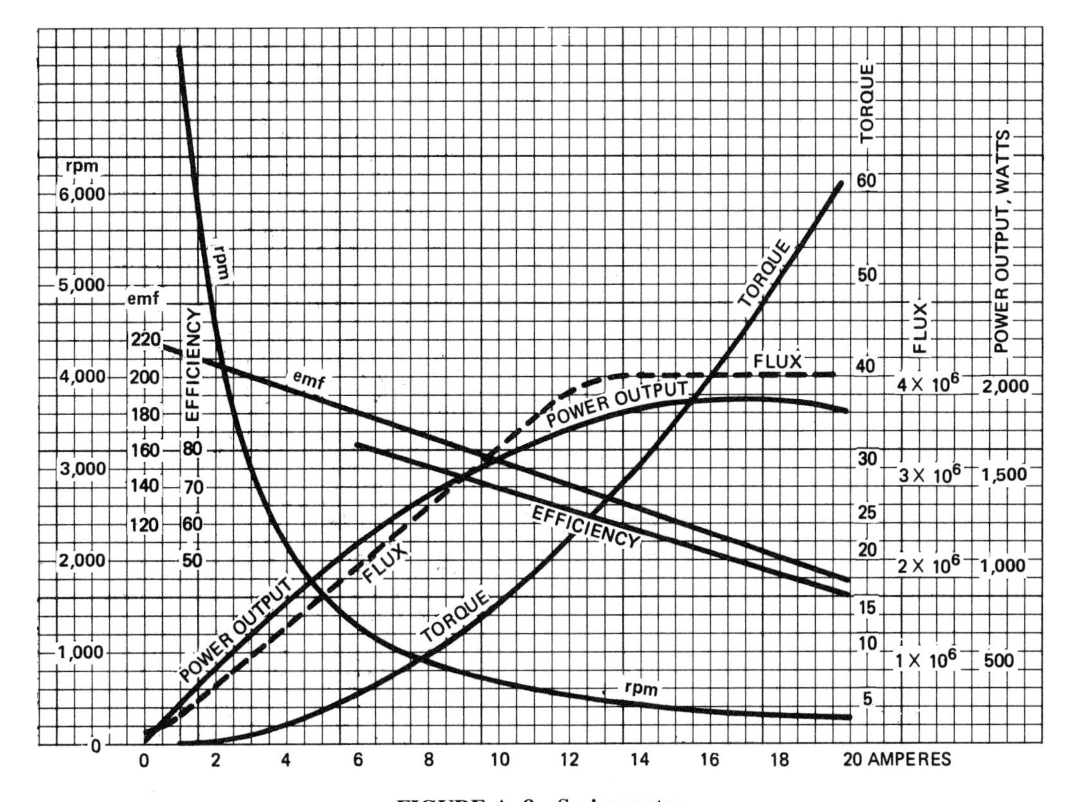

FIGURE A–8 Series motor

TABLE OF VALUES FOR SERIES MOTORS

I (amp)	Field Flux	Total *IR*	emf	Torque	rpm	Output (watts)	Input (watts)	Electrical Efficiency
0	Low	0	220.	0	—	0	0	
0.5	160,000	3.2	216.8	0.04	19,800	108	110	
1	320,000	6.4	213.6	0.15	9,800	214	220	
2	640,000	12.8	207.2	0.60	4,750	414	440	
3	960,000	19.2	200.8	1.38	3,180	602	660	
4	1,280,000	25.6	194.4	2.46	2,230	777	880	0.87
5	1,600,000	32.	188.0	3.84	1,725	940	1,100	0.855
6	1,920,000	38.4	181.6	5.53	1,390	1,090	1,320	0.825
6.25	2,000,000	40.0	180.0	6.0	1,320	1,125	1,375	0.82
7	2,240,000	44.8	175.2	7.5	1,150	1,226	1,540	0.795
8	2,560,000	51.2	168.8	9.8	970	1,350	1,760	0.77
9	2,880,000	57.6	162.4	12.4	830	1,462	1,980	0.74
10	3,200,000	64.0	156.0	15.4	716	1,560	2,200	0.71
11	3,520,000	70.4	149.6	18.6	607	1,646	2,420	0.68
12	3,840,000	76.8	143.2	22.1	550	1,718	2,640	0.65
14	4,000,000	89.6	130.4	30.1	480	1,826	3,080	0.59
16	4,000,000	102.	117.6	39.3	432	1,880	3,520	0.535
18	4,000,000	115.	104.8	49.8	385	1,885	3,960	0.475
20	4,000,000	128.	92.0	61.5	338	1,840	4,400	0.42

FIGURE A–9

IR = current × 6.4 ohms. Emf is found by subtracting each value of IR from 220. Torque is found from a × ϕ × I; rpm is found from emf = b × ϕ × rpm. Watts output is emf × amperes; watts input is 220 × amperes.

Our first calculation on the series motor shows 6 foot-pounds of torque produced with the motor operating steadily at 1,320 rpm, on 6.25 amperes. Suppose the torque load is suddenly increased to 12.4 foot-pounds. This increase slows the motor, reducing rpm and therefore emf. Reduced emf permits more current. A little more current causes a lot more torque, which carries the increased load. Notice that torque depends on current and flux; increasing the current also increases the flux. Therefore, torque is proportional to the *square* of the current, unless the field iron reaches saturation. Rpm is reduced from 1,320 down to 830 because of the increased flux. At this load, the motor can run no faster than 830; if it does, it generates so much emf that the necessary current cannot be maintained.

Now, reduce the torque load from 6 foot-pounds to 2.46 foot-pounds while the motor is operating at 1,320 rpm. Since this reduction makes the armature turn more easily, its first response is to speed up a little. Increased speed generates more emf. More emf reduces current input. Reduced current reduces flux. If the speed increases from 1,320 to 1,390, the current is reduced to 6 amperes, Figure A–9, which is enough to keep the motor accelerating (5.53 foot-pounds produced; load, 2.46 foot-pounds). When the speed is up to 2,230, emf is up to 194.4, which lets in 4 amperes to produce the 2.46 foot-pounds that carries the load at a steady 2,230 rpm. Notice that emf does not rise as fast as the speed does, because as the speed increases, the field flux decreases.

In the *shunt* motor, where flux is constant, a small increase in the rpm produces enough increase in emf to reduce the current to such a value that there is not enough torque to accelerate the motor further. (Compare tabulated values for series motor and shunt motor, Figures A–5 and A–9).

When the load is removed from a shunt motor, it accelerates a little, and emf rises to a value sufficient to shut off the torque-producing current. When the load is removed from a series motor, speed and emf also increase in the same way. However, due to decreased flux, emf does not increase rapidly enough to shut off the torque-producing current. The torque accelerates the motor further, theoretically without limit, Figure A–9. With the load removed, speed can rise to 9,800 rpm. At 9,800 rpm the 1-ampere input produces 0.15 foot-pounds of torque. If the friction in the motor is equivalent to less than 0.15, speed builds up further.

In actuality the top speed of series motors is limited.

- In small motors, friction from bearings, brushes, and windage limits the speed. At 10,000 rpm the entire power input can be expended on friction and there is no further increase in speed.
- In large motors, high speed produces inertial forces that burst the bands holding the armature coils in place. At 5,000 rpm, the surface of an 8-inch-diameter armature is travelling at about 2 miles per minute, and each ounce of copper wire in the slots requires a force of 180 pounds to hold it in place.

Efficiency at low-power output is not plotted on the preceding graphs. Values of electrical efficiency, calculated from emf divided by line volts, would be misleading. When current is 1 ampere, calculated electrical efficiency is 0.97, but if 50 watts is used on mechanical friction, actual efficiency is 0.745. Actual efficiency is not found by calculation, rather it is found by test on the motor.

MOTOR FORMULAS

From the fundamental definitions of current and magnetic field strength, the force on a current passing across a magnetic field is

$$\frac{B \times l \times I}{10}$$

where

B is the number of lines per square centimeters
l is the length of the wire in centimeters
I is the current in amperes
Force is measured in dynes

Since

$$\text{pounds} \times 445{,}000 = \text{dynes}$$

$$\frac{\text{Lines}}{\text{sq in.}} \times \frac{1}{6.45} = \frac{\text{Lines}}{\text{sq cm}}$$

$$\text{Inches} \times 2.54 = \text{cm}$$

$$\text{Force} \times 445{,}000 = \frac{B}{6.45} \times \frac{I \times l \times 2.54}{10}$$

This expression reduces to

$$F = \frac{B \times l \times I}{11{,}300{,}000}$$

where

F is the force in pounds on a single wire
l is the length in inches of wire in the field
B is the lines per square inch
I is the current in amperes

Torque on one wire is equal to the force (lb) × radius (ft), Figure A–10.

$$\text{Torque} = \frac{B \times l \times I \times r}{11{,}300{,}000}$$

Calling the total number of wires Z, the total torque on all of the wires is

$$\text{Torque} = \frac{Z \times B \times l \times I_a \times r}{11{,}300{,}000 \times \text{m}}$$

where I is the current in a wire and is $\frac{I_a}{m}$

I_a is the total armature current
m is the number of parallel paths through the armature

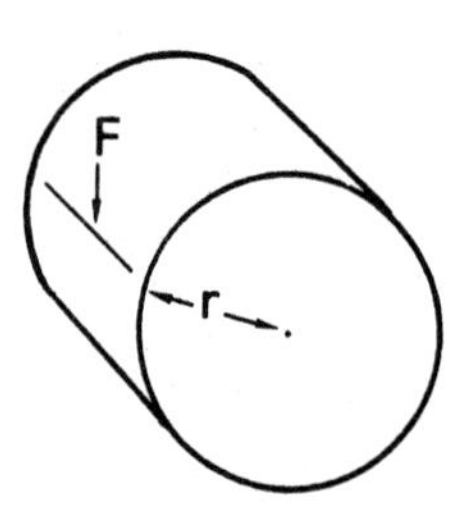

FIGURE A–10

The flux (Φ) passing through the armature can be found by multiplying the lines per square inch (B) by the area (in square inches) of half of the cylindrical armature surface (assuming that the motor is of compact construction so that the armature is surrounded by field poles).

The entire curved surface is $2\pi r \times l$ square inch; half of this surface is $\pi r l$, Figure A–11.

$$\phi = \pi r l B \text{ (r in inches)}$$

$$\phi = 12\pi r l B \text{ (r in feet)}$$

$$\frac{\phi}{12\pi} = r l B$$

Substituting $\phi/12$ for $4r \times l \times B$ in the previous torque formula,

$$T = \frac{Z \times I_a \times \phi}{11{,}300{,}000 \times m \times 12\pi}$$

$$T = \frac{Z \times I_a \times \phi}{425{,}000{,}000 \times m}$$

where

T is the torque in foot-pounds
Z is the total number of armature wires
I_a is the total armature current
m is the number of parallel paths through the armature
ϕ is the flux passing through the armature

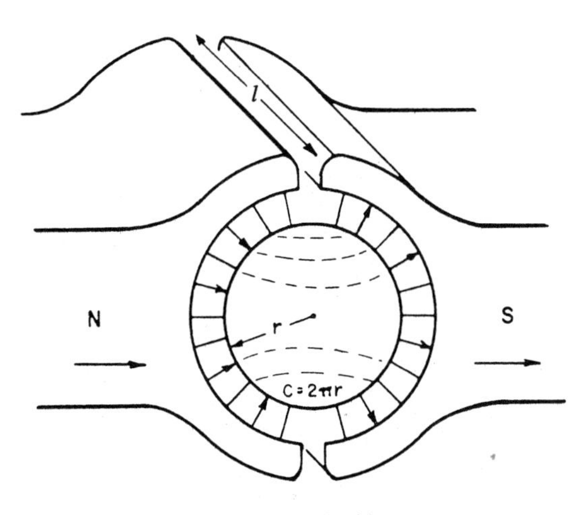

FIGURE A–11

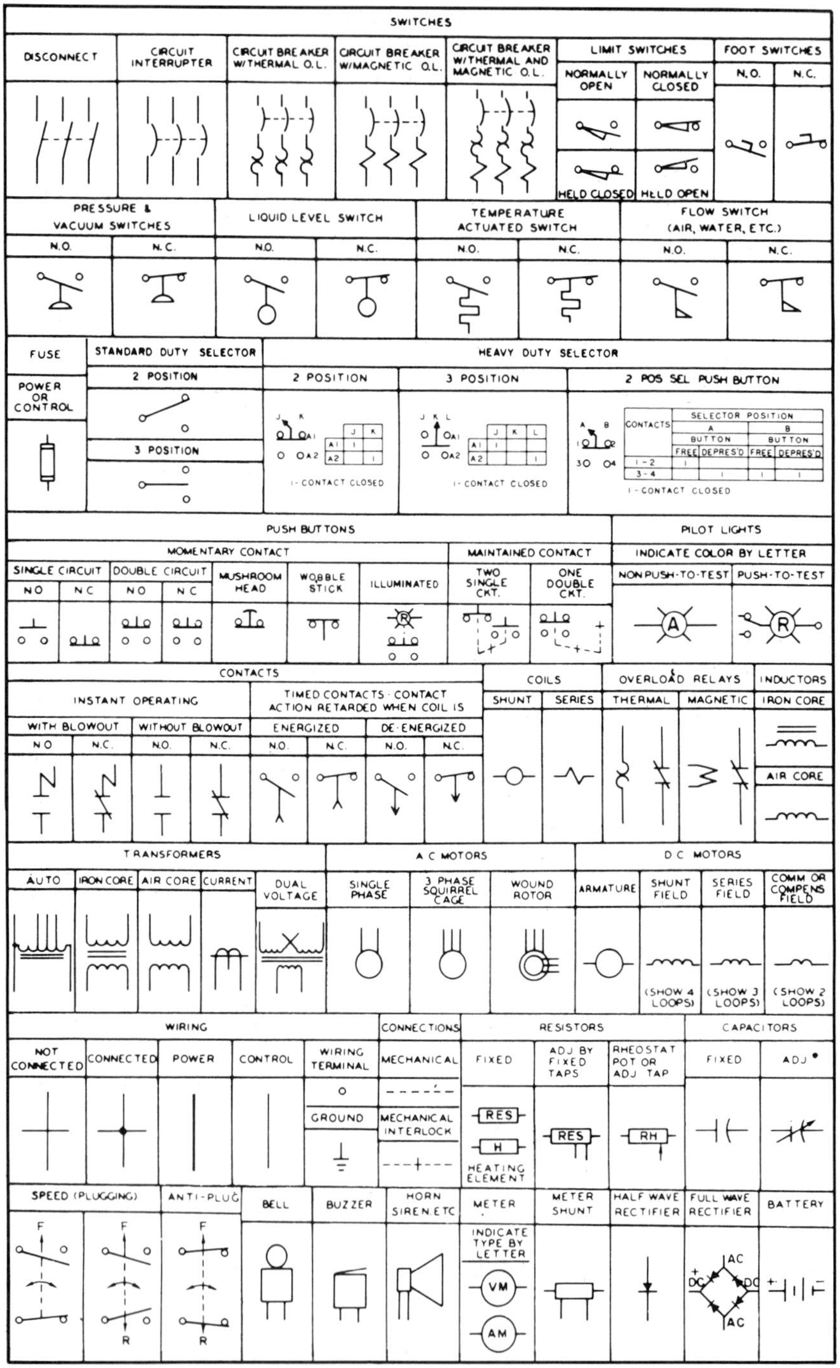

SWITCHES

DISCONNECT | CIRCUIT INTERRUPTER | CIRCUIT BREAKER W/THERMAL O.L. | CIRCUIT BREAKER W/MAGNETIC O.L. | CIRCUIT BREAKER W/THERMAL AND MAGNETIC O.L.

LIMIT SWITCHES: NORMALLY OPEN | NORMALLY CLOSED | HELD CLOSED | HELD OPEN

FOOT SWITCHES: N.O. | N.C.

PRESSURE & VACUUM SWITCHES: N.O. | N.C.

LIQUID LEVEL SWITCH: N.O. | N.C.

TEMPERATURE ACTUATED SWITCH: N.O. | N.C.

FLOW SWITCH (AIR, WATER, ETC.): N.O. | N.C.

FUSE: POWER OR CONTROL

STANDARD DUTY SELECTOR: 2 POSITION | 3 POSITION

HEAVY DUTY SELECTOR: 2 POSITION | 3 POSITION | 2 POS SEL PUSH BUTTON

2 POSITION: J K A1 A2 — 1 - CONTACT CLOSED

3 POSITION: J K L A1 A2 — 1 - CONTACT CLOSED

2 POS SEL PUSH BUTTON: A B 1 2 3 4 — CONTACTS 1-2, 3-4; SELECTOR POSITION A BUTTON FREE DEPRES'D, B BUTTON FREE DEPRES'D — 1 - CONTACT CLOSED

PUSH BUTTONS

MOMENTARY CONTACT: SINGLE CIRCUIT NO, NC | DOUBLE CIRCUIT NO, NC | MUSHROOM HEAD | WOBBLE STICK | ILLUMINATED

MAINTAINED CONTACT: TWO SINGLE CKT. | ONE DOUBLE CKT.

PILOT LIGHTS

INDICATE COLOR BY LETTER: NON PUSH-TO-TEST | PUSH-TO-TEST

CONTACTS

INSTANT OPERATING: WITH BLOWOUT N.O., N.C. | WITHOUT BLOWOUT N.O., N.C.

TIMED CONTACTS - CONTACT ACTION RETARDED WHEN COIL IS: ENERGIZED N.O., N.C. | DE-ENERGIZED N.O., N.C.

COILS: SHUNT | SERIES

OVERLOAD RELAYS: THERMAL | MAGNETIC

INDUCTORS: IRON CORE | AIR CORE

TRANSFORMERS: AUTO | IRON CORE | AIR CORE | CURRENT | DUAL VOLTAGE

A C MOTORS: SINGLE PHASE | 3 PHASE SQUIRREL CAGE | WOUND ROTOR

D C MOTORS: ARMATURE | SHUNT FIELD (SHOW 4 LOOPS) | SERIES FIELD (SHOW 3 LOOPS) | COMM OR COMPENS FIELD (SHOW 2 LOOPS)

WIRING: NOT CONNECTED | CONNECTED | POWER | CONTROL | WIRING TERMINAL | GROUND

CONNECTIONS: MECHANICAL | MECHANICAL INTERLOCK

RESISTORS: FIXED (RES, H — HEATING ELEMENT) | ADJ BY FIXED TAPS (RES) | RHEOSTAT POT OR ADJ TAP (RH)

CAPACITORS: FIXED | ADJ

SPEED (PLUGGING) (F, R) | ANTI-PLUG (F, R) | BELL | BUZZER | HORN SIREN ETC | METER (INDICATE TYPE BY LETTER: VM, AM) | METER SHUNT | HALF WAVE RECTIFIER | FULL WAVE RECTIFIER (AC, DC, +, −) | BATTERY (+, −)

FIGURE A–12

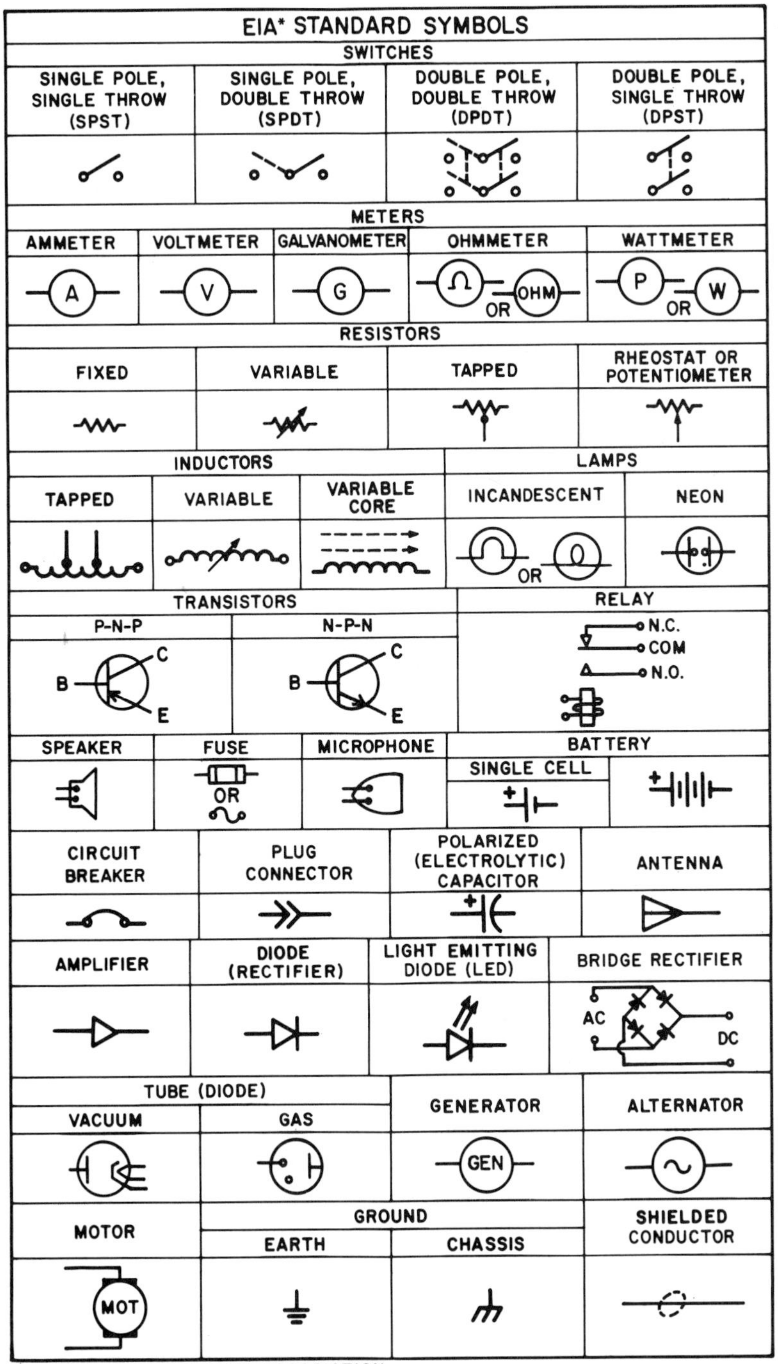

EIA* STANDARD SYMBOLS

SWITCHES

SINGLE POLE, SINGLE THROW (SPST)	SINGLE POLE, DOUBLE THROW (SPDT)	DOUBLE POLE, DOUBLE THROW (DPDT)	DOUBLE POLE, SINGLE THROW (DPST)

METERS

AMMETER	VOLTMETER	GALVANOMETER	OHMMETER	WATTMETER
A	V	G	Ω OR OHM	P OR W

RESISTORS

FIXED	VARIABLE	TAPPED	RHEOSTAT OR POTENTIOMETER

INDUCTORS			LAMPS	
TAPPED	VARIABLE	VARIABLE CORE	INCANDESCENT	NEON
			OR	

TRANSISTORS		RELAY
P-N-P	N-P-N	
B C E	B C E	N.C. COM N.O.

SPEAKER	FUSE	MICROPHONE	BATTERY	
	OR		SINGLE CELL	

CIRCUIT BREAKER	PLUG CONNECTOR	POLARIZED (ELECTROLYTIC) CAPACITOR	ANTENNA

AMPLIFIER	DIODE (RECTIFIER)	LIGHT EMITTING DIODE (LED)	BRIDGE RECTIFIER
			AC DC

TUBE (DIODE)		GENERATOR	ALTERNATOR
VACUUM	GAS	GEN	

MOTOR	GROUND		SHIELDED CONDUCTOR
	EARTH	CHASSIS	
MOT			

* ELECTRONICS INDUSTRY ASSOCIATION

FIGURE A–13

Glossary

alnico a permanent magnet made of aluminum, nickel, and cobalt
alternating current a current that reverses periodically due to reversal of polarity at the voltage source
American Wire Gauge a system of classifying the size of electrical conductors by gauge numbers (see Appendix)
ammeter an instrument for measuring electrical current
ampere the basic unit of electrical current. This is equal to 1 coulomb of electrons flowing past a given point in 1 second.
ampere-hour a unit describing the storage capacity of a battery (the product of current and time)
ampere-turns the basic unit of magnetomotive force (the product obtained by multiplying the current times the number of turns on the coil)
analog meter an instrument that indicates measurements by a pointer moving over a scale of numbers with continuous values
anode the terminal by which the electrons exit from a device
armature the moving element in an electromechanical device, such as the rotating part of a motor or the moving part of a relay or bell
armature reaction the undesirable distortion of the magnetic field of a motor or generator caused by its interaction with the flux developed by the armature
atom the smallest particle of an element

battery an assembly of cells maintaining a potential difference (voltage) between its terminals
bimetal a switching device made of two different metals, each having a different rate of expansion, welded together. It activates by bending when heated or cooled.
bleeder resistor a resistor used to stabilize the output from a voltage divider
brushes a conducting device, made of carbon or graphite, riding on the slip rings or commutators of rotating machines
Btu abbreviation for the term British thermal unit

cathode the terminal by which electrons enter a device

cathode-ray tube abbreviated CRT. A tube with a long neck behind a viewing screen (a picture tube). The screen is phosphorcoated to make visible a high-speed electron beam that is accelerated inside the neck of the tube.

cell a single-unit of a battery that produces a DC voltage by converting chemical energy into electrical energy

circuit breaker a resettable device, operating on thermal or electromagnetic principles, used to disconnect electrical circuits under overload.

circular mil a unit of measurement describing the cross-sectional area of a conductor

commutator the part of a motor (or generator) armature to which its coil ends are connected

compound generator a DC generator employing a series field as well as a shunt field

conductance the ability of a substance to carry electrical current. Conductance is also expressed as the reciprocal of resistance.

conductor a substance having many free electrons

contactor a switch that is closed or opened by the magnetic pull of an energized relay coil

conventional current electric current considered to flow from positive to negative through the external circuit

copper losses the power, or I^2R, losses of a resistor, dissipated in the form of heat

coulomb a great number of electrons equivalent to 6.25×10^{18} electrons

counter-emf the back-emf produced by an inductance, such as a motor circuit. Cemf opposes the applied voltage.

current the rate of flow of electrical charges (electrons). Electrons flow from the negative pole (electron surplus) to the positive pole (electron shortage).

cycle one complete reversal of an alternating current from positive to negative and back to the starting point

d'Arsonval meter a permanent-magnet-type instrument, used in the construction of voltmeters, ammeters, and ohmmeters

diamagnetic a material that has a permeability rating that is slightly less than that of air

dielectric heating a process in which the object to be heated is used as the dielectric of a capacitor in a high-frequency, alternating current circuit

direct current current that flows in one direction only

DPDT switch a double-pole, double-throw switch with six terminals

DVM an abbreviation for the term digital voltmeter

dynamic braking a method of quickly stopping an electric motor by dissipating the back-emf generated in its armature in a resistor

eddy current an undesirable current induced into and circulating within the core of magnetic devices operated with varying current

electrodynamometer a machine used for the measurement of torque developed by a motor for purposes of determining the horsepower output

electrolysis the decomposition of a chemical compound by means of an electric current

electrolyte a chemical solution capable of conducting electric current by its dissociation into ions

electron the negatively charged particle of an atom

electronics the application of electricity for the processing of informational signals, using semiconductor devices or electron tubes

electrostatics a term related to the force field that surrounds a stationary, charged object

element one of the substances that forms the basic building blocks of nature. Each element is identified by the structure of its unique type of atom.

emf an abbreviation for the term electromotive force (voltage)

ferromagnetic the quality of a substance that allows it to be easily magnetized, similar to that of iron (ferrus). A substance with a permeability rating much greater than that of air.

field coil the coils mounted within the frame of an electric motor or generator for the purpose of providing the magnetic field

flux the lines of force of a magnetic or electrostatic field

foot-candle a unit of measurement for surface illumination, defined as the illumination from a standard candle on a surface 1 foot away

frequency in alternating current, the number of cycles per second

fuse a protective device, operating on principles of heat, used to open a circuit under overload. Fuses, like circuit breakers, are available in different current ratings.

galvanometer a current-sensitive meter, usually with the 0 position at the center of the scale, indicating negative values as well as positive values

gauss a unit of magnetic flux density in the metric (cgs) system. One gauss equals 1 maxwell per square centimeter.

giga a metric prefix equal to 10°

gilbert a unit of magnetomotive force in the cgs system

hydrometer a device for determining the specific gravity of a fluid, such as the electrolyte in a battery

hysteresis a form of molecular friction caused by the fact that the formation of magnetic flux lags behind the magnetomotive force that produces it

incandescent emitting visible light as a result of heat

induction heating a process of heating a metallic object by inducing a high-frequency current into it

ion an unbalanced atom that has gained or lost electrons

joule the unit of energy in the International System of measurement. The work done when 1 ampere flows through 1 ohm in 1 second.

kilo a prefix used to designate the number 1,000

kilowatt-hour the unit of electrical energy used by power companies to bill their customers

linear pertaining to electrical devices, such as resistors, in which one variable is directly proportional to another
lumen the unit of luminous flux

magnetomotive force the magnetizing force produced by a current flowing through a solenoid
maxwell the unit of magnetic flux in the cgs system
mega a prefix used to designate the number 1,000,000
megohmmeter an instrument, often referred to as a "megger," used for measuring extremely high resistance, such as the insulation resistance of cables
mica a mineral silicate used as heatproof insulating material
micro a prefix designating one millionth of a unit (10^{-6})
mil a unit for measuring the diameter of a wire, equal to one thousandth of an inch
mil-foot a 12-inch-long conductor with a diameter of 1 mil, used as a standard for defining resistivity of a given material
milli a prefix designating one thousandth of a unit (10^{-3})

nano a prefix designating one billionth of a unit (10^{-9})
neutral wire a conductor grounded to the earth, generally colored white
nichrome short form for nickel-chromium, a special alloy used for resistance heating
normal speed the speed of a motor without any external resistance added to its circuitry
normally open (closed) the condition of a relay contact in the absence of electricity

ohm the unit of electrical resistance
oscilloscope an instrument for displaying electrical waveform on a screen

paramagnetic a material that has a permeability rating slightly larger than that of air
permalloy an alloy (approximately 78% nickel, 21% iron) used in the construction of electromagnets
permeability a measure denoting the ability of a material to establish, or concentrate, lines of magnetic flux
photoconductive the quality of a substance that changes its resistance under the influence of light
photovoltaic the quality of a substance producing a voltage from light energy
pico a prefix denoting 10^{-12}
piezoelectric referring to the production of a voltage by applying pressure to a crystal
pm motor a motor employing permanent magnets instead of field coils
potentiometer a variable resistor with three terminals, used to vary the voltage potential of a circuit
power the rate of doing work
primary cell a cell that cannot be recharged after it has run down
prime mover the engine or turbine used to drive a generator
programmable controller an electronic device, using microprocessors, for the control of electrical machinery
Prony brake a device used for determining the torque of a motor
proton a positively charged particle contained within the nucleus of an atom

rectifier an electrical device for changing alternating current (AC) into direct current (DC)
reluctance the opposition of a material to the establishment of magnetic flux. The reciprocal of permeability.
residual magnetism the magnetism retained by the core of an electromagnet after the coil has been deenergized
rheostat a variable resistor, utilizing only two of its three terminals, used to control the current in a circuit

saturation the condition of an electromagnet where an increase of current yields little or no further increase of magnetic flux
SCR an abbreviation for silicon-controlled rectifier, a semiconductor device used extensively for the control of electrical current
secondary cell a chemical cell that can be recharged after it has been run down
sensitivity a measure of quality for electrical meters, expressed as current flow necessary to force full-scale deflection of the pointer
shunt motor a DC motor in which the field coil is connected parallel to the armature
shunt resistor a resistor connected parallel to some other device, such as the shunt used in the construction of an ammeter
slip ring a metallic band mounted on but insulated from the shaft of a motor or generator. Used to conduct current to or from a rotating coil.
solenoid an electromagnet with a movable plunger
specific gravity the ratio of the weight of a given volume of a substance to the weight of an equal volume of water
speed regulation a number indicating the amount of change in speed of a motor as it goes from a condition of no load to full load. Generally expressed as a percentage.
stray losses energy losses in a machine other than I^2R losses, such as friction, windage, hysteresis, eddy currents

tesla the unit of flux density in the mks system of measurement. 1 tesla = 1 weber per square meter.
thermocouple a device made of two dissimilar metals, which produces a small-scale voltage when the junction of the two metals is heated
toroid a doughnut-shaped electromagnet
torque the rotational force on a wheel

voltage divider a series circuit of resistors designed to obtain specific voltage drops
voltage regulation a number expressing the change in voltage as a generator goes from a condition of no load to full load. Generally expressed as a percentage.

watt the unit of electrical power. The power expended when 1 ampere of current flows through a 1 ohm resistance.
weber a unit of magnetic flux in the mks system of measurement. One weber equals 100 million (10^8) lines of flux.

Selected Answers to Odd-Numbered Problems

Note: Only *numerical* solutions to problems are provided.

CHAPTER 5

1. 746
3. 84,900
5. 0.0235
7. 8.00
9. 569,000
11. 89.1
13. 11,600
15. 4,710
17. 0.426
19. 29.9
21. 0.299
23. 5,280,000
25. 7,380,000
27. 146
29. 0.00268

POWERS OF 10

Note: For these exercises only, every other answer is provided for *all* three problems.

1a. 10^2
1c. 10^5
1e. 10^9
1g. 10^3
1i. 10^{-1}
1k. 10^{-7}
1m. 10^{-12}
1o. 10^4
2a. 1
2c. 1,000
2e. 100,000,000
2g. 0.096
2i. 0.1
2k. 0.00001
2m. 0.000000000001
2o. 3,100,000
3a. 4.38×10^2
3c. 1×10^{-1}
3e. 7.7×10^5
3g. 8.2×10^{-5}
3i. 2×10^{-4}
3k. 3.81×10^4

Prefix Kilo

1. 1.2 kΩ
3. 68 kΩ
5. 220 kΩ
7. 18 kV
9. 0.15 kΩ
11. 2,700 V
13. 3.2 kΩ
15. 390 Ω
17. 500 kΩ

Prefix Milli

1. 2 mA
3. 370 mA
5. 52 mA
7. 1,700 mA
9. 2.9 mA
11. 0.025 A
13. 450 mA
15. 0.00064 A
17. 0.0038 A

Prefix Mega

1. 3.2 MΩ
3. 0.0025 MV
5. 210,000 V
7. 0.000018 MΩ
9. 0.0048 MV
11. 0.7 MΩ
13. 2,500,000 Ω
15. 10,000,000 Ω
17. 33 kΩ

Prefix Micro

1. 1 μV
3. 690 μV
5. 0.00025 A
7. 8,300 μA
9. 62,000,000 μV
11. 0.000058 A
13. 100,000,000 μA
15. 9,500 μV
17. 0.000083 A
19. 0.0836 V

ELECTRICAL UNITS, SCIENTIFIC NOTATION, CONVERSIONS AND ABBREVIATIONS

1. 1,000; 10^3
3. 0.001; 10^{-3}
7. 0.000000001; 10^{-9}
9. 1,000,000 Ω; 10^6
13. 0.5 W; 5×10^{-1}
17. 0.9 A; 9×10^{-1}
19. 0.000075 V; 7.5×10^{-5}
23. 47,000 Ω; 4.7×10^4
25. 0.00005 V; 5×10^{-5}
27. 10 V; 1×10^1
29. 0.0004 s; 4×10^{-4}
31. 3,000,000 V; 3×10^6
33. 0.0003 A; 3×10^{-4}
35. 100,000; 1×10^5
37. 0.000000011; 1.1×10^{-8}
39. 50,000 Ω; 5×10^4
41. 0.1 Ω; 1×10^{-1}
43. 0.001 s; 1×10^{-3}
45. 0.05 H; 5×10^{-2}

CONVERSIONS

47. 35,000
49. 1
51. 0.001
53. 200,000

ABBREVIATIONS

55. milliamperes
57. millivolts
59. microvolts
61. kilovolt
63. mega

CHAPTER 6

DC VOLTMETER READINGS

1. 3.2; 16; 32; 160; 320
3. 2.3; 11.5; 23; 115; 230
5. 2.7; 13.5; 27; 135; 270
7. 2.2; 11; 22; 110; 220
9. 2.5; 12.5; 25; 125; 250
11. 3.8; 19; 38; 190; 380
13. 1.4; 7; 14; 70; 140
15. 3.2; 16; 32; 160; 320
17. 4.4; 22; 44; 220; 440
19. 4.6; 23; 46; 230; 460

DC AMMETER READINGS

1. 2; 0.002
3. 200; 0.2
5. 0.66; 0.00066
7. 26; 0.026
9. 500; 0.5
11. 0.034; 0.000034
13. 34; 0.034
15. 0.26; 0.00026
17. 0.7; 0.0007
19. 4.8; 0.0048

METER SCALE READINGS

First Position Answers

A. 0.43
B. 7
C. (0–300) 65
C. (0–150) 32.5
D. (250–0–250) – 180
D. (50–0–50) – 36
D. (10–0–10) – 7.2

CHAPTER 7

COLOR CODE EXERCISES

Part A

1. 47 Ω ± 10%
3. 2.7 kΩ ± 10%
5. 1 MΩ ± 10%
7. 15 kΩ ± 10%
9. 10 Ω ± 5%
11. 22 kΩ ± 10%
13. 470 kΩ ± 10%
15. 220 Ω ± 5%
17. 220 kΩ ± 10%
19. 5.6 Ω ± 10%

Part B

1. Red, Violet, Yellow, Gold
3. Yellow, Violet, Green
5. Red, Violet, Red
7. Red, Red, Orange, Gold
9. Brown, Green, Red, Gold
11. Yellow, Violet, Gold, Gold
13. Brown, Green, Yellow, Silver
15. Brown, Black, Red
17. Brown, Green, Orange, Silver
19. Orange, Orange, Gold, Gold

CHAPTER 8

1. 30 Ω
3. 3 A
5. 80 Ω
7. 30 kΩ
9. 120 V
11. 5 mA
13. 33.75 V
15. 105.6 V
17. 90 V
19. 30 kΩ

CHAPTER 9

9. 600 W
11. 25 W
13. 240 Ω
15. $0.54
17. 0.25 A
19. 6.22 A
21. 18 V
23. 12 Ω
25. 298 W
27. 2.22 A
29. 9,336,000 ft-lb
31. 14,600 W
33. 4.57 hours
35. 20 minutes

CHAPTER 10

1. R_1 = 10 kΩ
R_2 = 2.5 kΩ
R_T = 12.5 kΩ
3. 4 mA
5. 108 V
7. 6 Ω and 10 Ω
9a. 214 V
9c. 16 V
9d. 1.33 Ω
9e. 1,600 ft
9f. 0.831 Ω
9g. AWG #9 would suffice but is commercially unavailable. Choose #8 instead.
11. 36 V and 96 W
13. 2 A and 48 Ω
15. 12 V and 24 Ω
17. 440 V and 968 W
19. 2.5 A and 50 W
21. 20 V, 120 V, 40 W, 240 W
23a. 681
23b. 0.25
23c. 0.25
23d. 0.25
23e. 12.5
23f. 2,730
23g. 0.25
23h. 12.5
23i. 0.25
23j. 0.25
23k. 50
23l. 25
23m. 250
23n. 6.25
23o. 6.25
23p. 170
23q. 0.25
23r. 200
23s. 62.5
23t. 200
23u. 0.25
23v. 100
23w. 50
23x. 1.56
23y. 681
23z. 170

CHAPTER 11

1. 3.75 Ω
3. 2.53 Ω
5. 200 Ω
7. 5 Ω
9. 43.7 Ω
11. 22.5 Ω
13. 136 V
15. 1.54 A

CHAPTER 12

1a. 20 Ω
1b. 20 Ω
1c. 60 Ω
1d. 300 V
1e. 66.7 V
1f. 33.3 W
1g. 500 W
1h. 111 W
1i. 333 W
1j. 55.5 W
1k. 1,500 W

3. 2,000 Ω
5a. 70 Ω
5b. 2 A
5c. 280 W
5d. 1.33 A
5e. 0.667 A
5f. 1 A
5g. 1 A
5h. 1 A
5i. 28 V
5j. 28 V
5k. 112 V
5l. 32 V
5m. 80 V
5n. 37.3 W
5o. 18.7 W
5p. 112 W
5q. 32 W
5r. 80 W
9. $R_T = 100\ \Omega$ and $I_T = 3$ A
11a. 72 V
11b. 90 V
11c. 6 Ω
11d. 3 A
11e. 18 V
11f. 36 V
11g. 6 Ω
11h. 1 A
11i. 36 Ω
11j. 2 A
11k. 18 V
11l. 6 Ω
11m. 210 V
11n. 210 V
11o. 210 V

CHAPTER 17

5. 0.404 Ω
7. 1 kΩ
9a. 0.025 V
9b. 0.025 V
11. 900 kΩ
13a. 12 kΩ
13b. 0.002 Ω
15a. 0.005 Ω
15b. 7,500 Ω

CHAPTER 18

5. 40,000 lines

CHAPTER 19

1. 3,750,000 lines
9a. 141.4 V
9b. 3,360 W
9c. 625 W
9d. 29 kW
11a. 87%
11b. 81%
13a. 45.8 A
13b. 234.6 V
13c. 241.5 V
13d. 1,054 W
13e. 16.3 hp

CHAPTER 20

7. 225 ft-lb
9. 180 lb

CHAPTER 21

1a. 18.8 V
1b. 105.2 V
1c. 5.64 hp
1d. 84.8%
1e. 752 W
9a. 209.5 V
9b. 3,142 W
9c. 4.21 hp
15a. 44,940 W
15b. 204.3 A
15c. 3,714 W
15d. 3,486 W
17a. 15.15 hp
17b. 51.1 A
19a. 1,167 A
19b. 81.8 A

CHAPTER 22

5a. 40.7 A **5b.** 150% of 30 A = 45 A
Yes, it meets requirements.

CHAPTER 23

1. 2,900 W **3.** 32 ft of #22 wire, or 20 ft of #24 wire **5.** 20.7 candlepower

CHAPTER 24

1. 951 mA from top to bottom **3.** 2.94 mA from left to right **5.** 1.67 Ω

Index